Errata

Robert G. Winch, *Telecommunication Transmission Systems*, **Second Edition**

The figure below replaces Figure 4.32 on page 177.

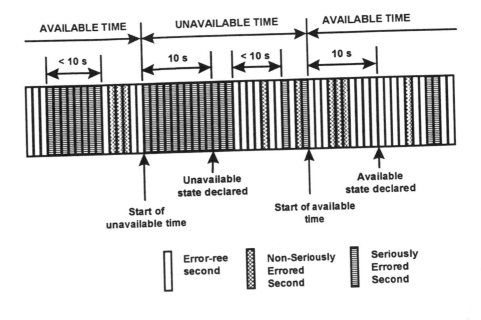

Telecommunication Transmission Systems

Other McGraw-Hill Telecommunications Books of Interest

Telecommunication Transmission Systems

Robert G. Winch

Second Edition

McGraw-Hill

New York San Francisco Washington, D.C. Auckland Bogotá
Caracas Lisbon London Madrid Mexico City Milan
Montreal New Delhi San Juan Singapore
Sydney Tokyo Toronto

Library of Congress Cataloging-in-Publication Data

Winch, Robert G.
    Telecommunication transmission systems / Robert G. Winch. — 2nd
ed.
        p.   cm.
    Includes index.
    ISBN 0-07-070970-X
    1. Telecommunication systems.   I. Title.
TK5101.W48   1998
621.382—dc21
                                                                    97-48537
                                                                    CIP

McGraw-Hill

A Division of The McGraw-Hill Companies

1 2 3 4 5 6 7 8 9 0   DOC/DOC   9 0 3 2 1 0 9 8

ISBN 0-07-070970-X

*The sponsoring editor for this book was Stephen S. Chapman, the
editing supervisor was David E. Fogarty, and the production supervisor
was Sherri Souffrance. It was set in Century Schoolbook by Michele
Bettermann, Joanne Morbit, and Michele Pridmore of McGraw-Hill's
Professional Book Group Hightstown composition unit.*

Printed and bound by R. R. Donnelley & Sons Company.

This book is printed on recycled, acid-free paper containing
a minimum of 50% recycled, de-inked fiber.

Contents

Preface

The telecommunications industry is proving to be a dynamic catalyst that is fueling the engines of economic growth in a manner the world has never previously experienced. The global implementation of digital telecommunications equipment has enabled the merger of the traditional telecommunications network designed for voice communications with data communications (computer information transfer). The resulting phenomenon we all know as "the Internet" is already changing the fabric of society and accelerating the globalization of commerce.

In today's world, technical advancement is occurring so rapidly that it is very difficult for most engineers and technicians to stay current with the enormous amount of literature being produced in each discipline. Most of us can only hope to keep up with developments occurring in a relatively narrow field. Telecommunications is a vast technical subject and it is the intention of this book to consider the most essential transmission systems information and focus on some specific aspects in considerable detail.

This text is designed for the *graduate engineer* or *senior technician*. However, the amount of mathematics has been purposely reduced to a minimum and emphasis is placed on the underlying concepts that shape telecommunications transmission equipment together with the practical application of the theory. It is intended that the material discussed be a shortcut to experience, to give the practicing engineer/technician both a better understanding of how existing telecommunication systems have evolved and an insight into their future development.

To ensure international compatibility, telecommunications equipment must be designed to conform to international standards. The design and performance specifications of some of the latest equipment is described with an acute awareness of the international standards that have been created to ensure compatibility. In this respect I am indebted to the International Telecommunication Union (ITU) for granting me authorization to reproduce information of which the ITU is the copyright holder. Due to the limitations of space, only parts of the ITU Recommendations are given or in some instances only a reference is made to the original source. For further information, the complete ITU publications can be obtained from the ITU General Secretariat, Sales and Marketing Service, Place des Nations, CH-1211 Geneva 20, Switzerland; e-mail Internet : Sales@itu.int.

The main themes of the book are multiplexing, microwave radio, satellite, optical fiber, wireless, and data communications. The term bandwidth, which relates to the amount of information that can pass through a system in a given time, is one of the most important parameters defining today's networks. The bandwidth attributes or restrictions of each transmission medium are treated in great detail.

Each topic starts from fundamental principles and therefore the material is also of interest to managerial staff who are new to the subject or may not have had time to keep up with all the latest technical advances published in journals.

Chapter 1 is an introduction which sets the scene for the rest of the book by describing the configuration for each of the main telecommunication systems: microwave radio, satellite, optical fiber, and cellular radio.

Chapter 2 describes the digitization of voice signals and how voice, data, or video channels can be combined by the time division multiplexing technique. The new and very innovative synchronous digital hierarchy (SDH) is also described together with its benefits over the traditional plesiochronous digital hierarchy (PDH). A brief description of television digitization is included as a prelude to future HDTV.

Chapter 3 discusses modulation techniques which are evolving to enhance bandwidth efficiency and error control to improve performance. The spread spectrum technique is described, highlighting its potential benefits for mobile communication systems.

Chapter 4 describes how the theory behind microwave communications leads to the physical appearance of today's systems. Attention is paid to the effects of terrain and atmospheric conditions on microwave propagation. The system performance is characterized and methods of enhancing performance are discussed.

Chapter 5 is a detailed study of digital microwave radio systems design and the measurements that are made to evaluate operational performance.

Chapter 6 traces the emergence of satellite communications from a fledgling telecom industry to a major player in the TV distribution and data networking businesses. The new global mobile phone systems using low and medium earth orbit satellites are also described.

Chapter 7 details the development of mobile radio (wireless) communications, with a comparison of the major systems presently in existence, namely, the digital TDMA (AMPS and PACS) and CDMA (IS-95) systems designed in

North America, and the TDMA (GSM) system from Europe. Other wireless topics are discussed such as cordless telephones and data over wireless.

Chapters 8 and 9 give an overview of the evolution of fiber optics from the early step-index fiber communication systems up to soliton transmission and coherent detection systems. Chapter 8 emphasizes the characteristics of the fibers and components used in the systems, and Chapter 9 deals with systems design for various lengths of optical communication links, from short distance local area networks to transoceanic distances.

Chapter 10 is a brief introduction to data communications, stating some of the international standards used to establish equipment compatibility and describing the basics of packet switching, LANs, and ISDN. The building blocks of the Internet are described and some of the networking bandwidth bottlenecks are highlighted, with possible solutions presented to overcome them.

I am grateful to authors, publishers, and companies who have granted permission to reproduce figures and photographs from previous publications. Finally, I cannot find adequate words to express my gratitude to my wife, Elizabeth, for her patience, dedication, preparation of figures, and expert proofreading of the manuscript, without which this project would not have been completed.

Robert G. Winch

1

Introduction

The telecommunications market is moving rapidly toward US $1 trillion per year. One tends to consider the subject of telecommunications in relation to the industrialized world without realizing the fact that this comprises only the minority of global population. The developing world is a potentially enormous additional market for the future.

In all regions of the world, there is a correlation between a country's per capita gross national product (GNP) and its telephone density (number of telephones per 100 people). Just as telecommunications has proved to be the fuel for the engine of growth in the developed world, it will surely be the same in the twenty-first century for what are presently developing nations. Already many Asian countries are testimony to this fact. The continued importance of telecommunications to development should ensure the telecommunications market a long life for decades to come.

Meanwhile, technology is in a state of positive feedback, accelerating at a pace so fast it is hard to keep up with the literature on the subject. International standards are essential to ensure that new equipment has compatibility on a global level with existing and future equipment. International standards organizations are having a hard time keeping current with technological advancements. The objective of the following chapters is to provide some insight into present and future technological trends, and also to give technical details of many aspects of present-day, high-usage, digital telecommunications transmission equipment.

1.1 Transmission Media

The telecommunications objective is to produce high-quality voice, video, and data communication between any pair of desired locations, whether the distance between locations is 1 or 10,000 km. The distance between the two locations determines the type of transmission equipment used for setting up the connection. First, communication over a distance on the order of a few meters, such as within a building, is done using metallic wires, optical fibers, or very small cell radios. Any routing of information within the building is done by a switch on the premises, a switch known as a private branch exchange (PBX).

When the distance is extended to a neighboring building or to span a distance within a village, town, or city, the local telephone network is usually used. This entails making a connection to the nearest switching exchange by a pair of copper wires or radio, routing the initiating party to the desired receiving party, and completing the connection on the recipient's pair of copper wires, or radio, which are also connected to the nearest exchange. The *switching exchange* is also known as the *central office,* or CO, and the terms are used interchangeably in the rest of this text. The connection between the CO and the customer is called the *local loop*, while the term *subscriber* is also used for customer; they are also used interchangeably. If the connection is within the same neighborhood, the two parties are connected via the same CO, but if the connection is across town, routing from one CO to another is necessary. It is at this stage that the choice of technology becomes important in the overall cost of the network. In the early days of telecommunications, all interexchange traffic was done using numerous pairs of copper wires (one pair for each interconnection). This was very cumbersome, because interexchange cables were required, and such connections required hundreds or thousands of copper pairs. A technique known as *multiplexing* was subsequently devised for passing multiple simultaneous telephone calls (referred to as *traffic*) down one pair of copper wires. More recently, optical fibers have been introduced to fill this role. Future networks will connect COs to customers using optical fibers in the local loop, but the manner in which this should be done is still being debated in many parts of the world. The mobile telephone also comes into the local loop category, and over the past few years the deployment of *cellular* mobile systems has experienced an explosive growth on a global scale. A cellular radio that is nonmobile (sometimes called *fixed wireless*), when used instead of a cable pair to the customer premises, is known as a *wireless local loop* (WLL).

The next stage of interconnection is intercity, or long-distance, connections. The contenders to fill this role are microwave radio, optical fiber, and satellite. Fierce competition has emerged among these three technologies. Microwave and satellite communications are far more mature technologies, but fiber optics technology has recently caught up and in many aspects has overtaken the other two. The rapid progress made by fiber optics over the past 10 years indicates that it is in a good position to "win the race" and become the dominant technology of the future. Many people see the impact of fiber optics on telecommunications, particularly all-optical systems, as being similar to the invention of the transistor and its effects on computer technology.

There are several advantages of geostationary satellite links for long-distance telecommunications media. First, the broadcast nature of satellites is very attractive, especially for television transmission. The information transmitted from a satellite can be received over a very large area, enabling it to serve a whole continent simultaneously. Also, the cost of satellite communications is *independent of the distance* between the source and the destination (e.g., transmission over 1 or 5000 km costs the same). However, the satellite system only becomes cost competitive with microwave radio and optical fiber systems when the distance is large (e.g., greater than 500 km). There are some situations that are ideally suited to satellite communications. A country like Indonesia, for example, consists of hundreds of small islands. It is cheaper to use a dedicated satellite for telecommunications than to interlink all of the islands by microwave or optical fiber systems. A similar situation exists in very mountainous regions where there are hundreds of villages within the valleys of a mountain range.

The technology used by satellite communications overlaps terrestrial microwave radio technology to a large extent. The radio nature and operating frequencies are the same. The main differences lie in the scale of the components. Because the satellite link is over 36,000 km long, high-power transmitters and very low-noise receivers are necessary. Also, the size and weight of the satellite electronics must be kept to an absolute minimum to minimize launch costs. Considerable attention has recently been devoted to very small aperture terminal (VSAT) satellite technology. As the definition of VSAT implies, these systems have earth station terminals that use antennas of only 1 to 4 m in diameter. This is a significant reduction from the 30-m-diameter antennas used in the original earth station designs of the 1970s. The use of such small-diameter antennas enables business organizations to use satellite communications cost effectively, because a complete earth station can be placed on company premises. Again, long distance and broadcast-type transmission produce the highest cost effectiveness.

Perhaps one of the main disadvantages of satellite communications is the propagation delay. It takes approximately a quarter of a second for the signal to travel from the earth up to the satellite and back down again. This is not a problem in two-way speech communication, provided echoes are removed from the system by sophisticated electronic circuits. If satellites are used for intercontinental communications, three geostationary satellites are needed for complete global coverage. In order to speak to someone at a place on the earth diametrically opposite, or outside the "vision" of one satellite, a double satellite hop is required, which produces a propagation delay of about half a second. Some user discipline is required in this situation because interruption of the speaker, as occurs in normal conversation, results in a very disjointed dialogue. This delay is totally unsatisfactory for many people. However, data communications and data over voice channels, such as Telefax, etc., are not adversely affected by this delay time. The double hop delay can be improved a little by satellite-to-satellite transmission, particularly if there are more than three satellites in the global system. The cost of this type of transmission is considerably less, as one of the earth stations is eliminated from the connection.

A major new application of satellite communications is the much-publicized global mobile telephone system. Many telecommunications organizations are already catering to the demands of the urban populations of many countries by offering a mobile telephone handset that uses UHF or microwave radio technology to make the interconnection between the CO and the customer. Already the enormous demand for these systems warrants the operation of a global mobile telephone network. Such satellite systems have been proposed by several consortia, the first of which was a project called *Iridium*. This project was proposed by Motorola, and comprises 66 satellites operating in low earth polar orbit to provide a global cellular network structure. The initial estimated cost of this project in 1990 was in excess of 2.5 billion U.S. dollars. Whether all of the proposed projects literally get off the ground remains to be seen. One major obstacle to the use of these systems in urban areas is the fact that penetration of signals from satellites through buildings to individual handsets is poor, particularly in high-rise buildings. Satellite communications and microwave mobile communications are inherently narrowband in nature compared to optical systems. The competition between satellite and optical fiber communications is extremely intense, and already the economic and performance advantages of optical fiber are allowing it to eat into satellite's market share.

Although the long-term future of satellite systems might be uncertain, voice, broadcast TV, and low-bit-rate data traffic (with occasional high-bit-rate traffic for alternative routing during fiber restoration) should keep satellite technology alive for many years to come. Microwave mobile cellular radio systems are excellent for voice transmission, but presently rather limited for data transmission. Optical fiber for the vast majority of wideband home services appears to be imminent in the foreseeable future for major cities. In rural areas and developing countries it is considerably further in the future. Wideband services can be provided by satellite and microwave radio, but the cost of this limited bandwidth resource is comparatively high. The question is "What is the highest data rate that can be offered at a competitive cost?" One excellent point in favor of satellite and microwave cellular radio is the mobility factor. The main disadvantage of optical-fiber-based networks is that the user is "tethered" and cannot be mobile.

The bulk of long- and medium-distance telephone traffic is currently transmitted over terrestrial-style microwave radio and optical fiber links, which at present are primarily digital electronics technologies. Chapter 2 describes the process of digital multiplexing, which is a means of combining voice, video, and data channels into one composite signal ready for transmission over the satellite, microwave radio, or optical fiber link. This composite signal is usually referred to as the *baseband* (or BB).

Chapters 3 to 9 describe how the three competing technologies have acquired their present-day capabilities. There are applications for which satellite or microwave radio systems might never be replaced by fiber optics. There are situations where all of them or combinations of each complement each other within the same network, and some situations where all are applicable (in which case a prudent choice has to be made based on economics). The choice is not always easy. Chapters 4 and 5 describe the theory and applications of microwave radio systems. Chapter 6 shows that the satellite communications link (which is mainly an extension of the terrestrial microwave radio link) has some unique features. Chapter 7 explains the natural progression of fixed microwave radio systems to cellular mobile radio systems for interconnecting the customer to the CO. Chapters 8 and 9 describe the theory and applications of optical fiber systems. Telecommunications involves not only voice telephone interconnectivity; data information transmission and networking have become an increasingly important telecommunications requirement and Chap. 10 is devoted to that subject.

Economics is the driving force that determines the fate of a new technology. No matter what the benefits might be, if the cost is too high, the new technology will have only limited application. Relatively low cost combined with improved performance will undoubtedly ensure that fiber optics will realize global connectivity. A large portion of the world has made a substantial investment in microwave communications systems. The introduction of fiber optics does not mean that the existing microwave radio equipment has to be scrapped. As higher capacities (more voice, video, or data channels) become necessary, the optical fiber systems can be installed and will work side by side with the microwave equipment. Many developing countries that are in the early stages of expanding their communications networks are in an excellent position to take immediate advantage of the new fiber optic equipment and consequently "leapfrog" the copper wire and microwave-based technologies.

Before entering into technical details, some obvious statements can be made about long-distance satellite, microwave radio, and fiber optic systems. Satellite and microwave links use radio wave propagation from point to point, whereas fiber optic links have a continuous cable spanning the distance from point to point. This obvious difference between the two systems automatically defines some applications for which both techniques are applicable and, conversely, indicates some applications from which each is excluded. For example, in a mountainous terrain, microwaves can "hop" from peak to peak across a mountain range effectively, unimpeded by intervening rocks, forests, rivers, etc. Similarly, microwave radio systems can interlink chains of islands whose distances are relatively close, without concern for underwater cabling techniques or the depth of water. When link security is a problem in unsettled parts of the world, microwave radio systems are usually the preferred choice. Cable, whether optical fiber, coaxial, or twisted copper pair, cannot be as well protected as a microwave station. Cable suspended between poles is particularly vulnerable to sabotage or severe weather conditions. Underground placement of optical fiber cable is considered to be a better arrangement. Unfortunately the cost is quite high (usually at least twice as much as overhead installation of new cables). Unintentional damage of cables by agriculture and construction activities is by no means rare. In fact, in some places such occurrences can be so frequent that the statistics are too embarrassing to publish. Over relatively flat terrain, optical fiber cable might seem to be the best option at first sight because of lower cost. However, if this terrain is mainly rock, installation costs rise, making cable a less attractive choice.

For interexchange traffic in cities, fiber cable can very easily replace old twisted-pair or coaxial cable in existing ducts. If the ducts are full and the twisted-pair or coaxial equipment is too new to retire, a microwave radio system would then be the preferred option. Several cities have excessively high water tables, causing severe electrical problems. Here is an excellent application for optical fiber cable, because it is not metallic. With little or no additional cost, fiber cable can be installed to eliminate deterioration by water. The nonmetallic nature of optical fiber is also useful in other applications. For example, in power-generating stations electromagnetic induction can play havoc with communications equipment that uses metallic cable. Optical fibers are almost impervious to electromagnetic interference. Furthermore, because the fibers are made of completely dielectric materials, there can be no short circuits. This is very desirable in areas where explosions could be caused by sparks from short-circuited wires.

For the local loop, there is an increasing trend toward cordless or cellular radio systems. These systems have the dual benefit of allowing mobility, which is highly desirable for many sectors of a country's population, and eliminating the need for costly cable installation. Cordless systems have limited mobility, whereas cellular systems are more mobile. In countries where major cities need to upgrade rapidly to improve or expand service, the installation of radio technology is orders of magnitude faster than laying new cable. For wideband services optical fiber will no doubt be used in the future, with customers using satellite or microwave radio handsets for their narrowband mobile voice and low-to-medium-rate data connectivity.

There are many other instances where the choice between microwave radio and optical fiber systems is not so clear. When considering a high-capacity, countrywide backbone route over hundreds or even thousands of kilometers, the days when telecommunications companies and authorities installed twisted-pair or coaxial cable are long gone. The decision might be in favor of satellite or microwave radio for very rugged terrain, or fiber optics for very flat terrain. When the region under consideration contains both very flat and very mountainous areas, a combination might be suitable. The problem of which to choose is compounded by dynamic economic conditions. Countrywide backbone (star) networks are evolving into highly reliable self-healing ring (mesh) structures using optical fiber wherever possible. Prediction of future advances in technology is highly desirable. Although this cannot be done with any certainty, close observation of research and development results can define the trends and make future projections possible. For example, if a "spur" route from a ring is required to supply a small village, present economics dictate that a microwave radio system would be cheaper than an optical fiber system. The number of channels required increases with the population to be serviced. There is a specific population size at which the price of microwave radio equipment equals that of optical fiber equipment (including cable installation). As time passes the price of optical fiber equipment is coming down. To what extent it will be reduced in the future is debatable. Also, the population of the village might increase or decrease depending on many factors. This apparently simple situation is already starting to develop into a complex problem. It also appears that any decision will involve at best a prediction based on present trends. One consoling fact is that both optical fiber and microwave radio systems can be upgraded in capacity.

For the expansion of a microwave radio system, additional transmitters and receivers are required, but the waveguides and antennas can often remain the same. Similarly, the fiber optic terminal equipment can be changed to increase the capacity. This is the case only if high-bandwidth (capacity) cable is installed to allow for future expansion. Microwave radio systems have the added flexibility of being able to redirect the path of a link to accommodate changing communication requirements. The equipment can be readily moved from one location to another, the free space propagation media being conveniently amenable and omnipotent. The installed cable, unfortunately, cannot be similarly moved without incurring significant additional cable costs. The atmosphere in which microwaves are propagated causes its share of problems, as Chap. 4 addresses in detail.

The high-quality cost effectiveness of the communication link is a prerequisite for a successful system. Cost and quality are, as usual, interrelated. The way in which they are related is very complex, and leads the discussion to a technical level.

1.2   Digitization

The major quality improvement obtained in digital transmission systems is due to the receiver signal recovery technique (regeneration). In analog transmission systems, each repeater retransmits the received signal and also retransmits the noise. The noise accumulates at each repeater, so after a certain transmission length the signal-to-noise ratio (S/N) is so poor that communication is impossible. In digital transmission systems, each repeater "regenerates" the original received stream of pulses (ones and zeros), and retransmits them free of noise. Theoretically, therefore, digital transmission has no transmission length limit. However, in reality there is a phenomenon called *jitter*. This is described later; it is a pulse position noise observed as small variations of the pulse zero crossing points of the digital bit stream from their precise positions (see Fig. 5.53). Jitter accumulates because of its introduction by several electronic circuits within a digital transmission system. Excessive jitter causes unacceptable bit errors to occur and therefore limits the maximum link length capability of the digital system. In analog systems the signal-to-noise ratio determined the quality of the link or channel. In digital systems, it is now the bit errors and their frequency of occurrence that determine the quality of the link or channel. To summarize, the advantages of digital systems over analog are:

- All subscriber services such as telephony, high-speed data, TV, facsimile, etc., can be sent via the same transmission medium. Consequently, the concept of the *integrated services digital network* (ISDN) can be realized.

- The bit error ratio (BER) in digital radio systems is unaffected by fading until the received RF level abruptly approaches the threshold value. These characteristics are discussed in Chap. 4 (see Fig. 4.35).

- High immunity against noise makes digital transmission *almost* independent of path length.

- Use of integrated circuits makes digital systems economical and alignment-free.

- Easy maintenance, based on go/no-go types of measurement.

- Synergistic integration of digital transmission systems such as optical fiber, digital satellite, and digital microwave radio systems with digital exchanges.

Interestingly, digital radio is a frequently used term, but it is rarely appreciated that all radio transmission is an analog phenomenon. In other words, the digital radio carrier is an analog wave and it is only the information superimposed on the analog carrier and the method of placing it on the carrier (modulation) that has a digital format. The terms S/N or C/N (carrier-to-noise ratio) are therefore still alive.

As networks have become more digitized, the combination of time-division multiplexing (TDM), time-division switching, digital radio, and optical fiber systems is considerably more economical and technically flexible than the corresponding analog networks.

1.3   Digital Microwave Radio System Configuration

Figure 1.1 shows a simplified microwave link incorporating just one regenerative repeater and two end terminal stations. The terminal stations house switching equipment that connects the customers to the long-distance paths. In this illustration, a large number of customer signals (around 2000) are multiplexed together into a single signal, ready for transmission over the microwave link. The signal is converted to the microwave frequency (around 6 GHz) and transmitted over a path of typically 30 to 60 km from station A to the receiving antenna at the repeater station. The repeater either (1) simply amplifies the signal and sends it off on its journey using a different microwave frequency to minimize interference, or (2) it completely regenerates the individual pulses of the bit stream before reconverting the signal back to a microwave beam for onward transmission. Station B receives the microwave signal, processes it, and unravels the individual channels ready for distribution to the appropriate customers at this end of the link.

Figure 1.2 is a simplified block diagram showing the major differences between analog microwave radio (AMR) and digital microwave radio (DMR) transmitters. At the intermediate frequency (IF) and above, the two systems are very similar. The IF-to-RF conversion shown here is done by the hetero-

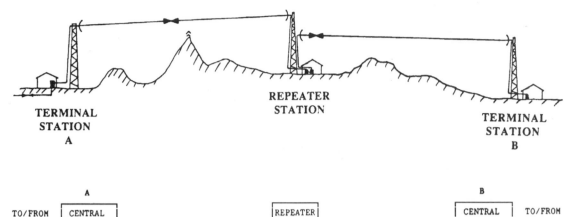

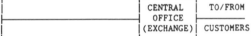

Figure 1.1 Basic microwave link incorporating a repeater.

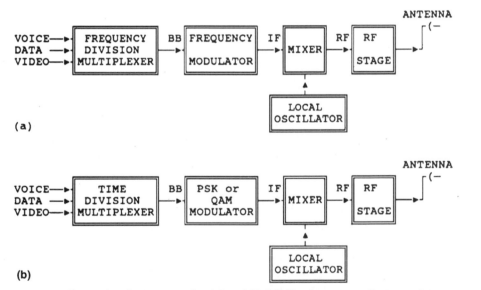

Figure 1.2 Comparison between analog (*a*) and digital (*b*) microwave radio transmitters.

dyne technique. The modulated signal is mixed with an RF local oscillator to form the RF signal, which is then amplified and filtered, and made ready for transmission from the antenna.

In both analog and digital systems, there are one or two variations on this theme. For example, the digital information can directly modulate the RF signal without going through an IF stage. This is called *direct RF modulation*. Another technique is to use a frequency multiplier to convert the IF signal to the RF signal. The advantages, disadvantages, and fine details of these systems are highlighted in Chap. 5.

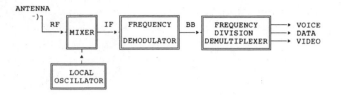

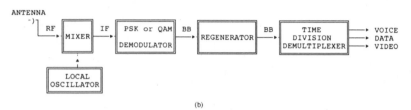

Figure 1.3 Comparison between analog (*a*) and digital (*b*) microwave radio receivers.

The major differences between analog and digital microwave radios lie in:

1. The composition of the baseband

2. The modulation techniques

3. The service channel transmission (not shown in Fig. 1.2)

The baseband is the combined multiple voice, data, and/or video channels that are to be transmitted over a telecommunications transmission system.

As indicated in the simplified diagram of Fig. 1.3, the receivers differ mainly in the demodulation technique and demultiplexing of the baseband down to the voice, data, or video channels. For the AMR receiver (Rx), the incoming RF signal is downconverted, frequency demodulated, and then frequency-division demultiplexed to separate the individual voice, data, or video channels.

For the DMR receiver, the incoming RF signal is similarly downconverted to an IF prior to demodulation. Coherent demodulation is preferred. However, for coherent demodulation, the exact transmitted carrier frequency and phase of the modulated signal must be obtained at the receiver. One way around this problem is to use differential encoding and decoding, as described in Chap. 5. The demodulated signal is subsequently restored to its original transmitted bit stream of pulses by the regenerator. Finally, the time-division demultiplexer separates the individual voice, data, or video channels for distribution to their appropriate locations.

Long-distance DMR link systems use regenerative repeaters. *Regenerative* is a term used when the signal goes through a complete demodulation-regeneration-modulation process (Fig. 1.4). Regenerators can interface with digital multiplex equipment for data insert and drop applications. Recently, add and drop multiplexer crossconnect equipment has been developed to simplify this process. In *regenerative* repeaters the noise and distortion are largely removed in the regeneration process and so the only type of noise accumulation is jitter accumulation. The occurrence of errors (transmitted 1s received as 0s, or transmitted 0s received as 1s) is also an important consideration, and is addressed in Chaps. 4 and 5.

1.4   The Satellite System Configuration

This is a vast subject, and one that has provided the material for many texts since its inception in 1962. An exhaustive coverage of satellite communications is impossible within a few pages, and in this text an attempt will be made to present only the main principles involved. Some of the basic engineering design aspects will be discussed, and how the laws of physics lead to specific equipment configurations. In many respects the satellite communication link can be viewed as a super-long-distance microwave link, and many of the calculations for terrestrial links can be extended to satellite paths. As with terrestrial telecommunication systems, the satellite industry is gradually becoming digitalized. Interestingly, the satellite is transparent to the passage of analog or digital information. Also, the baseband signals, whether analog or digital, are packaged on an analog radio carrier and only very recent satellites incorporate demodulation or demultiplexing. Some satellites already have onboard processing that is done at the individual channel (subbaseband) level, but the increased level of complexity and cost-effectiveness of future onboard processing is still being debated.

Regardless of whether satellite communications are international or domestic, there are several major categories of satellite users:

1. Individual voice band

2. Corporate data

3. TV broadcasting

4. Government and military

Satellite systems are very attractive for satisfying the wide area of coverage and the point-to-multipoint nature required for broadcasting. Video coverage of an event at one place on the globe can be sent up to a satellite and redistributed (broadcast) over large areas of the populated world in the form of clear television pictures. A network of satellites can provide global coverage so that telephone conversations can take place between individuals located at any of the remotest

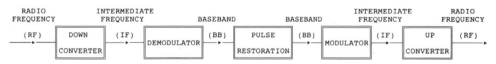

Figure 1.4   A regenerative repeater.

places in the world, with only a pocket handset required by the participants. Government applications are mainly to provide surveillance information, and military operations are coordinated and facilitated by satellite communications. Each application requires a significantly different system design approach.

Perhaps the most serious limitation of satellite communications is the *total available satellite bandwidth*. Although the information-carrying capacity of satellites has been expanding steadily over the years since its inception, the available bandwidth is still very small compared to optical fiber capabilities. For speech communications, there is ample available bandwidth, but for video transmission or high-speed data throughput there are severe limitations. Progress in digital compression techniques is gradually reducing the bandwidth needed for video transmission. Partial-motion videoconferencing is becoming feasible in the kilobit-per-second rather than megabit-per-second realm. Full-motion video, however, still requires several megabits per second, which is fine for broadcasting a few simultaneous TV programs but a problem for the individual requiring multimedia interaction facilities.

On the commercial side, one of the interesting aspects of satellite communications is that the cost of a single-satellite-hop telephone call is *almost independent of distance*. Whether calling a next-door neighbor or someone on another continent, the amount of equipment involved in the process is almost the same. The application determines the type of satellite system required. Satellites can be placed in three types of orbit:

1. Equatorial (geostationary)
2. Polar
3. Inclined

The type of satellite system determines the altitude at which the satellite is fixed. Figure 1.5 distinguishes between the three types. The geostationary orbit is the style most widely used for broadcasting, where the satellite is in an equatorial orbit and appears to be at a fixed point in the sky relative to an observer on the earth. Simple Newtonian mechanics shows the altitude of geostationary satellites to be an amazing 36,000 to 41,000 km, depending on the earth latitude of the observer. This vast distance makes life difficult for the

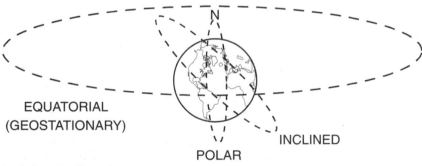

Figure 1.5 Satellite orbits.

designer, but good-quality voice and video communications via satellite have been commonplace for many years now. One of the major drawbacks of this large distance is the time taken for a round-trip signal to travel from the earth up to the satellite and back down to earth. Even at the speed of light the signal takes 0.24 to 0.27 s. This time delay can cause an annoying echo unless electronic circuitry is employed to minimize the effect. Another disadvantage of the geostationary orbit is the fact that at latitudes farther away from the equator, the geostationary satellite appears lower and lower toward the horizon. Eventually, at about 5° north or south, the satellite is too low on the horizon to receive a clear signal.

Satellites in polar or inclined orbits do not have this problem, but they are no longer geostationary. This would be a problem for broadcasting because an observer on earth would have to track the satellite as the earth moved beneath it, and if permanent transmission were required, several satellites would be needed with some mechanism for making a smooth transition between them before each one sequentially disappeared over the horizon. Polar orbits of about 800-km altitude are convenient for global coverage, so long as several satellites are moving in the same polar orbit and several polar orbits are used. These low-earth-orbit (LEO) satellite systems are the subject of potentially revolutionary constellations of satellites that are effectively global cellular systems. Inclined orbits of about 10,000 km provide global coverage with fewer satellites than LEO systems. These medium-earth-orbit (MEO) systems are consequently in direct competition with the LEO systems.

The satellite was traditionally used in the "bent pipe" mode. As the term suggests, the satellite acts like a slingshot to redirect the incoming signal to different locations on earth. In reality, as the simplified satellite link in Fig. 1.6

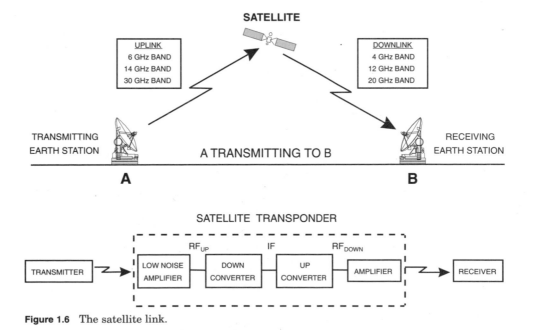

Figure 1.6 The satellite link.

indicates, the uplink microwave RF signal is amplified, downconverted to the IF, upconverted to a different frequency downlink microwave RF signal, amplified again, and retransmitted. That process is hardly the action of a bent pipe. However, this technique has been used since satellite communications began, and is still used on many of today's satellites. The addition of onboard processing in modern satellites allows separate parts of the incoming information package to be routed to different destinations by retransmission via different antenna beams.

1.5   Mobile Radio Systems

So far, the systems described are fixed-terminal or point-to-point in nature and, as far as the customer is concerned, the choice of system equipment used for an interconnection (be it microwave radio, optical fiber, or satellite) is largely irrelevant. However, the customer has recently become acutely aware of telecommunications equipment in the form of the portable radio telephone or the car telephone. These are the major mobile systems; others of importance include aircraft telephones and telephones to ships and trains. All of these systems are examples of radio technology being used directly *in the local loop*. Historically, mobile telephones were mainly installed in vehicles, and were initially rather heavy and cumbersome devices. With the acceleration of electronic circuit miniaturization over the past decade, the portable telephone market has mushroomed. In the industrialized world the day might not be too far away when city dwellers consider a portable telephone to be as much a part of their normal attire as a wrist watch. The term *city dweller* is used because portable telephones link up to the main national telephone network via very expensive equipment installed in the main city or urban areas. Gradually, coverage of entire countries is being accomplished, although the economics of such an enterprise might be prohibitive for many developing countries.

Radio waves have a large attenuation as they travel through the atmosphere, so in order to keep the required customer transmitter power to an acceptably low level, the distance between the customer and the nearest *base station* to link into the telephone network must be kept as small as possible. This has led to the need for numerous base stations arranged in a type of honeycomb, otherwise known as the cellular structure (see Fig. 1.7). In early mobile systems, each customer used a particular communication frequency for the duration of each call. Because of the limited frequency spectrum available, the number of subscribers able to use the systems simultaneously would appear to be quite small. However, the cellular system lends itself to frequency reuse, whereby subscribers in different areas use identical frequencies of communication and, by careful design, do not suffer any noticeable interference with each other. The intricacies of this and other advanced systems will be described in detail in Chap. 7.

1.6   The Optical Fiber System Configuration

The optical fiber link (Fig. 1.8) has some similarities to the microwave link. Both systems transmit the same output from the digital multiplexer (i.e., the baseband). The bit stream in the case of the optical fiber system can be used directly to turn a laser on and off to send light pulses down the fiber cable. Eventually, optical fiber systems might use a heterodyning or homodyning technique to improve the overall performance of the system. Regenerators are used at intervals to boost the signal, as in the microwave radio system. The distance between regenerators is gradually increasing for optical fiber systems as optical amplifier technology improves. Conversely, the line-of-sight microwave radio system regenerator spacing is limited by a physical, not a technological, constraint (i.e., the curvature of the earth). The increasing regenerator spacing of optical fiber systems is an important factor in enabling overall costs to be reduced. The spacing of optical regenerators is limited by the dispersive (pulse-broadening) characteristics of the fiber. As usual, the receiver incorporates a detector, an amplifier, and a means of restoring the original baseband bit stream ready for demultiplexing to voice, data, or TV signals.

The debate as to which system will become dominant in the future is gradually being won by the attractive optical fiber system economics. In addition,

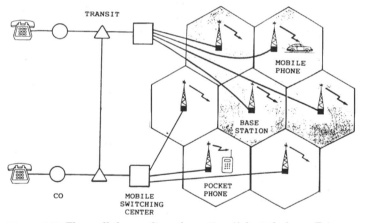

Figure 1.7 The cellular radio schematic. *(Adapted from Ericsson Review no. 3, 1987, Soderholm, G., et al., Fig. 1.)*

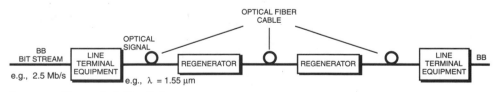

Figure 1.8 The optical fiber link.

the error performance is significantly superior to radio. It is widely believed that fiber optics is not only here to stay, but has the potential to transform our lifestyles in the decades to come.

1.7   Data Communications and the Network

Data communication is simply the transmission of a digital bit stream from one location to another. A bit stream is a sequence of millions of 1s and 0s that represent a combination of one or many voice, data, or video channels. The voice information to be communicated starts as analog information that is digitized for transmission, then reconverted to analog signals at the destination. There are inevitable analog-to-digital (A/D) and digital-to-analog (D/A) conversions in this type of network. While voice traffic through the telephone was (and still is) the primary aim of most telecommunications operators in the world, a rapid growth of data traffic has developed. While data traffic can mean a combination of digitized voice or video, it very often means computer-to-computer communication.

During the transition from an analog to a digital network, it has been necessary to accommodate a combination of analog and digital traffic. New transmission equipment installed in the network in both the developed and the developing world is digital in nature. In this case, as stated above, there is an unavoidable A/D conversion process for information that starts as analog signals. However, for digital computers the transmission path could be digital throughout. If the transmission involves, for example, the connection of a subscriber's PC to a computer mainframe in a downtown office, the subscriber loop should preferably have data transmission capability.

There is a fundamental problem that arises when trying to merge data transmission into the telephone network designed for the analog telephone traffic bandwidth. The telephone network was initially designed to have 300- to 3400-Hz bandwidth for a voice channel. This is not compatible with the megabit-per-second data rates (bit rates) ideally used by computers. The rates achievable at present are barely fast enough to allow a reasonable amount of information transfer, and the required data rates are increasing rapidly. The obvious solution that springs to mind is to use a wide-bandwidth transmission medium such as optical fiber cable for data transmission. Supplying optical fiber to every customer is a very expensive enterprise, but will probably become a reality for many countries near the year 2020. The bandwidth capability of the optical fiber would allow a variety of services to be provided, including videophone, video for TV, stereo music channels, computer networking, etc. Such a broadband integrated services digital network (B-ISDN) would incorporate broadcast facilities, which is already causing heated debate in several countries concerning the legal details. For example, how is a cable TV company affected if the telephone company has a better distribution system?

Before fiber-to-the-home (FTTH) becomes a reality and bandwidth is no longer a concern, an interim ISDN was proposed that extends a 64-kb/s path out to the customer premises. This is a considerable improvement over the early data bit rates of 9.6 or 14.4 using A/D and D/A converters, called *modems*, over an analog telephone network, 3.1-kHz-bandwidth circuit. However, recent modem improvement to 56 kb/s is providing competition in this area. Furthermore, some exciting innovations now allow megabit-per-second data transmission on regular twisted-pair telephone lines by what is called the *asymmetric digital subscriber line* (ADSL).

It is difficult to discuss data transmission without referring to the network as a whole. In particular, this encompasses the subject of switching, which plays an even greater major role in data communications than regular voice communications. For example, speech and video signals are real time in nature. Instantaneous processing and transmission are required so that there is no irritating delay as is sometimes experienced on satellite links. There is an important difference here when considering data that does not have this real-time constraint. Instead of using the conventional speech circuit switching, data can be split up into packets and transmitted when convenient time slots become available in a time-division multiple-access (TDMA) type of medium. This *packet switching* has become a major force in data communications today. The interconnection of thousands of local area networks (LANs) using packet switching technology has led to the evolution of the global network known as the *Internet*. This and other data communications subjects will be discussed in Chap. 10.

1.8   International Standards

Standards are essential. A simple example to illustrate the necessity for standards is the requirement for the receiver of a communications system in one country to operate on the same frequency as the message originator's system in another country. Obviously, if they are operating on different frequencies, there will be no communication. Without consensus from all member states, not only frequency problems but also a multitude of equipment incompatibilities and quality disparities would arise.

Throughout the rest of this text there are frequent references to the recommendations made by the International Telecommunication Union (ITU), which is a specialized agency within the United Nations. The ITU is an intergovernmental organization, and any sovereign state that is a member of the UN can be a member of the ITU. The ITU has been a global communications standards-setting body since 1865. Its standards are in the form of recommendations and clearly they cannot be forced on any country but, interestingly, almost all countries adhere to these standards as much as possible. Until recently the International Telegraph and Telephone Consultative Committee (CCITT) and the International Radio Consultative Committee (CCIR) were organizations that were part of the ITU. They have now been superseded by the Telecommunication Standards Bureau, known as ITU-T, and the Radiocommunications Bureau, known as ITU-R. These two committees have global representation. Since these forums are apolitical in nature, there has been considerable cooperation between their member nations over the past few decades, regardless of political bias.

The Institute of Electrical and Electronics Engineers (IEEE) is another global organization that is scientifically based, and which has also been very active in standards development.

There are standardization bodies other than the ITU that are regional or national rather than global in nature. There is a regional standards body called the European Telecommunications Standards Institute (ETSI). In the United States, the American National Standards Institute (ANSI) was created in 1984 at the time of the breakup of the Bell System, because de facto standards could no longer be expected. ANSI has many subcommittees and the Committee T1 on Telecommunications and the Telecommunications Industry Association (TIA) have considerable collaboration and cooperation with the ITU.

As the world shrinks, there is a growing need to harmonize the national communications standards of individual nations with the international standards of the ITU to achieve global standardization. If only one global body exists, there will be less duplication of work, fewer wasted resources, and no incompatible standards such as, for example, initially occurred with HDTV.

1.9   Telecommunication Systems' Driving Forces

Two widely differing forms of telecommunications have emerged: (1) narrowband connectivity for voice communications, and (2) wideband connectivity for computer networking.

1. First, the need for simple voice communications, characterized by narrow bandwidth, remains strong. As a percentage of total business, plain old telephone sets (POTS) remains the largest telecommunications revenue generator globally. Recently, this voice service requirement has become increasingly mobile and it is anticipated that mobile telephones will eventually become dominant over the next decade.

2. The second broad category of service is computer communications, requiring increasingly larger bandwidth per user compared to voice systems. Increasing numbers of computer users are creating a bandwidth appetite that can no longer be satisfied by the traditional wire-based transmission media. End-to-end *all-optical communications* are now foreseen to be the only long-term solution to this problem. The Internet and World Wide Web are demanding larger and larger bandwidths as more individuals want to be connected and multimedia-style presentations are expected. Eventually, as full-motion video is incorporated into networking, all-optical networks will be essential to cope with the hundreds of terabit-per-second (Tb/s) or even petabit-per-second (Pb/s) bandwidth requirements.

There is a growing sector of society that would like to see the benefits of mobility applied to networking. Many people already like to connect into the

Internet with a laptop from a mobile phone. The problem is that the normal telephone voice bandwidth connection has a severe limitation for high rates of data transfer. While there is a substantial research effort into the wider-bandwidth transmissions of megabits per second and above, radio technology inherently has a narrow-bandwidth problem. *There is a fundamental incompatibility between the fixed bandwidth resource available for radio transmission and the increasing number of users demanding an ever-increasing per-person bit rate.* Satellite radio systems fall within the same category as terrestrial radio systems in this respect. For this reason, regardless of how sophisticated the future electronic processing becomes for radio and satellite technology, it will always remain several orders of magnitude below the throughput capability of tethered all-optical networks. How these facts will translate into products and services over the next decade or two should prove to be an interesting story.

2

Digital Multiplexing

The digital baseband signal transmitted by digital microwave radios, satellite earth stations, or optical fiber line terminal equipment is formed nowadays by time-division multiplexing (TDM), instead of the analog frequency-division multiplexing (FDM) that was formerly used. Digital multiplexer equipment is considerably simpler and cheaper than its analog counterpart. While electronic processing might be digital, the physical world is analog. Unavoidable conversions must be made from analog (e.g., speech) to digital data, then eventually back to analog. The TDM technique involves periodically sampling numerous channels and interleaving them on a time basis. The resulting transmitted bit stream of 1s and 0s is unraveled at the receiving end so that the appropriate bits are allocated to the correct channels to reconstruct the original signal. The digital primary multiplexer packages channels in groups of 24 for North America and Japan, and 30 for Europe and elsewhere. The pulse code modulation (PCM) scheme is universally used for this purpose. For higher-capacity links, several 24- or 30-channel bit streams are digitally multiplexed to a higher-order bit rate (e.g., 45 Mb/s contains 672 channels in North America or 140 Mb/s provides 1920 channels in Europe). This higher-order multiplexing can be done on either a synchronous or nonsynchronous basis. Since the inception of PCM in the 1970s, the systems manufactured were almost completely the nonsynchronous type until the early 1990s. International standards were derived in 1988, based on the U.S. SONET standard, to ensure that the next generation of digital multiplexers would be the synchronous type. The term SONET means synchronous optical network, and although this scheme was originally intended for optical fiber systems, it is also applicable to microwave radio systems.

The term *synchronous* relates to the clocking system and the number of customer information pulses passing through the multiplexer per second. In a synchronous system this value is constant, but for the nonsynchronous multiplexers it varies plus or minus a few pulses per second. Because there are limits placed on the extent to which the number of pulses per second can vary, such systems are not totally asynchronous and a special name, *plesiochronous,* is given to them. Two terms have therefore evolved: the *synchronous digital hierarchy*, known as SDH, and the *plesiochronous digital hierarchy*, or PDH.

The SDH variety has some major benefits, as will become clear. However, because of the enormous quantity of PDH multiplexers still in existence, it will be several years before the transformation is completed in developed countries and a decade or more in developing countries. Both types will be outlined in the higher-order multiplexer section.

2.1 Pulse Code Modulation

Regardless of whether a multiplexer chain is part of a PDH or SDH system, the first stage of multiplexing requires the conversion of a voice or modem signal into a digital pulse stream and is the same for PDH or SDH. This process is called *pulse code modulation* or PCM. Pulse code modulation is simply an analog-to-digital conversion (A/D) process, and it was quite simple in its initial form. Because there is a direct link between cost and the number of bits per second required to transmit a good-quality voice signal, a large research effort is now being devoted to refining the already very complex algorithms that have been invented over the past few years. The starting point for bit rate reduction is 64 kb/s, which was used in digital multiplexers for many years before the new techniques started to gain acceptance in the early 1990s. For this reason, a chronological view of this subject will be taken.

2.1.1 Sampling

One of the first steps in the conversion of an analog signal into a digital one is the process of *sampling*. A sample is the magnitude of a modulating signal at a chosen instant, usually represented by the voltage of the modulating signal. Sampling is the process of measuring amplitude values at equal intervals of time (i.e., periodically, as shown in Fig. 2.1). The sampling rate for periodic sampling is the number of samples per unit time. For example, for telephony the sampling rate is 8000 samples per second, or 8 kHz. The sampling period is therefore 1/8 kHz (i.e., 125 µs). This sampling period is fixed by the *Shannon's sampling criterion,* which states that the sampling frequency must be at least double the highest frequency to be transmitted. Any rate lower than this has been mathematically shown to result in a loss of information transfer. While this is still true today, the loss of information does not necessarily reduce intelligibility. This fact is clear if one acknowledges that during speech there are pauses during which no information is transmitted. The extent to which information can be acceptably cut is subjective, and everyone has a minimum level of acceptability that is reached long before speech becomes unintelligible.

Within one 125-µs sampling period, samples of several telephone channels can be sequentially accommodated. This process is called TDM. The principle

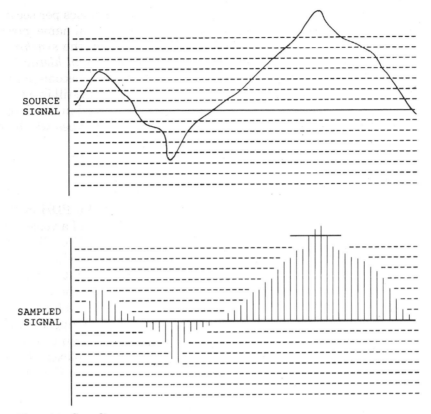

Figure 2.1 Sampling.

is used in all PCM systems. In the 24-channel PCM system, samples of 24 telephone channels are available for transmission in each sampling period.

A sampled signal contains complete and unambiguous information about the source signal as long as the sampling frequency is at least twice the highest frequency f_s of the source signal B. It can be shown mathematically that if the source signal has a spectrum as in Fig. 2.2a, the sampled signal will have a spectrum as in Fig. 2.2b. This consists of a number of subspectra, the first of which (number 1) lies in the frequency range 0 to B and is identical to the spectrum of the source signal. Subspectrum number 3 is identical to number 1 but moved f_s Hz in frequency (f_s = the sampling frequency). Number 2 is a reflected image of number 3 in f_s, and so on to infinity at a cycle of f_s Hz. Because the spectrum of the sampled signal coincides with that of the source signal in the $f_s + B$ band, we can conclude that the sampled signal contains all the information about the source signal, provided that f_s is greater than $2B$. If this condition is not satisfied, the situation in Fig. 2.2c will arise. The subspectra of the sampled signal will overlap each other, and this will result in the loss of information about the original source signal, called *aliasing distortion*.

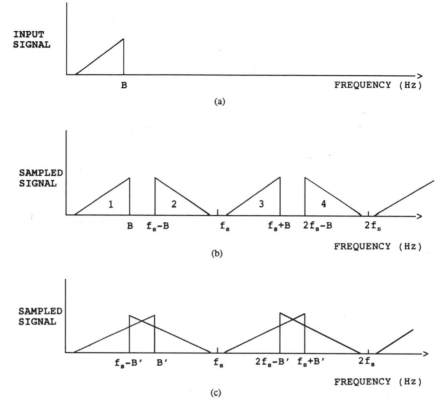

Figure 2.2   Spectra for sampled signals. (a) Input signal spectrum; (b) sampled signal spectrum for $f_s > 2B$; (c) sampled signal spectrum for $f_s < 2B$.

2.1.2 Quantization

Quantizing.   Quantizing is a process by which an analog sample is classified into one of a number of adjacent quantizing intervals. A sample whose amplitude falls anywhere within a particular interval is represented by a single value called the *quantized value*. At the receiver, the reconstructed sample is equal to the quantized sample. That means the reconstructed sample is almost always slightly different from the actual sample value, but the discrepancy is less as the chosen number of quantization values increases. Quantizing usually involves encoding as well. A quantizing interval is limited by two decision values at the extremities of the quantizing interval (see Fig. 2.3).

Uniform quantizing.   This is quantizing in which the quantizing intervals are of equal size. Figure 2.3 shows how the uniform quantizing scheme converts the incoming samples in the form of a pulse amplitude modulated (PAM) signal into 8-bit codewords containing the PAM data. In both the North

American and European PCM systems, the codeword has eight bits. The first bit (the most significant bit) gives the polarity of the PAM signal, and the next 7 bits denote the magnitude. Transmission of code words of 8 bits means that $2^8 = 256$ quantizing intervals are used.

Quantization noise. The distortion produced by quantization is significant. Because at the receiving end the analog signal is reconstructed from discrete amplitude samples, it is not an exact replica of the original signal. Because of the finite number of steps in the quantizing process, the input signal can only be approximated as shown in Fig. 2.4. For example, if there are 128 quantization steps and the maximum speech amplitude is, say, 1.28 V, each step has a width of 10 mV. All pulse amplitudes lying between 280 mV and 290 mV will therefore be encoded as 285 mV, the mean value between those steps. This difference between the input signal and the quantized output signal is called the *quantizing noise*. Clearly, the quantizing noise can be reduced by increasing the number of quantizing steps, but this would amount to increasing the number of bits in the code word designating the amplitude information. An increase in the number of bits per word can be accommodated only by reducing the pulse width. This means increasing the bandwidth, or reducing the number of multiplexed channels. By using a binary code word, the number of amplitude levels that can be encoded with n bits is 2^n.

The recovered analog signal-to-quantization-noise power ratio (S/N) is found to be approximately $6n$ dB, where n is the number of bits in the PCM word. Therefore, using uniform quantization, the PCM system S/N in decibels varies linearly with the number of bits in the word, and therefore with the bandwidth. In contrast, for frequency modulation (FM) systems the S/N in decibels varies with the logarithm of the bandwidth. Table 2.1 indicates the approximate S/N that can be achieved, depending on the number of bits in the code word.

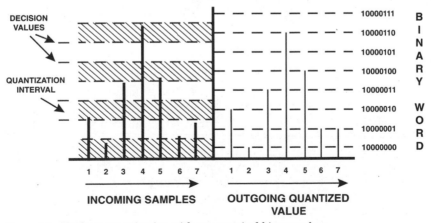

Figure 2.3 Uniform quantization with symmetrical binary values.

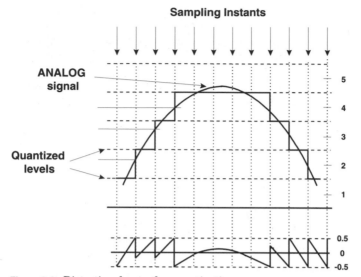

Figure 2.4   Distortion due to the quantization process.

TABLE 2.1   Quantization Details

Number of bits per code	Number of quantizing steps, 2^n	S/N, dB
7	128	42
8	256	48
10	1024	60
12	4096	72

A system transmitting speech signals must be able to accommodate signals of about 60 dB in dynamic range (i.e., a voltage range of 1000:1). To achieve this with uniform quantization, the number of bits per code word should be at least 10 (1024 quantization steps). To conserve bandwidth, 8-bit code words are used, but the quantizing steps are not equal and this nonuniform quantization improves the dynamic range.

Companding (nonuniform quantizing). With uniform quantization, the quantization distortion for signals with small amplitudes is greater than that for signals with larger amplitudes. Also, in telephony, the probability of the presence of smaller amplitudes is much greater than that of larger amplitudes. Low signal levels must therefore be amplified more than stronger signals to achieve a reasonably constant S/N. This is done by passing the analog speech signal through a nonlinear amplifier so that its dynamic range is *com*pressed at the transmitting end prior to being uniformly quantized. The reconstructed signal is *expanded* at the receiving end (hence the term *companded*) by passing the signal through a device having an inverse characteristic of the compressor. In PCM, this is effectively the same as making the quantization steps wider for higher-level signals and narrower for lower-level signals. By this process, the signal-to-quantization distortion ratio is nearly independent of the signal level.

Because the ear's response to sound is proportional to the logarithm of the sound amplitude, the compression curves used by all equipment manufacturers have an approximate logarithmic characteristic.

In North America and Japan the μ-law characteristic is used for companding (Fig. 2.5). This is represented by the function:

$$y = \text{sgn}(x)\left[\frac{\ln(1 + \mu x)}{\ln(1 + \mu)}\right] \tag{2.1}$$

where x = input amplitude
$\text{sgn}(x)$ = the polarity of x
μ = amount of compression and is chosen to be 255

Another compander characteristic used in Europe and many other parts of the world is the A-law curve (Fig. 2.5). According to ITU-T Recommendation G.711, this is defined by the equations:

$$Y = \frac{1 + \ln Ax}{1 + \ln A} \quad \textit{for } 1/A < x < 1 \tag{2.2}$$

$$Y = \frac{Ax}{1 + \ln A} \quad \textit{for } 0 < x < 1/A \tag{2.3}$$

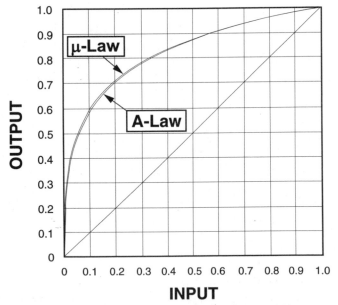

Figure 2.5   The μ-law and A-law compander characteristics.

where x = normalized input level
Y = normalized quantized steps
ln = natural logarithm
A = amount of compression and is chosen to be 87.6

As can be seen in Fig. 2.5, for μ = 255 and A = 87.6, the A-law curve is almost indistinguishable from that of the μ-law. In practice, the logarithmic curve of the A-law compander is approximated by a 13-linear-segment compander circuit.

Figure 2.6 illustrates how the *compressed* signal at the transmitting end, when combined with the *expanded* signal at the receiving end, produces a linear result. The signal heard at the receiving end is therefore a good reconstruction of the transmitted signal, except for a small amount of quantization noise.

2.1.3 Encoding and decoding

Encoding (or coding) is the conversion of an analog sample, within a certain range of values, into an agreed combination of digits. In PCM it is the generation of code words allocated to quantized intervals. These represent the quantized samples.

Encoder or coder.   In PCM, *quantizing* and *encoding* are very closely related. In actual implementation, the encoder or coder is usually one complete device providing both quantizing and encoding.

Operating principle of the iterative encoder.   The method used for encoding is as follows: At the beginning of each encoding period, the first bit is assumed to be a 1 if the sample is positive, and 0 if it is negative. The sample is then compared with an analog reference voltage corresponding to half of the maximum level. If the sample amplitude is greater than this reference voltage, this reference is maintained during the rest of the comparison (i.e., $B2 = 1$). If the sample amplitude is smaller than this reference voltage, this reference voltage is removed (i.e., $B2 = 0$). In this manner, the reference voltage corresponding to each of the code bits is tried in turn until all of the code bits

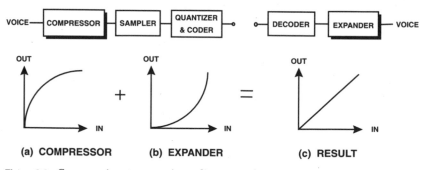

Figure 2.6 Compression + expansion = linear result.

have been determined. At the end of the comparison, the code word corresponding to the sample amplitude is directly available at the output of the circuit controlling the reference voltages. The output of this circuit is then stored, so that the encoder can process the next sample. The iterative A/D conversion decision sequence is shown in the "decision tree" of Fig. 2.7.

Decoding.   Decoding is the inverse process of encoding. In PCM it is a process in which a "reconstructed sample" is produced, corresponding to an 8-bit code word. Because this involves reading the word amplitude directly from a look-up table and does not involve an iterative process, decoding is much faster than encoding. Decoding at the receiver retrieves the encoded signals in the form of PAM samples. These PAM samples are passed through a low-pass filter and then through the expander part of the compander to form the reconstructed voice signal.

2.1.4   Recent coding developments

The standard bandwidth for voice channels in a 64-kb/s digital PCM transmission system has been set at the customary 300- to 3400-Hz value. The first important developments in speech coding techniques led to ITU-T Recommendation G.726, which allows a reduction of the transmitted bit rate to 32 kb/s while preserving audio quality. This means that, for a given transmission bit rate, the number of voice channels transmitted can be *doubled*. The technique that was developed for reducing the sampling rate to 32 kb/s was called *adaptive differential PCM* (ADPCM).

ADPCM works on the following principle: When successive samples are quantized, because of the gradual nature of the variation of the source signal (see Fig. 2.1), a lot of redundant information is transmitted. For example, when a sine wave signal is quantized, the successive sample quantization amplitudes near the peak of the sine wave vary only a small amount from one to the next. So, instead of transmitting the long digital word for each sample, the same information can be transmitted by forming a small digital word indicating the difference between one sample amplitude and the next. This more efficient method of quantization is the essence of differential PCM and leads to a lower bit rate per channel than the original method. The inclusion of an adaptive predictor circuit for tracking the trend of the sampled signal and statistically predicting its future amplitude provides further transmission efficiency, resulting in an even lower bit rate per channel.

Figure 2.8a shows the schematic circuit for the ADPCM encoder. The companded (log) PCM signal is linearized and a difference signal is formed by subtracting a previous estimate created from an earlier input. Every 125 µs, the difference signal is 4-bit coded in the 16-level adaptive quantizer to produce the 32-kb/s output. The inverse adaptive quantizer forms part of the feedback loop, in which a quantized difference signal is added to the signal estimate as part of the adaptive prediction process. Figure 2.8b shows the

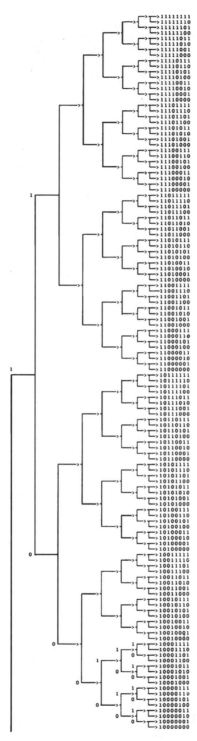

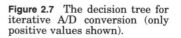

Figure 2.7 The decision tree for iterative A/D conversion (only positive values shown).

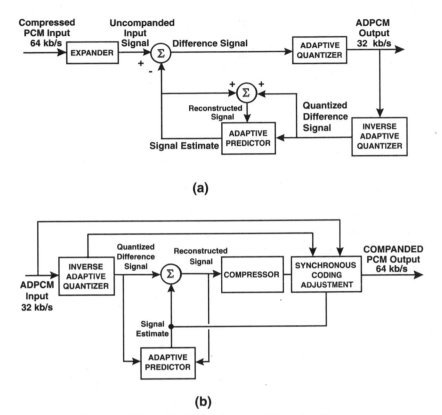

Figure 2.8   Adaptive differential PCM coder (*a*) and decoder (*b*).

ADPCM decoder, which simply performs the inverse process to revert the signal to a 64-kb/s companded PCM output. There are two interesting points to be made about this encoding-decoding process. First, the encoder contains the decoder circuit. Second, the ADPCM codec effectively introduces a doubling of the companding process, so that it can operate on a linear signal.

ITU-T Recommendation G.728 describes a 16-kb/s speech encoder using a lower-delay code-excited linear predictor (LD-CELP), and ITU-T Recommendation G.729 describes an 8-kb/s speech encoder using an advanced code-excited linear predictor (ACELP). Various CELP techniques are used to code speech signals down to 4.8 kb/s. For example, the IS-54 North American cellular system 7.95-kb/s speech coder uses vector-sum excited linear predictive (VSELP) coding. Rapid quality deterioration is experienced below this rate. Adequate modeling of the harmonic structure within the speech spectrum is just one of the problems. Considering the enormous effort and the astounding progress made in this field in the 1990s, it would be difficult to forecast a limit on the lowest bit rate possible. The objective of all these compression techniques is to maintain toll-quality speech as experienced in a regular PSTN connection. It is highly probable that bit rates less than 1 kb/s will be used in future cellular systems.

The main problem with all of these coding techniques that remove redundancies and/or predict the future amplitude of various frequency components of the speech spectrum is *delay time* caused by the algorithm. Excessive delay is often perceived as an echo by the listener. Today's transmission systems can include a mobile radio hop and a geostationary satellite orbit (GSO) or low earth orbit (LEO) satellite hop. It is the cumulative delay that must now be kept below the level that causes disjointed conversations. This is a very subjective matter, and an important question is "What is the maximum delay a customer will accept?" This subject is discussed in more detail in Sec. 3.5.

2.1.5 PCM primary multiplexer

A typical schematic diagram for a 24- or 30-channel PCM primary multiplexer is shown in Fig. 2.9.

Transmission path. The speech signal is first passed through the compressor part of the compander. It is then band-limited by the low-pass filter so that only the frequency band to 3400 Hz is transmitted. The speech signal is then sampled at the rate of 8 kHz to produce the PAM signal. This signal is temporarily stored by a *hold* circuit so that it can be quantized and encoded in the A/D converter. Samples from a number of telephone channels (24 in North America and Japan, or 30 in Europe) can be processed by the A/D converter within one sampling period of 125 µs. These samples are applied to the A/D converter via their respective gates, selected by the transmit timing pulses. At the output of the A/D converter, the speech samples exit as 8-bit PCM code words. These code words from the speech path are combined with the *frame alignment* word, service bits, and the signaling bits in the multiplexer to form *frames* and *multiframes*. The purpose of the formation of multiframes is to allow the transmission of signaling information for all 24 (or 30) channels during one complete multiframe. They are then passed on to the line encoder, which converts the binary signals into the bipolar line code known as alternate mark inversion (AMI) in North America, or the high-density bipolar 3 (HDB3) in Europe. The resulting line-coded signals are then ready for transmission of a meter or less to the next level of multiplexing or for many kilometers over a DMR link.

The transmission rate out of the primary PCM multiplexer is 1.544 Mb/s (2.048 Mb/s in Europe). This is controlled by the timing clocks in the transmission end, which control the processing of the speech, signaling, synchronizing, and service information.

Reception path. The 1.544-Mb/s AMI (or 2.048-Mb/s HDB3) signal that comes from the line is first decoded into a binary signal. This signal is then separated by the demultiplexer into the respective speech channels, together with supervisory information (signaling, etc.). The speech PCM signal is sent to the

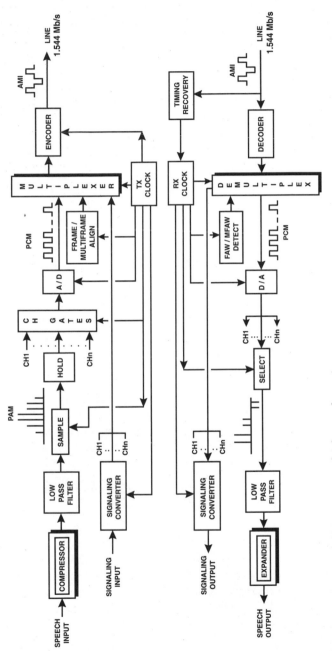

Figure 2.9 The basic PCM primary multiplexer schematic.

D/A converter, the signaling bits to the signaling converter, and the frame and multiframe alignment bits and service bits for alarms, etc., to the frame and multiframe alignment detectors and alarm unit. The timing signals for the receiver are recovered from the line codes, and are then processed in the receiver timing unit to generate the clock signals for processing the received signals. In this manner the receiver is kept synchronized to the transmitter. The codes belonging to the speech signal are then converted to PAM signals by the D/A converter. Next, they are selected by their respective gates and sent to their own channels via the respective low-pass filters and expanders, which reconstruct the original analog speech patterns. The bits belonging to signaling are converted into signaling information by the receive signaling converter and sent to the respective telephone channels.

2.1.6   Formats for 24-channel PCM systems

The 24-channel frame is 125 μs long (Fig. 2.10) and contains 24 time slots each having 8 bits. The first 7 bits are always used for encoding, and the eighth bit is for encoding in all frames except the sixth frame, where it is used for signaling. At the start of every frame, 1 bit is included for frame and multiframe alignment purposes. Each frame therefore contains $(24 \times 8) + 1 = 193$ bits. Because the sampling rate is 8 kHz, there are 8000 frames per second, giving $193 \times 8000 = 1.544$ Mb/s. The signaling bit rate is $(8000/6) \times 24 = 3200$ b/s.

The 24-channel system has a 1.5-ms multiframe consisting of 12 frames. The frame and multiframe alignment words are transmitted sequentially, by transmitting 1 bit at the beginning of each frame. They are sent bit by bit on the odd and even frame cycles, and their transmission is completed only after each multiframe has been transmitted. The frame and multiframe alignment words are both 6-bit words (101010 and 001110, respectively).

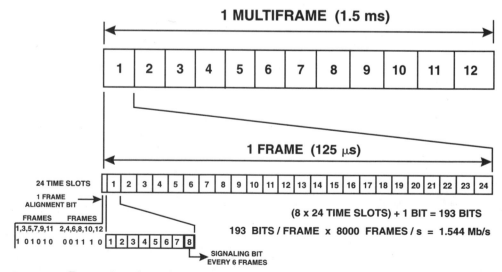

Figure 2.10 Twenty-four-channel frame structure.

2.1.7 Formats for 30-channel PCM systems

The 24-channel primary systems in operation in North America and Japan are not used in Europe and most of the developing world, where the design has 30 channels for the primary PCM system. The basic concept of A/D conversion is the same. The 30-channel frame is 125 μs long, the same as the 24-channel frame. However, the 30-channel frame contains 32 time slots each of approximately 3.9-μs duration. The time slots are numbered from 0 to 31. Time slot 0 is reserved for the frame alignment signal and service bits. Time slot number 16 is reserved for multiframe alignment signals and service bits and also for the signaling information of each of the 30 telephone channels. Each multiframe consists of 16 frames, so the time duration of one multiframe is 2 ms.

In European systems, the signaling information for each telephone channel is processed in the signaling converter, which converts the signaling information into a 4-bit code for each channel. These bits are inserted into time slot 16 of each PCM frame except frame number 0. The 16 frames in each multiframe are numbered 0 to 15. In each frame signaling information from two telephone channels is inserted into time slot 16, so signaling information from the 30 telephone channels can be transmitted within one multiframe. The construction of the frame and multiframe is shown in Fig. 2.11.

In time slot 0 of each frame, the frame alignment word (0 0 1 1 0 1 1) is sent on every *even* frame, and the service bits (Y 1 Z X X X X) are sent on every *odd* frame. In time slot 16 of frame 0 only, the multiframe alignment word (0 0 0 0) is sent. In time slot 16 of frames 1 to 15, the signaling information of channel pairs 1/16, 2/17, etc., are sent in their specific frame allocations.

A comparison of the construction of the frame and multiframe for the U.S./Japan 24-channel and the European 30-channel primary PCM systems is summarized in Table 2.2.

Frame alignment. As mentioned earlier, a time slot of 8 bits per frame is available for frame alignment. This means that out of the 2.048 Mb/s transmitted, 64 kb/s are reserved for frame alignment. The basic principle of frame alignment is that the receiver identifies a fixed word and then checks its location at regular intervals. This makes it possible for the receiver to organize itself to the incoming bit flow and to distribute the correct bits to the correct channels. In addition to frame alignment, the assigned time slot is also used for transmission of information concerning the alarm states in the near-end terminal and the remote-end terminal. Spare capacity is also available for both national and international use. The 16 frames are numbered 0 to 15. The words in time slot 0 in frames with even numbers are often called *frame alignment word 1*, while those in odd frames are called *frame alignment word 2*.

When the receiver reaches the frame alignment state, its only function is to make sure that frame alignment word 1 recurs where it should occur, and at regular intervals. *If the frame alignment word is incorrect four consecutive*

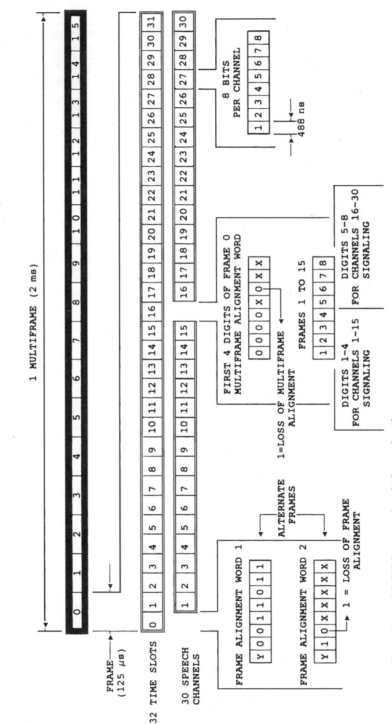

Figure 2.11 Thirty-channel PCM frame and multiframe details.

TABLE 2.2   Comparison of 24- and 30-Channel Systems

	24-Channel system	30-Channel system
Sampling frequency, kHz	8	8
Duration of time slot, μs	5.2	3.9
Bit width, μs	0.65	0.49
Bit transfer rate, Mb/s	1.544	2.048
Frame period, μs	125	125
Number of bits per word	8	8
Number of frames per multiframe	12	16
Multiframe period, ms	1.5	2
Frame alignment signal in	Odd frames	Even frames
Multiframe alignment signal in	Even frames	TS16 of frame
Frame alignment word	101010	0011011
Multiframe alignment word	001110	0000

times, the frame alignment is considered lost, and the search process is started again. In practice it might happen that a bit in the frame alignment word becomes distorted in transmission, and it would be quite unnecessary to resynchronize the system every time this happens. By waiting for four consecutive incorrect alignment words before taking action, the result is a very stable synchronizing system with a high degree of insensitivity to disturbances. In fact, realignment will seldom be required in normal operation.

The strategy for the frame alignment alarm is shown in Fig. 2.12. This diagram illustrates the electronic decision-making process that results in the frame alignment alarm being indicated if four consecutive frame alignment

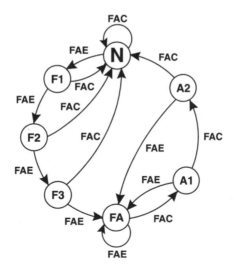

Figure 2.12   Frame alignment strategy.
FAC = frame alignment correct;
FAE = frame alignment error;
N = normal state;
FA = alarm state;
F1, F2, F3 = prealarm state;
A1, A2 = postalarm state.

errors (FAEs) exist. It also shows the decision process for the alarm state (FA) to be normalized only when three consecutive correct frame alignments (FACs) are detected. For example, if the system is in the normal state N and one incorrectly received frame alignment word occurs, the system is in the prealarm state F1. If the next FAW is correct, the system is back to the normal state N, but if it is incorrect, the system is in the prealarm state F2. Another incorrect word takes it to F3. At this point a correct frame alignment word would take the system back to the normal state N, but an incorrect word would be the fourth one, taking the system into the alarm state FA.

Multiframe alignment.   Multiframe alignment might seem more complicated than frame alignment, because the multiframe alignment word occurs only once every 16 frames and should therefore be harder to find. However, the system first performs frame alignment and then multiframe alignment. The multiframe alignment logic receives information about the starting point of the frame from the frame alignment logic, via the so-called 64-kb/s interface. If the starting point of the frame is known, it is easy to establish the location of time slot 16 and then just wait for the frame that contains the multiframe alignment word (i.e., frame number 0).

The multiframe alignment process is quite simple. The system is multiframe aligned as soon as a multiframe alignment word is found (b1 b2 b3 b4 = 0 0 0 0). The reason for this is that the risk of imitation is practically nonexistent, because the starting point of the frame is known and the combination 0000 never occurs in either the first half or the second half of time slot number 16 in any frame except frame number 0. This leads to the requirement that the combination 0000 never be used for signaling. *The multiframe alignment is considered as lost if two consecutive incorrect multiframe alignment words have occurred.* This means there is an element of inertia, which makes it possible to avoid unnecessary realignment in the event of isolated bit errors.

2.2   Line Codes

The PCM signal is made up of bit sequences of 1s and 0s. Even when transmitting this information a meter or two from one multiplexer to the next higher-order multiplexer, to the microwave radio rack, or to the optical fiber line terminal rack, the 1s and 0s can be detected incorrectly if they are not transmitted in the correct form. There are several types of pulse transmission, which are called *line codes*. The two main categories of line code are the *unipolar* and the *bipolar*. Also, the pulses in the line code are categorized as either *non return to zero* (NRZ) or *return to zero* (RZ). NRZ (as in Fig. 2.13) means that the 1 pulses return to 0 only at the end of one full clock period, whereas the RZ signal (Fig. 2.14) has 1 pulses that return to 0 during the clock period.

It is usually necessary to convert from one code to another in the PCM transmission equipment. These conversions are done by appropriate circuits called

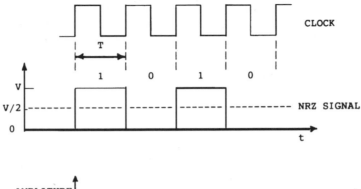

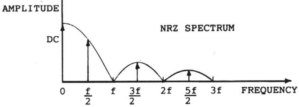

Figure 2.13   The NRZ code and its spectrum.

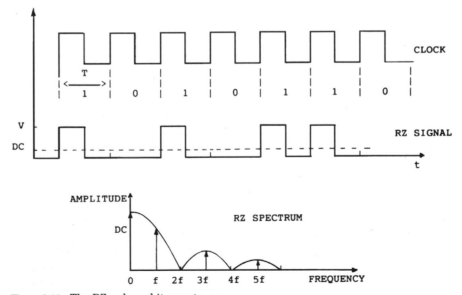

Figure 2.14   The RZ code and its spectrum.

code converters. The important features that affect the choice of transmission code are dc content, timing for the transmitter and receiver synchronization, bandwidth, and performance monitoring. Usually, a code must fulfill the following conditions:

1. There must be no significant direct current (dc) component, because an alternating current (ac) transformer or capacitive coupling is used in most wire-line transmission systems, including telephone lines, to eliminate ground loops.

2. The energy at low frequencies must be small; otherwise physically large circuit components will be needed for the equalization circuitry.

3. A significant number of zero crossings must be available for timing recovery at the receiving end (i.e., for clock frequency recovery).

4. The coded signal must be capable of being uniquely decoded to produce the original binary information signal (with no ambiguity).

5. The code must provide low error multiplication.

6. Good coding efficiency is necessary to minimize bandwidth.

7. Error-detection or correction capability is necessary for high-quality performance.

The most common codes will be examined in the following subsections.

NRZ code (100 percent unipolar).   From the circuitry point of view the NRZ code is the most common form of digital signal, because all logic circuits operate on the ON-OFF principle, and so the NRZ code is used inside the equipment (e.g., multiplexers, DMR, optical fiber line terminals, etc.). By observing the signal in Fig. 2.13, one can see that all 1 bits have a positive polarity, and so its spectrum has a dc component, the average value of which depends on the 1/0 ratio of the signal stream. If the signal consists of a 10101010 sequence, the dc component will be $V/2$. Depending on the signal, the dc component can assume any value from 0 (all 0) to V volts (all 1). The spectrum of the NRZ code is as shown in Fig. 2.13. The fundamental frequency occurs at half the clock frequency f, and only the odd harmonics are present. Furthermore, there is no signal amplitude at the clock frequency, so it is difficult to extract the clock frequency at the receiving end. Also, during transmission via wire-cable, if the noise peaks are summed up so that a 0 is simulated as 1, it is impossible to detect the error. These disadvantages make the use of the NRZ unsuitable for transmission via cable, unless accompanied by scrambling. This line code is used for high-bit-rate, scrambled, SONET/SDH bit streams at 155 Mb/s and above (as described in Sec. 2.5).

RZ code (50 percent unipolar).   This is similar to the NRZ signal, but with a pulse duration reduced to one-half. This code is also convenient for equipment circuitry, because all logic circuits operate on the ON-OFF principle. One can see from Fig. 2.14 that it also produces a dc component in the spectrum. However, the fundamental frequency is now at the frequency of the clock signal, with only the odd harmonics existing as in Fig. 2.13. This makes it possible to extract the clock frequency at the receiving end provided long sequences of 0s are not present. Detection of errors, as explained earlier, is not possible. This code is therefore rarely used at any stage of the system. However, a bipolar version of the RZ code is widely used, as discussed shortly.

Alternate mark inversion (AMI) code (bipolar code).   By observing the signal in Fig. 2.15, one can see that the 1s are alternately positive and negative, so there is no dc component in the spectrum. The apparent absence of the clock frequency in the spectrum can be overcome by simply rectifying the received signal to invert the negative 1s and making the resulting signal similar to the RZ signal. In this case, the received AMI signal has already passed through the transmission line. Consequently, the appearance of the dc component is of no interest, and the clock frequency can be extracted from the new spectrum of the signal. Another advantage of the AMI signal is that it can correct errors. If, during line transmission, noise peaks are summed up to simulate a 1 instead of 0, there would be a violation of the code that necessitates that the 1s be alternately positive and negative. Unfortunately, the recovery of the clock frequency is not easy with AMI coding if a long sequence of 0s is present.

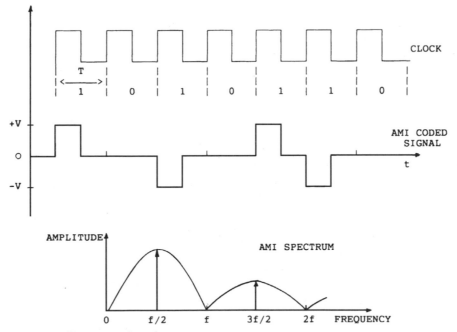

Figure 2.15 The AMI code and its spectrum.

The AMI code is recommended by the ITU-T (Rec. G.703) for the 1.544-Mb/s interface. Its modified version, as discussed shortly, is used for 2.048 Mb/s.

Alternate digit inversion (ADI) code (unipolar 100 percent duty cycle). In this form of coding, every second, or alternate, digit or bit is inverted. The example in Fig. 2.16 shows how the 8-bit PCM words are coded in ADI code. The speech code from the A/D converter contains a relatively high number of 0s for speech levels close to zero, while the occurrence of 1s increases as the level is raised. The probability of the speech signal being near zero is high. First, when a given channel is not seized at all, the level is zero. Second, when a channel is seized, only one subscriber at a time speaks, which means that the other subscriber has zero level in his or her direction of transmission. ADI is very useful because even with large strings of 0s or 1s, extraction of the clock signal at the receiver is possible when the transmit signal is ADI-encoded.

Binary N zero substitution codes. Binary N zero substitution line codes (BNZS) are used in North America. BNZS is a type of AMI code that ensures that long strings of zeros do not occur. For any incoming long strings of zeros, the lack of zero crossings can lead to the loss of synchronization and increased timing jitter (pulse position noise). For each sequence of N zeros, a special code is substituted.

The binary three-zero substitution (B3ZS) line code is specified by ITU-T Recommendation G.703 for the 44.736-Mb/s (DS3) interface. Every string of three zeros is substituted by either 00V or B0V, where V is a 1 (which violates the AMI rule), and B is a 1 (which obeys the AMI rule). The choice of 00V or B0V depends on the recent bit-stream history as shown in Table 2.3. The substituted code words (00V and B0V) are selected in a manner that maintains

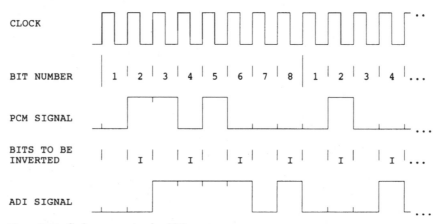

Figure 2.16 Code conversion by ADI.

TABLE 2.3   B3ZS Substitution Sequence

Number of B pulses since last V pulse	Polarity of last B pulse	Substitution sequence	
Even	+	− 0 −	(BOV)
Even	−	+ 0 +	(BOV)
Odd	−	0 0 −	(OOV)
Odd	+	0 0 +	(OOV)

positive and negative pulse balance, so there is no resulting dc component. Also, any errors occurring during the transmission through a line can be detected because the received sequence would not comply with the B3ZS coding rules. In other words, an error would cause the number of 1s between violation pulses to be even instead of odd.

Example. Figure 2.17 shows the data stream to be transmitted and in this case the first pulse is considered positive.

a. Data to be transmitted   1 1 0 0 0 0 1 1 0 0 0 0

b. B3ZS coded data   1 1 <u>0 0 V</u> 0 1 1 <u>B 0 V</u> 0

    (if last V positive)   + −(0 0 −) 0 + −(+ 0 +) 0

c. B3ZS coded data   1 1 <u>B 0 V</u> 0 1 1 <u>B 0 V</u> 0

    (if last V negative)   + −(+ 0 +) 0 − +(− 0 −) 0

For the first string of three zeros, either 00V or B0V must be substituted. If the last V pulse were positive (case *b*), the next V would be negative to preserve the AMI rule for the violation pulses. The last data pulse before the V was negative so the negative V is in violation, and 00V is substituted. For the next three zeros the substituted V must be positive. Because the last data pulse was negative, a B pulse must be included with positive polarity to make V a true violation with respect to the previous pulse. A different result would be attained if the last violation before this sequence started were positive (case *c*). An analysis could also achieve the same outcome using the rules of Table 2.3.

Another code used in North America is the binary six-zero substitution (B6ZS) at the 6.312-Mb/s rate for *one symmetric pair*. Bipolar violations are introduced in the second and fifth bit positions of a six-zero substitution as follows:

$$\text{Last pulse } (+) \rightarrow \text{code to substitute is } 0 + - 0 - +$$

$$\text{Last pulse } (-) \rightarrow \text{code to substitute is } 0 - + 0 + -$$

i.e., each block of six successive zeros is replaced by 0VB0VB.

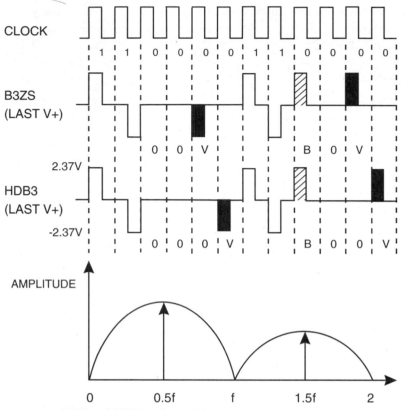

Figure 2.17 B3ZS and HDB3 codes and their spectra.

Example. Assuming the last pulse before the first block of six zeros to be positive:

Data to be transmitted: 1 <u>0 0 0 0 0 0</u> 1 0 1 1 <u>0 0 0 0 0 0</u> 1 0 1
B6ZS coded data: + (0 + − 0 − +) 0 − 0 + − (0 − + 0 + −) + 0 −

The B8ZS line code is specified by ITU-T Recommendation G.703 for *one coaxial pair* at 6.312 Mb/s. In this code, each block of eight successive zeros is replaced by 000VB0VB.

HDB3 code. The purpose of the HDB3 code is also to limit the number of zeros in a long bit stream, and in this case the limit is three zeros. This assures clock extraction in the regenerator of the receiver. This code is recommended by the ITU-T (Rec. G.703) for the input and output bit streams of 2-, 8-, and 34-Mb/s systems. An example is shown in Fig. 2.17. Longer sequences of more than three zeros are avoided by the replacement of one or two zeros by pulses according to specified rules. These rules ensure that the receiver recognizes that these pulses are replacements for zeros and does not confuse them with code pulses. This is achieved by selecting the polarity of the pulse so as to violate the alternate mark inversion polarity of the AMI code. Also, the replacement pulses themselves must not introduce an appreciable dc component. The rules for HDB3 coding can be summarized as follows:

1. Invert every second *1* for as long as a maximum of three consecutive *zeros* appear.

2. If the number of consecutive *zeros* exceeds three, set the *violation pulse* in the fourth position. The violation pulse purposely violates the AMI rule.

3. Every alternate violation pulse shall change polarity. If this rule cannot be applied, set a *1* according to the AMI rule in the position of the first *zero* in the sequence.

By observing the polarities of the preceding data pulse and the violation pulse, Table 2.4 summarizes the substitution sequence.

The HDB3 code can also be summarized as follows:

1. Apply the three rules step by step.

2. 000V and B00V generation:
   a. 000V is substituted if there is an odd number of 1s since the last violation pulse.
   b. B00V is substituted if there is an even number of 1s since the last violation pulse. The B pulse follows the AMI rule for its polarity.

The HDB3 code is very similar to the B3ZS code. The only difference is that for B3ZS it is the third zero that is substituted, whereas for HDB3 it is the fourth.

Manchester coding (biphase). This line code is not usually associated with multiplexers, but it is an important code because it is used in Ethernet and token-ring local area networks. In Manchester coding (Fig. 2.18), a binary 0 is represented by a pulse that is negative for the first half of the bit duration

TABLE 2.4   HDB3 Substitution Sequence

a	b	c	d	e	
Polarity of last V pulse	Polarity of preceding data	Number of B pulses since last V pulse	Polarity of last B pulse	Substitution sequence	
+	+	Even	+	− 0 0 −	(B 0 0 V)
−	−	Even	−	+ 0 0 +	(B 0 0 V)
+	−	Odd	−	0 0 0 −	(0 0 0 V)
−	+	Odd	+	0 0 0 +	(0 0 0 V)

Either (a + b = e) or (c + d = e).

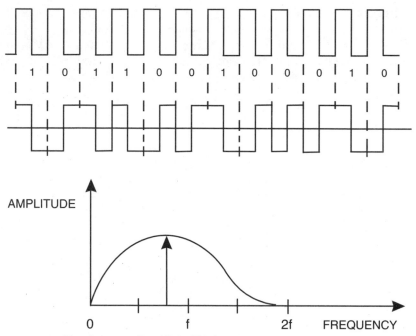

Figure 2.18 Manchester coding (digital biphase).

and positive for the second half. Conversely, a binary 1 is represented by a pulse that is positive for the first half of the bit duration and negative for the second half. The advantages of this code are (1) it has no dc component, and (2) it always has mid-bit transitions (so it is easy to extract timing information). It has error-rate performance identical to polar NRZ codes. The main disadvantages are (1) it has larger bandwidth than other common codes, and (2) it does not have the error-detecting capability of BNZS or HDB3.

Coded mark inversion (CMI) encoding.   The CMI is a line code in which the 1 bits are represented alternately by a positive and a negative state, while the 0 bits are represented by a negative state in the first half of the bit interval and a positive state in the second half of the bit interval. The CMI code specifications are given in ITU-T Recommendation G.703 as the interface code for signals transmitted at 139.264 Mb/s. It is also the electrical interface for the SDH rate of 155.52 Mb/s, known as the STM-1e (Sec. 2.5). An example of CMI encoding is shown in Fig. 2.19.

Comparison of codes.   To summarize, Fig. 2.20 is a comparison of codes used for the transmission of a digital bit stream. Note that in reality the clock frequency would not be the same for all codes, because they are designed for different operating bit rates. This diagram provides an easy comparison of the way in which the codes described above differ from each other.

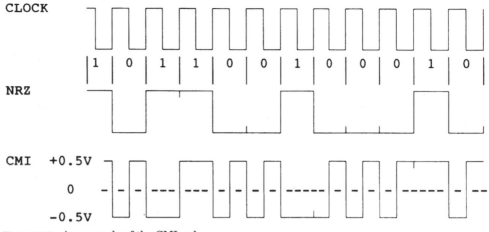

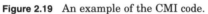

Figure 2.19 An example of the CMI code.

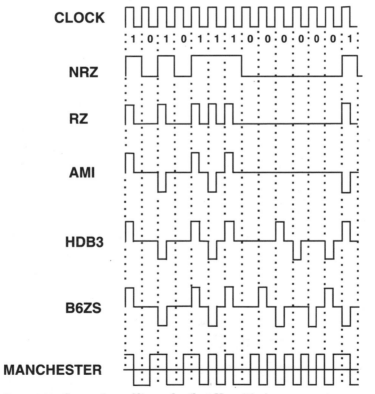

Figure 2.20 Comparison of line codes (last V positive).

2.3   Plesiochronous Higher-Order Digital Multiplexing (North America)

2.3.1   PDH multiplexing (North America)

The 24-channel PCM system is only the first, or primary, order of digital multiplexing. If it is necessary to transmit more than 24 channels, the system is built up as in the *plesiochronous digital hierarchy* (PDH) diagram of Fig. 2.21.

In the North American hierarchy, four primary systems are combined (multiplexed) to form an output having 96 channels. This is the second order of multiplexing. Seven 96-channel systems can be multiplexed to give an output of 672 channels (third order of multiplexing). Six 672-channel systems are multiplexed to give an output of 4032 channels (fourth order). Higher orders of multiplexing are now available, but all new multiplexers from the fourth order upward are *synchronous digital hierarchy* (SDH) multiplexers. Details of these higher-order PDH multiplexers are given in Table 2.5.

Plesiochronous multiplexing, as described elsewhere, is a method of multiplexing that is nonsynchronous. Although this is now the older technology, it will be beyond the year 2000 before it is all replaced by the newer synchronous multiplexing equipment.

The first level of multiplexing was described in Sec. 2.1.5. The line code used for transmission of this 1.544-Mb/s bit stream is the AMI code with 50 percent duty cycle (i.e., the pulse width is one-half of one bit interval). Notice

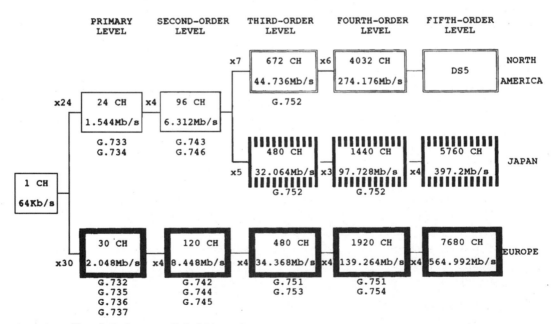

Figure 2.21   The plesiochronous digital hierarchy.

TABLE 2.5   PDH for North American Systems

Level	No. of channels	Bit rate, Mb/s	Line code
DS1	24	1.544	Bipolar
DS2	96	6.312	Bipolar (B6ZS)
DS3	672	44.736	Bipolar (B3ZS)
DS4	4032	274.176	Polar bipolar

that there is no simple relationship between the bit rates for this hierarchy and the European hierarchy. This has been the source of considerable difficulty in interfacing international traffic between different regions of the world. The problem will be resolved satisfactorily only when there is global adoption of the *synchronous* multiplexing hierarchy as described in Sec. 2.5. This new hierarchy has been created with a view to incorporating the existing plesiochronous bit rates from the North American, European, and Japanese hierarchies.

2.3.2   Second-order (DS2) PDH multiplexing (1.544 to 6.312 Mb/s)

The 6.312-Mb/s output of a second-order (DS2) multiplexer is created by multiplexing *four* first-order (DS1) multiplexer outputs. This is done by interleaving the bit streams of the four primary systems. Each individual bit stream is called a *tributary*. The main problem to overcome in this process is the organization of the four incoming tributaries. As just mentioned, there are two categories of digital multiplexers:

1. *Synchronous* digital multiplexers

2. *Plesiochronous* digital multiplexers

   Synchronous digital multiplexers have tributaries with the same clock frequency, and they are all synchronized to a master clock. *Plesiochronous digital multiplexers* have tributaries that have the same nominal frequency (that means there can be a small difference from one to another), but they are not synchronized to each other. The difference between the two types becomes apparent when one imagines the situation at a point where the four tributaries merge. For the synchronous case, the pulses in each tributary all rise and fall during the same time interval. For the plesiochronous case the rise and fall times of the pulses in each tributary do not coincide with each other. The plesiochronous designs have been globally deployed, but the synchronous systems are rapidly gaining acceptance and will become prominent in the near future. The multiplexing of several tributaries can be achieved by either:

1. Bit-by-bit multiplexing (bit interleaving)

2. Word-by-word multiplexing (byte interleaving)

Figure 2.22*a* and *b* illustrates the difference between bit and byte interleaving. The terms are self-explanatory. For example, in Fig. 2.22*a* there are four bit streams (tributaries) to be multiplexed. One bit is sequentially taken from each tributary so that the resulting multiplexed bit stream has every fifth bit coming from the same tributary. In Fig. 2.22*b* a specific number of bits (usually 8), forming a word, are taken from each tributary in turn.

Byte interleaving sets some restraints on the frame structure of the tributaries and requires a greater amount of memory capacity. Bit interleaving is much simpler because it is independent of frame structure and also requires less memory capacity. Bit interleaving is used for North American and European PDH systems. A typical 6.312-Mb/s plesiochronous multiplexer has four primary (DS1) multiplexers, each having an output of 1.544 Mb/s, bit interleaved to form the next level in the hierarchy. Note that this output bit rate of 6.312 Mb/s is not exactly four times the tributary bit rates of 1.544 Mb/s. This is a result of the nonsynchronous nature of the system, and will now be

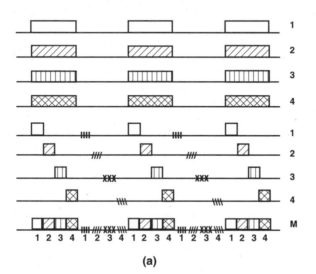

(a)

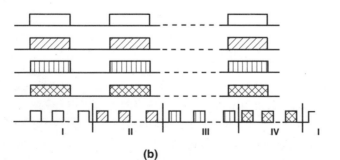
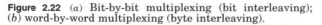

(b)

Figure 2.22 (*a*) Bit-by-bit multiplexing (bit interleaving); (*b*) word-by-word multiplexing (byte interleaving).

studied in detail. Every tributary has its own clock. Every tributary is timed with a *plesiochronous frequency*, that is, a nominal frequency about which the shifts around it lie within prefixed limits. For example, the primary multiplexer output is 1.544 Mb/s ± 50 ppm. To account for the small variations of the tributary frequencies about the nominal value when multiplexing four tributaries to the next hierarchical level, a process known as *positive stuffing* (also known as *positive justification*) is used.

2.3.3 Positive pulse stuffing (or justification)

Pulse stuffing involves intentionally making the output bit rate of a channel higher than the input bit rate. The output channel therefore contains all the input data plus a variable number of "stuffed bits" that are not part of the incoming subscriber information. The stuffed bits are inserted, at specific locations, to pad the input bit stream to the higher output bit rate. The stuffed bits must be identified at the receiving end so that "destuffing" can be done to recover the original bit stream. Pulse stuffing is used for higher-order multiplexing when each of the incoming lower-level tributary signals is unsynchronized, and therefore bears no prefixed phase relationship to any of the others. The situation is shown by the simplified diagrams of the two-channel multiplexer in Fig. 2.23a and b. The input bit rates, in Fig. 2.23a, are *exactly* the same and the pulses arrive synchronized. The output subsequently has a perfect byte-interleaved bit sequence. However, in Fig. 2.23b, the input bit rates are *not* identical so the pulses do not arrive in a synchronized manner. The difference in bit rates is exaggerated for the purpose of the example. The situation is similar for a plesiochronous multiplex system, where the input bit rates are very close but not exactly the same, and there is a timing offset due to the unsynchronized arrival of the bit streams. The output of the system in Fig. 2.23b requires some additional (stuffed) pulses in order to make up the difference in input bit rates. Each frame in the output signal of Fig. 2.23b contains a fixed number of time slots. Each frame contains a stuffing control bit (C) and a stuffing bit (S). When a control bit is a 0, the (S) bit contains real data information. When a (C) bit is a 1, the respective (S) bit is a stuffed bit (i.e., a 1).

The higher-order plesiochronous multiplexed outputs are composed of frames. In the formation of the DS2 frame from four DS1 tributaries, each tributary signal is written into a memory by the 1.544-Mb/s clock extracted from the incoming signal. The data is read out by a multiplexing frequency clock at 1.578 Mb/s. Writing of customer data is inhibited at specific time slots for the inclusion of frame alignment bits, service bits, and stuffing control bits. The stuffing control message, requesting a stuffing bit, is generated when the memory reading and writing phases reach a predetermined threshold.

At the receiving end, the writing clock has the same characteristics as those of the transmit reading clock. That is, it has a frequency that is on average the same as that of the tributary, but it presents periodic spaces for the frame structure and random spaces for the stuffing process. A *phase-locked loop* (PLL) circuit is used to reduce:

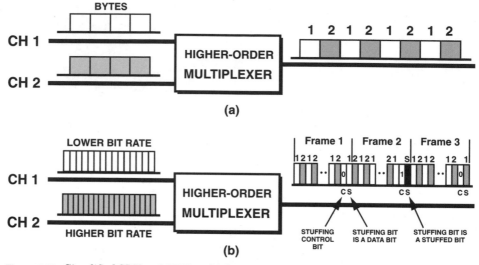

Figure 2.23 Simplified SDH and PDH multiplexing. (*a*) Synchronous byte interleaving (no stuffing needed); (*b*) plesiochronous bit interleaving (stuffing needed).

1. Jitter caused by the frame structure
2. High-frequency jitter components (waiting time) caused by stuffing
3. Tributary signal jitter
4. Jitter introduced by the 6.312-Mb/s link

In summary, the positive stuffing method involves the canceling of a clock pulse assigned to a particular tributary in some of the frames in order to coordinate the timing of the plesiochronous tributaries into a multiplexed output. Random spaces are therefore created in the frame, as well as periodic spaces. In the periodic spaces frame alignment word bits, service bits, and stuffing control bits are inserted. The tributary information bits are inserted in the random spaces in the absence of stuffing, or a logic 1 is used when stuffing takes place. Remember, the stuffing pulses carry no subscriber information.

2.3.4   The 6-Mb/s PDH frame structure

Figure 2.24 displays the structure of the frame for the second-order, DS2, 6.312-Mb/s multiplexer formed by the plesiochronous multiplexing technique. Each frame contains:

1. The frame alignment word
2. The multiframe alignment word
3. Alarm control bits
4. Stuffing control bits (stuffing message)
5. Stuffing bits
6. Tributary bits

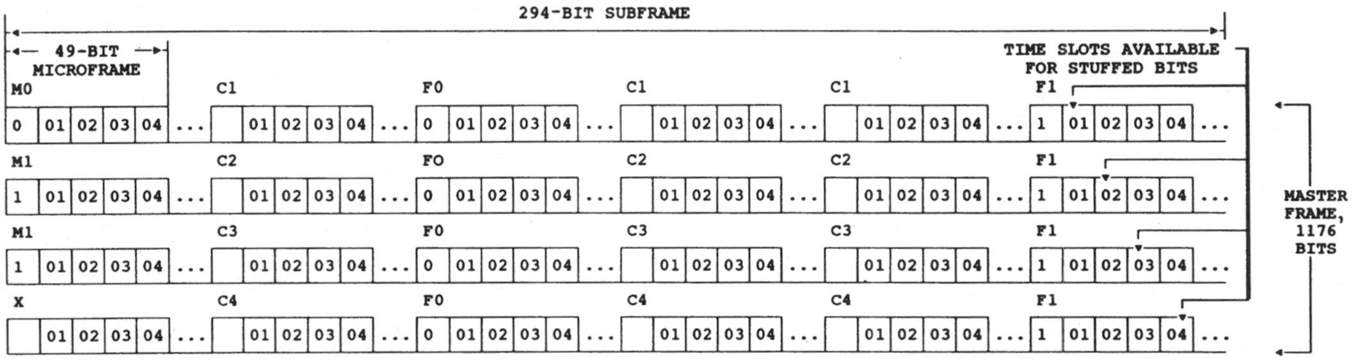

Figure 2.24 The 6.132-Mb/s frame format.

Notes: (1) The frame alignment signal is F0 = 0 and F1 = 1.

(2) M0, M1, M1, X is the multiframe alignment signal and is 011X, where X is an alarm service digit. The normal (no alarm) state is X = 1.

(3) C1, C1, C1 and C2, C2, C2, and C3, C3, C3 and C4, C4, C4 are the stuffing (indicators) message words for DS1 input channels 1, 2, 3, and 4, respectively, where 000 indicates no stuffing and 111 indicates stuffing is required.

Every frame has 1176 bits. Each of the four subframes contains 294 bits, and each microframe within the subframes has 49 bits. Every microframe starts with an "overhead" (housekeeping) bit (e.g., frame alignment) followed by bits taken sequentially from the four input tributaries on a bit-by-bit basis (by bit interleaving). Each frame has a 4-bit multiframe alignment word 011X, where X is the alarm service digit (X = 1 means no alarm). Each subframe contains the frame alignment signal 01, where 0 is always the first bit in the third microframe and 1 is always the first bit in the sixth microframe. Each subframe also has a 3-bit stuffing message (indicator word) formed from the first bit of the second, fourth, and fifth microframes (e.g., subframe 1 has the stuffing indicator word for tributary 1, etc.). The word 000 indicates no stuffing is required, whereas 111 denotes stuffing is required. The stuffing bits, if required, are inserted in the first time slot designated for each tributary following the frame alignment bit in the sixth microframe of each subframe. Finally, of the 1176 bits in each frame, 1148 to 1152 bits are information bits, depending on how many stuffing bits are required. Note that the B6ZS line code is used at the output from this level of multiplexer.

It is important to notice that the output frame structure of all higher-order multiplexers is unrelated to the frame structure of the lower levels making up the input tributaries. As far as the higher-order multiplexer is concerned, each input tributary is merely a bit stream, and it does not care that it is composed of a highly sophisticated bit structure, within which frame alignment bits, multiframe alignment bits, and signaling bits are all transmitted along with the information bits containing telephone calls, computer data, or video images.

2.3.5   PDH DS3 multiplexing (6 to 45 Mb/s)

For the next level of multiplexing, seven 6.312-Mb/s input tributaries are plesiochronously multiplexed to give an output of 44.736 Mb/s. The frame format shown in Fig. 2.25 indicates a 4760-bit frame containing seven subframes, each having 680 bits. Each subframe comprises eight microframes, each having 85 bits. Every microframe starts with an overhead bit followed by bits taken sequentially from the seven input tributaries on a bit-interleaving basis. The multiframe alignment signal in this case appears only in the fifth, sixth, and seventh subframes, and is designated 010. This time slot for the other four subframes is for alarm and parity information. Subframes 1 and 2 might be used for the alarm service channel, where XX must be identical, and are usually 1s (i.e., X = 1 for no alarm). The two Ps in subframes 3 and 4 are parity bits formed by considering parity over all *information* bits in the preceding frame. If the digital sum of all information bits is 1, PP = 11. If the sum is 0, PP = 00. The frame alignment signal is F0 = 0 and F1 = 1, and each pair appears twice in each subframe. The stuffing indicator words are the same as previously discussed except that here they appear in microframes 3, 5, and 7. The stuffing bits, if required, are placed in the final microframe of each subframe in the first time slots for their respective tributaries. The B3ZS line code is used at the output from this level of multiplexer.

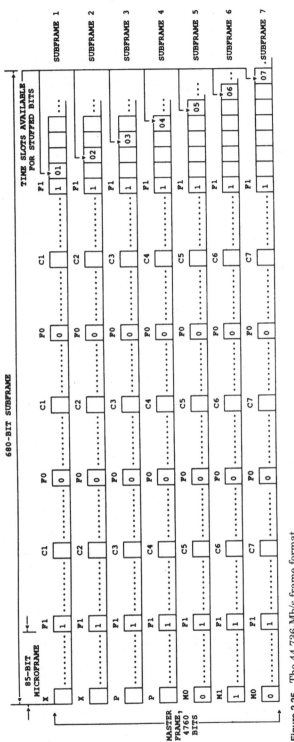

Figure 2.25 The 44.736-Mb/s frame format.

Notes: (1) The frame alignment signal is F0 = 0 and F1 = 1.

(2) M0, M1, M0 is the multiframe alignment signal and appears in the fifth, sixth, and seventh subframes. M0 = 0 and M1 = 1.

(3) PP is parity information taken over all message time slots in the preceding M frame. PP = 1 if the digital sum of all message bits is 1 and PP= 0 if the sum is 0. These two parity bits are in the third and fourth M subframes.

(4) XX is for alarm indication. In any one M frame the two bits must be identical. Presently XX = 1.

(5) C1, C1, C1 to C7, C7, C7 are the stuffing (indicators) message words for DS2 input channels 1 to 7, respectively, where 000 indicates no stuffing and 111 indicates stuffing is required.

55

2.3.6 PDH DS4 multiplexing (45 to 274 Mb/s)

For the next level of multiplexing, six 44.736-Mb/s input tributaries are plesio-chronously multiplexed to give an output of 274.176 Mb/s having a two-level binary line code. The frame format shown in Fig. 2.26 indicates a 4704-bit frame containing 24 subframes, each having 196 bits. Each subframe comprises two microframes each having 98 bits. In this scheme, the first two bits of each microframe are devoted to overhead, followed by bits taken sequentially from the six input tributaries on a bit-interleaving basis. The multiframe alignment (M), alarm (X), and stuffing indicator (C) bits occur in complementary pairs (i.e., the second bit is the inverse of the first). The multiframe alignment bits appear in the first three subframes and follow the sequence 101. The alarm bits occur in the second three subframes. The stuffing indicator words occur over the remaining six sets of three subframes. As previously, the word 111 indicates stuffing is required, whereas 000 indicates stuffing is not required. The time slot for stuffing is the eighth slot for a given tributary after the last bit of the stuffing indicator word has appeared for that particular tributary. The P bits appear as identical pairs of bits, where P1 indicates even parity over the 192 previous *odd*-numbered information bits and P2 indicates even parity over the 192 previous *even*-numbered information bits.

2.4 Plesiochronous Higher-Order Digital Multiplexing (Europe)

2.4.1 PDH multiplexing (Europe)

Europe and many other regions of the world use the PDH hierarchy in Fig. 2.21. In this hierarchy, four primary systems are combined (multiplexed) to form an output having 120 channels (second order of multiplexing). Similarly, four 120-channel systems can be multiplexed to give an output of 480 channels (third order). Four 480-channel systems are multiplexed to give an output of 1920 channels (fourth order). Four 1920-channel systems are multiplexed to give an output of 7680 channels (fifth order). All new multiplexers from the fourth order upward are now *synchronous digital hierarchy* (SDH) multi-plexers. The approximate bit rate for each PDH multiplexer level is shown in Table 2.6.

TABLE 2.6   PDH for European Systems

Level	No. of channels	Bit rate, Mb/s	Line code
First	30	2.048	HDB3
Second	120	8.448	HDB3
Third	480	34.368	HDB3
Fourth	1920	139.264	CMI
Fifth	7680	565.992	CMI

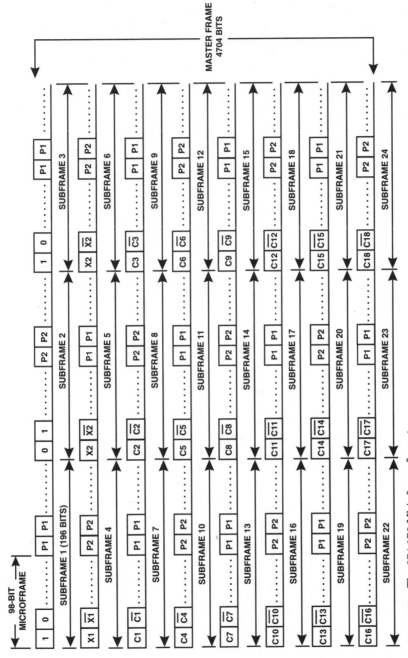

Figure 2.26 The 274.176-Mb/s frame format.
Notes: (1) M, X, and C bits appear as complementary pairs. (a) M appears in the first three subframes and must be 10, 01, 10. (b) X appears in the fourth, fifth, and sixth subframes. (c) C1, C1, C1 to C6, C6, C6 are the stuffing (indicators) message words for DS3 input channels 1 to 6, respectively, where 000 indicates no stuffing and 111 indicates stuffing is required. The time slot for stuffing is the eighth message bit position following the last C1 bit.
(2) The P bits appear as identical pairs and are for even parity over the 192 previous odd-numbered message bits.
(3) All other bits are message (information) bits.

57

2.4.2   The PDH 8-Mb/s frame structure

The frame structure is shown in Fig. 2.27. Each frame contains 848 bits and is divided into four subframes, each containing 212 bits. The first 12 bits in every frame contain 10 frame alignment bits (1111010000) and 2 service bits. The service bits relay information concerning alarms, synchronization errors, etc. A 3-bit stuffing control word for each tributary occupies a total of 12 bits (4 bits after each of the first three subframes). Finally, the stuffing bit for each tributary is inserted (if required) before the start of the fourth subframe. Each tributary signal is written into a memory by the 2.048-Mb/s clock extracted from the incoming signal. The data is read out by a multiplexing frequency clock at 2.112 Mb/s. Writing is inhibited at specific time slots for the inclusion of frame alignment bits, service bits, and stuffing control bits. The stuffing control message, requesting a stuffing bit, is generated when the memory reading and writing phases reach a predetermined threshold.

When a stuffing pulse is to be inserted, the stuffing control message is formed for the particular tributary concerned. This consists of setting the stuffing control message to 111 (i.e., a 1 in each of the three periodic spaces following each tributary information). The three sets of four periodic spaces allow the formation of the 3-bit stuffing control message for each of the four tributaries. When stuffing does not take place, the stuffing control message is 000. This is illustrated in Fig. 2.28 where the first C subscript denotes the *tributary* number and the second subscript denotes 1 of the 3 bits in the stuffing message word for that tributary. At the receiving end, if a majority of 1s (two out of three) is detected in the stuffing control message, the stuffing pulse is canceled by nulling the receive writing clock for that pulse period (i.e., the stuffing bit is not part of the required received information). Conversely, if a majority of 0s (two out of three) is received for the stuffing control message, the stuffing pulse time slot is not canceled, thus allowing a data bit (1 or 0) to be received.

The receive writing clock has the same characteristics as those of the transmit reading clock. That is, it has a frequency that is on average the same as that of the tributary, but it presents periodic spaces for the frame structure and random spaces for the stuffing process. Again, a PLL circuit is used to reduce jitter.

2.4.3   PDH third-order multiplexing (8 to 34 Mb/s)

There are many similarities between the third order and the second order of multiplexing. Four 8.448-Mb/s tributaries are bit interleaved to produce a 480-channel output at a bit rate of 34.368 Mb/s. The frame structure is shown in Fig. 2.29. Each frame consists of 1536 bits. The frame is divided into four subframes each containing 384 bits. As in the second-order multiplexer, the first 12 bits of the frame are reserved for the 10-bit frame alignment word and the 2 service bits. Also, stuffing control and stuffing bits are organized as in the second-order system.

SUBFRAME	SIGNAL	NUMBER OF BITS	BIT NUMBERING
1	-Alignment word 1111010000 -Service bits -Tributary bits	10 2 200	1 to 10 11 to 12 13 to 212
2	-Stuffing message -Tributary bits	4 208	213 to 216 217 to 424
3	-Stuffing message -Tributary bits	4 208	425 to 428 429 to 636
4	-Stuffing message -Stuffing bits -Tributary bits	4 4 208	637 to 640 641 to 644 645 to 848

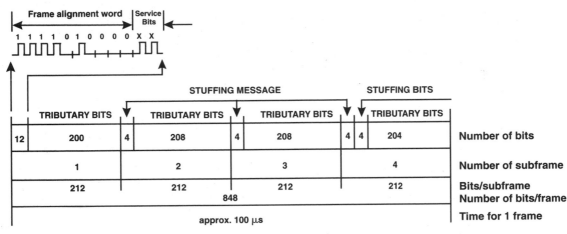

Figure 2.27 The 8 Mb/s-frame structure (positive stuffing/justification).

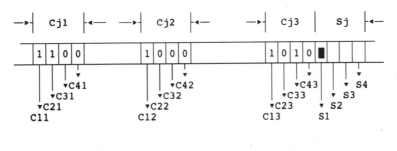

C11, C12, C13 = 1,1,1 ----▸ S1 IS A STUFFING BIT

C21, C22, C23 = 1,0,0 ----▸ S2 IS AN INFORMATION BIT

C31, C32, C33 = 0,0,1 ----▸ S3 IS AN INFORMATION BIT

C41, C42, C43 = 0,0,0 ----▸ S4 IS AN INFORMATION BIT

Figure 2.28 The formation of the stuffing message word.

SUBFRAME	SIGNAL	NUMBER OF BITS	BIT NUMBERING
1	-Alignment word 1111010000	10	1 to 10
	-Service bits	2	11 to 12
	-Tributary bits	372	13 to 384
2	-Stuffing message	4	385 to 388
	-Tributary bits	380	389 to 768
3	-Stuffing message	4	769 to 772
	-Tributary bits	380	773 to 1152
4	-Stuffing message	4	1153 to 1156
	-Stuffing bits	4	1157 to 1160
	-Tributary bits	376	1161 to 1536

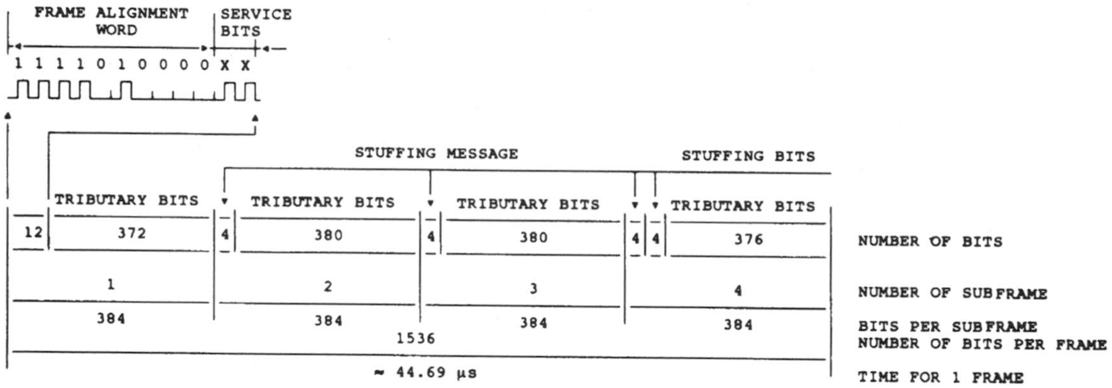

Figure 2.29 The 34-Mb/s frame structure (positive stuffing/justification).

2.4.4 PDH fourth-order multiplexing (34 to 140 Mb/s)

Again, the fourth-order multiplexer has a similar structure to the second- and third-order multiplexers, but it also has some important differences. In this case four third-order outputs at 34 Mb/s are bit interleaved to give a 1920-channel output at a bit rate of 139.264 Mb/s. The frame structure is shown in Fig. 2.30. The 2928-bit frame is divided into *six* subframes each containing 488 bits. Note that this is different from the second- and third-order multiplexers, which each had only *four* subframes. The FAW is different from the second- and third-order systems (i.e., 111110100000), occupying the first 12 bits in this case. Also, the service bits occupy 4 bits here instead of the 2 bits in the other systems, and the stuffing control message has 5 bits instead of the 3 bits in the other systems. The actual stuffing bits are the same as the other systems (i.e., 1 bit for each tributary per frame). Finally, the line code used for the transmission of 140 Mb/s has been designated by the ITU-T to be the CMI code instead of the HDB3 code used for the first-, second-, and third-order transmission codes.

2.4.5   PDH fifth-order multiplexing (140 to 565 Mb/s)

The fifth order of multiplexing is a natural progression from the previous levels. Four 139.264-Mb/s tributaries are bit interleaved to give a 7680-channel output at 564.992 Mb/s. The frame structure is shown in Fig. 2.31. The 2688-bit frame is divided into *seven* subframes each containing 384 bits. The frame alignment word occupies the first 12 bits. The service bits occupy 4 bits in the seventh subframe. The stuffing control message has 5 bits, composed of 1 bit at the beginning of subframes 2, 3, 4, 5, and 6, as in the case of the fourth level of multiplexing. The actual stuffing bits are the same as the other systems, that is, 1 bit for each tributary per frame. In this case the actual stuffing bits are in the seventh subframe. Finally, the line code used for the transmission of 565 Mb/s has been designated by the ITU-T to be the CMI code.

SUBFRAME	SIGNAL	NUMBER OF BITS	BIT NUMBERING
1	-Alignment word 111110100000	12	1 to 12
	-Service bits (1 for alarm)	4	13 to 16
	-Tributary bits	472	17 to 488
2	-Stuffing message	4	489 to 492
	-Tributary bits	484	493 to 976
3	-Stuffing message	4	977 to 980
	-Tributary bits	484	981 to 1464
4	-Stuffing message	4	1465 to 1468
	Tributary bits	484	1469 to 1952
5	-Stuffing message	4	1953 to 1956
	-Tributary bits	484	1957 to 2440
6	-Stuffing message	4	2441 to 2444
	-Stuffing bits	4	2445 to 2448
	-Tributary bits	480	2449 to 2928

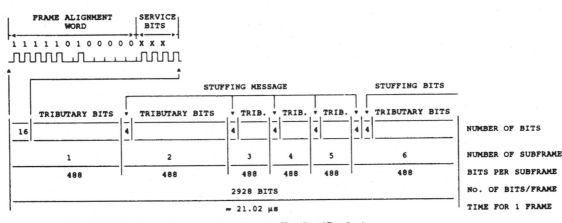

Figure 2.30 The 140-Mb/s frame structure (positive stuffing/justification).

SUBFRAME	SIGNAL	NUMBER OF BITS	BIT NUMBERING
1	-Alignment word -Tributary bits	12 372	1 to 12 13 to 384
2	-Stuffing message -Tributary bits	4 380	385 to 388 389 to 768
3	-Stuffing message -Tributary bits	4 380	769 to 772 773 to 1152
4	-Stuffing message Tributary bits	4 380	1153 to 1156 1157 to 1536
5	-Stuffing message -Tributary bits	4 380	1537 to 1540 1541 to 1920
6	-Stuffing message -Tributary bits	4 380	1921 to 1924 1925 to 2304
7	-Service bits -Stuffing message -Tributary bits	4 4 376	2305 2308 2309 2312 2313 2688

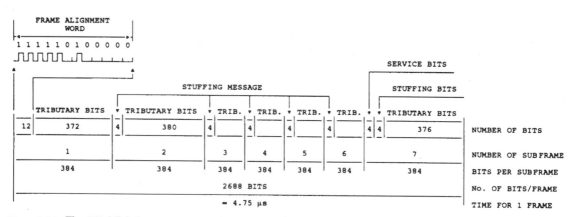

Figure 2.31 The 565-Mb/s frame structure (positive stuffing/justification).

2.5 Synchronous Digital Hierarchy (SDH) Multiplexing

Plesiochronous multiplexers have the benefit of operating independently without a master clock to control them. Each PDH multiplexer has its own independent clock. This so-called plesiochronous transmission has small differences in frequency from one multiplexer to another, so that when each provides a bit stream for the next hierarchical level, bit stuffing (justification) is necessary to adjust for these frequency differences. Despite the attractive aspects of plesiochronous multiplexing, there is one major drawback. If, for example, a 274-Mb/s system is operating between two towns, it is not possible to identify and gain access to individual channels at locations en route. In other words, add and drop (drop and insert) capability requires a complete demultiplexing procedure. In contrast, the synchronous multiplexing technique does allow this easy add and drop facility, and also allows multiplexing of tributaries that have different bit rates.

In 1988, the ITU-T reached an agreement on a worldwide standard for SDH, in the form of Recommendations G.707, 708, and 709. In addition to being a technical milestone, this agreement also unified the bit rates so that this new synchronous system did not have the existing interface problems between North America and Japan and the rest of the world. The resulting Recommendations were intended for application to optical fiber transmission systems and were originally called the synchronous optical network (SONET) standard. Although SDH now supersedes the SONET description, they both refer to the same subject matter and are used interchangeably in the literature. During the development stages of the new SDH, it was essential to establish a system which allowed both North American and European hierarchical bit rates to be processed simultaneously. This led to extensive discussions as to how the synchronous system should be constructed. Each negotiating team proposed solutions that favored its own existing systems, and eventually both sides had to make a compromise to ensure a successful outcome.

2.5.1 Network node interface

For international communications and the growing demand for broadband services, it has become increasingly essential to specify a universal *network node interface* (NNI). The NNI is acquiring increasing importance with respect to digital telecommunication systems. Within a communications system a transport network has two elementary functions: first, a transmission facility, and second, a network node. The transmission facility has various media such as optical fiber, microwave radio, and satellite. The network node performs terminating, crossconnecting, multiplexing, and switching functions. There can be various types of node such as a 64-kb/s-based node or broadband nodes. The NNI is the point at which the transmission facility and the network node meet. The objectives or requirements for a NNI are that the network should have:

- Worldwide universal acceptance
- Unique interface for transmission, multiplexing, crossconnecting, and switching of various signals
- Improved operation and maintenance capabilities
- Easy interworking with existing interfaces
- Easy future provision of services and technologies
- Application to all transmission media

An example of NNIs within a network is presented in Fig. 2.32. This diagram shows typical communication between two network users. In this case, there are two digital switches. The input to or the output from a digital switch

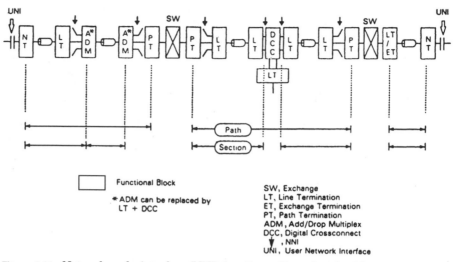

Figure 2.32 Network node interface (NNI) locations. *(Reproduced with permission from IEEE. Asatani, K., K. R. Harrison, and R. Ballart, CCITT Standardization of Network Node Interface and Synchronous Digital Hierarchy, IEEE Communications Magazine, August 1990, pp. 15–20. © 1990 IEEE.)*

is via a path termination (PT), line termination (LT), or CO termination. The equipment between the path terminations is referred to as the *path*. Each transmission line, which could be an optical fiber or PCM cable, is terminated by line termination equipment. At the end of a *section* of transmission line there might be add/drop multiplexers (ADMs) to enable local traffic to enter or leave the network, or there could be a digital crossconnect (DCC) between line terminals. In this example, it is the points between the LT/ADM, ADM/PT, PT/LT, and LT/DCC that are the NNIs. The implementation of the SDH is a very significant milestone in moving toward the achievement of a worldwide, universal NNI.

2.5.2 Synchronous transport signal frame

From the North American point of view, the basic building block and first level of the SDH is called the *synchronous transport signal–level 1* (STS-1) if electrical, or *optical carrier–level 1* (OC-1) if optical. The STS-1 has a 51.84-Mb/s transmission rate and is synchronized to the network clock. The STS-1 frame structure has 90 columns and 9 rows (Fig. 2.33). Each column has an 8-bit byte, so SDH multiplexing is a byte-by-byte multiplexing (byte-interleaving) process. The 8-bit bytes are transmitted row by row from left to right, and one complete frame is transmitted every 125 μs. Because there are $90 \times 9 = 810$ bytes in one frame, 810×8 bits are transmitted in 125 μs, so in one second $6480/125 \times 10^{-6} = 51.84$ Mbits are transmitted. The first three columns of the frame contain *section* and *line* overhead (housekeeping) bytes. The remaining 87 columns and 9 rows are used to carry the STS-1 *synchronous payload envelope* (SPE).

The SPE also includes 9 bytes of *path* overhead (Fig. 2.33). The STS-1 can carry a DS3 channel (44.736 Mb/s) or a variety of lower-order signals at DS1 (1.544 Mb/s), DS1C (3.152 Mb/s), and DS2 (6.312 Mb/s) rates.

Overhead.   The overhead columns are divided into *section*, *line,* and *path* layers as in Fig. 2.34. The SONET standard differs slightly from the SDH standard here. Because SONET was designed specifically for optical fiber equipment, the term *line* was introduced. SDH, on the other hand, is not media-specific, so it just uses the *section* and *path* overhead terminology, and some bytes in the section overhead are reserved for media-specific functions. The overhead is structured in this manner so that each network element (terminal, repeater, and ADM) needs only to access the necessary information. This enables cost savings by allowing engineers to design each network element so that it accesses only the *necessary* overhead instead of all

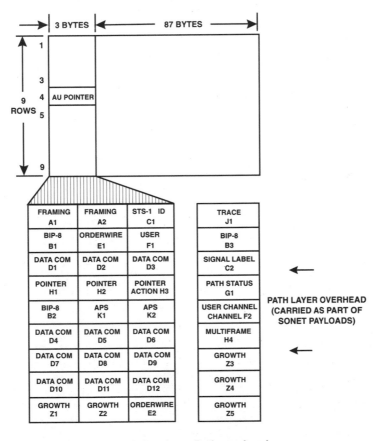

Figure 2.33 Synchronous transport signal—level 1 (STS-1) and overhead bytes.

of it. Figure 2.33 shows the position of the overhead bytes in the STS-1 frame. The section overhead has:

- Two bytes that show the start of each frame
- A frame identification byte (STS-1 ID)
- An 8-bit bit-interleaved parity (BIP-8) check for monitoring section errors
- An order-wire channel for maintenance purposes
- A channel for operator applications
- Three bytes for data communications maintenance information

When a SONET/SDH signal is scrambled, the only bytes that are not included in the scrambling process are the framing and identification bytes. The line overhead information is processed at all equipment regenerators. It includes:

- Pointer bytes
- A BIP-8 line error monitoring byte
- A 2-byte automatic protection switching (APS) message channel
- A 9-byte line data communications channel
- A line order-wire channel byte
- Bytes reserved for future growth

The path overhead bytes are processed at the terminal equipment. The path overhead includes:

- A BIP-8 for end-to-end payload error monitoring
- A byte to identify the type of payload being carried
- A byte to carry maintenance information signals
- A multiframe alignment byte

This very extensive array of overhead bits allows greatly enhanced maintenance, control, performance, and administrative capability of a network incorporating SONET/SDH equipment.

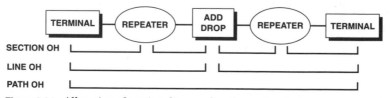

Figure 2.34 Allocation of section, line, and path overhead.

SONET/SDH bit rates. Higher-bit-rate synchronously multiplexed signals are obtained by word-by-word (byte) interleaving N frame-aligned STS-1s into an STS-N as in Fig. 2.35. In this manner the ITU-T standard level (155.52 Mb/s) or any other of the SDH levels is constructed as in Table 2.7.

Whereas all of the section and line overheads in the first STS-1 of an STS-N are used, many of the overheads in the remaining STS-1s are unused. Only the section overhead framing, STS-1 ID, BIP-8 channels, and the line overhead pointer and BIP-8 channels are used in *all* STS-1s in an STS-N. The STS-N is then scrambled and either converted to an optical carrier (whose line rate is one of the above hierarchical levels) or it could be transmitted over DMR. Note: the line rate in megabits per second for OC-N is exactly N times OC-1.

Fixed location mapping. The bit-stuffing technique for plesiochronously multiplexing several tributaries has already been discussed. The *fixed location mapping technique* has been used in the early SDH multiplexing equipment.

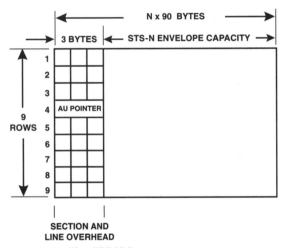

Figure 2.35 The STS-N frame.

TABLE 2.7   SONET / SDH Levels

	Level		Line rate, Mb/s
OC-1	STS-1	STM-0	51.84
OC-3	STS-3	STM-1	155.52
OC-12	STS-12	STM-4	622.08
OC-24	STS-24	STM-8	1244.16
OC-48	STS-48	STM-16	2488.32
OC-96	STS-96	STM-32	4976.64
OC-192	STS-192	STM-64	9953.28
OC-768	STS-768	STM-256	39813.12

This technique uses specific bit positions in a higher-rate SDH signal to carry lower-rate synchronous signals. Each frame position is dedicated to information for a specific tributary, and there is no pulse stuffing. However, there is no guarantee that the high-speed signal and the tributary will be phase aligned with each other. Small frequency differences between the high-speed signal and the tributary can occur because of small network synchronization deficiencies. To account for such variations, 125-μs buffers are incorporated at the multiplexer interfaces to phase align or even slip the signal. A slip is a repeat or deletion of a frame of information in order to correct any frequency differences. These buffers are undesirable, because slipping causes signal delay and subsequently signal impairment.

Payload pointer. The SDH/SONET standard contains a very innovative technique known as the *payload pointer*. It is used to frame-align STS-N or STM-N signals, and is also used for multiplexing synchronization of PDH signals. The payload pointer allows easy access to synchronous payloads without requiring 125-μs buffers. The payload pointer is a number carried in each STS-1 line overhead that indicates the location of the first byte of the STS-1 SPE payload within the STS-1 frame. These are bytes H1 and H2 in Fig. 2.33. The payload is therefore not locked to the STS-1 frame structure, as in the case of fixed location mapping, but floats within the STS-1 frame. The positions of the STS-1 section and line overhead bytes define the STS-1 frame structure. Figure 2.36 shows how the 9-row by 87-column SPE payload fits into two 125-μs STS-1 frames. If the STS-1 payload has any small variations in frequency, the pointer value increases or decreases to account for the variations. Figures 2.37 and 2.38 illustrate the mechanism. Consider the payload bit rate to be high compared to the STS-1 frame bit rate. The payload pointer is decreased by 1 and the H3 overhead byte is used to carry data information for one frame. This is equivalent to the negative stuffing process. If the payload bit rate is slow compared to the STS-1 frame bit rate, the data information byte immediately following the H3 byte is nulled for one frame, and the pointer is increased by 1. This is equivalent to a positively stuffed byte. By this mechanism, slips (lost data) are avoided. The phase of the synchronous STS-1 payload is always known by checking the pointer value. This technique combines the advantages of bit stuffing and fixed location mapping, resulting in a minimal cost for pointer processing. This is easily done by integrated circuit technology.

Concatenation. When several STS-1s are combined into an STS-N, frame alignment is facilitated by grouping the STS-1 pointers together for easy access at the receiving end, which therefore permits using a single framing circuit. For example, if three STS-1 payloads are required, the phase and frequency of the three STS-1s must be locked together, and the three signals are considered as a single signal that is transported through the network. This

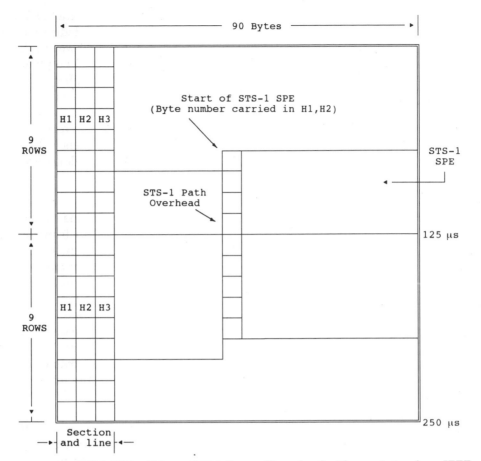

Figure 2.36 STS-1 SPE within an STS-1 frame. *(Reproduced with permission from IEEE. Ballart, R., and Y.-C. Ching, SONET: Now It's the Standard Optical Network, IEEE Communications Magazine, March 1989, pp. 8–14. © 1989 IEEE.)*

is called *concatenation*, and is achieved by using a *concatenation indication* in the second and third STS-1 pointers. The concatenation indication is a pointer value that notifies the STS-1 pointer processor that the present pointer being considered should have the same value as the previous STS-1 pointer. The STS-N signal that is locked by pointer concatenation is called an STS-Nc signal (c for concatenation). In North America an STS-3c signal would be considered three STS-1s, but Europe and other countries consider the STS-3c as the basic building block for the new synchronous digital hierarchy.

2.5.3 Synchronous transport module frame

Countries other than those in North America and Japan call the STS-3c the *synchronous transport module–level 1* (STM-1), and it has a bit rate of

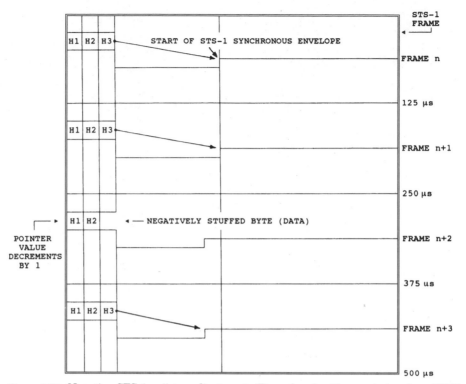

Figure 2.37 Negative STS-1 pointer adjustment. *(Reproduced with permission from IEEE. Ballart, R., and Y.-C. Ching, SONET: Now It's the Standard Optical Network, IEEE Communications Magazine, March 1989, pp. 8–14. © 1989 IEEE.)*

155.52 Mb/s. The STM-1 has a 9-row by 270-column frame structure as shown in Fig. 2.39. The STM-1 is also known as the 155.52-Mb/s administrative unit (AU). The overhead and payload pointers for the STM-1 frame have a similar format to the STS-1 frame.

The diagrams in Fig. 2.40 compare the STM-1 frame format with that of the STS-1 frame. The compatibility, or equivalence, is self-evident. Notice how the pointers are used to indicate the beginning of the SPE. The STS-1 is also known as a 51.84-Mb/s AU, and this is easily expanded to form the STS-3. To harmonize the North American and European-based notations, STS-1 is also called STM-0.

Nesting. As already stated with respect to international acceptance, it is essential that one type of AU (e.g., 155.52 Mb/s) in one country can be accepted by a country that has another type of AU (e.g., 51.84 Mb/s). This is where nesting is useful. A nested signal is a set of AUs contained within the STM-1 (i.e., 155.52-Mb/s AU), and this is then transported through the network. Figure 2.41 shows how a nested signal such as three AU-3s (at 45 Mb/s) can be carried in a European country in the STM-1 format without constructing a special network to manage it. Conversely, in North America, four European 34-Mb/s signals can be transported in an STS-3c (STM-1). Nested signals are used to carry bulk traffic through the network, whereas lower-bit-rate traffic such as DS1s, 2.048 Mb/s, etc., would be transported in virtual containers (virtual tributaries) as described later on.

2.5.4 Comparison of PDH and SDH/SONET interfaces

It is important to note that PDH signals can be synchronously multiplexed to the next hierarchical level without difficulty. However, it must be appreciated that the add and drop feature applies only to the complete PDH signal. Subportions of the PDH signal (e.g., one of the 45-Mb/s tributaries of a 274-Mb/s stream) can be accessed only by plesiochronously demultiplexing the whole PDH signal.

Compatibility of the present PDH with the new SDH is one of the main successes of the latest SDH Recommendations. Although there will no doubt eventually be a completely SDH network, a significant amount of time will elapse

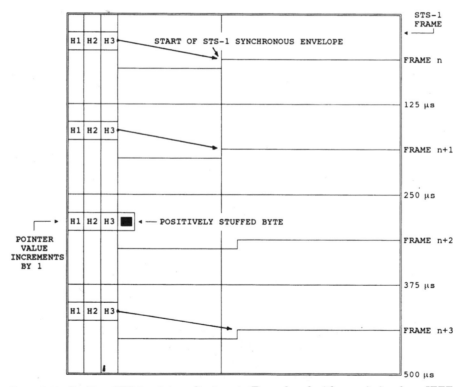

Figure 2.38 Positive STS-1 pointer adjustment. *(Reproduced with permission from IEEE. Ballart, R., and Y.-C. Ching, SONET: Now It's the Standard Optical Network, IEEE Communications Magazine, March 1989, pp. 8–14. © 1989 IEEE.)*

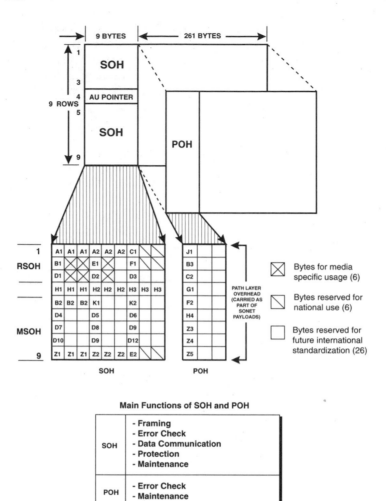

Figure 2.39 The STM-1 frame structure. AU = administrative unit; RSOH = regenerator section overhead; POH = path overhead; MSOH = multiplexer section overhead.

before this can be achieved. Present estimates indicate that it will be around the year 2000 in the developed world and much later in the developing world. The transition period requires the existing PDH network elements to be multiplexed into the SDH format. Figure 2.42 illustrates the current interfaces compared to the new SDH interfaces that will eventually be the worldwide, universal NNI. Note that even in the PDH system the primary multiplexers at 1.544 Mb/s and 2.048 Mb/s are described as synchronous. The plesiochronous nature is apparent only when multiple primary multiplexers are grouped together to form a higher-order multiplexer, or when these resulting bit streams are multiplexed to the next hierarchical level.

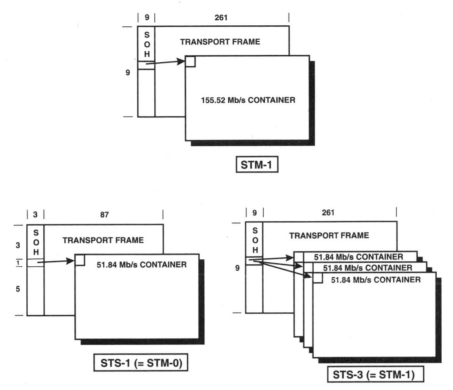

Figure 2.40 Comparison of the North American and European SDH frames.

So, considering the new SDH interfaces, the existing PDH networks of North America and Japan allow the basic primary 1.544 Mb/s to be synchronously multiplexed up to the STS-1 at 51.84 Mb/s, and eventually to the STM-1 or STS-3c at 155.52 Mb/s. The European format is a direct synchronous multiplexing of 2.048 Mb/s up to 155.52 Mb/s. Any further higher-order synchronous multiplexing would be integer multiples of 155.52 Mb/s.

2.5.5 SDH/SONET multiplexing structure summary

Having discussed the major differences between SDH and PDH multiplexing, particularly with respect to the frame format, the formal ITU-T specifications for the SDH scheme will now be summarized. The complete SDH multiplexing structure in Fig. 2.43 shows all the possible ways of forming an STS-1 and subsequently an STM-1. To accompany this diagram there are several new definitions associated with SDH multiplexing that need to be clarified.

Administrative unit. An AU is simply a "chunk" of bandwidth that is used to manage a telecommunications network. In North America and Japan, this

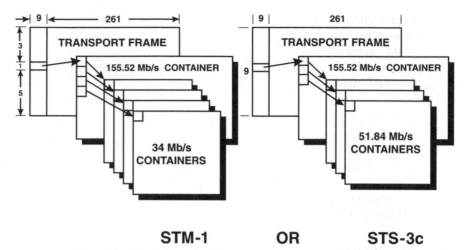

Figure 2.41 Nested signals. *(Adapted with permission from IEEE. Boehm, R.J., Progress in Standardization of SONET, IEEE Lightwave Communications Systems Magazine, May 1990, pp. 8–16. © 1990 IEEE.)*

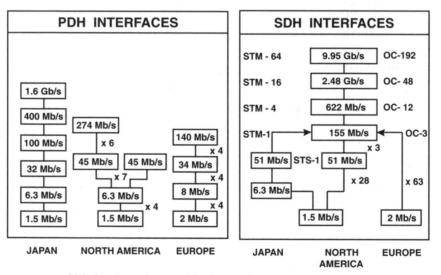

Figure 2.42   Global universal network node interfaces.

value has been set at 51.84 Mb/s (DS3 size), whereas most of the rest of the world has the 155.52-Mb/s AU.

Container.   The first block in Fig. 2.43 is called a *container* and is denoted as C-nx (where n = 1 to 4, and x = 1 or 2); n refers to the PDH level and x indicates the bit rate (i.e., x = 1 is for 1.544 Mb/s, and x = 2 is for 2.048 Mb/s). Note that C-11 is stated as C-one-one and not C-eleven.

Virtual container. The next block is the *virtual container* denoted as VC-*n* (where *n* = 1 to 4). This consists of a single container or assembly of tributary units together with the path overhead so that the virtual container is a unit that establishes a path in the network. Each of the containers is said to be *mapped* into a virtual container.

Tributary units. The next block is the *tributary unit* denoted as TU-*nx* (where *n* = 1 to 3, and *x* = 1 or 2). The tributary unit consists of a VC together with a pointer and an AU. The pointer specifies the phase of the VC. The VCs are said to be *mapped* or *aligned* with respect to the TUs. The TUs and the AUs therefore contain sufficient information to enable crossconnecting and switching of the VC and its pointer. In North America the VC is called a *virtual tributary* (VT) and sub-DS3 signals are placed in virtual tributary containers.

Tributary unit group. The next block is the *tributary unit group*, denoted as TUG-*n* (where *n* = 2 or 3). The TUG is a national grouping of TUs formed by the multiplexing process. For example, TUG-2 has four TU-11s, three TU-12s, or one TU-2. The TUG-2 can then be multiplexed to either a TUG-3 or VC-3, and the TUG-3 can be multiplexed to VC-4.

In North America the virtual tributary group was established, which is simply a set of virtual tributaries that have been grouped together to carry similar virtual tributaries. For example, a virtual tributary group could contain four DS1 (1.544-Mb/s) signals or three 2.048-Mb/s signals, and they would be contained in a VT6. This duplication of definitions often causes some confusion because they are used interchangeably in the literature.

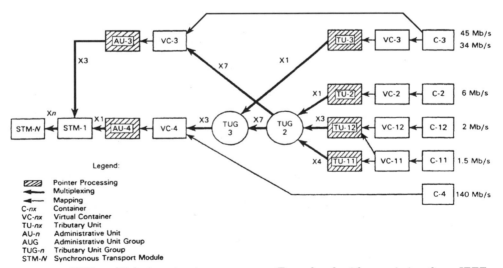

Figure 2.43 SDH multiplexing structure summary. *(Reproduced with permission from IEEE. Shafi, M., and B. Mortimer, The Evolution of SDH: A View from Telecom New Zealand, IEEE Communications Magazine, August 1990, pp. 60–66. © 1990 IEEE.)*

STM-*N*. At this stage the VC-3 and VC-4 can be aligned (mapped) to the respective administrative units (i.e., AU-3 or AU-4 as illustrated in Figs. 2.41 and 2.44). The VC floats in phase with respect to the AU, and the AU pointer value denotes the number of bytes offset between the pointer position and the first byte of the VC. Any frequency offset between the frame rate of the section overhead and that of the VC results in stuffing bytes in the AU pointer.

The STM-1 is then formed by multiplexing either three AU-3s or an AU-4 together with the *section* overhead information. The STM-1 can be multiplexed into an STM-*N* by synchronously byte-interleaving *N* STM-1s.

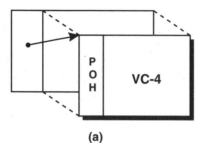

(a)

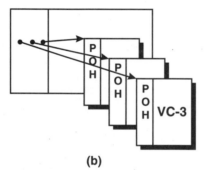

(b)

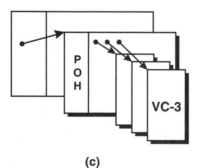

(c)

Figure 2.44 Mapping methods.
(*a*) VC-4 mapping into an STM-1.
(*b*) VC-3 direct mapping into an STM-1.
(*c*) VC-3 nested mapping into an STM-1.

Sub-STS-1 payloads.   The success of the SDH specifications in satisfying North American, Japanese, and European equipment bit rates can be seen by observing the composition of the STS-1 signal. The STS-1 SPE is divided into VT (TU) payload structures. Four sizes of VT/TU have been specified: VT1.5/TU-11, VT2/TU-12, VT3, and VT6/TU-2. These have bandwidths large enough to carry DS1 (1.544-Mb/s), primary (2.048-Mb/s), DS1C (3.152-Mb/s), and DS2 (6.312-Mb/s) signals, respectively. Each VT/TU occupies several nine-row columns within the SPE as follows:

VT1.5   occupies   3 columns   (27 bytes)

VT2   occupies   4 columns   (36 bytes)

VT3   occupies   6 columns   (54 bytes)

VT6   occupies   12 columns   (108 bytes)

A VT *group* (TUG-2) is defined as a 9-row by 12-column payload structure that can carry four VT1.5s, three VT2s, two VT3s, or one VT6. The STS-1 SPE therefore contains seven VT groups (84 columns), one path overhead column, and two unused columns.

There are two different methods of transporting the payloads within a VT/TU:

1. Floating mode

2. Locked mode

The floating mode uses a VT/TU pointer to establish the starting byte position of the VT/TU SPE within the VT/TU payload structure. This is the mode used for plesiochronous mapping of nominally PDH signals. The locked mode does not use the VT/TU pointer. Instead, the VT/TU payload structure is directly locked to the STS-1 SPE (which still floats with respect to the STS-1 frame). Byte-synchronous mapping in both the locked mode or floating mode can be used for transporting unframed SDH signals.

International compatibility is an important achievement of SDH/SONET. Also, on a network level SDH offers many benefits in addition to the add and drop capability. Its ability to handle PDH multiplexed signals within the synchronous multiplexed hierarchy is a major achievement. The implementation of SDH is discussed further in Chaps. 5, 9, and 10.

2.5.6 Comparison of PDH and SDH/SONET equipment

The strong growth in telecommunications traffic in all parts of the network and particularly in the trunk network has caused service providers to change quickly to SONET/SDH equipment at the OC-48 level of 2.488 Gb/s, and to OC-192 (9.952 Gb/s). The SONET/SDH equipment is more compact, and is a lower-cost investment compared to the equivalent PDH installations. This has arisen from a combination of new and more flexible devices, together with a trend toward using microprocessor-driven equipment components. The key SONET/SDH components are:

- Compact terminal equipment
- Versatile add/drop multiplexers
- New digital crossconnects

This SDH/SONET equipment and the excellent foresight of the SDH/SONET pioneers in creating their 1988 global standard has given the management of transmission equipment a major leap forward. The power and benefits of this inherent management capability become more and more evident as SDH/SONET deployment moves closer to 100 percent.

The PDH multiplexer "mountain" is eliminated by SDH/SONET multiplexers. Figure 2.45 shows the classic PDH terminal equipment for multiplexing up to 140 Mb/s (or three DS3s). The equivalent STM-1 terminal has up to 84×1.544-Mb/s (or 63×2.048-Mb/s) inputs and a 155.52-Mb/s output. This same terminal when equipped with additional circuit cards can be expanded to an add/drop multiplexer. This device is very efficient. In extreme cases, it can branch a T1 (1.544 Mb/s) to a village or individual customer premises, or it can split traffic equally into two directions for route diversity or for two destinations. Figure 2.46 shows the PDH multiplexer mountain again being replaced by the compact drop/insert multiplexer. Notice the 1.544- (or 2.048-) Mb/s block is now the smallest container or channel grouping to be dropped or inserted.

A new network element called the digital crossconnect has been introduced by the SDH/SONET scheme, which opens up new dimensions for network architecture. This powerful and very flexible element was not available in PDH systems. Figure 2.47 illustrates again how the old PDH multiplexer equipment stack in a CO can be replaced by a single digital crossconnect. In this case, it is not only a reduction in the amount of equipment. Reconfiguring the PDH 1.544- (or 2.048-) Mb/s destinations means manually rerouting traffic at the digital distribution frame (DDF), whereas the SDH/SONET digital crossconnect uses software to manage the reconfiguration. This microprocessor-driven approach also has other major benefits. Route protection and restoration can be done very rapidly by automatic, hands-off, computer-controlled rerouting. Reconfiguration for leased-line allocation can also be done very easily and quickly.

Telecommunications management network. The overhead incorporated into the STS or STM frames forms the basis of the modern *telecommunications management network* (TMN). The information carried in these frames is processed and presented on monitors at various locations around the network to facilitate the reconfiguration of traffic as well as maintenance functions. The equipment sections discuss this in more detail (e.g., Sec. 5.2.3).

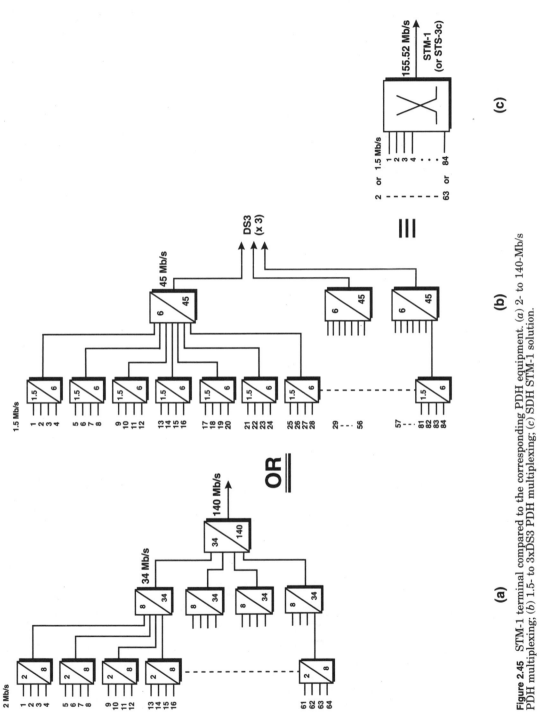

Figure 2.45 STM-1 terminal compared to the corresponding PDH equipment. (*a*) 2- to 140-Mb/s PDH multiplexing; (*b*) 1.5- to 3xDS3 PDH multiplexing; (*c*) SDH STM-1 solution.

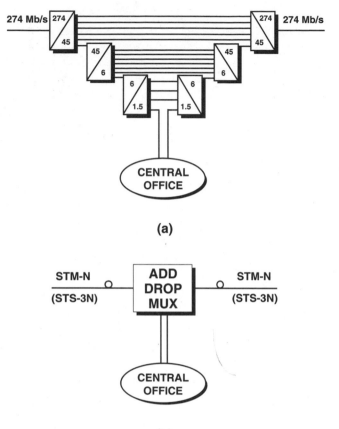

(a)

(b)

Figure 2.46 The add/drop multiplexer. (*a*) PDH solution; (*b*) SDH solution.

2.6 Multiplexing Digital Television Signals

So far, the discussion has centered around the multiplexing of voice telephone channels into various hierarchical bit rates. In addition to voice traffic, data and video are also important media for incorporation into the multiplexing structures. The 24- or 30-channel primary multiplexers allocate voice channels for data transmission. The bandwidth required for television signals is relatively large, and a lot of work is currently being devoted to determining specifications for digital coding of TV signals. This subject is still in a considerable state of flux, particularly because high-definition television (HDTV) is emerging in the marketplace in some countries, and specifications for the three main types of signal construction need to be harmonized. This is very difficult, because the debate is still raging as to which type of HDTV system will evolve as the international standard.

The term *HDTV* is frequently heard to describe the new, improved TV format. Also, video telephones in the ISDN or on the computer screen are emerging, and videoconferencing is already available. The subject of video/TV is a very broad area of electronics that can be placed in either of two categories, communications or broadcasting. Many books have been written solely on this subject. Here, the intention is to address the process of converting analog video signals to digital bit streams. The resulting bit streams are multiplexed with voice and/or data traffic and transmitted over medium- or long-distance routes using microwave, optical fiber, or satellite media as required. Coding standards have emerged from the motion picture experts group (MPEG) that have become globally recognized. A brief description of their key features will also be described (Sec. 2.6.3).

2.6.1 Digitization of TV signals

Until recently, TV signals were transmitted from the TV studio to the public in analog form, using approximately 6 MHz of bandwidth. If the signal is translated to a digital bit stream by an A/D process, the resulting digital signal is in excess of 155 Mb/s without any compression. Using compression techniques, the bit rate can be reduced to 1 to 10 Mb/s while maintaining MPEG-1 quality. For video telephone or even videoconferencing, this is still excessive bandwidth. The cost to the subscriber would be prohibitively large. For video telephones one might argue that the video bandwidth must not exceed the voice frequency range of 4 kHz, which presents a formidable problem to design engineers. Although it is not impossible, the question is "Can the minimum level of acceptable quality be achieved in a 4-kHz bandwidth?" For international transmission,

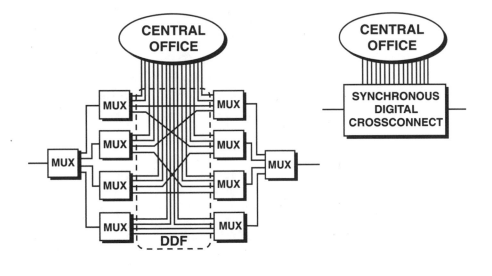

(a) (b)

Figure 2.47 The new digital crossconnect (*b*) replaces the PDH multiplexer "mountain" (*a*).

the problems are compounded by system incompatibilities. ITU-T Recommendation H.261 specifies methods of video telephony communication.

For the digital transmission of video/TV signals, the bandwidth must be reduced to a minimum while maintaining high picture quality. It is this last proviso that makes this subject very interesting. There is a trade-off between picture quality and transmission bit rate. To appreciate this problem better, it is necessary to understand the basic principles of analog color television. The subject is large and complex, and only the important features will be highlighted.

2.6.2 Analog color TV

The color TV signal has two major components:

1. Black and white intensity, known as *luminance*

2. Color variation, called *chrominance*

The luminance signal (Y) can be expressed as a combination of varying intensities of the three primary colors—red, green, and blue:

$$Y = 0.299R + 0.587G + 0.114B$$

For the chrominance signal, instead of transmitting all three primary colors, the bandwidth is reduced by transmitting two color difference (chrominance) signals:

$$U = R - Y \quad \text{and} \quad V = B - Y$$

The picture to be transmitted is formed by allocating a certain number of vertical lines to the picture and horizontally scanning these lines from left to right and top to bottom. During the "flyback" time from the end of one line to the beginning of the next, and from the bottom of the picture back to the top, no picture information is obtained, and the signal is said to be *blanked*. This blanking time is wasted time in the analog system, but can be used in the digital TV system for signal processing.

Two types of scanning have evolved: (1) *interlaced,* and (2) *progressive,* also known as *sequential.* For the interlaced type, the video picture (called the frame or field) is divided into two fields in which the scan lines of the first field fall between those of the second. In other words, every alternate (odd) line is scanned during the first half of a frame scan while the other alternate (even) lines are scanned during the second half of the frame scan. Two full field scans are therefore necessary to form a complete picture. The progressive type of scanning involves scanning the total number of lines contained in one field in a sequential manner. This technique provides higher vertical resolution but requires more transmission bandwidth than interlaced scanning.

One of the major problems for creating a global standard for HDTV is the fact that the NTSC system uses 525 lines for scanning and the power-line frequency of 60 Hz is used to define the rate of scanning, that is, 60 fields per second. The PAL and SECAM systems use 625 lines and 50 Hz for scanning parameters.

2.6.3 Video compression techniques

The standard A/D conversion for voice channels is also used as the basis for video digitization, after which compression is necessary to minimize the transmission bit rate. Compression is possible because there are redundancies in video transmission and also because the human visual perception has peculiar facets that can be exploited. For example, taking the analogy of motion picture projection, it is well known that the motion is actually an illusion created by viewing a series of still pictures, each having a slightly different position of the moving components. This same principle is used in TV broadcasting. The number of still pictures required is determined by the human visual perception. If there are not enough stills, the picture appears to flicker, which occurs at about 16 stills per second or less.

When considering a motion picture, there is only a certain portion of the picture that changes from one still to the next. For a fixed camera position, only the parts of the picture that move need to be updated, because the background remains the same. This means a differential PCM (DPCM) type of processing will greatly reduce the required transmission bandwidth. In a video telephone application, it is mainly the facial movements (and to some extent the motion of the whole head) that need to be periodically updated. DPCM is a very powerful tool when used together with some predictive signal processing, so that past motion that causes differences in successive picture frames can be extrapolated to predict future frames.

Other aspects of the human visual system can be electronically exploited. For example, differences in brightness (luminance) are perceived much more prominently than differences in color (chrominance). This leads to the luminance being sampled at a higher spatial resolution. After sampling, each line in the frame contains a number of picture elements called pixels, or PELS. The luminance signal might have, for example, 720×480 pixels compared to 360×240 pixels for color signals. Also, the eye is more sensitive to energy with low spatial frequency than with high spatial frequency. This characteristic is exploited by using more bits to code the high-frequency coefficients than low-frequency coefficients. Combining all of these techniques results in bit-rate-compression ratios of 20:1 to 100:1. At 20:1 the reconstructed video signal is almost indistinguishable from the original signal. Even at a 100:1 compression ratio, the reconstructed signal is similar to analog videotape quality.

Inevitably, high compression ratios lose some information, so that the received video signals are not exactly the same as those transmitted. The amount of compression possible and the effects of loss caused by compression vary depending on the application. For example, in a video telephone or conferencing situation there is relatively little motion.

The amount of compression obtainable depends, to a large extent, on the video coding technique chosen. The two main categories of video coding are source coding and entropy coding.

Source coding usually results in lossy, or degraded, picture quality and is classed as either intraframe or interframe coding. Intraframe coding exploits the redundancies in the spatial domain caused by small differences between neighboring pixels within each frame. Interframe coding eliminates redundancies in the temporal domain resulting from the small differences in adjacent frames necessary to depict motion.

There are several techniques that are now being applied to video coding to achieve high compression ratios. The *discrete cosine transform* (DCT) has emerged as a powerful tool for intrafield and interframe source coding. It is a process of mapping pixels of images from the time domain into the frequency domain. Images are separated into, for example, 8×8 blocks, meaning eight lines of eight samples. Each 8×8 block is transformed to produce another 8×8 block, whose coefficients are then quantized and coded. A large portion of the DCT coefficients are zero. A completely zero transform would mean no picture motion. Inverse DCTs are used to recover the original block. Quantization noise control is very important for the quality of a digitally compressed video picture. Mixed coding might evolve as the best solution, where motion-compensated interframe coding is used for lower-order DCT coefficients and intrafield coding for higher-order DCT coefficients.

Entropy coding is another valuable method that encodes frequent events with more bits than infrequent events, and is theoretically lossless. Huffman coding is a form of entropy coding that uses predetermined variable code words.

MPEG Standards. Video compression standards evolved in the late 1980s and early 1990s from the participants of the motion picture experts working group, known as the MPEG. The MPEG-2 has now become a de facto standard for video compression.

MPEG-1, formally known as the ISO/IEC IS 11172 and published in 1993, is a three-part Standard called video, audio, and system. It initially allowed full-motion TV video compression down to 1.5 Mb/s with a resolution of 352×240 pixels per frame and progressive scanning, for applications such as computer-to-computer communications, digital storage and retrieval systems, and CD-ROMs. MPEG-1 has been updated to include up to 10 Mb/s for TV broadcasting and distribution with a resolution of 480×720 pixels. The audio compression algorithms chosen were Dolby AC-2 and Musicam, which both compress a stereo channel into 256 kb/s.

MPEG-2 standardization, which started in 1991, was intended to build on MPEG-1 by allowing interlaced scanning to achieve ITU-R Recommendation 601 video quality of 480×720 pixels per frame within the transmission range of 2 to 10 Mb/s, depending on program content. It can also be extended to lower resolution to be downward compatible with MPEG-1. For higher-resolution signals such as HDTV, for example, to 1080 × 1440 or 1250 × 1920 pixels per frame, the bit rate can be increased to 25 to 30 Mb/s or higher.

In 1995, the quality of 6-Mb/s MPEG-2 coding exceeded the older systems such as NTSC, PAL, or SECAM, and MPEG-2 coding at 9 Mb/s satisfied ITU-R Recommendation 601. Because this subject is still rapidly evolving, no further details will be discussed here. A typical existing digital TV transmission system that uses some of the above compression techniques will be described.

2.6.4 A typical digital TV transmission system

While direct satellite broadcasting (DBS) and community antenna TV (CATV) via cable have become important modes of distributing existing TV signals, there is still the need to transport the TV signal from the studio to terrestrial transmitters or to link a TV switching center to a satellite earth station. Whether the telecommunications network or the broadcasting network organizes the link, the technical problems remain the same. High-quality transmission is essential. Digital transmission of high-quality TV signals at the 45- or 34-Mb/s interface was devised to fit smoothly into the North American or European PDH systems.

A TV signal A/D converter can usually be used for digitizing either NTSC, PAL, or SECAM signals. Encoding is performed according to ITU-R Recommendation 601, which refers to luminance signal sampling at 13.5 MHz and chrominance sampling at 6.75 MHz with linear encoding into 8-bit words. This ratio is known as the 4:2:2 digital code studio standard. A ratio of 2:1:1 is not used, because the Electronic News Gathering (ENG) organization already uses some frequencies that make 4:2:2 more appropriate. The 13.5- and 6.75-MHz rates are equivalent to 864/858 samples per line for the luminance signal and 432/429 samples per line for the chrominance signal. The number of active samples per line are 720 and 360 for luminance and chrominance, respectively. The number of active lines per frame is 485 for the 525-line system and 575 for the 625-line system.

Luminance processing. The luminance signal can be processed to reduce the bit rate significantly. Suppression of line and field blanking intervals reduces the bit rate by 25 percent. Adaptive DPCM coding and adaptive quantization reduce the number of bits per sample from 8 to 4. Figure 2.48 shows the feedback loops of intra- and interframe predictors used in the adaptive DPCM coding process. Switching between the two predictors is determined by evaluating the instantaneous picture movement inside the frame. Statistical coding then reduces the number of bits per sample to an average of 2.3.

Chrominance processing. Coding and bit-rate reduction of the chrominance signals U and V are achieved by a similar sequence of steps. U and V are sampled alternately at the transmit side, and the omitted color picture elements

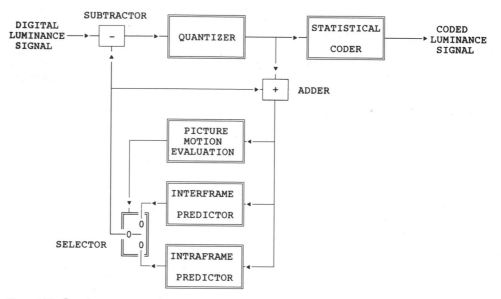

Figure 2.48 Luminance processing.

are reconstructed in the receiver by interpolation. The line and frame blanking intervals are suppressed. A 2:1 reduction is obtained by sampling only every other color picture element. DPCM coding with *fixed* prediction and adaptive quantizing gives an 8-to-4 reduction in the number of bits per sample. Statistical encoding then further reduces this to an average of 2.4 bits per sample.

Error correction.   Error correction circuits are included to compensate for the increased error sensitivity created by compression (bit-rate reduction). A typical forward error correction (FEC) mechanism for correcting single errors and error bursts would have a 248-bit block length, containing 234 information bits, allowing the correction of an error burst up to 5 bits. Interleaving four blocks of 248 bits improves the correctable error burst length to 20 bits and individual errors to 5 bits. ITU-T Recommendation 723 advises error protection by using a Reed-Solomon (255,239) code with an interlacing factor of 2.

2.6.5   High-definition television

The implementation of HDTV is such a huge commercial venture that the potential profits are enormous. Consequently, the major players are all trying to outmaneuver their competitors. As a result, the progress of HDTV has been plagued by the inability of the interested parties to converge on a global standard. The evolution of analog TV resulted in three main categories of system, namely, PAL, SECAM, and NTSC, with several variations within these categories. The necessity for HDTV to be compatible with the existing national analog systems has led to a divergence of research and a rather rigid stance of HDTV design from one group to another. At present, it appears inevitable that two studio standards will emerge, and much work is being done to minimize the differences between the two. The main debate has centered on the number of lines scanned per second and the proposed field rates of either 50 or 60 Hz.

There are some areas of universal acceptance. The picture aspect ratio (length to width) for HDTV will be 16:9 instead of the present TV value of 4:3. This will take advantage of the psychophysical characteristics of human visual perception. In other words, the picture will appear to be clearer and easier on the eye, particularly for programs that contain rapid movements.

ITU-R Recommendation 601, which was initially created in 1982 for extended definition TV using digital technology, has become the basis for HDTV standards. As two systems evolve, having field rates of 50 and 60 Hz, converters are necessary to ensure compatibility. This incurs extra expense and a small deterioration of the converted signal. Even if a single system were designed to satisfy North American and European requirements, the resulting system would still be incompatible with the Japanese systems, which have already had substantial amounts of capital invested in development of the 1125/60 configuration. The number of lines is also a problem, with the NTSC group proposing 1050 lines, with 2:1 interlacing, and the European group proposing 1250 lines and 50-Hz progressive scanning. The Japanese are proceeding with their multiple sub-Nyquist encoding system (MUSE), which has 1125 lines at 60 fields per second.

Regardless of the incompatibility problems concerning the transmission and broadcasting of HDTV, the bit rate will be approximately 1 Gb/s or higher before compression. The high postcompression bit rate places HDTV in the category of the broadband integrated services digital network, or B-ISDN (see Chap. 10). The mode of transporting such bit rates will be by the asynchronous transport mode (ATM) as described in Chap. 10. The ITU-R has defined three main categories of HDTV transmission:

1. *Contribution*, which will be for interstudio and intrastudio transmission. This will be at 622 Mb/s, which corresponds to the SDH levels STS-12 or STM-4. The compression technique will probably be the lossless subband DPCM.

2. *Primary distribution*, for interconnecting broadcast stations, video theaters, and CATV suppliers. The transmission rate of 155 Mb/s will probably use subband DCT compression.

3. *Secondary distribution*, which will be for directly transmitting to customers at 30 to 50 Mb/s. The techniques necessary to achieve this high level of compression are still being studied.

Although the exact form of HDTV is still developing, there is an emerging overlap between digital telecommunications and broadcasting. The customer benefits of this new and upcoming technology should be truly astounding.

3

Signal Processing for Digital Communications

There are some digital signal processing techniques that are common to satellite, DMR, and optical fiber technology. For example, error detection and correction is a major feature used by all. In the quest to pack more and more voice channels into a limited satellite, fixed radio or mobile radio frequency band, modulation schemes have evolved that are becoming increasingly bandwidth efficient. Unfortunately, the problem of transmission delay is created by the error correction and speech encoding process described in Chap. 2. This chapter addresses these topics.

3.1 Modulation Schemes

The methods of modulation used in previous analog communication systems could be broadly categorized as amplitude modulation (AM) and frequency modulation (FM). Digital communication systems follow a different approach. The signal to be transmitted in a digital system is a stream of 1s and 0s. There are only two amplitude levels, ON or OFF. At first sight this appears to be a much simpler form of transmission than that of an analog signal whose amplitude varies in a very complex manner. Unfortunately, a pulse is composed of a fundamental tone plus an infinite number of harmonics. Theoretically, that requires an infinite bandwidth for the transmission of a single pulse. Any communication system is limited in available bandwidth, and it is this constraint that causes a considerable complexity in the design of digital modulators and demodulators (modems).

3.1.1 Bandwidth efficiency

In the days of analog microwave radio (AMR), 1800 voice channels could be transmitted in a 30-MHz bandwidth in the 6-GHz band. In the early days of

DMR, cost-effectiveness made it essential for DMR systems to be able to transmit at least the same number of voice channels within the same analog radio RF bandwidth. This competition between AMR and DMR was clearly won by DMR, and the improvement in the quality of service of digital systems was a major factor. There is still a fierce competition for the limited frequency spectrum resource, not only from fixed microwave system service providers but now also from mobile radio service providers. It is useful to define the term *bandwidth (or spectrum utilization) efficiency*, which is the number of transmitted *bits per second per Hertz* (b/s/Hz). In the synchronous digital hierarchy, 155 Mb/s contains 84 DS1s, totaling 2016 voice channels. If the bandwidth efficiency is 1 b/s/Hz, a transmission bandwidth of 155 MHz would be required. Because this is an excessive value, some system modifications must be applied to improve the bandwidth efficiency. This is done during the modulation process. There are several modulation techniques, which can be broadly categorized as follows:

1. Pulse amplitude modulation (PAM)

2. Frequency shift keying (FSK)

3. Phase shift keying (PSK)

4. A mixture of phase and amplitude modulation, called quadrature amplitude modulation (QAM)

Before discussing each type of modulation in detail, it is informative to indicate briefly the effects of *filters* used in the modulation process on the bandwidth efficiency of a digital signal. Because the bandwidth required to transmit pulses perfectly is infinite, if the available bandwidth is comparatively narrow, there will be a significant effect on the shape of pulses emerging from a bandwidth-limiting device such as a filter.

3.1.2 Pulse transmission through filters

The use of low-pass or bandpass filters in the modulation or upconversion processes of a communication system is unavoidable. Passing the pulses through a low-pass or bandpass filter will eliminate some components of the pulses, resulting in output pulses having very "rounded" corners instead of sharp, right-angle corners. Eventually, if the cut-off frequency of a low-pass filter reaches a low enough value, the pulses become so rounded that they do not reach their full amplitude. *The Nyquist Theorem states that if pulses are transmitted at a rate of f_s b/s, they will attain the full amplitude value if passed through a low-pass filter having a bandwidth $f_s/2$ Hz.* This is the minimum filtering requirement for pulse transmission without performance degradation (i.e., *no intersymbol interference*).

Figure 3.1 illustrates this *ideal* Nyquist filter, which allows pulses to reach their maximum amplitudes. Unfortunately, this type of ideal Nyquist filter does not exist. If it did, it would require an infinite number of filter sections (which would therefore have an infinite cost). The filter characteristics for

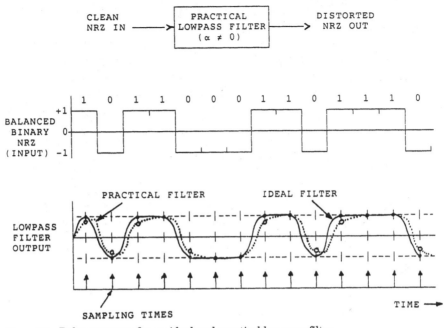

Figure 3.1 Pulse response for an ideal and practical low-pass filter.

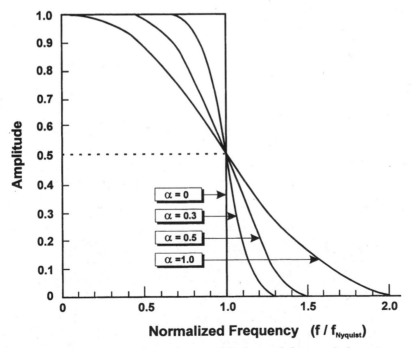

Figure 3.2 Filter characteristics. $f_{\text{Nyquist}} = f_s/2; f_s = $ symbol transmission rate.

transmission of impulses (approximately the same as for very narrow pulses) are shown in Fig. 3.2. The value of α = 0 is the ideal filter case. A more practical value of α = 0.3 requires a bandwidth of 30 percent in excess of the Nyquist bandwidth. This means that instead of transmission at an ideal bandwidth efficiency of 2 b/s/Hz, the value is 2/1.3 = 1.54 b/s/Hz. Figure 3.1 also shows the output response for this type of nonideal filter. At the sampling instants, the signal does not always reach its maximum value, so the imperfect filter introduces intersymbol interference. Many types of complex filters have recently been designed to overcome the intersymbol interference problem without significantly reducing the bandwidth efficiency.

The surface acoustic wave (SAW) filter has been introduced into some DMR equipment. It has some qualities that allow it to be designed very closely to the ideal Nyquist filter. SAW filters have cosine roll-off characteristics and "saddle"-shaped passband characteristics. Figure 3.3 shows a typical frequency response for a SAW filter.

3.1.3 Pulse amplitude modulation

There are several linear modulation techniques available to improve the bandwidth efficiency. The first, PAM, is simple; Fig. 3.4 shows a conversion from a binary NRZ signal to multilevel PAM. Here, a two-level NRZ signal is converted to four levels. Each four-level symbol contains 2 bits of information. Based on the Nyquist theorem, it is theoretically possible to transmit, without intersymbol interference, two symbols per second per hertz. Because each symbol contains 2 bits of information, four-level PAM should ideally be able to transmit 4 b/s/Hz. Each eight-level PAM contains 3 bits of information, ideally allowing a transmission of 6 b/s/Hz. Unfortunately, the

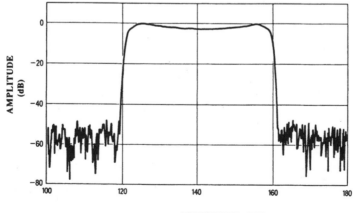

Figure 3.3 Saw filter passband characteristics. (*Reproduced with permission from Siemens Telcom Report: Special "Radio Communication," vol. 10, 1987, p. 242, Fig. 3.*)

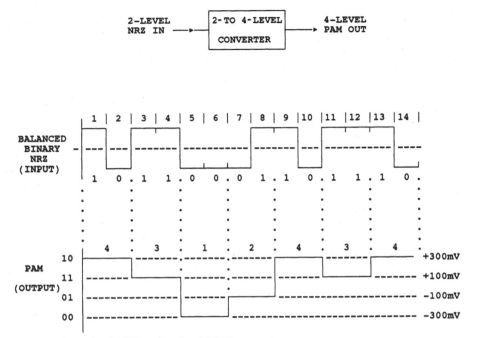

Figure 3.4 Two-level NRZ to four-level PAM conversion.

error performance (BER) of digital AM is inferior to other forms of digital modulation. However, there is a very important use of AM combined with phase modulation: QAM. Its use in DMR systems is discussed in detail in Sec. 3.1.6.

3.1.4 Frequency shift keyed modulation, minimum shift keying (MSK), and Gaussian MSK

FSK is simply the allocation of one fixed frequency tone for 0s and another tone for 1s. The input data bit sequence is used to switch back and forth between these two frequencies in sympathy with the changes from 1 to 0 or 0 to 1 (Fig. 3.5). From a circuit point of view, this can be accomplished by feeding the input data into a voltage controllable oscillator (VCO) for the modulation process and using a PLL for demodulation. For FSK it is useful to define a modulation index m as

$$m = \frac{f_0 - f_1}{f_r} \tag{3.1}$$

where f_1 = 1s frequency
f_0 = 0s frequency
f_r = symbol rate frequency

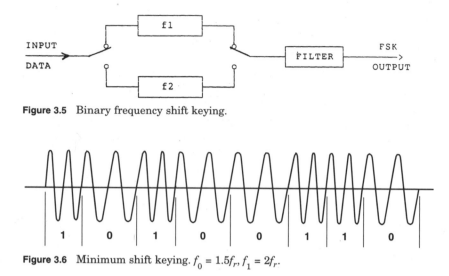

Figure 3.5   Binary frequency shift keying.

Figure 3.6   Minimum shift keying. $f_0 = 1.5f_r, f_1 = 2f_r$.

The error performance of FSK is generally worse than PSK. FSK is cheap to implement, and there are variations of FSK that have some other performance benefits. For example, fast frequency shift keying (FFSK) increases the rate of transmission without increasing the occupied bandwidth. In other words, it improves bandwidth efficiency. This can be done by making the value of m quite low. For example, if $f_1 = 1200$ Hz, $f_0 = 1800$ Hz, and $f_r = 1200$ b/s, $m = 0.5$. For a bit sequence of 010, the 1200-Hz oscillator is connected to the output for one cycle, then the 1800-Hz oscillator is connected for 1.5 cycles, followed by the 1200-Hz oscillator for one cycle. Notice there is oscillator phase continuity at the transition from one bit interval to the next. From communications theory, it has been shown that there is equivalence between FFSK and MSK. MSK is binary FSK, with the two frequencies selected to ensure that there is exactly a 180° phase shift difference between the two frequencies in a single bit interval. MSK therefore produces a maximum phase difference at the end of the bit interval using a *minimum* difference in frequencies (Fig. 3.6) and maintains phase continuity at the bit transitions. MSK is attractive because it has a relatively compact spectrum with out-of-band characteristics that are better than FSK. Low out-of-band emission is necessary for low adjacent-channel interference. One method of improving the out-of-band emission is to preshape the data stream with a filter prior to MSK modulation. Several types of filter have been used for this purpose. The classic Nyquist raised-cosine filter allows ISI-free transmission, but a Gaussian-shaped filter that accepts about 1 percent ISI has considerably better out-of-band performance. This is consequently called Gaussian MSK, or GMSK, modulation, and it is used in the GSM digital cellular mobile radio systems (Sec. 7.8.1).

3.1.5   Phase shift keyed modulation

PSK modulation is widely used in DMR technology today. There are several levels of PSK. The simplest is two-phase PSK as shown in Fig. 3.7a. In this type of modulation, the incoming bit stream is given a phase reversal of 180° every time a 1 changes to a 0 or vice versa. Figure 3.7b shows the waveform changing between 0 and 180°. The next level of PSK is four-phase PSK, otherwise known as quadrature or quaternary PSK (QPSK or 4-PSK). As indicated in Fig. 3.8a, the incoming bit stream is divided into two parallel bit streams by using a *serial-to-parallel converter*. These two bit streams are known as the in-phase (or I) and the quadrature-phase (or Q) bit streams. The transmit oscillator generates the unmodulated carrier frequency, which is passed through a 0 and 90° phase splitter. The I and Q baseband NRZ bit streams are time-domain multiplied by the carrier signals using a mixer. The summed output is then passed through a bandpass filter, resulting in the waveform of Fig. 3.8b. The phase diagram shows that there are four phase states, 90° apart from each other at 0, 90, 180, and 270°. The next level of PSK, as shown in Fig. 3.9, is 8-PSK. Here the incoming bit stream is divided into three before the carrier is modulated. The output waveform has eight states, spaced 45° apart. Each time the level is raised, the *theoretical* bandwidth efficiency is increased:

$$2\text{-PSK} \rightarrow 1 \text{ b/s/Hz}$$

$$4\text{-PSK} \rightarrow 2 \text{ b/s/Hz}$$

$$8\text{-PSK} \rightarrow 3 \text{ b/s/Hz}$$

In reality, the nonideal practical filters reduce these theoretical maximum values.

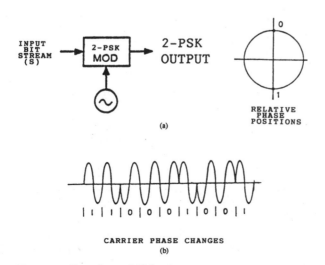

Figure 3.7   Two-phase shift keying.

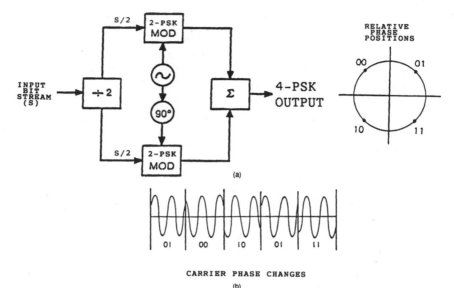

Figure 3.8 Four-phase shift keying.

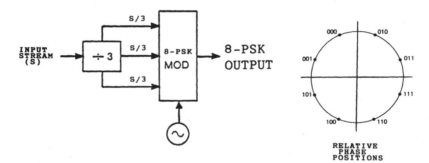

Figure 3.9 Eight-phase shift keying.

π/4-QPSK.   2-PSK and 4-PSK modulated carriers sometimes suffer 180° phase jumps during which the instantaneous carrier amplitude crosses the zero point. This is not good, especially when the signal passes through an amplifier operating in the nonlinear region close to saturation, as can be the case in systems such as VSATs and some mobile cellular radio systems. Conventional 4-PSK simply has four possible transmitted phases (or states). π/4-QPSK has eight possible phases, as shown in Fig. 3.10. This figure also shows the allowable transmission between states, which can be summarized as ±π/4 and ±3π/4 transitions. This type of modulation has a transmission rate of 2 bits per symbol. Any phase changes that occur do not pass through the origin, which makes the signals less affected by amplifier nonlinearities.

This π/4 shifted, differentially encoded 4-PSK type modulation is used in the North American digital cellular radio system IS-54/IS-136. The spectral efficiency of π/4-QPSK is typically 1.6 b/s/Hz after filtering. The incoming bit

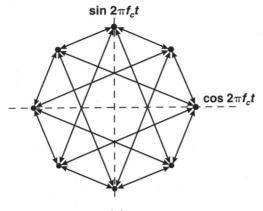

(a)

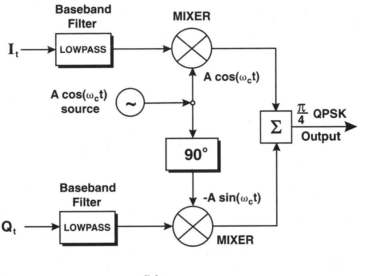

(b)

Figure 3.10 (*a*) π/4-QPSK constellation; (*b*) circuit diagram for π/4-QPSK.

stream is serial-to-parallel converted into odd and even bit streams X and Y. The I and Q streams are formed from the odd and even streams according to the differential encoding as follows:

$$I_t = I_{t-1} \cos \Delta\phi - Q_{t-1} \sin \Delta\phi$$

$$Q_t = I_{t-1} \sin \Delta\phi + Q_{t-1} \cos \Delta\phi$$

where I_{t-1} and Q_{t-1} are amplitudes one pulse time earlier (see Sec. 5.2.1 for more details on the differential encoding process). The X_t and Y_t symbols are related to $\Delta\phi$ by:

X_t	Y_t	$\Delta\phi$
0	0	$\pi/4$
0	1	$3\pi/4$
1	0	$-\pi/4$
1	1	$-3\pi/4$

The I_t and Q_t provide the inputs to a regular 4-PSK modulator. The resulting constellation is the combination of two 4-PSK constellations, where one is rotated by $\pi/4$ with respect to the other. One 4-PSK four states corresponds to the even symbols, and the other 4-PSK four states to the odd symbols. The relative phase shift between two successive symbols is therefore $\pm\pi/4$ or $\pm3\pi/4$.

3.1.6   Quadrature amplitude modulation

The next level of PSK, 16-PSK, is not used very much. In preference, a modulation having both PSK and amplitude modulation has evolved—QAM. The reason for this is improved error performance. QAM can be viewed as an extension of PSK. In the special case of 4-QAM, where two amplitude levels are used as inputs to a 2-PSK modulator, the system is identical to 4-PSK. However, higher-level QAM systems are distinctly different from the higher-level PSK systems. Figure 3.11a shows the generalized QAM modem. Figure 3.11b shows the modem for 16-QAM together with the respective waveforms. The bit stream to be modulated is split into two parallel bit streams that are then each converted into a four-level PAM signal. One of the signals is mixed with a carrier from a local oscillator (LO), while the other signal is mixed with the same LO after undergoing a 90° phase shift. The two signals are then added. The result is a 16-QAM signal; Figure 3.12a shows the signal state-space diagram for 16-QAM. The *signal state-space diagram* is often referred to as the *constellation diagram*. Notice that, in comparing this to the 16-PSK signal (Fig. 3.12b), the 16-QAM signal does not have a constant envelope, whereas the 16-PSK does. This has some interesting noise implications, which are discussed later.

The prime motive for moving to higher levels of QAM is simply improvement in bandwidth efficiency. For example, a 64-QAM digital microwave radio system has a bandwidth (spectral) efficiency that allowed it to be a direct substitute for old analog systems of similar capacity. It operates in the same channel arrangement as analog systems. Also, analog and digital systems could operate on adjacent channels without mutual interference during the transition period. Figure 3.13 is a comparison of QAM configurations for 16-, 64-, 256-, and 1024-QAM, etc., together with their respective constellation diagrams. In each case, the input bit stream is split into four, six, eight, etc., bit streams by a series-to-parallel converter. A PAM process then provides two bit streams (I and Q) each with 4, 8, 16, etc., amplitude levels. Each

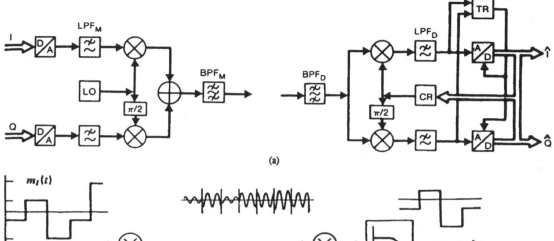

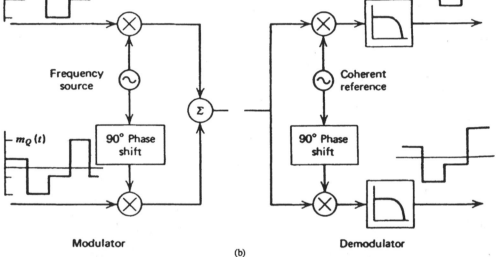

Figure 3.11   (*a*) Generalized QAM modem. (*Reproduced with permission from IEEE © 1986. Noguchi, T., Y. Daido, and J. A. Nossek, Modulation Techniques for Microwave Digital Radio, IEEE Communications Magazine, vol. 24, no. 11, September 1986, pp. 21-30.*) (*b*) 16-QAM modem. (*From Bellamy, J., Digital Telephony, © 1982. Reprinted by permission of John Wiley & Sons, Inc.*)

I and Q bit stream is then mixed with the IF oscillator (directly and 90° phase shifted, respectively). The modulated outputs are then added to form the QAM signal.

3.1.7   Comparison of modulation techniques

The spectral shapes of PSK and QAM signals having the same number of states (e.g., 16-PSK and 16-QAM) are identical. The error performance of any digital modulation system is fundamentally related to the distance between points in the signal constellation diagram. Figure 3.14 illustrates how the distance between points is the same for 2-PSK as for 4-PSK. The next level,

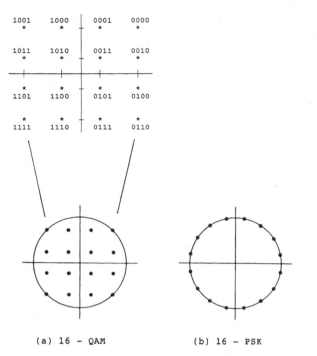

Figure 3.12   The signal state-space diagram for (*a*) 16-QAM and (*b*) 16-PSK.

8-PSK, has points more closely spaced, so the error performance is expected to be worse than 2- or 4-PSK. The reason for this is that, as the states become closer together, noise that causes them to occupy a broader area around the required points gives a higher probability of states overlapping (as in Fig. 3.15). This simple result has important implications for coded modulation, as described in the next sections. Overlapping of the states indicates interference, which causes errors. This constellation representation has a very useful application in DMR maintenance methods, as discussed in Chap. 5. It is quite evident that for 16-QAM, the distance between the points is greater than for 16-PSK, thus providing better error performance. Notice, as indicated in Fig. 3.14, *PSK is a double-sideband suppressed-carrier (DSB-SC) modulation.* This is an important feature that must be taken into consideration when choosing demodulation circuitry.

Spectrum-power trade-off.   Having established the desired value of bandwidth efficiency, it is important to assess how much received power is required to produce a specified BER (that is, to maximize power efficiency). Unfortunately, the higher levels of modulation require higher values of carrier-to-noise ratio (C/N) to achieve a given BER. Figure 3.16 is a comparison of the error rates for several PSK and QAM systems. For a required BER, the C/N must be increased significantly as the level of modulation increases. The probability of error is

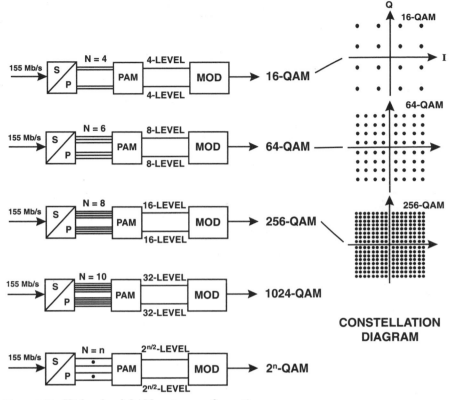

Figure 3.13 Higher-level QAM system configurations.

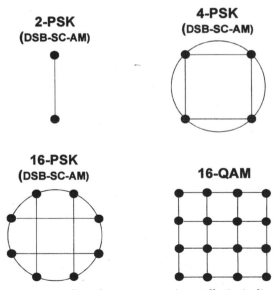

Figure 3.14 Signal state-space (constellation) diagrams for several types of modulation.

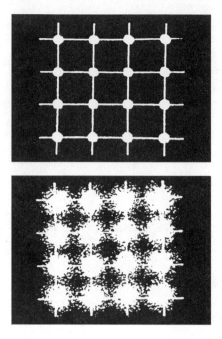

Figure 3.15 The constellation diagram showing noise.

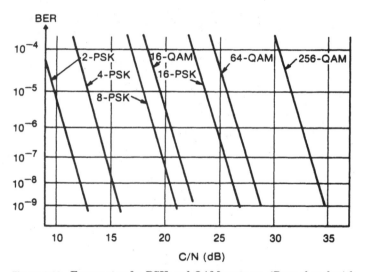

Figure 3.16 Error rates for PSK and QAM systems. (*Reproduced with permission from IEEE © 1986. Noguchi, T., Y. Daido, and J. A. Nossek, Modulation Techniques for Microwave Digital Radio, IEEE Communications Magazine, vol. 24, no. 11, September 1986, pp. 21-30.*)

plotted against the C/N. This graph stresses the following important point: *Improving the bandwidth efficiency by employing a higher level of modulation requires a greater C/N to maintain a good BER.*

The spectral density (the same as bandwidth efficiency) η is simply the ratio of the bit rate to bandwidth W, the width of each channel. For M-PSK and M-QAM the transmitted bit rate is

$$\frac{1}{T} \log_2 M \tag{3.2}$$

where T is the symbol rate in bauds. Therefore

$$\eta = \frac{1}{WT} \log_2 M \quad \text{b/s/Hz} \tag{3.3}$$

$WT = 1$ only for the case of $\alpha = 0$, which as already stated is the ideal (impractical) filter. For practical filtering in the transmitter and receiver, the total RF bandwidth is therefore $(1 + \alpha)/T$. This means that because $\alpha \neq 0$, the transmitted bandwidth exceeds the allotted channel bandwidth W. It is usual to organize filtering so that half of the cosine roll-off shaping is done in the transmitter and half in the receiver. Figure 3.17 shows the transmitted spectrum for cosine roll-off shaping for various values of α. Note that only $\alpha = 0$ keeps the transmitted energy within the channel region of ± 0.5. Studies by the FCC and CEPT (now ETSI) have resulted in the transmitted spectrum emission masks of Fig. 3.18, which must not be exceeded by microwave radio operators. With $\alpha = 0.33$ and $1/WT = 0.75$, the FCC and ETSI masks for 4-, 6-, and 11-GHz operation are satisfied. The spectral efficiency in that case is

$$\eta = \frac{1}{(1 + \alpha)} \log_2 M \tag{3.4}$$

$$\eta = 0.75 \log_2 M \tag{3.5}$$

so for various levels of QAM

$$\text{4-QAM} \rightarrow \eta = 1.5 \text{ b/s/Hz}$$

$$\text{16-QAM} \rightarrow \eta = 3.0 \text{ b/s/Hz}$$

$$\text{64-QAM} \rightarrow \eta = 4.5 \text{ b/s/Hz}$$

$$\text{256-QAM} \rightarrow \eta = 6.0 \text{ b/s/Hz}$$

$$\text{512-QAM} \rightarrow \eta = 7.5 \text{ b/s/Hz}$$

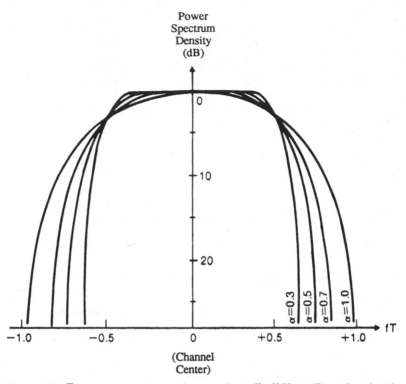

Figure 3.17 Frequency spectrum using a cosine roll-off filter. (*Reproduced with permission from IEEE © 1986. Noguchi, T., Y. Daido, and J. A. Nossek, Modulation Techniques for Microwave Digital Radio, IEEE Communications Magazine, vol. 24, no. 11, September 1986, pp. 21-30.*)

Impairments. The higher-level modulation systems are more vulnerable to equipment and atmospheric propagation impairments. Both impairments cause degradation in transmission quality, but whereas the radio designer has no control over the propagation and can only take countermeasures to minimize the effects, there is a choice concerning modem design. To take deliberately the option of, say, 256-QAM means not only accepting an unavoidable power penalty (degradation in C/N) compared to lower levels of QAM, but also stipulating that the design tolerances must be extremely tight; otherwise additional C/N degradation will occur. The major factors that cause this further C/N degradation are:

1. Amplitude distortion (linear and nonlinear)

2. Delay distortion (linear and nonlinear)

3. Timing error

4. Recovered carrier phase error

Items 3 and 4 include jitter in addition to fixed errors. These problems are addressed in more detail in Chap. 5. At this juncture, it is important to stress

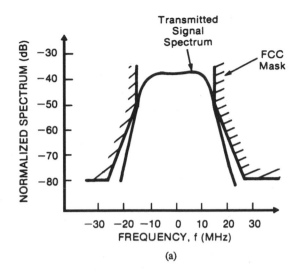

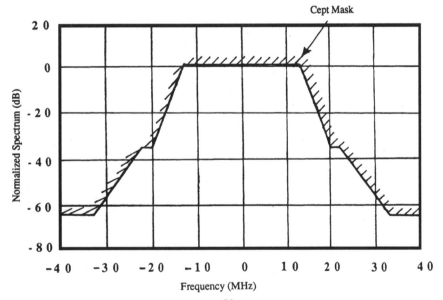

Figure 3.18 (a) FCC frequency spectrum limits for lower 6-GHz DMR. (*Reproduced with permission from IEEE © 1986. Noguchi, T., Y. Daido, and J. A. Nossek, Modulation Techniques for Microwave Digital Radio, IEEE Communications Magazine, vol. 24, no. 11, September 1986, pp. 21-30.*) (b) CEPT (now ETSI) frequency spectrum limits for the lower 6-GHz DMR. (*Reproduced with permission from Terrestrial Digital Microwave Communications by Ivanek, F. (ed.), Artech House, Inc., Norwood, MA, 1989. http://www.artech-house.com.*)

that when comparing QAM to 4-PSK modulation, as the level of modulation increases (16-QAM, 64-QAM, etc.), any of the above impairments will very significantly increase the C/N required to achieve a specific BER. The C/N degradation is *in addition* to the values shown in Fig. 3.16.

Conclusion. Finally, the improvement of bandwidth efficiency at the expense of the C/N is shown in Fig. 3.19. The choice of modulation scheme would appear to be simply a trade-off between bandwidth efficiency and power efficiency. There is also the constraint of staying within the required transmitted power spectrum mask. The tolerance to impairments is also important. As usual, cost is the prime deciding factor, and it is recognized that the cost per channel is lower for the higher levels of modulation. Already DMR systems for SHD comfortably pack a 155.52-Mb/s signal into 40 MHz using 64-QAM with trellis-coded modulation (TCM), often called 64-TCM. Designs for 2 × 155.52-Mb/s systems into 40 MHz using 512-TCM are appearing. The bandwidth efficiency of these systems is 3.888 and 7.776, respectively. The digital signal processing needed to achieve these high values has increased dramatically over recent years. It might be that at 1024-QAM and above the system complexity required to overcome the technical impairments to ensure a satisfactory performance will be prohibitively expensive. Time will tell.

3.1.8 Demodulation

Demodulation is the inverse process of modulation. This rather obvious statement is made evident by the demodulation schematic in Fig. 3.11. However,

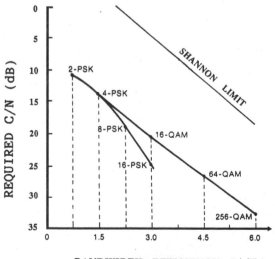

Figure 3.19 Graph of bandwidth efficiency against C/N for various types of modulation (BER = 10^{-6}; WT = 1.33).

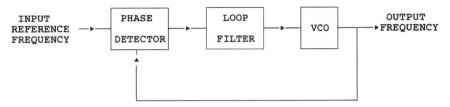

Figure 3.20 The basic phase lock loop.

demodulation is not so simple in reality. The major difference between the modulation and demodulation diagrams lies in the carrier recovery and timing recovery circuits.

A transmitted QAM signal, for example, is a suppressed-carrier signal. The reinsertion of a locally generated carrier is necessary for recovery of the transmitted information. Furthermore, *coherent detection* is essential. This means the inserted carrier must have the same phase and frequency as the transmitted carrier.

Two successful methods of carrier recovery are (1) the *Costas loop* and (2) the *decision-directed* method, as described in the following.

The Costas loop.   The Costas loop is an extension of the famous PLL method of carrier recovery and demodulation. The PLL has been used extensively in AMR circuits for establishing:

1. A stable oscillator frequency

2. Carrier recovery of a transmitted double-sideband suppressed-carrier (DSB-SC) signal

3. Demodulation

To refresh the memory, the PLL circuit of Fig. 3.20 has an output that locks onto the input reference signal and then tracks any changes that occur in the input signal. The phase lock is done by using the phase comparator to compare the phase of the output signal with the phase of the input reference. The phase difference produces an error voltage that is used to modify the frequency (and therefore phase) of the VCO. As this error voltage tracks the phase or frequency of the input reference signal, it effectively demodulates the input reference signal. An equation can be derived for the error voltage required to maintain phase lock as follows:

$$V_f = K \sin (\theta_i - \theta_o) \tag{3.6}$$

where K is a constant and $\theta_i - \theta_o$ is the phase error.

The PLL circuit is applicable to both analog and digital demodulation. For an analog 2-PSK signal, the circuit for the Costas loop is shown in Fig. 3.21. One can see that this is merely an extension of the PLL concept. The loop filters are

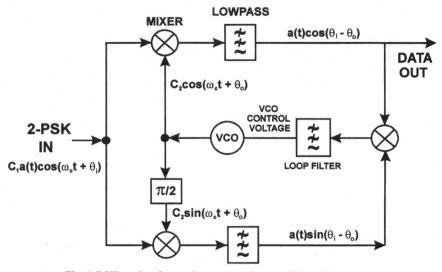

Figure 3.21   The 2-PSK analog Costas loop. $a(t)$ = data and has values ±1; θ_i = input signal phase; θ_0 = VCO output phase; $(\theta_i - \theta_0)$ = phase; error, ω_c = angular carrier frequency; C_1 and C_2 are constants.

included to remove the harmonic frequency terms. The error voltage, which is the input to the loop filter, is found to be

$$e_f = C \sin 2(\theta_i - \theta_o) \tag{3.7}$$

where C is a constant. This result differs from the basic PLL equation only in the fact that the Costas loop error voltage is zero for a phase difference of 0° and 180°, instead of just 0° as for the basic PLL. This means that phase lock can occur for two different phase angles between the VCO and the input signal. Data can therefore appear in the upper or lower half of the circuit, depending upon the angle onto which the loop locks. The data are a faithful reproduction of the modulated transmitted signal.

The differences between the analog and digital 2-PSK Costas loops are (1) the multiplier in the feedback loop is replaced by an EXCLUSIVE OR gate and (2) the low-pass filter outputs are followed by hard limiters for the digital 2-PSK circuit; 4-PSK requires a slightly more elaborate Costas loop for demodulation, as indicated in Fig. 3.22. Because a 16-QAM signal requires the demodulation of a 4-PSK prior to demodulation of the PAM signal, this type of Costas loop is widely used in 16-QAM demodulators.

The 4-PSK input in this case is

$$A_1 a(t) \sin \omega_c t + A_2 b(t) \cos \omega_c t \tag{3.8}$$

where $a(t)$ and $b(t)$ are the data signals, which can each have values of ±1; A_1 and A_2 are the amplitudes of the PSK signals; and ω_c is the angular carrier frequency. The input to the upper phase detector is

$$2 \sin (\omega_c t + \theta_e) \qquad\qquad (3.9)$$

where θ_e is the phase error (i.e., $\theta_i - \theta_o$). The input to the lower phase detector is

$$2 \cos (\omega_c t + \theta_e) \qquad\qquad (3.10)$$

The data I output is

$$\text{sign} [a(t) \cos \theta_e + b(t) \sin \theta_e] \qquad\qquad (3.11)$$

The data Q output is

$$\text{sign} [b(t) \cos \theta_e - a(t) \sin \theta_e] \qquad\qquad (3.12)$$

The data outputs I and Q are subsequently processed at the baseband level to recover the originally transmitted bit stream at, for example, 155 Mb/s. For a 155-Mb/s system, a 155-MHz VCO is used during the modulation process to produce the 16-QAM signal.

Other techniques can be used for carrier recovery. Remodulation is a favorable method, especially because it has a faster acquisition time than the 4-PSK Costas loop demodulator. The main purpose of the 4-PSK remodulation technique is to demodulate the signal into I and Q bit streams, after which the two I and Q signals are remodulated. The remodulated output is fed back into the loop filter together with the 4-PSK input signal, to produce the loop error signal.

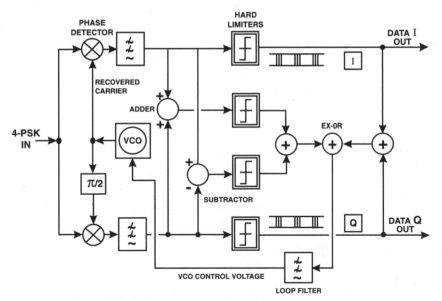

Figure 3.22 The 4-PSK digital Costas loop demodulator.

Decision-directed method.   The decision-directed method of timing recovery is yet another variation of the PLL, and it has been applied successfully to high-level QAM demodulation circuitry. Because this method is emerging as the overall best technique for carrier recovery of QAM signals, it is worthy of extensive coverage. However, this subject is unavoidably highly mathematical so, in this text, just a few details will be highlighted.

In the demodulation diagram of Fig. 3.11, the outputs of the A/D converters are the I and Q digital data streams. From these two streams, the constellation plot could be observed by feeding the I and Q streams into a constellation analyzer. The existence of noise in the DMR causes the constellation points to be spread, as in Fig. 3.15. In addition, other impairments of the DMR system cause the constellation states to be offset from their precise locations.

Figure 3.23 is a 64-QAM constellation, superimposed on a template of 256 small square regions (four for each state). The coordinates of each of these 256 squares are stored in a memory in the carrier recovery circuit. Concentric circles are also drawn on this diagram. If noise and/or equipment impairment causes a constellation state to be moved from its ideal position to a position in a "black" square, the state needs to be rotated clockwise to correct its position. Counterclockwise rotation would be necessary for states displaced into the shaded square areas. This information of control increments required to correct the position of the QAM states is applied to a VCO in a loop. The control increments derived from the template effectively replace the phase detector function of the familiar PLL.

More advanced decision-directed circuits incorporate the baseband adaptive equalizer (see Sec. 5.1.5) and automatic gain control (AGC) circuits within the loop. Sophisticated algorithms are applied in such circuits, which result in excellent carrier recovery even in the presence of very hostile propagation conditions.

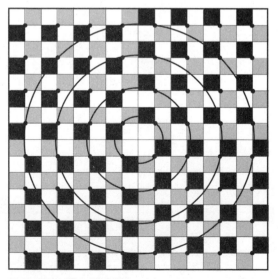

Figure 3.23 The 64-QAM carrier recovery template.

3.2   Error Control (Detection and Correction)

The BER and *errored second ratio* (ESR) are perhaps the most important quality factors to observe when evaluating a digital transmission system. First, it is instructive to make a statement about the noise, which causes errors. The worst type of noise is random noise. Error calculations are often based on the type of random noise described as *additive white Gaussian noise* (AWGN). If the noise has a structure, that structure can be used to the system designer's advantage. Short error bursts are not random, in the sense that there are definite reasons for and locatable sources of such errors. The probability of receiving errors for a specific modulation scheme depends on the S/N at the receiver and the transmission bit rate. In practical systems, there is usually no opportunity to change those two factors, so one must resort to *error control coding*, also called *channel coding*, or broadly speaking, *FEC*. During the past 10 years, developments in incorporating error control into the modulation process have produced some exciting results. This technique is called *trellis-coded modulation* (TCM).

A second category of error control technique is *automatic request for repeat* (ARQ). In this technique, redundancy is also added prior to transmission. The philosophy here is different from FEC, where the redundancy is used to enable a decoder at the receiver to correct errors. ARQ uses the redundancy to detect errors. Once detected at the receiver, a request is made to the transmitter for a repeat of the transmitted bit sequence. In this technique, a return path is essential. Unfortunately, the delay introduced by retransmission would be unsatisfactorily high for voice and video systems. However, ARQ is used extensively for data communications where the delay is less important.

3.2.1   Forward error correction

The objectives of error correctors are to reduce the residual BER by several orders of magnitude and also to increase the system gain. This is achieved by encoding the bit stream prior to modulation by a process of adding extra bits to the bit stream according to specific rules. The additional bits do not contribute information to the message transmitted. They serve only to allow the decoder in the receiver to recognize and therefore correct any errors that might arise during the transmission process. As a result, *error correction deliberately introduces redundancy* into a transmission system. In other words, the improvement in residual BER is achieved at the expense of an increase in the transmission bit rate. FEC does not need a return path. This is analogous to a person speaking over a noisy telephone line. If that person includes some repetitions from time to time (adding redundancy), the message is clarified without any need for the receiving party to ask for a repeat. Increased bit rate implies that the bandwidth efficiency suffers some degradation. Although this is true for most types of coding, recent advances in TCM achieve improved noise immunity (low residual BER) with a minimal reduction in bandwidth efficiency.

Error correction codes can be broadly categorized into (1) *block* codes, (2) *convolutional* codes, or (3) a combination of both block and convolutional, called *concatenated* codes.

Block codes.   In Fig. 3.24, the input bit stream is read into the FEC encoder in *blocks* of information bits containing k bits. The output from the FEC encoder has the k bits plus r check bits, so an n-bit *code word* is transmitted. This would be called an (n,k) encoder, having a *coding rate* of k/n. The FEC encoder adds check bits that are used in the receiver to identify and possibly correct any errors. In its simplest form, there is only 1 check bit, which is a parity check bit. The larger the number of check bits, the better the residual BER. Clearly there is a trade-off here between BER and excess bandwidth required because of the check bits increasing the bit rate. When comparing coded to uncoded signals, the reduction in the value of coded signal C/N which is required to yield a specific residual BER (e.g., 1×10^{-10}) is called the *coding gain*. Using a block code having a coding rate of 81/84, which means there are only 3 check bits per code word, a coding gain in excess of 3 dB is easily achieved for a 64-QAM system at a BER of 10^{-6}.

An important family of block codes is the *Reed-Solomon* (RS) codes. They are a subclass of the famous Bose-Chaudhuri-Hocquenghem (BCH) codes that were derived for correcting errors that occur independently. RS codes are effective in correcting combinations of both random errors and error bursts. They operate on multibit symbols instead of individual bits. For example, an RS(64,40) code with 6-bit symbols would have the information data grouped into blocks of 240 bits (6 × 40) in the encoder. The encoding process would then expand the 40 symbols into 64 symbols (384 bits), of which 24 symbols (144 bits) constitute redundancy. There is quite a lot of choice when considering RS codes. This is illustrated by the general inequality that states that RS(n,k) codes having b bits per symbol exist for all values of n and k, where $0 < k < n < 2^b + 2$. It is common to use $b = 8$ so each symbol is 1 byte. RS codes have a powerful correction capability for a given percentage of redundancy. An RS(64,62) code would have 3.2 percent redundancy, which is typical for a 64-QAM DMR system. As indicated in Fig. 3.25, this would provide a coding gain of about 4.4 dB (at BER = 10^{-8}). The term *asymptotic coding gain* (ACG) is often used to describe the coding gain at BER = $10^{-\infty}$ (zero errors). Caution must be observed when using this term because it can be

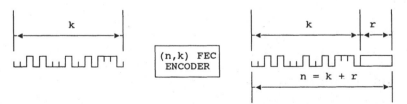

Figure 3.24 A block encoder. k = message bits, r = check bits, n = code word.

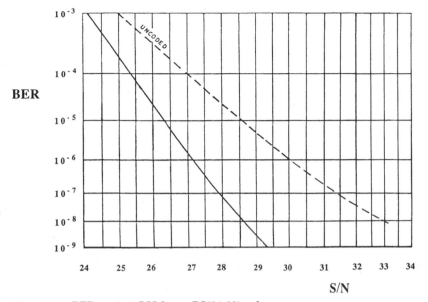

BER

S/N

Figure 3.25 BER against S/N for an RS(64,62) code.

somewhat misleading. For example, the coding gain at BER = 10^{-10} for some codes could be considerably less than the ACG value.

The general corrective power is described as follows: If there are $r = n - k$ redundant symbols, t symbol errors can be corrected in a code word so long as t does not exceed $r/2$. For RS(64,62), this implies that only singly occurring symbol errors are correctable in each block of 61 symbols [i.e., $(64 - 62)/2 = 1$]. A highly redundant code such as RS(64,40) would allow every pattern of up to 12 symbol errors to be corrected in a block of 40 symbols.

The ability of a block code to control random errors is dependent upon the minimum number of positions in which any pair of encoded blocks differs. This number is called the *Hamming distance* for the code. For example, a received sequence of 110111 differs from the transmitted sequence of 111111 by a Hamming distance of 1, whereas 101101 would have a Hamming distance of 2.

Convolutional codes. The convolutional code was the first type of coding to receive widespread acceptance, and it was initially used in satellite transmission applications.

A convolutional code differs from a block code in that the convolutional code word exiting the encoder depends on the block of message bits (k) *and* the previous ($n-1$) blocks of message bits. This rather complex process is performed by shift registers and adders. Figure 3.26 illustrates a convolutional encoder using only three shift registers. The number of shift registers forming the encoder is called the constraint length K, so Fig. 3.26 has $K = 3$. The input

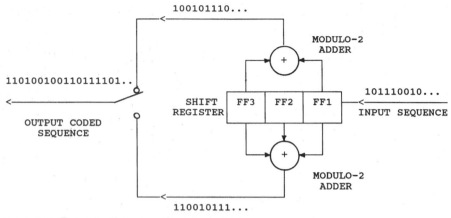

Figure 3.26 Rate 1/2, $K = 3$ convolution encoder.

data is gradually shifted through the registers, and the outputs are fed into the two modulo-2 adders as indicated. A modulo-2 adder, by definition, produces a 1 output for an odd number of 1 inputs. A simple truth table check for two inputs to a modulo-2 adder confirms it to be the familiar EXCLUSIVE OR function. The switch moves back and forth to take alternate bits from each modulo-2 adder output to produce the encoded output. In this case there are 2 coded bits for every input data bit. This is called a rate 1/2 *(not half-rate)* encoder. Note this is different from the half-rate encoder terminology used in mobile communications (Chap. 7). The code rate in general is $1/n$, where n is the number of coded output bits per input data bit. For example, consider an input data sequence of 101110010. The outputs of the modulo-2 adders are shown in Table 3.1 (considering the shift registers initially set at 000 and the sequence was followed by zeros). The switch then takes alternate bits to form the encoded output:

110100100110111101...

TABLE 3.1   Convolutional Encoder Bit Sequences

Output from 3-input adder	Shift register values	Output from 2-input adder
1	001	1
1	010	0
0	101	0
0	011	1
1	111	0
0	110	1
1	100	1
1	001	1
1	010	0

It is now instructive to represent the encoding process in the form of a tree (Fig. 3.27) and subsequently the *trellis* of Fig. 3.28. This will be presented here as an introduction to TCM. The tree is a decision process diagram. Notice how the coded sequence can be derived from this tree diagram. The first five input bits to the coder were 10111. This would lead us through the path A-C-F-I-O-S, during which the coded sequence becomes 11 01 00 10 01. Notice that the tree repeats itself after three splits. LMNOLMNO is the same for the top half of the diagram as for the bottom half. This is because $K = 3$. The output is influenced only by the 3 bits that are currently in the three flip-flops of the shift register. As the fourth bit enters FF1, the first bit drops out of FF3 and no longer influences the output. Therefore, if $K = 4$, the tree would repeat itself after four splits. For a nonmathematical appreciation, there is another visual aid that is very helpful in clarifying this apparently obscure subject; that aid is the trellis diagram of Fig. 3.28a. The trellis is formed by plotting all possible paths through the tree. Solid lines indicate a path due to a 0, and dotted lines indicate that of a 1. It is interesting to note that there are only four possible states in the trellis. State 1 is ABDHLP, state 2 is CEIMQ, state 3 is FJNR, and state 4 is GKOS.

The trellis diagram gives some insight into a means of decoding the convolutional-coded sequence. The repetitive aspect of the trellis is the key to the operation of the decoding process known as the Viterbi algorithm. A block code has a formal structure, and this property is exploited to decode precisely the bit sequence in the receiver. Unfortunately, this is not the case for convolutional decoding, and a considerably more complex mechanism is necessary.

First, the terms *hard* and *soft* need to be defined with respect to decoding. Previously, the output from a demodulator was considered to be a hard decision. For QAM, the output was a specific state in the constellation diagram. In the case of a soft decision, the output from the demodulator would be quantized into several levels (e.g., eight levels for a 3-bit quantization). A decoding algorithm can then operate on this output information to enhance the overall BER. Stated another way, a soft distance would be the Euclidean distance instead of the Hamming distance.

The Viterbi algorithm is known as a *path of maximum likelihood* process. The paths through the trellis represent all possible transmitted sequences. The repetitive nature of the tree allows the number of computations required to home in on the correct sequence to be reduced to a manageable quantity. For example, consider decoding the received sequence 1101001001. Taking the digits in pairs (11 01 00 10 01), the path on the trellis is traced, noting the Hamming distance for all paths. Because 11 is the path AC, the Hamming distance for AC is 0, whereas the path AB is a Hamming distance of 2. This leads to the new trellis diagram of Fig. 3.28b, where the Hamming distances are noted for each branch of the tree. The cumulative distances are the important values and these are noted in brackets, the first number being the upper best route

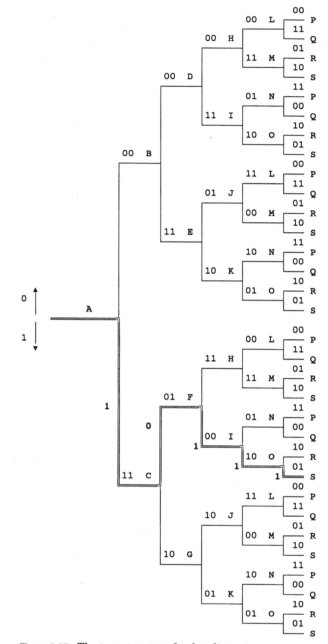

Figure 3.27 The tree structure for decoding.

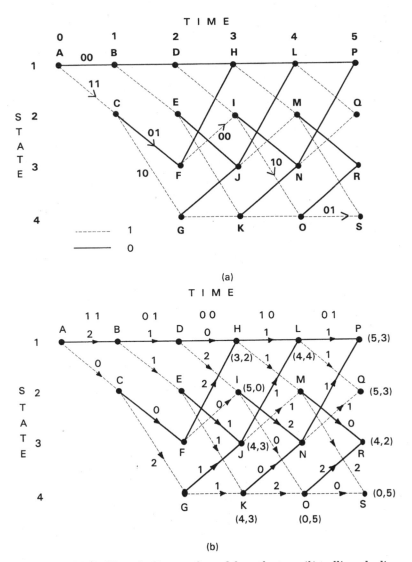

Figure 3.28 (a) Trellis code diagram formed from the tree; (b) trellis code diagram with a Hamming distance example.

and the second number being the bottom route. At point E, the paths EK and EJ both have equal total distances of (4,3). One path is rejected while the other is the "survivor." This is the key to streamlining the calculation that is characteristic of the Viterbi algorithm. The least-distance path is ACFIOS, which is a coded sequence of 1101001001. The received sequence therefore has a Hamming distance of 0. The corresponding transmit sequence of 10111 is therefore the "most likely" sequence to have been transmitted.

All of the above analysis has been for a value of $K = 3$. The complexity clearly increases rapidly for higher values of K. Instead of the four states for $K = 3$, there would be eight for $K = 4$, sixteen for $K = 5$, etc., with correspondingly large increases in the number of possible paths in the trellis. The technological advancements of fast *very large-scale integration* (VLSI) circuits able to handle such large numbers of computations have been essential to the success of convolutional coding.

A convolutional decoder requires a certain number of computations per decoded information bit. In general, the greater the number of computations the better the BER. This implies that computation time becomes a limiting factor.

The performance of convolutional coders depends on the constraint length K. There is a trade-off between BER and bit rate. The higher the K value the better the BER performance, but the lower the highest operating bit rate. Typical design values for K range between 5 and 11, with 7 being popular. A value of 11 would be more desirable from a BER improvement point of view, but this would limit the operation to very slow channels. For a data throughput of several hundred megabits per second, K would be restricted to about 5. In general, convolutional codes tend to be superior to RS codes for *random error* correction.

Convolutional codes with a high code rate can be *punctured* by periodically deleting code symbols from a low-rate code. Punctured codes produce good performance, and the decoder has a relatively low complexity. Finally, the complexity of convolutional coders decreases as the redundancy increases. Conversely, for block codes, the complexity increases as the redundancy increases.

Code interleaving. Code interleaving is a countermeasure against short error bursts. Figure 3.29 illustrates how a simple block code deals with an error burst. Interleaving means that N blocks of data are transmitted bit by bit from each successive block instead of block by block. The number of blocks involved in the interleaving process is called the interleaving depth. Suppose the burst causes errors in N2, A3, B3, ..., etc., of the transmitted bit stream. Because of the interleaving, these errors will show up in the receiver as isolated errors in blocks A, B, and N.

If the burst of errors is less than the interleaving depth, the errors can be corrected in the same manner as if they were random errors. The greater the interleaving depth, the longer the error burst that can be corrected but the greater the added redundancy and system delay. Interleaving is used on both RS and convolutional codes, with the RS interleaving being preferable for long error burst (large interleaving depth) situations.

Interleaving tends to convert error bursts into semirandom errors, which is contrary to the philosophy that random errors are the most difficult to correct. Nevertheless, interleaving does have an important part to play in the practical implementation of error correction.

Concatenated codes. Concatenated codes can provide the benefits of both RS codes and convolutional codes to produce a performance that is synergistic in nature. In other words, the concatenated-coded BER is better than the sum of

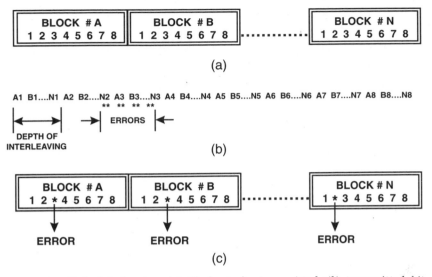

Figure 3.29 Code interleaving. (a) Blocks to be transmitted; (b) transmitted bit sequence; (c) blocks received.

the two individual improvements of RS coding and convolutional coding. The two coders and decoders are essentially in series, as indicated in Fig. 3.30. If the first (inner) coder and decoder is used to improve the performance of a system degraded by Gaussian noise conditions, the second (outer) coder and decoder is then used to clean up short error bursts due to interference and fading. For relatively low bit rates (a few hundred Mb/s), the convolutional coder with Viterbi decoder is a good option for the inner codec. Block coding would be more appropriate for higher bit rates and also if the noise is not Gaussian. The most suitable outer codec is usually the RS type. Figure 3.31 illustrates the dramatic performance improvement of a concatenated RS and convolutional code compared with the convolutional code alone and the uncoded signal. Admittedly, the redundancy of the RS(255,223) coder and (2,1), $K = 7$ convolutional coder in this case is high, but even with a small percentage of redundancy, the performance is significantly improved.

Summary. To summarize, the following statements can be made about the use of coding to counter Gaussian noise:

1. Convolutional coding is good for relatively low bit rate systems requiring BER values in the 10^{-3} to 10^{-7} region.

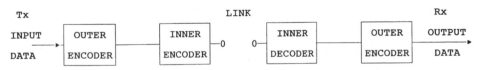

Figure 3.30 Concatenated coding.

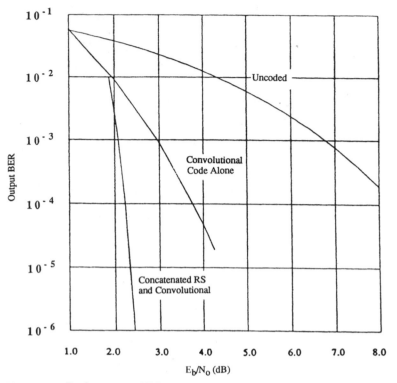

Figure 3.31 Performance of RS(255,223) and (2,1), $K = 7$ convolutional code. *(Reproduced with permission from IEEE © 1987, Berlekamp, E. R., et al., "The Application of Error Control to Communications," IEEE Communications, April. 1987, Fig. 10.)*

2. Concatenated convolutional and block coding is applicable to relatively low bit rate, excellent BER (10^{-10}) systems.

3. Block codes are useful for high bit rate, excellent BER requirement applications (better than BER = 10^{-10}).

3.2.2 Cyclic redundancy check

The cyclic redundancy check (CRC) is a very powerful error-detecting technique that uses simple circuitry and suffers minimal delay or additional overhead. The CRC-16 is recommended by ITU-T for use in high-speed modems (e.g., ITU-T Rec. V.42). It can detect all single and double bit errors; all errors having an odd number of bits; all burst errors of 16 bits or less; and 99.997 percent of 17-bit error bursts.

The CRC principle is to modify the transmitted message so that it is always divisible by a predetermined number at the receiver. For example, if the number 43 is to be transmitted and it is decided that the received number must be divisible by 7, then 43 must first be altered before transmission. It could be multiplied by, say, 9 to give 387, and then 5 must be added to make it divisible by 7. So, 392 is transmitted, and if at the receiver the number is divided by 7 exactly with no remainder, error-free transmission is indicated.

The CRC-6 is used in the primary multiplexer (1.544-Mb/s) frame and the CRC-4 for the 2.048-Mb/s frame. The error correction code is the remainder of the message polynomial $P(x)$ divided by a generating polynomial $G(x)$. This might sound complex, but in the binary system it is very simple. The division is done by the modulo-2 division using a digital circuit containing only EX-OR gates and shift registers.

Consider the CRC-6 in the 1.544-Mb/s multiplexer as an example. The first bit (F-bit) in the 193-bit frame is for frame alignment, to identify where each frame is located within the 24 frames, and also for the 6 CRC check bits. A 6-bit multiframe alignment word is formed by designating every fourth frame with bits that make up the word 001011. Also, the F-bits in frames 2, 6, 10, 14, 18, and 22, which have positions within the multiframe at bits 194, 966, 1738, 2510, 3282, and 4054, respectively, form the CRC-6 check bits. There are a total of $24 \times 193 = 4632$ bits making up the multiframe.

The CRC-6 generator polynomial $G(x)$ is $x^6 + x + 1$, which in binary terms is 1000011 (derived from $x^6 + x^5 + x^4 + x^3 + x^2 + x^1 + x^0$). First, the F-bits of frame N are replaced by 1s. The 4632 message block is multiplied by x^6 and then modulo-2 divided by the generator polynomial. The remainder of this calculation is the CRC-6 message block transmitted. The CRC-6 check bits generated for each multiframe are used as the CRC-6 message block for the next multiframe.

At the receiver, the same procedure is followed. That is, the F-bits are all replaced by 1s, the 4632 bits are multiplied by x^6, and the result is modulo-2 divided by $x^6 + x + 1$ to give the remainder. These 6 bits are compared with the CRC-6 check bits received in the next multiframe. If they are identical, the frame was received with no errors.

A 4632 (13-bit) modulo-2 division is too lengthy to show here, but for illustration purposes a simple example for a 5-bit message, a 3-bit generator, and a 2-bit CRC word follows:

The message polynomial is $P(x) = x^4 + x^3 + x$, which in binary is 11010.

The generator polynomial is $G(x) = x^2 + 1$, which in binary is 101.

The message is multiplied by x^2, because there are two CRC bits. So,

$$x^2(x^4 + x^3 + x) = x^6 + x^5 + x^3, \text{ which in binary is } 1101000.$$

This is modulo-2 divided by $x^2 + 1$ as in the following calculation:

Transmit	Receive
11101	11101
101)1101000	101)110100001
101	101
111	111
101	101
100	100
101	101
100	100
101	101
01	101
	101
	000

The CRC bits are therefore 01 and the transmitted sequence is:

$$1101000 \ 01$$

After the receive side process, the remainder is 000, indicating no errors.

3.2.3 Automatic request for repeat

First, there are positive and negative acknowledgment methods for ARQ. The negative scheme requires the data to be retransmitted *only* when a request is made. No request means there are no errors detected. In the positive acknowledgment, the transmitting end requires confirmation of every correctly received block of data. The negative method is clearly superior in terms of redundancy, but the positive method boasts superior data security. Because of the poor throughput of the positive method, it is used only in special circumstances, and the negative method is predominant. Two main types of ARQ are:

1. Go back N

2. Selective repeat

In the *go back N* system, negative acknowledgment is usually used when an error is received, in which case the receiver sends a request to the transmitter to repeat the most recent N blocks of data. N is typically 4 or 6. The advantage of this system is that the blocks do not have to be labeled, which minimizes redundancy. The main disadvantage is that the retransmission is greater than desirable. This is necessary to ensure the error is eliminated. Furthermore, the maximum round-trip delay for the link must not exceed the time to transmit N blocks; otherwise the block containing errors might not be repeated.

As the name *selective repeat* suggests, the receiver requests the transmitter to repeat only specific blocks of data. While this method has the advantage that only the blocks containing errors are repeated (less redundancy than go back N), the blocks have to be individually numbered (more redundancy than go back N). Clearly, long blocks reduce the overhead, but they are also more likely to contain errors.

To summarize, ARQ is useful for data channels that are mainly error-free but suffer rare error bursts of limited duration. While the output from an ARQ system has a predictable quality, the throughput of data is heavily dependent on the channel conditions. These characteristics make ARQ particularly well suited to packet-switched networks. For example, in a packet-switched network having selective ARQ, if one or more packets experience a noisy or failed path, the packet is repeated on a different path. The additional overhead inherent to the ARQ technique is easily absorbed by the packet-switched network.

Conversely, FEC systems have a constant throughput, and the data quality depends on the channel noise conditions. The complementary nature of ARQ and FEC leads one to believe that a combination of the two would be extremely effective for data traffic error control. In certain noise circumstances this is true. In a hybrid scheme where FEC is followed by ARQ, the idea is for the FEC to be dominant and for the ARQ to clear up any errors not detected by FEC. This combination is very effective for radio air interfaces that are almost uniformly noisy, but have infrequent error bursts caused by interference or fading.

3.2.4 Trellis-coded modulation

There are several techniques for improving the BER of a digital radio communications system. As elaborated in more detail in Chap. 5, some techniques analyze and modify the received bit sequence, whereas others modify the received IF characteristics. Some techniques operate on the transmitter and receiver sides of the system. FEC, as just discussed, is successfully employed by sending some extra (redundant) bits to improve the BER. This has the disadvantage of increasing the bit rate and therefore the bandwidth. Because higher levels of modulation have evolved to improve bandwidth efficiency, it seems a retrograde step to lose some of this valuable and hard-earned bandwidth efficiency by using an error-detecting code. *TCM was developed to improve the BER without increasing the bit rate.* Shannon's original work on information theory more than 45 years ago predicted the existence of such coded modulation schemes that could fulfill this function. Ungerboeck subsequently produced definitive papers (1976, 1982, etc.) outlining the implementation of TCM.

TCM can achieve a coding gain of up to 6 dB in the E_b/N_o, depending on the modulation scheme. Note that *asymptotic* coding gain is implied here when the term *coding gain* is used. The susceptibility of a modulation scheme to ISI depends on the distance between states in the signal state-space diagram. The trellis coding technique obtains its advantage by effectively maximizing the Euclidean (physical) distance between the states of a modulation scheme by expanding the signal set while avoiding bandwidth expansion. Notice this is distinctly different from convolutional coding and Viterbi decoding that seek to minimize the Hamming distance. The term *trellis* refers to the state transition diagram that is similar to the binary convolutional coding trellis diagrams. The difference in the case of TCM is that the branches of the trellis represent redundant nonbinary modulation signals instead of binary code symbols.

Trellis coding for QAM.   One technique central to TCM is called *set partitioning*. Consider the 16-QAM constellation diagram at the top of Fig. 3.32. If this is now displayed as eight subsets each containing 2 of the 16 states, it is clear that the distance between each pair in any subset is larger than the minimum distance (between adjacent states) in the full 16-QAM constellation. Although this is a simplified explanation, it illustrates the essence of expanding the signal set to achieve coding gain while maintaining a fixed bandwidth efficiency.

This is called eight-state trellis coding, and simply refers to the number of possible states in which an encoder can be. If it is applied to higher levels of QAM, the number of states in the eight subsets would increase. For m number of bits per symbol, each subset contains 2^{m-2} states, and the number of states in the new constellation is 2^{m+1}, as in Table 3.2 (where 16-QAM-TCM is highlighted).

The smallest distance between states, called the *free distance* (d_{free}), is an important parameter in establishing the coding gain. In a QAM signal this is simply the distance between adjacent states. The misinterpretation of any of the states in the constellation, caused by noise affecting the precise location of states, is referred to as the transition of one state to another. These transitions form the basis of the trellis diagram for TCM. In the 16-QAM diagram of Fig. 3.32, transitions are most likely to occur between adjacent states whose

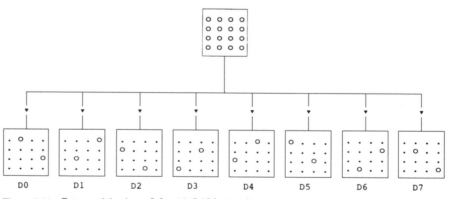

Figure 3.32   Set partitioning of the 16-QAM signal sets.

TABLE 3.2   Eight-State Trellis Coding

m	QAM	2^{m+1}	2^{m-2}
3		16	2
4	16	32	4
5	32	64	8
6	64	128	16
7	128	256	32
8	256	512	64

free distance is Δ_0. However, transitions between states other than adjacent states can occur. In the eight subsets of Fig. 3.32, the free distance between states within each subset is $\Delta_0\sqrt{8}$. Taking into account the possibility of transitions occurring between states in different subsets, the free distance for this code is $\Delta_0\sqrt{5}$, which is still considerably better than Δ_0.

Practical implementation.   When TCM is included in 16-QAM, the constellation is expanded from 16 states to a minimum of 32 states, as shown in the 32-cross QAM diagram of Fig. 3.33. This eight-state TCM encoding is done using the convolutional coding process illustrated in Fig. 3.34. Block coding could also be used, but convolutional coding is simpler and more effective. Each state is increased from four to five bits and the extra bit is also used for FEC protection. Note that the symbol *rate* does not increase. Of the four incoming bit streams, two are mapped directly and the other two are convolutionally encoded into a third bit stream. The third (extra) bit is used to FEC-protect those 3 bits. The 3 bits have eight states, and so this process has the name *eight-state TCM encoder*. Because the 3 bits are protected by a rate 2/3 FEC code, they effectively have almost the same error probability as the two unprotected bits. Also, the unprotected bits are mapped into the two widely spaced states of the four-state subset. The resulting five bit streams are therefore mapped into the eight groups of four states of the 32-cross QAM constellation (Fig. 3.33) to provide very robust protection against intersymbol interference.

Figure 3.35 illustrates the eight-state trellis code for any eight-state QAM signal. For a specific code sequence of D0-D0-D3-D6, four error paths are traced at a distance of $\Delta_0\sqrt{5}$ from that code sequence. The paths all start at the same state and reconverge after three or four transitions. There will

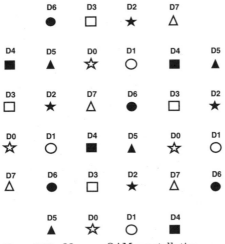

Figure 3.33   32-cross QAM constellation.

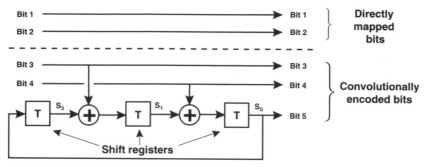

Figure 3.34   16-QAM TCM.

always be four error paths for this eight-state coding, regardless of the code sequence. Any error occurrences will most likely correspond to these paths, resulting in decision error bursts of length 3 or 4. It can be shown mathematically that the coding gain for eight-state coding for all relevant m values of QAM is around 4 dB. The coding gain can be increased to about 5 dB with 16 states, and nearly 6 dB with 128 states. This is clearly a diminishing-returns situation, with considerable increase in circuit complexity as the number of states increases. Eight-state TCM is a good compromise between performance and complexity (and the cost that complexity creates).

Figure 3.36a and b graphically presents the theoretical performance curves for eight-state TCM. Figure 3.36a shows E_b/N_o values plotted against BER for several levels of QAM-TCM. Figure 3.36b compares the E_b/N_o values required to give a BER of 10^{-3} and 10^{-10} for coded and uncoded TCM for many QAM values. In fact, the x axis plots the information rate (bits per symbol) which is related to the bandwidth efficiency or is a measure of the QAM level. Notice that the approximate asymptotic coding gain, denoted by BER = 10^{-10}, is nearly 4 dB.

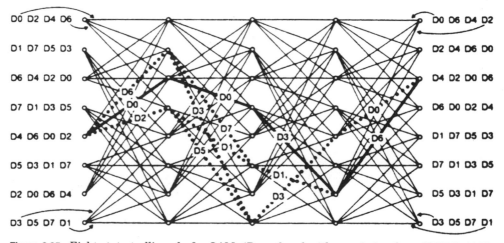

Figure 3.35   Eight-state trellis code for QAM. (*Reproduced with permission from IEEE © 1987, Ungerboeck, G., Trellis Coded Modulation with Redundant Signal Sets, Part 1—Introduction, IEEE Communications Magazine, vol. 25, no. 2, February 1987, pp. 5-11.*)

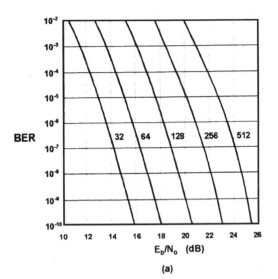

(a)

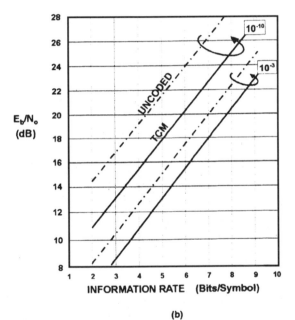

(b)

Figure 3.36 Eight-state TCM performance curves. (a) BER against E_b/N_o); (b) E_b/N_o versus information rate.

3.3 Spread-Spectrum Techniques

Initially, FM was used extensively for analog systems. Spread spectrum is now the subject of intense research and development for application to digital cellular mobile radio, fixed microwave radio, and LEO satellite communication systems. Spread-spectrum techniques theoretically promise to improve the bandwidth efficiency and interference characteristics of future radio systems in general and, in particular, cellular radio systems.

Spread spectrum was a technique initially envisioned for use in military communications, because it provided the militarily desirable characteristic of its signals having a low probability of detection. Today's mobile communication systems suffer from severe spectral congestion, and spread-spectrum techniques promote bandwidth efficiency. So, what is the link between these two diverse types of communication, both of which benefit from the spread-spectrum technique?

From the military perspective, the object of a spread-spectrum system is to ensure that it is difficult to intercept or to "jam" the communication signal. Once a signal is detected, the enemy can transmit noise at the frequency of operation, thereby successfully jamming (disrupting) the communication. *Spread spectrum can be broadly defined as a mechanism by which the bandwidth occupied by the transmitted signal is much greater than the bandwidth required by the baseband information signal.* A spread-spectrum receiver operates over such a large bandwidth that the jamming noise is spread over that bandwidth in a manner that does not significantly interfere with the desired transmission. It is this interference-resistant feature of spread-spectrum systems that is attractive for efficient spectrum utilization in commercial applications, such as microcellular radio. Spread-spectrum signals are not secure, in that they still require message encryption to disguise message content.

The principle of operation of spread-spectrum techniques can be separated into two categories:

1. Direct-sequence spread spectrum (DSSS).

2. Frequency-hopping spread spectrum (FHSS).

The essence of the two techniques is to spread the transmitted power of each user over such a wide bandwidth (preferably more than 10 MHz) that the power per unit bandwidth, in watts per hertz, is very small (see Fig. 3.37). This means that if the desired signal can be extracted, the power transmitted from unwanted, interfering sources and received by a typical narrowband receiver is a small percentage of the usual received power. The unwanted interference is therefore negligible.

In practice, a spread-spectrum receiver spreads the energy of an interfering signal over its transmitted wide bandwidth while compressing the energy of the desired received signal to its original *prespread* bandwidth. For example, if the desired signal is compressed to its original bandwidth of 4 kHz, and all other interfering signals were spread over 1.25 MHz in the receiver, the power in each interfering signal will be reduced by 1,250,000/4,000 = 312.5, or

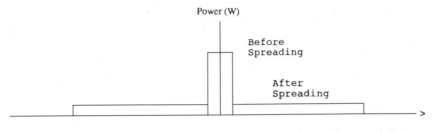

Figure 3.37   Spectrum spreading.

24.9 dB. This is a measure of the spectrum spreading, and is defined as the processing gain N:

$$N = \frac{\text{bandwidth of the signal after spreading}}{\text{bandwidth of the unspread signal}} \qquad (3.13)$$

As a background to the cellular mobile systems design discussed in Chap. 7, several associated techniques need to be clearly defined.

3.3.1 Pseudorandom noise generation

Both FHSS and DSSS use the principle of spreading the spectrum by using pseudorandom generated bit sequences. As the name suggests, these bit sequences are not truly random, but periodic. They are called pseudorandom because the periodicity is so large that usually more than 1000 bits occur before the sequence repeats itself. The simple pseudorandom noise generator circuit is shown in Fig. 3.38. The periodicity depends on the number of flip-flops in the circuit n, with the length of the sequence being given by $2^n - 1$. This is often called a *maximal-length* sequence generation. In this case, because $n = 6$, the sequence repeats itself after every 63 bits.

Truly random sequence generators are available, and the performance of spread-spectrum systems would be improved by purely random bit streams. Unfortunately, a system receiver must have knowledge of the transmitted bit sequence if it is to be effective in despreading and subsequently receiving the signal that is buried in background noise. The pseudorandom sequence is therefore the crucial element in the practical implementation of spread-spectrum systems.

Correlation is an important concept used in connection with pseudorandom sequences. An observation of two purely random sequences would result in the conclusion that there is zero correlation between the two sequences. When comparing two sequences, the correlation can be defined as

$$R = \text{number of agreements} - \text{number of disagreements}$$

If a sequence is compared (correlated) to itself, the outcome is called autocorrelation, whereas correlation with another sequence results in cross-correlation.

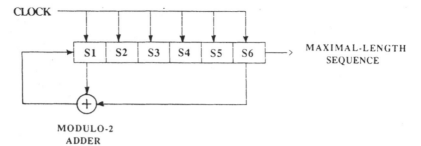

Figure 3.38 Pseudonoise generator.

Comparison of two identical pseudorandom sequences would have perfect cross-correlation only for the in-phase or zero time-shift condition, and a very small correlation for all other phase relationships. Two sequences are said to be *orthogonal* if the degree of cross-correlation (the term "correlation" is more commonly used) is close to zero over the entire sequence comparison.

Figure 3.39 illustrates the line spectrum nature of a pseudorandom sequence. The null points occur at multiples of the shift-register clock frequency. The longer the sequence (defined by the number of shift registers), the more lines there are in the spectrum. For 10 shift registers, the sequence would be 1023 bits long, and the lines in the spectrum would be so close together that the spectrum would appear to be almost continuous.

3.3.2 Frequency-hopping spread spectrum

The difference between FHSS and DSSS will now be discussed (see Fig. 3.40). The frequency-hopping systems achieve their processing gain by avoiding interference, whereas the direct-sequence systems use an interference attenuation technique. In either technique, the objective of the receiver is to be able to pick out the transmitted signal from a wide received signal bandwidth in which the signal is below the background noise level. This sounds impossible, especially when considering the S/N is typically *minus* 15 to 30 dB. The receiver must have a certain minimum amount of information in order to perform this technological stunt. It must know the carrier signal frequency, type of modulation, pseudorandom noise code rate, and phase of the code. Only the phase of the code is a problem. The receiver must be able to establish the starting point of the code from the received signal. This process, which is referred to as *synchronization*, is essential to despreading the required signal while spreading all unwanted signals.

Frequency hopping, as the name suggests, periodically changes the frequency of the transmitted carrier signal so that it spends only a small percentage of its total time on any one frequency (Fig. 3.40a). For example, if the total available bandwidth of 12.5 MHz is partitioned so that carriers are spaced apart by, say, 10 kHz, there are 1250 possible carrier frequencies to which the transmitted signal can hop. If a second user, occupying 12.5 MHz of bandwidth, also hops from one frequency to another, interference caused by the two simultaneously occupying the same frequency occurs only $1/1250 \times 100 = 0.08$ percent of the total time. It is evident that quite a number of users can simultaneously use such a system before the interference becomes noticeable. The main limitation of FHSS systems is the inability of the frequency synthesizer to change frequency quickly without generating unwanted noise signals. If the bandwidth were reduced to 1.25 MHz from 12.5 MHz, the interference would be 0.8 percent. This means there would have to be an order of magnitude fewer users for a given probability of hopping onto the same frequency.

Figure 3.41 shows a schematic diagram of a frequency-hopping system. A pseudorandom noise code generator directly feeds a frequency synthesizer to produce the hopping. This allows the transmission of each bit or part of a bit (known as a *chip*) on a different carrier. The term *chip* arises because the

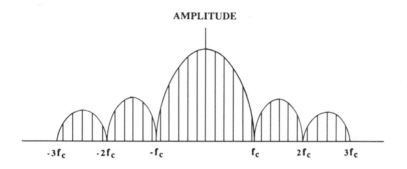

Figure 3.39   Practical spread-spectrum signal.

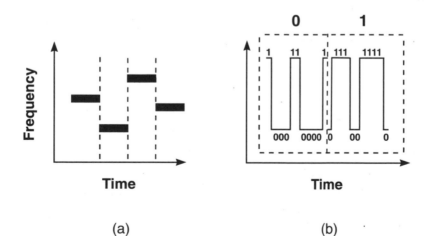

Figure 3.40   (*a*) Frequency-hopping spread spectrum (FHSS). (*b*) Direct-sequence spread spectrum (DSSS).

frequency-hopping rate is not equal to the transmission bit rate. The number of different carriers that can be used depends on the length of the pseudorandom bit sequence. At the receiving end the signal is dehopped by using a pseudorandom bit sequence identical to the one used in the transmitter. First, the RF signal is downconverted by mixing the received signal with the pseudorandom code-generated frequency. After passing through the IF bandpass filter, the signal is demodulated and the original bit stream recovered. A major advantage of frequency hopping is that the receiver does not need to have phase coherence with respect to the transmitter over the full spread-spectrum bandwidth. This means faster acquisition, because the receiver does not have to waste time searching and locking onto the correct phase of the transmitted signal. This noncoherent technique is, however, more susceptible to noise than a coherent method (e.g., direct-sequence modulation).

3.3.3 Direct-sequence spread spectrum

The DSSS technique (Fig. 3.40b) achieves superior noise performance compared to frequency hopping, at the expense of increased system complexity. The spectrum of a signal can be most easily spread by modulating (multiplying) it with a wideband pseudorandom code-generated signal (Fig. 3.42). It is essential that the spreading signal be precisely known so that the receiver can demodulate (despread) the signal. Furthermore, it must lock onto and track the correct phase of the received signal within one chip time (partial or subinteger bit period). At the receiving end, a serial search circuit is shown. There are two feedback loops, one for locking onto the correct code phase and the other for tracking the carrier. For code phase locking, the code clock and carrier frequency generator in the receiver are adjusted so that the locally generated code moves back and forth in time relative to the incoming received code. At the point that produces a maximum correlation at the correlator output, the two signals are synchronized, meaning the

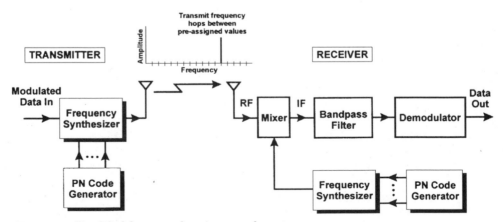

Figure 3.41 Simplified frequency-hopping spread spectrum.

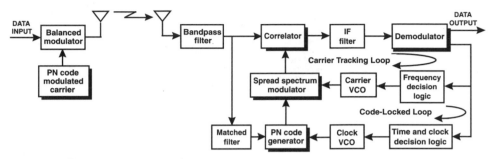

Figure 3.42 Direct-sequence spread spectrum.

correct code phase has been acquired. The second loop (carrier tracking loop) then tracks the phase and frequency of the carrier to ensure phase lock is maintained.

The input bandpass filter is designed for the full spread-spectrum bandwidth. In special circumstances, the IF filter must be designed to have additional width to cope with small Doppler shifts in frequency caused by high-velocity mobile stations (e.g., airplanes).

The success of DSSS depends largely on the chip rate. The spectrum should be spread at the highest possible chip rate. Complementary metal oxide semiconductor (CMOS) circuits can operate at approximately 100 megachips per second, and gallium arsenide field-effect transistor (GaAs FET) circuits can achieve up to several gigachips per second.

3.4   Access Techniques for Satellite and Mobile Communications

Both frequency division multiple access (FDMA) and TDMA are widely used for digital transmission, and these subjects are covered in most introductory texts on telecommunications. For completeness and to refresh the memory of those who might have forgotten, a very brief distinction between the two will now be clarified for the satellite and mobile communications application.

3.4.1   Frequency division multiple access

Broadly speaking, FDMA (Fig. 3.43a) simply means splitting up an available frequency band into a specific number of channels, and the bandwidth of each channel depends on the type of information signals to be transmitted. One pair of channels is used for full duplex operation. Information to be transmitted is superimposed on a carrier at the channel center frequency. The information can be a composite of several information signals, which are multiplexed prior to being superimposed on the carrier, or a single information signal can be placed on the carrier. This would be called a single channel per carrier (SCPC) system, which has been widely used in satellite technology.

Years ago, the analog information was superimposed on the carriers using FM. More recently, the analog signals have been converted to digital pulse streams and the PSK and QAM techniques employed.

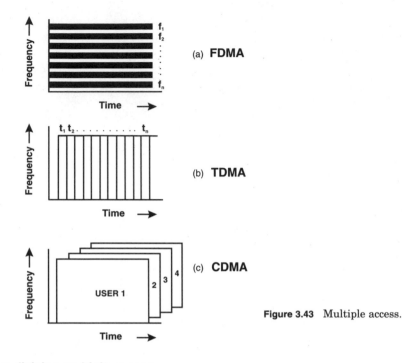

Figure 3.43 Multiple access.

3.4.2 Time division multiple access

By contrast, the TDMA scheme (Fig. 3.43*b*) uses only one frequency band, and many channels are created by transmitting information for each channel in allocated time slots. This mechanism creates the illusion that many users are accessing the radio system simultaneously. In reality, the time sequencing means only one user is occupying the system at any time.

In a TDMA mobile radio system, each base station is allocated a 25- or 30-kHz channel, and users share this same channel on a time-allotted basis. The *maximum number* of users of each channel depends on how many bits per second are required to digitize the voice of each user. As indicated in Chap. 2, the conventional voice A/D conversion process requires 64 kb/s. ADPCM has been accepted for several years as a means of reducing this to 32 kb/s. By incorporating several other digital processing techniques, such as linear predictive coding, 8 kb/s is possible without significantly noticeable degradation in quality, and 4 kb/s can be used if some noticeable degradation is tolerated. If a digital modulation technique such as 4-PSK is used, each voice channel can be digitized with a bandwidth efficiency of at least 1 b/s/Hz (i.e., 8.5 kHz of bandwidth is required for each digital voice signal). This means that for a 30-kHz TDMA channel, there can be three users per channel (or twelve users per channel with degradation).

3.4.3 Code division multiple access

Although the previous two access methods will already be known to most engineers and technicians, CDMA (Fig. 3.43*c*) has only recently received widespread interest because of its potential benefits in satellite and cellular mobile radio telecommunication systems. The objective of CDMA is to allow many users to occupy the exact same frequency band without interfering with each other. It sounds too good to be true. However, with our spread-spectrum knowledge, it is quite a reasonable statement. Each user is assigned a unique orthogonal code.

In a CDMA system, all signals from all users will be received by each user. Each receiver is designed to listen to and recognize only one specific sequence. Having locked onto this sequence, the signal can be despread, so the message stands out above the other signals (which appear as noise in comparison). Interference does become a limiting factor, because as more users occupy the same frequency band, the noise level rises to a point where despreading does not provide an adequate S/N.

TDMA requires precise network synchronization. In comparison, the non-synchronous nature of CDMA gives it the advantage of not requiring network synchronization, although synchronization between individual transceiver pairs is necessary. Second, it is relatively easy to add users to a CDMA system. Third, CDMA is more tolerant of multipath fading than TDMA or FDMA. In conclusion, CDMA has the potential to enable a more efficient use of the frequency spectrum than other techniques, and its use in satellite and mobile radio systems is examined in Chaps. 6 and 7.

3.5 Transmission Delay

The transmission delay in telecommunications circuits is becoming a serious problem, especially for satellite systems. Delay is caused by:

1. Transmitter to receiver signal transit time
2. Speech encoder digital processing
3. Error correction coding techniques

The transit time delay is clearly unavoidable and is significant only on GSO satellite links where the propagation delay is about a quarter of a second. In comparison, the earth's half-circumference of 20,000 km of optical fiber has only about 100 ms of delay. Speech encoding is becoming more important for packing ever-larger numbers of calls into cellular radio bands. Satellite systems are also caught up in this process of improving bandwidth efficiency in order to remain competitive. Error correction is now an essential part of all transmission systems, and even the less noisy environment of optical fiber systems employs error-correction techniques.

Time delay caused by the speech coding and error correction processes has been created as a direct result of the bandwidth restrictions imposed by the limited radio frequency resource. GSO satellite systems suffer from all three of the above time delay contributors. Even worse, when two cellular phone users talk through a satellite link, the delays are all additive.

Echo is often associated with transmission delay, but care must be taken to distinguish between the two. Audible echo is caused primarily by voice frequency signal reflections from a two-wire to four-wire interface known as the hybrid circuit in COs (exchanges); this is due to an impedance mismatch. Any reflections will be more noticeable as the transmission delay increases. In the early days of satellite technology, propagation delay was often mistaken to be the echo itself rather than the factor that makes echoes more noticeable. Echo cancellers at earth station receivers remove the echo but, of course, they cannot remove the propagation delay.

In analog connections there is a two-wire to four-wire interface in the CO, at the end of the local loop and at a concentrator if present. Acoustic echoes can be created between the loudspeaker and the microphone on the telephone set. For an ISDN connection four-wire lines are end to end from the caller's telephone to the called party's telephone, so only acoustic noise can be present. Delay on telephone circuits is a subjective problem that is more annoying to some people than others. The tolerance to echoes was investigated many years ago. The overall loudness ratio (OLR) is a term used in ITU-T Recommendation G.131 to indicate sensitivity to echo. For a complete digital connection having a one-way signal delay of 25 ms, the echo perceived by an OLR of 26 dB was detectable by 10 percent of those questioned. At a 33-dB OLR, only 1 percent noticed an echo. 25 ms is the maximum delay recommended as being acceptable without using echo suppression. A one-way delay up to 150 ms is stated by ITU-T Recommendation I.114 to be acceptable provided echo canceller circuits are used. This Recommendation adds that one-way delay should ideally not exceed 300 ms, and 400 ms is the limit that should definitely not be exceeded.

A GSO satellite connection one-way delay of 270 ms is already quite high. A double hop with a 540-ms delay should clearly be avoided. Echo cancellers eliminate the echo, but delays of this magnitude disrupt conversations. The normal "interrupt" characteristic of everyday conversations is not possible with such large delays. Table 3.3 shows some typical delay times. Notice that the GSM customer talking through a GSO satellite link would experience a one-way delay of about $90 + 270 = 360$ ms, which exceeds the recommended limit. Again, such connections should be avoided.

TABLE 3.3   Transmission Delay for Several Types of System

System type	Approximate one-way delay (ms)
GSO satellite at the equator (no coding)	240
GSO satellite at 5° elevation angle (no coding)	275
GSO satellite at the equator + VSELP	360
LEO satellite (no coding)	5
LEO satellite + VSELP	125
DECT cordless telephone	10
GSM mobile telephone	90
MPEG II video codec	150
Primary PCM system	0.75
Digital switch between analog interfaces	1.5
Digital switch between 1.544-Mb/s interfaces	0.5

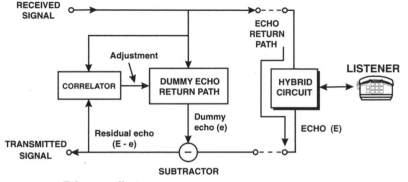

Figure 3.44 Echo cancellation.

However, for a GSM customer talking via a LEO satellite link, the delay is reduced to 95 ms, which is acceptable, provided the satellite signal processing delay is not too large.

Echo cancellers.   Echo suppressors were used before echo cancellers became widely available. Their effectiveness was not always satisfactory and initial syllable clipping was often a problem. A typical echo canceller works on the principle shown in Fig. 3.44. The received signal generates an echo signal E in the hybrid, which would be transmitted if cancellation were not effective. Part of the received signal is passed through a circuit to generate a dummy echo e. The subtractor prevents the echo from penetrating into the transmit signal. When the listener interrupts, the voice is mixed with the residual echo $(E - e)$. The received signal is therefore correlated with the residual echo to remove the effect of the listener's voice. Continual adjustment of the dummy-echo return path setting is made to keep the output of the correlator at zero as the transmission path conditions change.

The Microwave Link

In its simplest form the microwave link can be one *hop*, consisting of one pair of antennas spaced as little as 1 or 2 km apart, or can be a *backbone*, including multiple hops, spanning several thousand kilometers. Typically, a single hop is 30 to 60 km in relatively flat regions for frequencies in the 2- to 8-GHz bands. When antennas are placed between mountain peaks, a very long hop length can be achieved. Hop distances in excess of 200 km are in existence. The "line-of-sight" nature of microwaves has some very attractive advantages over cable systems. *Line of sight* is a term that is only partially correct when describing microwave paths. Atmospheric conditions and terrain effects modify the propagation of microwaves so that even if the designer can see from point A to point B (establishing a true optical line of sight), it may not be possible to place antennas at those two points and achieve a satisfactory communication performance.

The objective of microwave communication systems is to transmit information from one place to another without interruption, to be clearly reproduced at the receiver. Figure 4.1 indicates how this is achieved in its simplest form. The voice, video, or data channels are combined by the multiplexing technique to produce a BB signal. This signal is frequency modulated to an IF and then upconverted (heterodyned) to the RF for transmission through the atmosphere. The reverse process occurs at the receiver. The microwave transmission frequencies presently used are within the approximate range of 2 to 38 GHz. The frequency bands used for digital microwave radio are recommended by the ITU-R and summarized in Table 4.1. The mobile radio systems have gained some spectrum at the expense of DMR, namely by sharing the 1.9- to 2.3-GHz band with fixed radio. Until the mid-1990s, each Recommendation clearly defined the frequency range, the number of channels that could be used within that range, the channel spacing, the bit rate, and the polarization possibilities. The competition of DMR with optical fiber transmission systems has led to some remarkable liberalization of the radio spectrum utilization.

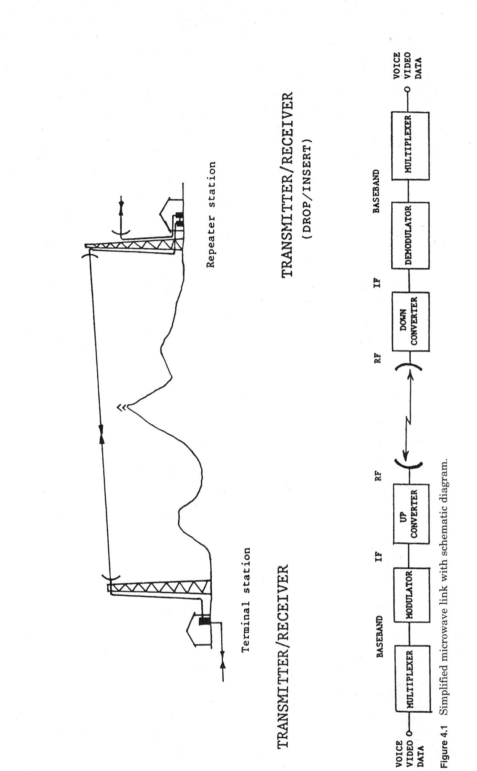

Figure 4.1 Simplified microwave link with schematic diagram.

TABLE 4.1   Frequency Bands and Channel Capacity for Digital Microwave Radio Systems

(a) Frequency Bands for DMR Systems			
Band (GHz)	Frequency range (GHz)	ITU-R Recommendation F-series	Channel spacing (Mb/s)
1.5	1.427 – 1.53	746 Annex 1	0.5; 1; 2; 3.5
2	1.427 – 2.69	701	0.5 (pattern)
	1.7 – 2.1; 1.9 – 2.3	382	29
	1.7 – 2.3	283	14
	1.9 – 2.3	1098	3.5; 2.5 (patterns)
	1.9 – 2.3	1098 Annexes 1, 2	14
	1.9 – 2.3	1098 Annex 3	10
	2.3 – 2.5	746 Annex 2	1; 2; 4; 14; 28
	2.5 – 2.7	283	14
4	3.8 – 4.2	382	29
	3.6 – 4.2	635	10 (pattern)
	3.6 – 4.2	635 Annex 1	90; 80; 60; 40
5	4.4 – 5.0	746 Annex 3	28
	4.4 – 5.0	1099	10 (pattern)
	4.4 – 5.0	1099 Annex 1	40; 60; 80
	4.54 – 4.9	1099 Annex	40; 20
L6	5.925 – 6.425	383	29; 65
	5.85 – 6.425	383 Annex 1	90; 80; 60
U6	6.425 – 7.11	384	40; 20
	6.425 – 7.11	384 Annex 1	80
7	7.425 – 7.725	385	7
	7.425 – 7.725	385 Annex 1	28
	7.435 – 7.75	385 Annex 2	5
	7.11 – 7.75	385 Annex 3	28
8	8.2 – 8.5	386	11.662
	7.725 – 8.275	386 Annex 1	29.65
	7.725 – 8.275	386 Annex 2	40.74
	8.275 – 8.5	386 Annex 3	14; 7
10	10.3 – 10.68	746 Annex 4	20; 5; 2
	10.5 – 10.68	747 Annex 1	7; 3.5 (patterns)
	10.55 – 10.68	747 Annex 2	5; 2.5; 1.25 (patterns)
11	10.7 – 11.7	387 Annexes 1, 2	40
	10.7 – 11.7	387 Annex 3	67
	10.7 – 11.7	387 Annex 4	60
	10.7 – 11.7	387 Annex 5	80
12	11.7 – 12.5	746 Annex 5	19.18
	12.2 – 12.7	746 Annex 5	20 (pattern)
13	12.75 – 13.25	497	28; 7; 3.5
	12.75 – 13.25	497 Annex 1	35
	12.7 – 13.25	746 Annex 5	25; 12.5
14	14.25 – 14.5	746 Annex 6	28; 14; 7; 3.5
	14.25 – 14.5	746 Annex 7	20
15	14.4 – 15.35	636	28; 14; 7; 3.5
	14.4 – 15.35	636 Annex 1	2.5 (pattern)
	14.4 – 15.35	636 Annex 2	2.5

TABLE 4.1   Frequency Bands and Channel Capacity for Digital Microwave Radio Systems (Continued)

Band (GHz)	Frequency range (GHz)	ITU-R Recommendation F-series	Channel spacing (Mb/s)
18	17.7 – 19.7	595	220; 110; 55; 27.5
	17.7 – 21.2	595 Annex 1	160
	17.7 – 19.7	595 Annex 2	220; 80; 40; 20; 10; 6
	17.7 – 19.7	595 Annex 3	3.5
	17.7 – 19.7	595 Annex 4	13.75; 27.5
23	21.2 – 23.6	637	3.5; 2.5 (patterns)
	21.2 – 23.6	637 Annex 1	112 to 3.5
	21.2 – 23.6	637 Annex 2	28; 3.5
	21.2 – 23.6	637 Annex 3	28; 14; 7; 3.5
	21.2 – 23.6	637 Annex 4	50
	21.2 – 23.6	637 Annex 5	112 to 3.5
	22.0 – 23.6	637 Annex 1	112 to 3.5
27	24.25 – 25.25	748	3.5; 2.5 (patterns)
	24.25 – 25.25	748 Annex 3	56; 28
	25.25 – 27.5	748	3.5; 2.5 (patterns)
	25.25 – 27.5	748 Annex 1	112 to 3.5
	27.5 – 29.5	748	3.5; 2.5 (patterns)
	27.5 – 29.5	748 Annex 2	112 to 3.5
	27.5 – 29.5	748 Annex 3	112; 56; 28
31	31.0 – 31.3	746 Annex 8	25; 50
38	36.0 – 40.5	749	3.5; 2.5 (patterns)
	36.0 – 37.0	749 Annex 3	112 to 3.5
	37.0 – 39.5	749 Annex 1	140; 56; 28; 14; 7; 3.5
	38.6 – 40.0	749 Annex 2	50
	39.5 – 40.5	749 Annex 3	112 to 3.5
55	54.25 – 58.2	1100	3.5; 2.5 (patterns)
	54.25 – 57.2	1100 Annex 1	140; 56; 28; 14
	57.2 – 58.2	1100 Annex 2	100

(b) Channel Capacity for DMR Systems

Channel bandwidth (MHz)	Channel capacity (minimum roll-off factor α)				
	4-QAM, 4-PSK	16-QAM	64-QAM	256-QAM	512-QAM
North American[*]					
2.5	2DS1 (0.5)	4DS1 (0.5)	6DS1 (0.5)	10DS1 (0.2)	10DS1 (0.35)
5	4DS1 (0.5)	8DS1 (0.5)	12DS1 (0.5)	20DS1 (0.2)	20DS1 (0.35)
10	8DS1 (0.5)	16DS1 (0.5)	1DS3 (0.25)	1 DS3 (0.65)	1DS3 (0.9)
				1STS1 (0.45)	1STS1 (0.6)
20	16DS1 (0.5)	1DS3 (0.65)	2DS3 (0.25)	2DS3 (0.65)	3DS3 (0.25)
			1STS1 (1.0)	2STS1 (0.45)	2STS1 (0.6)
40	1DS3 (0.65)	2DS3 (0.65)	4DS3 (0.25)	5DS3 (0.35)	6DS3 (0.25)
		2STS1 (0.45)	3STS1 (0.45)	4STS1 (0.45)	5STS1 (0.3)
			1STM1 (0.45)	1STM1 (0.9)	1STM1 (1.0)

TABLE 4.1 Frequency Bands and Channel Capacity for Digital Microwave Radio Systems (Continued)

Channel bandwidth (MHz)	Channel capacity (minimum roll-off factor α)				
	4-QAM,4-PSK	16-QAM	64-QAM	256-QAM	512-QAM
European[†]					
3.5	2E1 (0.6)	4E1 (0.6)	8E1 (0.2)	8E1 (0.6)	12E1 (0.2)
7	4E1 (0.6)	8E1 (0.6)	12E1 (0.6)	1E3 (0.5)	1E3 (0.7)
14	8E1 (0.6)	1E3 (0.5)	1E3 (1.0)	2E3 (0.5)	2E3 (0.7)
				1STS1 (1.0)	1STS1 (1.0)
28	1E3 (0.5)	2E3 (0.5)	2E3 (0.5)	5E3 (0.2)	5E3 (0.35)
		1STS1 (1.0)	2STS1 (0.5)	3STS1 (0.35)	3STS1 (0.5)
				1STM1 (0.35)	1STM1 (0.5)
56	2E3 (0.5)	5E3 (0.2)	7E3 (0.3)	10E3 (0.2)	10E3 (0.35)
	1STS1 (1.0)	3STS1 (0.35)	4STS1 (0.5)	6STS1 (0.35)	7STS1 (0.3)
		1STM1 (0.35)	1STM1 (1.0)	2STM1 (0.35)	2STM1 (0.5)

[*] DS1 is 1.544 Mb/s; DS3 is 44.736 Mb/s; STS-1 is 51.84 Mb/s.

[†] E1 is 2.048 Mb/s; E3 is 34.368 Mb/s; STM-1 is 155.52 Mb/s.

SOURCE: Part (a) adapted from ITU-R, Rec. F.746-2, Table 1, Geneva 1995, by permission from ITU.

The frequency bands have largely remained the same for fixed radio service, except for the inclusion of additional spectrum in some places, particularly at the higher frequency bands above 20 GHz. Many frequency bands can now have several types of configuration. The heavily used 4- and 6-GHz bands now have enhanced flexibility. As an example, the major changes for these bands can be described as:

■ Variable channel widths

■ Cochannel operation

To elaborate, the 3.6- to 4.2-GHz band can have channel spacings of 10, 40, 60, 80, or 90 MHz with either alternating polarization (interleaved) arrangements, or copolar. Modulation choices vary from 4-PSK through to 256-QAM. A wide variation of capacity is therefore now available. Also, two-carrier transmission is possible for counteracting difficult propagation paths.

Some new bands have been opened up. For example, the 4.4- to 5.0-GHz band (ITU-R Rec. 1099-1) has even greater flexibility. In addition to channel widths of 20, 40, 60, and 80 MHz, with cochannel operation usable as desired (see Fig. 4.2), there is also a multicarrier mode option on the 60- and 80-MHz channel arrangements. This means each of the 60-MHz-wide channels can have three or six carriers. Again, the multicarrier systems here are a strong measure to counteract very severe propagation conditions known as multipath fading that are described later. Using 256-QAM allows eight GO and eight RETURN channels, each carrying 311 Mb/s (2 × STM-1), plus two GO and two RETURN at about 200 Mb/s. The spectrum utilization efficiency is about 10 b/s/Hz for this configuration.

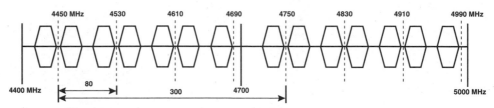

Figure 4.2　Frequency plans for the 4400- to 5000-MHz band.

Some of the factors influencing and leading to those recommendations for point-to-point microwave radio transmission are discussed in the following.

4.1　Antennas

Because antennas play a central role in microwave communications, they will be considered first. There are several shapes of antenna available for transmitting microwaves. Microwave telecommunication systems almost always use the parabolic type, and sometimes the horn type. These antennas are highly directional. The microwave energy is focused into a very narrow beam by the transmitting antenna and aimed at the receiving antenna, which concentrates the received power by a mechanism analogous to the telescope. Figure 4.3 shows how the microwave energy is transmitted by a parabolic antenna, by placing the microwave guide opening at the focus of the parabola. For the simplest style of antenna feed, the waveguide opens in the form of an enlarging taper, which is designed to match the impedance of the waveguide to that of free space. This system is analogous to the searchlight or flashlight beam at optical frequencies. Both light and microwaves are electromagnetic waves, so they have similar qualities. Because we can see light (or light allows us to see), it is often helpful to use the analogous optical mech-

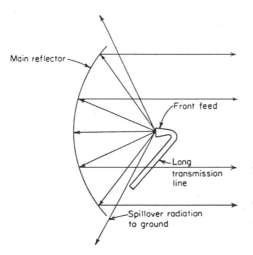

Figure 4.3　Front-feed paraboloid antenna. *Note:* Microwaves follow most of the rules of optics. (*Reproduced with permission from McGraw-Hill. Panter, P. F., Communication Systems Design, Fig. 11-4, McGraw-Hill, 1972.*)

anism for the purpose of shedding some light on the subject of microwaves. Parabolic antennas are available in sizes ranging from about 0.5 to 36 m in diameter.

4.1.1 Antenna gain

The most important characteristic of an antenna is its *gain*. This is a measure of an antenna's ability to transmit waves in a specific direction instead of all directions. It is a measure of directionality. An antenna radiating energy equally in all directions is called an *omnidirectional,* or *isotropic,* antenna (Fig. 4.4). For a point-to-point system, as in microwave communication systems, it is desirable to have a high degree of directionality. In other words, the isotropic antenna is not efficient because energy is wasted. The gain of an antenna describes the extent to which an amount of isotropically radiated energy can be directed into a beam. The narrower the beam, the more highly directional the antenna and therefore the higher the gain.

Mathematically,

$$\text{Gain } (G) = 10 \log_{10} \left(\frac{4\pi A e}{\lambda^2} \right) \quad \text{dB} \tag{4.1}$$

where A = effective area of the antenna aperture (m^2)
 e = efficiency
 λ = wavelength (m)

An isotropic antenna, by definition, has a gain of 1 (or 0 dB). For a parabolic antenna the efficiency is not 100 percent, because some power is lost by "spillover" at the edges of the antenna when it is illuminated by the waveguide fixed at the focus. Also, the antenna dish is not fabricated in a perfectly parabolic shape. The waveguide feed at the focus causes some reduction of the transmitted or received power because it is a partial blockage to the

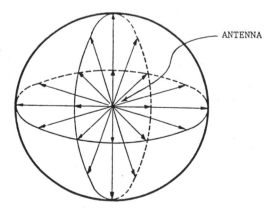

ANTENNA

Figure 4.4 Radiation pattern— isotropic antenna.

microwaves. Commercially available parabolic antennas have efficiencies in the region of 50 to 70 percent. For an efficiency of approximately 60 percent, Eq. (4.1) can be rewritten

$$G = 20 \log_{10} (8.1Df) \ \text{dB} \tag{4.2}$$

where D = antenna aperture (diameter) (m)
       f = frequency (GHz)

The graph in Fig. 4.5 shows the relationship between antenna diameter, frequency, and gain for a conservative antenna efficiency of 50 percent. As the equation suggests, the gain increases with frequency and also with antenna diameter. The most frequently used antenna for the 4- to 6-GHz band is the 3-m-diameter antenna, which has a gain in the region of 40 dB. The diagram in Fig. 4.6 illustrates the manner in which the microwave power is radiated from a parabolic antenna. The measured radiated power level is mapped over a full 360° range. The maximum power is transmitted in the direction called the *boresight*. The boresight field strength is designated as 0 dB, so the field strength in any other direction is referred to that maximum value. The major lobe (main beam) is quite narrow, and the sidelobes and backward radiation are more than 25 dB below the major lobe. Ideally, the transmitted energy should be only in the main beam, but imperfect illumination of the parabolic antenna and irregularities in its reflecting surface cause the sidelobes.

An important characteristic of antennas is the *front-to-back ratio*. It is defined as the ratio of maximum gain in the forward direction to the maximum

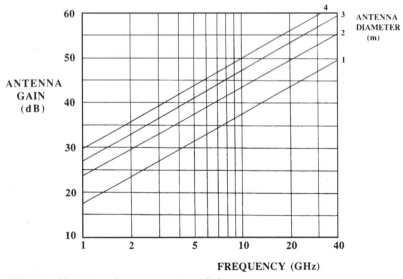

Figure 4.5 Variation of antenna gain with frequency.

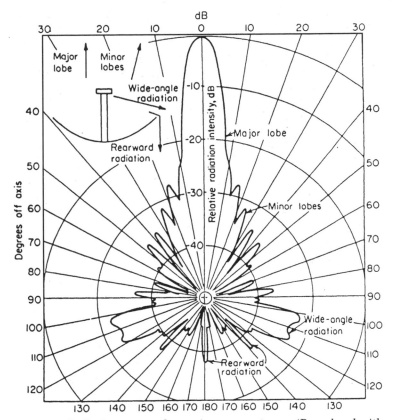

Figure 4.6 Radiation pattern for a microwave antenna. (*Reproduced with permission from McGraw-Hill. Panter, P. F., Communication Systems Design, Fig. 11-1, McGraw-Hill, 1972.*)

gain in the backward direction (backlobes). To illustrate the point, Fig. 4.6 shows a rather poorly designed antenna because it has a front-to-back ratio of only 32 dB. This backward radiation has great significance with respect to noise and interference. Signals reach the receiver after reflection from the ground behind the antenna. This introduces unwanted noise. Secondly, at repeaters where the same frequency is used for incoming and outgoing signals, coupling occurs between the receiver and transmitter antennas. Obviously, the same frequency is avoided wherever possible. Because transmitter power levels are usually at least 60 dB higher than receiver levels, it is evident that in same-frequency cases there must be a high degree of isolation between antennas in order to avoid severe interference.

4.1.2 Beamwidth

The beamwidth is another important characteristic of antennas, and for parabolic antennas the beamwidth is

$$\phi = \frac{21.3}{fD} \tag{4.3}$$

where ϕ = beamwidth measured at the half-maximum power points (3-dB
   down points)
   f = frequency (GHz)
   D = antenna diameter (m)

Figure 4.7 shows the antenna beamwidth plotted against the parabolic antenna dish diameter for several frequencies. Note that the beamwidth for a 3-m antenna is very narrow: less than 2° in the 4- to 6-GHz range. If larger antennas are used, the beamwidth is further reduced. Interference from external sources and adjacent antennas is minimized by using narrow beam antennas. Although large-diameter antennas provide desirably high gain, the decrease in beamwidth can cause problems. The two antennas in each hop must be aligned very precisely; the narrower the beamwidth, the higher the alignment precision required. A very small movement in either antenna will cause degradation of the received signal level. This problem can be serious, particularly when large antennas are used on high towers in very windy regions.

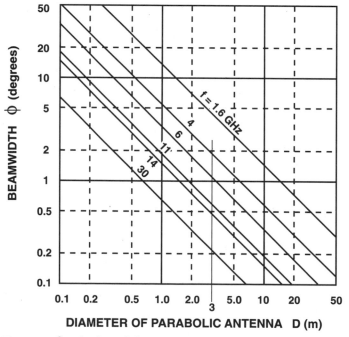

Figure 4.7 Graph of parabolic antenna gain against beamwidth.

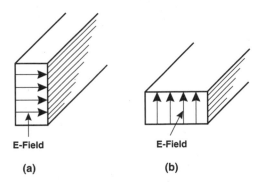

E-Field

(a)

E-Field

(b)

Figure 4.8 Waveguide representation of (a) horizontal and (b) vertical polarization.

4.1.3 Polarization

If a feed horn is used to illuminate the antenna, it is oriented in one of two positions. The narrow dimension of the rectangular opening is placed either horizontally or vertically with respect to the ground. If it is vertical, the electric field is in the horizontal plane and the antenna is said to be *horizontally polarized* (Fig. 4.8a). If the narrow dimension of the horn is horizontal, the electric field is in the vertical plane and the antenna is *vertically polarized* (Fig. 4.8b). If the signal is transmitted, for example, in the horizontal polarization, a small amount of power is unavoidably transmitted in the vertical polarization. The signal in the unwanted polarization is typically 30 to 40 dB lower than the intended polarization. This is referred to as a *cross-polarization discrimination* (or XPD) of 30 to 40 dB. Often two radio channels having opposite polarizations are placed on the same antenna at the same frequency. This is known as *frequency reuse*, and is used for *frequency diversity systems,* as discussed in detail in Sec. 4.7 and Chap. 5.

4.1.4 Antenna noise

Noise entering a system through the antenna can set a limit to the performance of the system. A communications system performance is designed to have the lowest possible noise figure, often stated in terms of noise temperature. These parameters are used interchangeably in most texts, so the relationship between noise figure and noise temperature is given in the graph of Fig. 4.9. It is useful to become familiar with both types of notation, and to memorize the approximate equivalent values for, say, 1-, 3-, 5-, and 10-dB noise figures (i.e., 1 dB $\equiv$ 75 K, 3 dB $\equiv$ 290 K, 5 dB $\equiv$ 627 K, and 10 dB $\equiv$ 2610 K, respectively). The noise temperature of the antenna, T_a, depends on:

- The loss between the antenna and the receiver input (i.e., waveguide run)
- Sky noise from the galaxy, sun, and moon

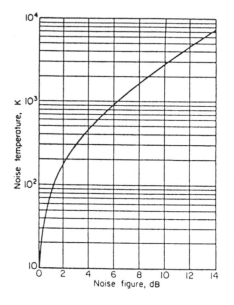

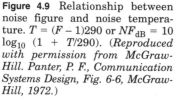

Figure 4.9 Relationship between noise figure and noise temperature. $T = (F - 1)290$ or $NF_{dB} = 10 \log_{10} (1 + T/290)$. (*Reproduced with permission from McGraw-Hill. Panter, P. F., Communication Systems Design, Fig. 6-6, McGraw-Hill, 1972.*)

- Absorption by atmospheric gases and precipitation
- Radiation from the earth into the backlobes of the antenna
- Interference from man-made radio sources

Suppression of the sidelobes is an important aspect of reducing antenna noise. Although the main beam might not be directed towards a "hot" part of the sky (e.g., the sun), one of the sidelobes might be, causing T_a to be too large. A countermeasure used to minimize sidelobe problems is to place a shroud (shield) around the edge of the parabolic dish. This is in the form of a metallic rim, as shown in Fig. 4.10. This improves the front-to-back ratio of a 3-m antenna operating at 6 GHz from the unshielded value of approximately 50 dB to greater than 70 dB. This is sufficient to allow transmission of the same frequency in both directions of a back-to-back repeater station with negligible interference between signals. A thin plastic cover (with negligible attenuation) called a *radome* is often attached to the shroud to provide weather protection.

4.1.5 High-performance antennas

Even higher performance can be obtained from a *horn reflector* antenna (Fig. 4.11). This design has a section of a very large parabola placed at an angle so that the energy fed from the waveguide is both focused and reflected from the parabolic surface. This antenna has excellent sidelobe suppression, and a 3-m antenna at 6 GHz has a front-to-back ratio of about 90 dB. It also has very low voltage standing-wave ratio (VSWR) characteristics, low noise temperature, and substantially wider bandwidth than ordinary parabolic antennas. Although the electrical characteristics are superior to the usual style of parabolic antenna, the horn reflector type is more expensive, larger, heavier, and more difficult to mount. Nevertheless, in circumstances where very high performance is warranted, it is indispensable.

4.1.6 Antenna towers

The towers that are built for microwave antennas significantly affect the cost of the microwave link. The higher the towers are built, the longer the hop and therefore the cheaper the overall link. Calculations show that for a 48-km hop, over relatively flat terrain, towers of about 75 m are required at each end. If there are obstacles such as trees or hills at some points on the path, towers of 100 m or more might be necessary to provide adequate clearance. Towers of this height require guy wires to support them. This is because the cost of self-supporting towers is too high; the cost of such towers increases almost exponentially with height, whereas the cost of guyed towers (having constant cross section) increases linearly with height. In both cases, the amount of land required for the towers increases considerably with tower

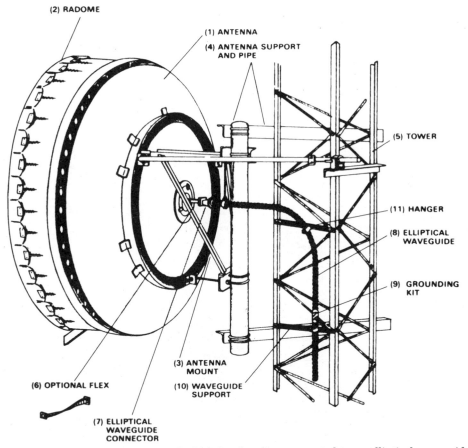

Figure 4.10 Parabolic antenna with shroud and radome connected to an elliptical waveguide system. (*Reproduced with permission from Andrew Corporation.*)

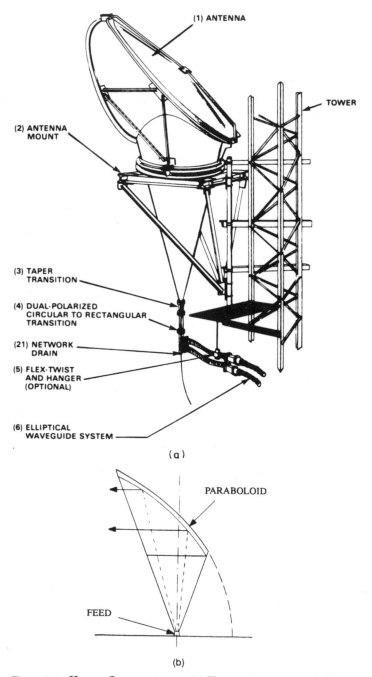

Figure 4.11 Horn reflector antenna. (*a*) Horn antenna connected to an elliptical waveguide system. (*Reproduced with permission from Andrew Corporation.*) (*b*) Schematic diagram of the microwave beam propagating from the antenna.

height. As an example, using Fig. 4.12, the land area required for a guyed tower is determined by the area occupied by the guy wires. They extend outward from the tower a distance of about 80 percent of the tower height. The minimum ground area required for a 91-m guyed tower is 1.47 hectares, or 3.64 acres. Although this might not be a problem in the countryside where land is usually available, in towns or cities there is often limited space at the terminal exchanges for building high towers. This sometimes necessitates mounting the towers on the roofs of CO buildings, in which case the structural adequacy of the roof must be carefully evaluated. In addition, local building codes or air-traffic control regulations can impose restrictions on the height of

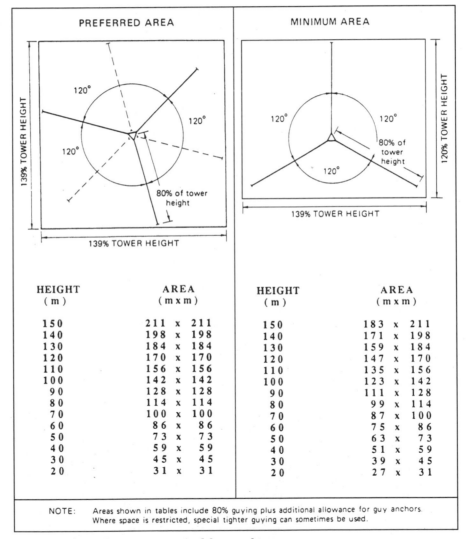

HEIGHT (m)	AREA (m x m)	HEIGHT (m)	AREA (m x m)
150	211 x 211	150	183 x 211
140	198 x 198	140	171 x 198
130	184 x 184	130	159 x 184
120	170 x 170	120	147 x 170
110	156 x 156	110	135 x 156
100	142 x 142	100	123 x 142
90	128 x 128	90	111 x 128
80	114 x 114	80	99 x 114
70	100 x 100	70	87 x 100
60	86 x 86	60	75 x 86
50	73 x 73	50	63 x 73
40	59 x 59	40	51 x 59
30	45 x 45	30	39 x 45
20	31 x 31	20	27 x 31

NOTE: Areas shown in tables include 80% guying plus additional allowance for guy anchors. Where space is restricted, special tighter guying can sometimes be used.

Figure 4.12 Approximate area required for guyed towers.

towers. Local soil conditions must also be taken into account. Extra cost can be incurred in areas with hard rock that must be moved or in very soft soil areas where extra-large concrete bases need to be built. Also, wind loading must be considered, or movement of the tower will cause outages. Even the antennas themselves can cause problems if too many are placed on one tower in an unbalanced configuration.

4.2 Free-Space Propagation

Because microwave energy is electromagnetic energy, it will pass through space in a manner similar to that of light. Nobody knows how it propagates through "nothingness," but we know it does propagate. The atmosphere and terrain have modifying effects on the loss of microwave energy, but first only the loss as a result of free space will be considered. Free-space loss can be defined as the loss between two isotropic antennas in free space, where there are no ground or atmospheric influences. The isotropic antenna by definition radiates energy equally in all directions. Although this is a hypothetical ideal that cannot be realized physically, it is a useful concept for calculations. As stated previously, it allows the gain of an antenna to be described relative to this omnidirectional reference, the isotropic antenna.

When imagining how energy emanates from an isotropic antenna, it is easy to appreciate how a fixed amount of energy emitted by the antenna will spread out over an increasing area as it moves away from the antenna. It is analogous to considering the spherical surface area of an expanding balloon. The energy loss due to the spreading of the wavefront as it travels through space is according to the inverse-square law. As the receiving antenna occupies only a small portion of the total sphere of radiation, it is reasonable to assume only a minute fraction of the total emitted radiation is collected. Mathematically, the free-space loss has been derived as

$$\alpha_{fs} = \left(\frac{\lambda}{4\pi d} \right)^2 \tag{4.4}$$

where λ = wavelength
d = distance from source to receiver

If the path distance is in kilometers and the frequency of the radiation is in gigahertz, the previous equation can be written as

$$\alpha_{fs} = 92.4 + 20 \log_{10} f + 20 \log_{10} d \tag{4.5}$$

(If d is in miles, 92.4 changes to 96.6.) This equation is illustrated in the form of a graph (Fig. 4.13) where free-space loss is plotted against distance for several frequencies. For a typical repeater spacing of approximately 48 km, the free-space loss is 132 dB at 2 GHz, and 148 dB at 12 GHz. This is an extremely large value of attenuation. If only 1 W of microwave power is transmitted, only

10^{-13} W of power are received. These figures are for the case of isotropic antennas at both the transmitter and receiver. Fortunately, the gain of each high-performance directional antenna used in microwave communication systems is approximately 40 dB. The received power is therefore improved from 10^{-13} W to about 10^{-5} W. Equation 4.5 is presented in a different form in Fig. 4.14 by plotting free-space loss against frequency for various distances. This graph gives a

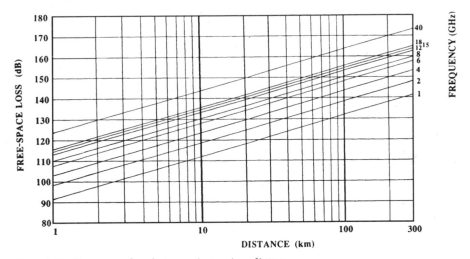

Figure 4.13 Free-space loss between isotropic radiators.

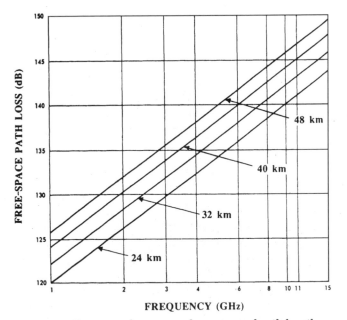

Figure 4.14 Free-space loss versus frequency and path length.

good feel for how the free-space loss increases with distance for a particular operating frequency.

4.3 Atmospheric Effects

4.3.1 Absorption

True free-space propagation exists for a portion of a satellite communication path, but terrestrial microwave communication paths always require propagation through the atmosphere. The oxygen in the atmosphere absorbs some microwave energy, as shown in Fig. 4.15. Fortunately, this attenuation is relatively small in the frequency range used for microwave communication. It is approximately 0.01 dB/km at 2 GHz and increases to 0.02 dB/km at 26 GHz.

The effect of rain on microwave radio propagation is quite significant, especially at the higher frequencies. As the graph in Fig. 4.15 shows, the attenuation increases rapidly as the water content of the atmosphere in the microwave path increases. At 6 GHz the attenuation due to water vapor in the air is only 0.001 dB/km. As the water content increases to fog and then to light rain, the attenuation increases to 0.01 dB/km, and for a cloudburst (very heavy rain) the attenuation is about 1 dB/km. The microwave energy is absorbed and scattered by the raindrops. For a 40-km hop, this would cause

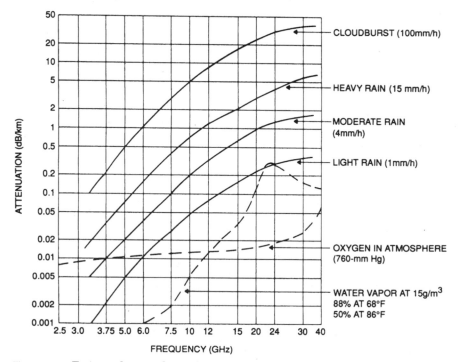

Figure 4.15 Estimated atmospheric absorption.

a 40-dB increase in the attenuation, which is enough to degrade the quality of transmission. Usually, cloudbursts do not cover a distance of 40 km, so the attenuation would not be as high as 40 dB. It can be concluded that rain is not a serious problem below 6 GHz, although it can impair quality in tropical regions that frequently have very heavy rainfall.

At higher frequencies, especially above 10 GHz, rainfall can cause severe transmission problems. For example, at 12 GHz, the attenuation can reach almost 10 dB/km. An extensive cloudburst can cause a temporary break in transmission. In a DMR system, that might cause several hundred calls to be dropped. In circumstances where very high reliability is required, it might be necessary either to reduce the length of hops or to use a lower transmission frequency.

4.3.2 Refraction

In addition to the attenuation effects of the atmosphere, there are also difficulties caused by refraction. The effect of refraction is to cause the microwave beam to deviate from its line-of-sight path. Its effect on transmission can be very serious, causing outages of a fraction of a second up to several hours. Predicting its occurrence can be done only on a statistical basis. Methods of protecting a link against refraction effects are discussed later.

Refraction is a bending of the radio waves due to changes in the characteristics of the atmosphere. The atmosphere changes in temperature, density, and humidity with increasing altitude from the earth's surface. The change in density affects the velocity of microwaves traveling through the atmosphere:

$$\text{Velocity }(v) = \frac{c}{n} = \frac{c}{\sqrt{\varepsilon_r}} \tag{4.6}$$

where n = refractive index
c = speed of light
ε_r = relative permittivity

Figure 4.16 recalls to mind the simple physics diagram of how the high velocities at the top of a wavefront, in thinner air, cause the microwave beam to bend. The bending in these circumstances is downward, toward the earth, which allows the microwaves to be transmitted farther than the direct straight-line path (Fig. 4.17). This is equivalent to increasing the radius of the earth's curvature. On average, atmospheric conditions cause the propagation path to have a radius of curvature that is approximately 1.33 times the true earth radius. In practice, this allows a propagation path length of approximately 15 percent longer than the shortest distance path. It is common practice to plot the radio path on profile paper that is corrected to four-thirds the earth's radius, so that the microwave beam can be plotted as a straight line. The change of earth curvature caused by refraction is denoted by the *k-factor*, which is defined as the ratio of the effective earth radius to the true earth radius:

$$k = \frac{\text{effective earth radius}}{\text{true earth radius}} \quad (4.7)$$

The effective earth radius is often misunderstood. It is not the radius of the microwave beam. For a given atmospheric condition, it is the radius of a fictitious earth that allows the microwave beam to be drawn as a straight line.

A more mathematically precise definition of the effective earth radius is $1/[(1/a) + (dn/dh)]$, where a is the true earth radius and dn/dh is the gradient of the refractive index n, with respect to the altitude h. So,

$$k = \frac{1}{1 + (adn/dh)} \quad (4.8)$$

The standard value of k is taken to be $\frac{4}{3}$. Variations in atmospheric conditions occur daily and also hourly, depending upon geographical location. On a larger time scale, significant seasonal changes can affect transmission. In general, k is larger than $\frac{4}{3}$ in warm temperature areas and less than $\frac{4}{3}$ in cold temperature areas, perhaps lying between 1.1 and 1.6 for most countries, depending upon latitude and season.

In addition, k can have values of less than 1 to ∞ or can even be negative. For example, when a temperature inversion occurs, a warm layer of air traps a cooler surface layer of air. The k-factor in this case is less than 0. Other conditions, such as a steep change in temperature of the air from the earth's surface to an altitude of several hundred meters, can produce $k = \infty$. This special case denotes that a microwave beam precisely follows the curvature of the earth. If this condition were present around the complete surface of the earth, it would be possible to transmit across the Atlantic Ocean with only one hop. Of course, this situation never occurs and, even if it did, the received signal level would be too low to be useful. Nevertheless, $k = \infty$ does occur over short distances under the appropriate atmospheric conditions. Further downward bending beyond the $k = \infty$ condition produces a negative k-factor. Figures 4.18

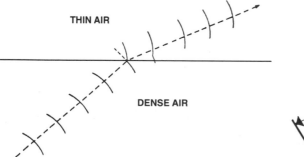

Figure 4.16   Refraction of a microwave beam.

Figure 4.17   Transmission distance increased by refraction.

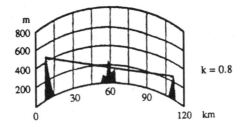

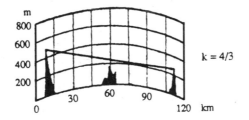

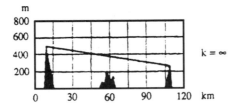

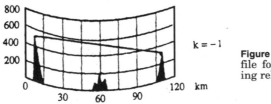

Figure 4.18 Effective earth profile for several k-factors, showing relative obstacle clearance.

and 4.19 illustrate the k-factor effects on radio paths. Figure 4.18 shows a straight-line microwave beam and how the effective earth radius causes a "bulging" effect that can enhance or reduce the path clearance, depending on the k- factor. Figure 4.19a shows the true situation of the real earth curvature and how a microwave beam transmitted at 90° to the earth's surface is curved, depending on the k-factor. Figure 4.19b shows how the path length is altered for various values of k-factor. *Note that the antenna has to be tilted to a different angle to the earth for each value of k-factor in order to achieve the maximum path length for a given k-factor.* This means that if the hop is designed for a $k = 4/3$ and the antennas are set on a day when $k = 4/3$, if temperature changes cause k to change to $2/3$, the transmitted beam is bent *upward,* and some power will be lost because the reception will be off the peak of the major lobe (off boresight). Maximum received power will similarly be unachievable if the temperature causes k to change to, say, 2.0.

In this case, the beam will be bent *downward,* with antennas set for $k = 4/3$. These are very extreme changes in the value of k, and the power loss would still only amount to about 1 dB for a 50-km-length link using 3-m-diameter antennas. For longer hops and larger antennas the loss is worse. Furthermore,

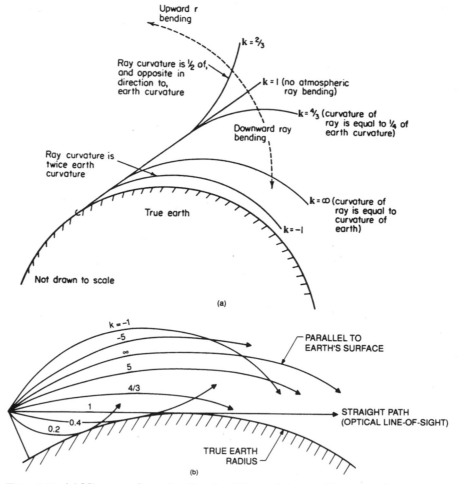

Figure 4.19 (*a*) Microwave beam bending for different k-factors. (*Reproduced with permission from McGraw-Hill. Panter, P. F., Communication Systems Design, Fig. 13-4, McGraw-Hill, 1972.*) (*b*) Antenna oriented for maximum path length for various k-factor values.

the downward bending of the $k = 2.0$ case could lead to signal loss problems due to inadequate clearance from obstacles.

When designing a microwave hop, it is often more convenient to plot a profile of the path using a straight-line microwave path rather than a curved line. For this purpose, an equivalent earth profile template is available, as in Fig. 4.20. This is derived using the equation

$$h = \frac{0.079d_1d_2}{k} \tag{4.9}$$

where h is the vertical distance (in meters) between the flat earth ($k = \infty$) and the effective earth at any given point, and d_1 and d_2 are the distances (in kilometers) from a given point to each end of the path. The main disadvantage of this technique is that if several values of the k-factor are to be considered, multiple plots are necessary, each on different profile paper. Conversely, if the flat earth method is used, several microwave beams (one for each k-factor) can be plotted on the same diagram.

4.3.3 Ducting

Atmospheric refraction can, under certain conditions, cause the microwave beam to be trapped in an atmospheric waveguide called a *duct*, resulting in severe transmission disruption. Ducting is usually caused by low-altitude, high-density atmospheric layers, most frequently occurring near or over large expanses of water or in climates where temperature or humidity inversions occur. Figure 4.21 gives an example of how the transmitted beam becomes trapped in a duct. When the beam enters the duct and it reaches the other interface between the two different density layers, the critical angle is exceeded so that internal reflection occurs. Subsequently, the beam bounces back and forth as it travels along the duct, and the receiving antenna loses the signal.

4.4 Terrain Effects

Propagation of microwave energy is affected by obstacles placed in its path. As described previously, the earth curvature (or effective curvature) is a dominant factor in determining hop length. Previously, a smooth earth was considered. Now the effects of obstacles such as rocks, trees, and buildings will be discussed. The shape and material content of any obstruction must be taken into account when surveying a microwave path. Unfortunately, objects close to the direct-line path can cause problems even though they are not obstructing the line of sight.

4.4.1 Reflections

The raylike beam that has been drawn in previous illustrations is a convenient tool for describing several physical concepts. However, it must be

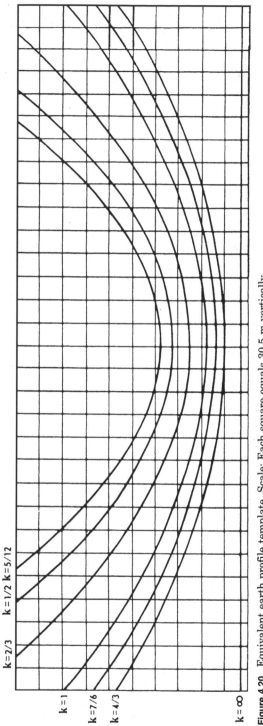

Figure 4.20 Equivalent earth profile template. Scale: Each square equals 30.5 m vertically and 3.2 km horizontally; or, each square equals 122 m vertically and 6.4 km horizontally.

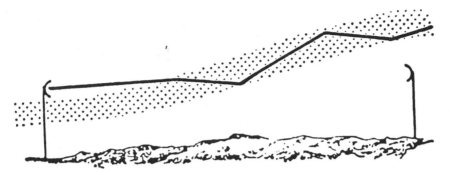

Figure 4.21 A microwave beam trapped in a duct.

remembered that, although a microwave beam might be only 1 or 2° in half-power width, this still represents a large area of energy spread at a distance of 40 km from the transmitter. Simple geometry indicates the half-power cone to have enlarged to a circle of approximately 1.4 km diameter for a 2° beam, or 0.7 km for a 1° beam. This means that some of the energy off boresight will be reflected from the ground (Fig. 4.22) or other nearby objects at either side of the direct line. Figure 4.23 shows how a wave undergoes a 180° phase reversal when it is reflected. At the receiving antenna, energy arrives from the direct and reflected paths. If the two waves are in phase, there is an enhancement of the signal, but if the waves are out of phase, a cancellation occurs that can disrupt transmission. This 180° phase shift occurs for horizontally polarized waves at microwave radio operating frequencies, whereas for vertically polarized waves the phase shift can be between 0 and 180° depending on the ground conditions and the angle of incidence.

4.4.2 Fresnel zones

The microwave energy that arrives at the receiving antenna 180° (or $\lambda/2$) out of phase with the direct wave determines the boundary of what is called the *first Fresnel zone*, as illustrated in Fig. 4.24. For a specific frequency, all points within a microwave link from which a wave could be reflected with a total additional path length of one half-wavelength (180°) form an ellipse that defines the first Fresnel zone radius at every point along the path (Fig. 4.26).

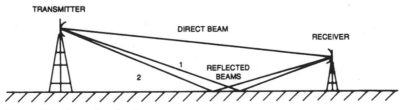

Figure 4.22 Ground reflection.

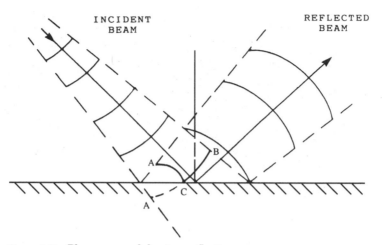

Figure 4.23 Phase reversal due to a reflection.

The second and third Fresnel zones are defined as the boundary consisting of all points from which the additional path length is two half-wavelengths and three half-wavelengths, respectively. So, at any point along the path, there is a set of concentric circles whose centers are all on the direct line-of-sight path line, denoting all of the Fresnel zone boundaries (Fig. 4.24). The distance F_n (in meters) from the line-of-sight path to the boundary of the nth Fresnel zone is approximated by the equation

$$F_n = 17.3 \sqrt{\frac{nd_1 d_2}{fD}} \tag{4.10}$$

where d_1 = distance from one end of the path to the reflection point (km)
d_2 = distance from the other end of the path to the reflection point (km)
$D = d_1 + d_2$
f = frequency (GHz)
n = number of Fresnel zone (1st, 2d, etc.)

Figure 4.25 is a graph formed from Eq. (4.10), and it indicates the first Fresnel zone radius at specific distances from either end of a link of known total length. This family of curves is drawn for a frequency of 6.175 GHz. This is a very popular frequency band for microwave radio operation. To convert the values for the first Fresnel zone radius calculated at 6.175 GHz to values at other frequencies, the factor $\sqrt{6.175/f}$ is used. For example, a link of 40 km in total length would have a first Fresnel zone radius (F_1) of about 12 m at a distance of 3.2 km from either end antenna at 6.175 GHz. At a distance of 12.8 km, F_1 increases to about 21 m at 6.175 GHz, and a change of frequency to 4 GHz further increases F_1 to 26 m.

These Fresnel zone radii have significant consequences when obstacles such as trees or hills within the microwave path approach the first Fresnel zone

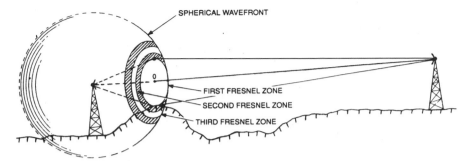

Figure 4.24 Fresnel zones.

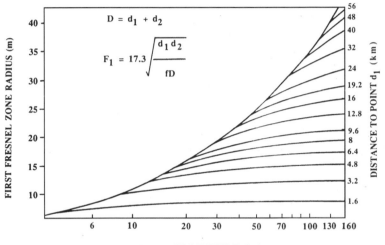

$$D = d_1 + d_2$$

$$F_1 = 17.3 \sqrt{\frac{d_1 d_2}{fD}}$$

Figure 4.25 First Fresnel zone radius (6.175 GHz).

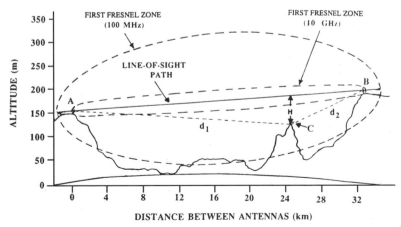

Figure 4.26 Typical profile plot showing first Fresnel zones for 100 MHz and 10 GHz.

radius. For determining the higher-order, nth, Fresnel zone radii when the first zone radius is known, the equation $F_n = F_1 \sqrt{n}$ is used.

When designing a microwave link, it is useful to obtain information relating to all obstacles in the region between the transmitter and receiver and draw them on the profile plot (as in Fig. 4.26). This diagram also has the first Fresnel zone plotted for 100-MHz and 10-GHz waves (note the ellipses defining these zones). This figure gives an indication of how the size of these zones varies with frequency. The radius of the first Fresnel zone at the center of the path in this case is about 17 m at 10 GHz, but it is about 170 m at 100 MHz. The rocky peak (point C) is outside the first Fresnel zone at 10 GHz, but at some lower frequency, it will be within the first Fresnel zone. However, an important point arises here. When the wave is reflected from point C for a small incident angle, a 180° phase reversal occurs (for horizontal polarization). Therefore, when AB and ACB differ by an odd multiple of a half-wavelength, the received signals add instead of canceling. Also, if the two paths differ by an even number of half-wavelengths, cancellation occurs. This means that signal energy at the second Fresnel zone distance from the reflection point and the direct signal produce cancellation. The loss in signal strength due to this cancellation can be very large. Atmospheric changes can cause refraction variations that make the reflected waves intermittently pass through the cancellation conditions. This is one form of fading called *multipath fading*, and it can occur over periods of a few seconds to many hours.

Considering reinforcement of waves arriving in phase, a signal level up to 6 dB higher than that obtained by direct free-space loss can be achieved. This should be considered a bonus, and hop designs never include this as an expected signal enhancement.

Because the vertical polarized wave can have a 0° phase shift on reflection, this represents a worst-case situation because reflection from an obstacle at the first Fresnel zone distance causes cancellation. Hop design experience has shown that to achieve a transmission unaffected by the presence of obstacles, the transmission path should have a clearance from these obstacles of at least 0.6 times the first Fresnel zone radius, F1 for $k = \frac{2}{3}$. Note that the $k = \frac{2}{3}$ obstacle clearance condition is worse than the normal condition of $k = \frac{4}{3}$ (*provided the antennas are set on a day when $k = \frac{2}{3}$*). For conservative design, one or more first Fresnel zone distances are chosen for clearance of obstacles, particularly when operating in the higher-frequency bands. This clearance condition applies to objects at the side of the path as well as below the path (ground reflections).

4.4.3 Diffraction

So far, discussion has been confined to perfectly reflecting surfaces. In practice, this applies only to paths that pass over surfaces such as water or desert. Such highly efficient reflective surfaces are often labeled *smooth sphere diffraction* paths. The majority of microwave paths have an obstacle clearance in the cat-

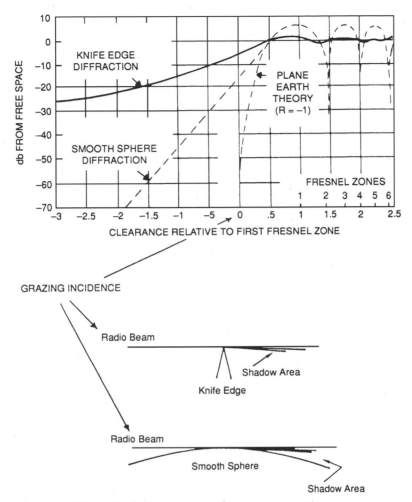

Figure 4.27 Effect of path clearance on radio wave propagation.

egory known as *knife-edge diffraction*. These paths traverse terrain that is moderately to severely rough, with brush or tree covering. Diffraction is a characteristic of electromagnetic waves that occurs when a beam passes over an obstacle with *grazing incidence* (i.e., just touching the obstacle; Fig. 4.27). The beam energy is dispersed by an amount that depends on the size and shape of the obstacle. *Shadow loss* is a term often used to describe the loss in an area behind an obstacle. This loss is dependent on frequency. The higher-frequency waves suffer higher diffraction loss. This simple fact explains the relative success of VHF radio technology in mountainous or rough terrain environments.

If microwave hop antennas are placed at a low height compared to the Fresnel zone clearance, then there will be a small angle of incidence for an obstacle near the line-of-sight path, and there will be some shadow loss (or diffraction loss) due to the grazing incidence. Figure 4.27 shows the loss compared to the expected free-space value plotted against clearance from the obstacle, for the two terrain extremes: (1) the smooth sphere case and (2) the knife-edge case. The worst loss situation is for smooth sphere grazing, which can be up to 15 dB depending on clearance. The knife-edge grazing causes approximately 6 dB of loss. This loss is not small. Most microwave paths have reflections occurring at one or several points along the path. The height of antennas must be sufficient to prevent the reflected signals from causing high losses due to varying propagation conditions. Particular attention should be paid to this matter for paths crossing water or desert.

Figure 4.28 shows the resulting computer plot of a microwave hop with a beam plotted for a k-factor of $^4/_3$. The plot also includes a one Fresnel zone radius factored into the calculations; 85-m antennas at each end of the hop are needed for the beam to clear the hill peak at the 10-km distance and the trees at the 27-km point. Notice the situation is not reciprocal, meaning that the A to B path is different from the B to A path. The question must be asked: Is this configuration adequate? The answer is yes, provided there are no significant atmospheric changes. In these circumstances it is useful to plot the situation on the same diagram for a k-factor of 2.0 to see if there is still adequate obstacle clearance in the event that the atmospheric conditions change for the antennas set at $k = ^4/_3$. As one can see, the clearance is now inadequate, so it would be preferable to increase the antenna height at one or both ends of the hop so that there is a little more clearance margin to allow for any atmospheric changes. The problem is more complicated in reality, because the

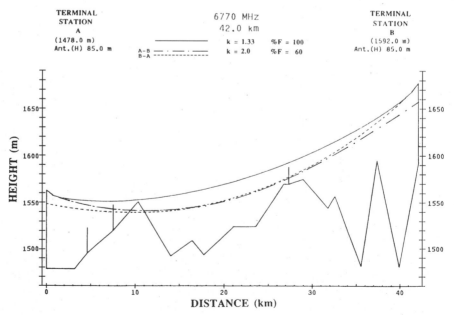

Figure 4.28 A typical hop using the flat earth presentation.

curves in Fig. 4.28 consider a homogeneous atmosphere, which is only an approximation of real conditions. Nevertheless, these graphs do provide a good engineering starting point from which modifications can be made to optimize each individual situation.

4.5 Fading

There are two main categories of fading:

1. Flat fading (frequency independent)
2. Frequency selective fading

Neither type of fading can be predicted accurately because each is caused by variations in atmospheric conditions. Experience has shown that some climates and terrain surfaces are more likely to cause fading than others, but in all circumstances fading can only be defined statistically. In other words, one can only say that based on probability theory, the microwave system will be inoperative for a certain percentage of the year because of fading. In some regions this percentage is too large to be tolerated. Fortunately, there are techniques that can be used to improve the outage time. Before discussing the statistics, a more detailed view of fading will be examined.

4.5.1 Flat fading

Two forms of flat fading were indicated earlier: ducting and rain attenuation fading. A more frequently occurring flat fading is due to beam bending. As discussed in the previous section, the microwave beam can be influenced by a change of the refractive index (dielectric constant) of the air; $k = \frac{4}{3}$ is considered to be the *standard* atmospheric condition, in which the microwave beam has one-fourth of the true earth curvature. Transmitting and receiving antennas are placed so that under standard conditions a full-strength signal will be received from the transmitter. When the density of the air subsequently changes such that the refractive index of the atmosphere is different from standard, the beam will be bent upward or downward, depending on the k-factor. When k is less than $\frac{4}{3}$, often called a *subrefractive* or *substandard* condition, it causes upward bending, and when k is greater than $\frac{4}{3}$, known as *superrefractive* or *superstandard* conditions, it causes downward bending (Fig. 4.29). Depending on the severity of the bending, either type can cause a considerable reduction in the received signal strength to the point of disrupting service.

For downward bending, provided the beam is not bent so far that some energy from the beam is reflected from an obstacle, the fading is wideband compared to the relatively narrow microwave frequency band (i.e., flat fading or non-frequency-selective fading). However, if some energy is reflected from an obstacle and interferes with the direct path energy, the fading becomes frequency selective. Similarly, for upward bent beams (provided

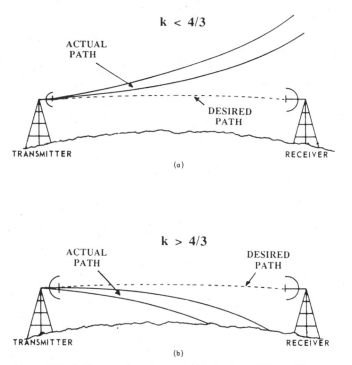

Figure 4.29 (a) Inverse bending, substandard conditions; (b) positive bending, superstandard conditions.

there is no energy reaching the receiver other than the direct path), as the beam is bent away from the receiver, flat fading occurs.

4.5.2 Frequency selective fading

Atmospheric multipath fading. When the atmospheric conditions are such that layers or stratifications of different density exist, as indicated earlier, ducting can occur. If the composition of the layers is such that the microwave beam is not trapped, but only deflected, as in Fig. 4.30, the microwave energy can reach the receiving antenna by paths that are different from the direct path. This multipath reception produces fading because the two waves are rarely received in phase. If they arrive in complete antiphase, for a few seconds a drop in received power, which can be 30 dB or more, is observed.

Ground reflection multipath fading. As indicated in Sec. 4.4.1, ground reflection can cause a multipath reception that will be observed as fading if the waves are received in antiphase. When ground reflection and atmospheric multipath fading occur simultaneously, short-term fades as deep as 40 dB can occur. If corrective action is not taken, this will cause service disruption. Multipath fading is frequency selective because, for antiphase cancellation, the different waves must reach the receiver after traveling distances that differ by one half-wavelength. Because the size of a half-wavelength varies significantly from 1 to 12 GHz, fading conditions that exist at one frequency might not exist at another.

4.5.3   Factors affecting multipath fading

Experience has shown that all paths longer than 40 km can be subject to multipath fading for frequencies of operation above 890 MHz. Atmospheric multipath fading is most pronounced during the summer months or, more specifically, when the weather is hot, humid, and wind-free. It has been found that fading activity most frequently occurs after sundown and shortly after sunrise. At mid-day, the thermal air currents usually disturb the atmosphere so that layers do not form and fading is not a problem. In some regions, there are certain times of the year when the atmospheric conditions produce multipath fading outages every day. In general, frequency selective fading is "fast fading." The average duration of a 20-dB fade is about 40 s, and the average duration of a 40-dB fade is about 4 s.

As the length of a microwave path is increased, there is a rapid increase in the number of possible indirect paths by which the signal might be received. For microwave hops in desert regions or over water, it is often necessary to reduce the hop path length to 35 km or less to avoid serious ground reflection multipath fading. Fortunately, multipath fading does not occur during periods of heavy rainfall, so fading is usually flat fading *or* frequency selective fading.

4.6   Overall Performance Objectives

The measure of system reliability is usually referred to as its *availability*. Ideally all systems should have an availability of 100 percent. Because this is

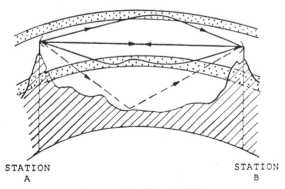

STATION
A

STATION
B

Figure 4.30 Mechanics of multipath fading.

not possible, the engineer strives to ensure the availability is as high as possible. Table 4.2 shows the average outage time expected as the availability (reliability) degrades from 100 percent. The system becomes unavailable for two main reasons. First, *person-made faults* are caused during maintenance, or failures occur because of inadequate equipment design or fabrication. Equipment failure due to old age can also be included in this category, because it should be taken out of service before its rated lifetime expires. The second category can be called unavoidable or non-person-made faults, and these are primarily caused by *changing atmospheric conditions*. This can be controlled to some extent by equipment and route design. Usually, service interruption due to fading can account for up to about half of the total outage. There is a third category, which includes *disasters* such as earthquakes, fire, terrorism, etc., which fortunately amounts to a very small percentage of the total unavailability, although at the time the effects can be temporarily devastating.

Because the fading effect on availability is closely linked to route design, it will be studied in more detail. Because atmospheric changes occur over a period of time, the depth of multipath fades is variable. Fortunately, 40-dB fades occur for only a very small portion of the total operating time (approximately 0.01 percent). It is therefore not possible to state exactly how much time a system will be interrupted each month or year, but one can calculate the statistically derived average times. As can be seen from the table of availabilities in Table 4.2, a value of 99 percent reliability might sound impressive, but closer evaluation shows that this is equivalent to an average system outage of 14.4 min each day.

For an analog system carrying only voice traffic, this might not be too serious. However, it depends on how this 14.4 min is distributed throughout the day. It is statistically improbable that the 14.4 min will occur as a continuous time interval. At the other extreme (which is more probable) each subscriber could suffer a break in conversation for 1 percent of the time of a call. In real-

TABLE 4.2   Relationship between System Reliability and Outage Time

Availability or reliability (%)	Outage time (%)	Outage time per		
		Year	Month (avg.)	Day (avg.)
0	100	8760 h	720 h	24 h
50	50	4380 h	360 h	12 h
80	20	1752 h	144 h	4.8 h
90	10	876 h	72 h	2.4 h
95	5	438 h	36 h	1.2 h
98	2	175 h	14 h	29 min
99	1	88 h	7 h	14.4 min
99.9	0.1	8.8 h	43 min	1.44 min
99.99	0.01	53 min	4.3 min	8.6 s
99.999	0.001	5.3 min	26 s	0.86 s
99.9999	0.0001	32 s	2.6 s	0.086 s

TABLE 4.3   Interrupt Time for Various Types of Traffic

Type of traffic	Maximum tolerable interrupt time	Effect if tolerance is exceeded
Voice circuits	100 ms	Seizure of exchange switching equipment
Video	100 µs	Loss of synchronization (rolling)
Data (64 kb/s)	10 µs	Error
Data (2.488 Gb/s)	1 ns	Error

ity, certain times of the day have higher incidences of fading than others, and during these times subscribers would experience several instances where the speaker's voice fades to a level that is inaudible. In a digital system, 99 percent reliability would be nothing short of disastrous for a telephone company. An outage in a digital system means that all calls are dropped for as much as several minutes until the system resets itself. This happens if the deep fade duration is as short as a few hundred milliseconds. Table 4.3 shows the order of magnitude of the maximum tolerable interrupt times for several types of information transmission. Video and data transmission are the most intolerant, but even 100 ms is enough to drop voice traffic. Obviously, if numerous short fades occur throughout the day, the system could be out of service for a large part of that day. A reliability of 99.999 percent would be an excellent target to achieve, and most common carriers in North America design their protected radio system paths for this level of availability. Even this high value causes an average outage of about 0.86 s per day, or 26 s per month, which can cause problems for data transmission. This might not seem to be many seconds per month, but if the equipment is designed to drop calls every time the BER increases above, say, 1×10^{-3} for a few milliseconds, calls could be dropped *thousands* of times each month. This is why the new performance specification standards for availability are described in terms of *errored seconds*. Errored seconds gives an indication of the distribution of errors (i.e., bursts or isolated events).

Although this chapter specifically addresses microwave links, performance objectives for all transmission media (including microwave) have been incorporated into ITU-T Recommendations G.826 and G.827, as follows.

4.6.1   Error definitions

Recommendation G.826 was derived to prepare for broadband services. Specifically, SONET/SDH equipment should now offer a higher quality of service than previous PDH equipment. For many years the BER has been the measurement primarily used to assess digital link performance. It defined transmission quality and acceptability. In ITU-T Recommendation G.821, if the BER is too high, the system is declared to have failed or be unavailable. The familiar bit error ratio (BER) is defined as the ratio of false bits to the total number of received bits:

$$BER = \frac{\text{false bits}}{\text{received bits}} \qquad (4.11)$$

Example. If a 1.544-Mb/s link produces 1 false bit per second, the error ratio will be

$$BER = \frac{1}{1.544 \times 10^6} = 6.5 \times 10^{-7} \qquad (4.12)$$

For PDH equipment, BER measurements were usually made in the DMR station terminal equipment. Manufacturers used the BER measurements to provide a *maintenance required* alarm if the BER became worse than, say, 1×10^{-5} or 1×10^{-6}, and an *out of service* alarm for worse than 1×10^{-3}.

There are some significant changes between the *bit error* ratio-based G.821 and the *block error* ratio-based G.826 Recommendations. Also, whereas Recommendation G.821 considered a bit rate of 64 kb/s, G.826 considers constant bit-rate paths at or above the primary rate (that is, 1.544 or 2.048 Mb/s or higher). With more signal processing incorporated into transmission equipment, and especially in SONET/SDH equipment, in-service monitoring of errors is becoming a well-established method of performance monitoring.

In general, PDH transmission equipment is designed to pass specific *bit error ratio* measurement tests, whereas SONET/SDH equipment is designed to conform to *block error ratio and errored second* measurement standards. Complications can arise when PDH bit streams are incorporated into SDH equipment because the Recommendation G.826 standards are more stringent than G.821. There is some latitude for equipment manufacturers to use G.821 as the basic minimum for PDH equipment, with G.826 being the norm for SONET/SDH equipment and, wherever possible, PDH equipment also. Microwave radio equipment will likely encounter the most difficulty in conforming to the new standard because of propagation-related error bursts.

Bit error ratio has been a convenient parameter used to describe digital equipment performance since it emerged in the early 1980s. The causes, types, and distribution of errors have become better understood since those early years, and the improvement in quality of service has demanded better error performance. Furthermore, it has become very clear that error performance analysis is not the simple subject it was once thought to be. This becomes evident when trying to reconcile the new Recommendation G.826 parameters with the simpler bit error ratio values. A few definitions are first necessary, as follows:

- *Errored block* (EB) is a predetermined number of consecutive bits that has 1 or more bits in error.

- *Errored second* (ES) is a 1-s period that contains one or more errored blocks.

- *Seriously (or severely) errored second* (SES) is a 1 s period that contains ≥30 percent errored blocks or at least one severely disturbed period (SDP).

- *Out-of-service measurement.* In this case, an SDP is when all of four contiguous blocks have a binary error density of 10^{-2}, or there is a loss of signal information.
    - *In-service monitoring.* In this case, an SDP is when there is a network defect such as a loss of signal, an alarm indication signal, or a loss of frame alignment.

- *Background block error* (BBE) is an errored block not occuring as part of a severely errored second.

- *ES ratio* (ESR) is the ratio of errored seconds to the total seconds of a measurement made during available time.

- *SES ratio* (SESR) is the ratio of severely errored seconds to the total seconds of a measurement made during available time.

- *BBE ratio* (BBER) is the ratio of the errored blocks to the total blocks measured during available time. Any blocks occuring during seriously errored seconds are not included.

There is no simple direct conversion factor of the G.826-recommended SES to G.821-recommended BER, because SES now depends on distance and on the bit rate as well as the distribution of errors. For a 50-km repeater span 155-Mb/s signal experiencing randomly occuring errors, an SES event occurs when the BER becomes worse than about 1.7×10^{-5}. This is considerably more stringent (by about two orders of magnitude) than the older G.821 Recommendation.

4.6.2 Availability

Previous objectives. This subject is in such a state of change that it is useful to compare the older, better-established performance criteria with the newly evolving criteria. In the late 1980s, optical fiber was starting to challenge microwave, but most countrywide backbone systems were still microwave. By 1990, the ITU had outlined a *hypothetical reference path* (HRP) to characterize a typical, realistic, end-to-end transmission across international borders. The HRP given in Recommendation G.821 was a 27,500-km path that had an international portion, and national portions with international gateways at the border crossing points. The aim of Recommendation G.821 was to be media-independent, so any part of the 27,500-km path could be optical fiber, microwave radio, or satellite. The national transmission network was subdivided into high-, medium-, and local-grade sections depending on their position in the network. Apportionment rules are shown in Table 4.4.

In ITU-T Recommendation G.821 the criterion for unavailability was when the BER became greater than (worse than) 1×10^{-3} for 10 consecutive seconds.

New objectives. ITU-T Recommendation G.827 concerns the availability objectives of international constant bit rate digital paths at or above the

TABLE 4.4   ITU-T Rec. G.821 Apportionment Rules

	Percent of error performance objective		
	Local grade	Medium grade	High grade (international section)
Error seconds	15% to each originating/terminating country	15% to each originating/terminating country	40%

A satellite HRDP (if applicable) is allocated 20% of the total error seconds.

primary rate (1.544 or 2.048 Mb/s). It takes into account that the paths are based on PDH, SDH/SONET, or cell-based transport. Two types of path are addressed: (1) those between international switching centers (international only) and (2) those that include national as well as international portions.

Figure 4.31 illustrates the various points (path elements) that make up an international path between customer premises (type 2 above). Three types of path element are defined:

1. Intercountry path core element

2. International path core element

3. National path element

Availability is allocated to each path element depending on its length. For a path less than 10,000 km there are 20 length portions, in increments of 500 km, so:

$$500\,(i - 1) \leq L < 500i \qquad \text{where } i = 1, 2, ..., 20$$

Beyond 10,000 km is considered as a separate category.

The new definition of unavailable time is as follows: Unavailable time starts at the beginning of a block of 10 consecutive severely errored seconds. Those 10 s are declared to be *unavailable time*. The available time returns when, for

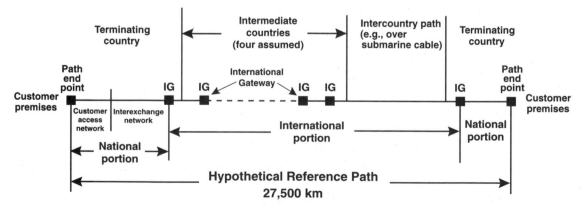

Figure 4.31 Hypothetical reference digital path for G.826. (*Adapted by permission from the ITU, ITU-T Rec. G.826.*)

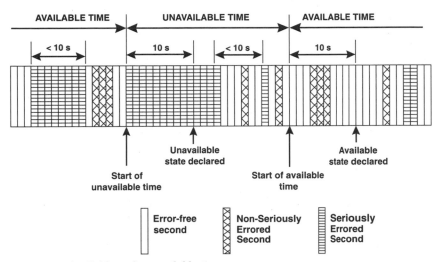

Figure 4.32   Available and unavailable time.

both directions of transmission, there is a new period (onset) of 10 consecutive nonseverely errored seconds. Those 10 s are considered to be *available time*. Figure 4.32 illustrates the definition of available and unavailable time.

Having defined the criterion for unavailable time in terms of severely errored seconds, the availability ratio is used to quantify end-to-end paths. *Availability ratio* is defined as the proportion of time that a path is available compared to the total observation period. This is the same definition that has been used for many years.

$$AR = \frac{\text{total time} - \text{outage time}}{\text{total time}} \qquad (4.13)$$

Apportionment is now as follows:

$$AR_j = \begin{cases} 1 - [b_{jn} + ix_{jn}] & \text{for } L < 10{,}000 \text{ km} \\ 1 - [b_{jn} + 21x_{jn}] & \text{for } L \geq 10{,}000 \text{ km} \end{cases}$$

where j = path elements (1, 2, 3 above)
b_{jn} = block allowance for path element j; length range n
i = length category
x_{jn} = distance based allowance for PE type j; length range n

At the time of writing, the values for b_{jn} and x_{jn} used to calculate the availability ratio objectives are still under evaluation. The mean time between outages is related to the availability ratio by:

$$M_0 = \frac{1}{1 - AR} = \frac{\text{total time}}{\text{outage time}} \qquad (4.14)$$

4.6.3   Microwave radio performance objectives

ITU-R Recommendation F.1189 specifies error performance objectives for international digital paths at a constant bit rate at or above the primary rate (1.5 or 2 Mb/s).

Radio relay paths form part of the national section of the 27,500-km HRP. Error performance objectives of the digital radio paths are based on block error allowances and distance-based accumulation of block errors. The ITU-R subdivides the national digital portion into three categories: (1) long haul, (2) short haul, and (3) access network, where long haul is defined as the interexchange network section including connections between the primary, secondary, or tertiary centers and the international gateway. Short haul includes the interexchange network section, including connections between a local exchange and the primary, secondary, or tertiary center. Access includes the connections between the path end-point (customer premises) and the corresponding local access switching center (local exchange). It recognizes long haul to contain part of the fixed errored block allowance and also a distance-based errored block allocation as shown in Table 4.5. Note that the errored block contributions for the short haul and access networks are about the same and no distance component is included.

This national portion fixed block allowance, as indicated in Note 1, has a combined (A_1 + B + C) allocation for ESR, SESR, and BBER amounting to 17.5 percent of the total path-end-point to path-end-point objective (see Fig. 4.31). The distance-based allowance is added to the fixed block allowance. For example, a 2000-km national microwave (or fiber) route would have the value, $A = 0.01 + (0.01 \times 2000/500) = 0.05$ or 5 percent, giving a total national portion of 22.5 percent. A satellite hop receives 35 percent allocation regardless of the distance spanned, whether it is counted within the national or international portion.

An international path could contain one or more national portions. When considering a path that could include a combination of microwave, satellite, and optical fiber routes, the apportionment can become complex. It is also apparent that there is some flexibility in the components that total 100 percent apportionment, as well as some restrictions. For example, a path having two national portions (17.5 + 17.5), a national satellite path (35 percent, and a 10,000-km fiber link (21 percent, totals only 91 percent leaving some room for reallocation. However, two satellite hops are not allowed in any one end-to-end path.

In-service path performance monitoring for PDH systems is based on the cyclic redundancy check (CRC-6 or CRC-4) for the 1.544- or 2.048-Mb/s primary rates, respectively, and parity for the DS3. For SONET/SDH, in-service monitoring uses the bit interleaved parity (BIP-n) in the path overhead, which corresponds to one block. The block is assumed to be an errored block if any one of the n separate parity checks fails.

TABLE 4.5   ITU-T Recommendation G.826 Details for Error Performance Objectives of Microwave Radio Paths

Bit rate (Mb/s)	ESR			SESS			BBER		
	Long haul	Short haul	Access	Long haul	Short haul	Access	Long haul	Short haul	Access
1.5 - 5	0.04A	0.04B	0.04C	0.002A	0.002B	0.002C	$3A \times 10^{-4}$	$3B \times 10^{-4}$	$3C \times 10^{-4}$
> 5 - 15	0.05A	0.05B	0.05C	0.002A	0.002B	0.002C	$2A \times 10^{-4}$	$2B \times 10^{-4}$	$2C \times 10^{-4}$
> 15 - 55	0.075A	0.075B	0.075C	0.002A	0.002B	0.002C	$2A \times 10^{-4}$	$2B \times 10^{-4}$	$2C \times 10^{-4}$
> 55 - 160	0.16A	0.16B	0.16C	0.002A	0.002B	0.002C	$2A \times 10^{-4}$	$2B \times 10^{-4}$	$2C \times 10^{-4}$
> 160 - 3500	*	*	*	0.002A	0.002B	0.002C	$A \times 10^{-4}$	$B \times 10^{-4}$	$C \times 10^{-4}$

$A = A_1 + (0.01 \times L/500)$

where A_1 is provisionally in the range of 1 to 2%

$\quad$ L is rounded up to the next multiple of 500 km

$B = $ is provisionally in the range of 7.5 to 8.5%

$C = $ is provisionally 8%

$* = $ for further study

1. The sum of the percentages $A_1\% + B\% + C\%$ should not exceed 17.5%, in accordance with the allocations to the national portion of an international CBR path given in ITU-T Recommendation G.826.

2. The provisional values agreed for $B\% + C\%$ are in the range 15.5% to 16.5%.

3. Depending on national network configurations, the $A\%$, $B\%$, and $C\%$ block allowances may be reallocated among the sections of the national portion of a radio path.

4. The suggested evaluation period is 1 month for any parameter.

4.6.4   Comments on the old and new error objectives

There are two main motives for moving to a new block-based error standard: (1) to improve network quality and (2) to take into account today's signal processing, such as FEC and scrambling, which tends to create errors in groups or blocks (bursts) rather than individual, random bit errors. Recommendation G.826 was derived to prepare for broadband services. Specifically, SONET/SDH equipment should now offer a higher quality of service than previous PDH equipment. PDH equipment might have difficulty in meeting the new standard, but some latitude for exemption is acknowledged during the transition from PDH to SDH/SONET.

The transition to a block-error-based measurement from the more familiar BER will no doubt take a long time. In the interim, many observers are trying to relate the new standard (G.826) to BER figures. In the past, a BER of 10^{-3} has been the failure point for most system specifications. The new block-based standard tightens this up to a BER of about 10^{-5}. There is no simple direct relationship between the two. It can be said that when the BER drops to 10^{-5}, there is a high degree of certainty that a severely errored second has occurred.

The background block error ratio (BBER) should not be confused with the ordinary BER. The BBER is an errored block not occurring as part of an SES. In other words, this is a measure of isolated error events rather than more serious long error bursts. In general, if the ESR objective is met, the BBER requirement will also be satisfied. However, the ESR objective is stringent. A BER of about 10^{-12} is needed to meet the ESR objective for a 50-km, 155-Mb/s span. A BER floor of 10^{-13} is therefore needed and both fiber and radio systems can achieve this these days. Certainly, radio will have a difficult task to meet a BER of 10^{-12}, but with FEC, transversal equalizers, diversity, etc., it is possible. Severe propagation conditions would upset this picture and could lead to a shortening of some radio paths to improve performance. It is estimated that equipment designed to meet the previous BER = 10^{-3} criterion for availability will have an outage increase of about 30 percent based on the new standard.

4.7   Diversity

Diversity is the simultaneous operation of two or more systems or parts of systems. It can be described as equipment redundancy or duplication. It is a means of achieving an improvement in the system reliability. A microwave path that has been expertly designed with respect to fade margin, path clearance, and elimination of reflections might still suffer from poor performance. Multipath fading can cause temporary failure in the best-designed paths. The system designer does not usually have the luxury of choosing the climate or terrain over which the microwave path must be designed. In regions where multipath fading conditions exist, it is necessary to incorporate diversity into the system design. The two types of DMR equipment diversity are

1. Space diversity

2. Frequency diversity

   Note: Neither space diversity nor frequency diversity provides any improvement or protection against rain attenuation.

4.7.1   Space diversity

In this mode of operation (Fig. 4.33a), the receiver of the microwave radio accepts signals from two or more antennas vertically spaced apart by many wavelengths. The signal from each antenna is received and then simultaneously connected to a diversity combiner. Depending upon the design, the function of the combiner is either to select the best signal for its output or to add the signals.

For a space diversity protected system, the direct signal travels two different path distances from the transmitter to the two receiver antennas, as indicated in Fig. 4.34. In addition, there might be reflected paths, where the signal entering each antenna has also traveled different distances from the transmitter. Experience has shown that when the reflected path causes fading by interference with the direct signal, the two received signals will not be simultaneously affected to the same extent by the presence of multipath fading, because of the different path lengths. Although the path from the transmitter

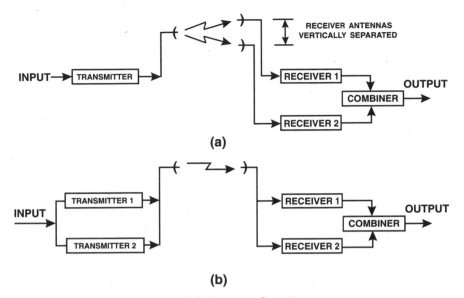

(a)

(b)

Figure 4.33 (*a*) Space diversity and (*b*) frequency diversity.

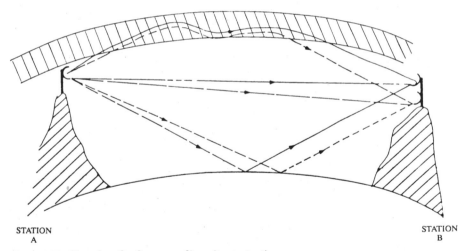

STATION
A

STATION
B

Figure 4.34 Signal paths for space diversity protection.

to one of the receiver antennas might cause phase cancellation of the direct
and reflected path waves, it is statistically unlikely that multiple paths to the
other antenna will cause phase cancellation at the same time.

Statistical analysis has shown the improvement in reliability (or reduction
of fading-caused outage) to be in the range of 10 to 200. This is a surprisingly
large improvement. It is further enhanced by an increase in frequency, fade
margin, antenna vertical spacing, and a decrease in path length. Typical

antenna spacings are at least 200 wavelengths (at least 10 m in the 6-GHz band). Mathematically, the improvement factor is:

$$I = \frac{T}{T_d} \tag{4.15}$$

where T_d and T are the outage times with and without diversity, respectively. Several diversity analyses have been made, all of which provide only an approximation to reality because of the enormous complexity and variability of the problem. The space diversity improvement on an overland link with negligible ground reflections can be approximated by the old, but reasonably accurate Vigants equation:

$$I_s \approx \frac{1.2 \times 10^{-3} \; \eta s^2 f \, 10^{(F - V)/10}}{d} \tag{4.16}$$

where η = effectiveness of the diversity switch
s = vertical spacing of the antennas (m) center to center ($5 \text{ m} \leq s \leq 15$ m)
f = frequency (GHz)
F = fade depth, usually taken as fade margin (dB)
V = difference in gain between two antennas
d = hop length (km)

Space diversity is usually the first choice for system protection. For a single channel, it is cheaper than the frequency diversity described in the next section. Also, it does not use extra bandwidth, whereas the frequency diversity system does.

4.7.2 Frequency diversity

System protection is achieved with this type of diversity by effectively *operating two microwave radios* between the same transmit and receive antennas. The information signal is simultaneously transmitted by two transmitters operating at different frequencies. They are coupled to waveguides, which run to the antenna, and then transmit by the same antenna (usually with opposite polarization). At the receiving end, the antenna collects the information and passes it through waveguides to a filter that separates the two signals and separate receivers extract the voice, video, or data information. As in space diversity, a combiner is used to provide the maximized output (Fig. 4.33b). If the separation in frequencies of the two transmitters is large, the frequency selective fading will have a low probability of affecting both paths to the same extent, thereby improving the system performance. A frequency separation of 2 percent is considered adequate, and 5 percent is very good. This means a separation of at least 120 MHz in the 6-GHz band.

The major disadvantage of frequency diversity is the extra bandwidth that the system occupies. In uncongested regions this is of no concern, but in large cities the number of channels available for new radios is limited, and the additional channels for system protection are even more difficult to acquire. The improvement gained by frequency diversity is considerably less than that of space diversity.

Calculations show at least a factor of 10 improvement for frequency diversity over the nondiversity system. Several elaborate mathematical models have recently been proposed, but the improvement factor as indicated in ITU-R Report 338 provides satisfactory results. The improvement factor is approximated by

$$I_f \approx \frac{80 \, \Delta f \, 10^{F/10}}{f^2 d} \qquad (4.17)$$

where Δf = frequency separation (GHz)
F = fade depth (dB)
f = carrier frequency (GHz) $(2 \leq f \leq 11)$
d = hop length (km) $(30 \leq d \leq 70)$

As mentioned earlier, depending upon conditions, space diversity can provide an improvement greater than a factor of 100. A combination of both space and frequency diversity is used in cases of extremely high multipath fading problems.

Maintenance technicians and engineers like frequency diversity because they can proceed with repairs on one of the two radios without interrupting service. Of course, during the repair time there is no diversity protection. On nondiversity systems maintenance has to be performed at the lowest-traffic periods, usually between 2 and 4 A.M.

To ensure that the transmission quality is not degraded by the switching process itself, a *hitless switching* technique has been devised. The received bit streams in the regenerated baseband are compared on a bit-by-bit basis. If an error is detected in one of the bit streams, by using an elastic store, a decision is made to switch to the bit stream devoid of errors. By this method switching can be error-free, or *hitless*.

4.8   Link Analysis

4.8.1   Hop calculations

A microwave route can span a distance of a few kilometers to several thousand kilometers. Each hop is surveyed for a line-of-sight antenna path having the necessary clearance, as mentioned above. The size of the antennas, transmitter output power, minimum acceptable receive power, and hop length are all interrelated. The minimum receivable power is the hop design starting point. This is determined by barriers created by the fundamental laws of physics and the state of the art of technology. The receive power level has a threshold value below which satisfactory communication is not possible. As Fig. 4.35 indicates, for an analog signal a gradual reduction of communication quality is experienced to the point where the noise is intolerably high. In practice, this effect is observed as a severe background hissing or crackling noise that eventually makes the talker's speech inaudible to the listener. In contrast, the threshold receive power level for digital systems is approached very abruptly. This received power level in dBm is

$$P_r = P_t - \alpha \qquad (4.18)$$

where P_t = transmitted power level (dBm)
       α = path loss (dB)

The transmitted power level is limited primarily by cost and reliability considerations. The present-day value of P_t is in the region of 1 W or less. The net path loss is

$$\alpha = \alpha_{fs} + \alpha_b + \alpha_f - G_t - G_r \qquad (4.19)$$

where α_{fs} = free-space loss
       α_b = RF branching network loss
       α_f = antenna feeder (waveguide) loss
       G_t = transmitter antenna gain
       G_r = receiver antenna gain

The free-space loss, as indicated earlier, increases as both the operating frequency and path (hop) length increase. The branching network has an unavoidable loss due to filters and circulators. This loss is only about 2 dB in a nondiversity system. The feeder loss is due to the waveguide run from the radio transceiver in the exchange or repeater building up to the antenna(s) on the tower. This loss can be in the range 3 to 9 dB/100 m at 6 GHz, depending on the type of waveguide used. The gain of the antennas increases with diameter of the antenna and also as the operating frequency increases. To minimize the net loss, there are some conflicting requirements. The hop length is the most controversial parameter here. Obviously, the shorter the hop, the lower the free-space loss. However, the whole objective of this exercise is to make hops as large as possible so that the number of expensive repeater stations is minimized. At the same time, the hops cannot be so long that the quality of performance and availability are compromised. In flat regions, the curvature of the earth will limit the hop length. The distance can be increased by building very high towers. Towers are expensive, but it is often cost effective to have towers even as high as 100 m so that less repeaters are necessary. In mountainous regions where long hops are possible, the free-space loss is the limiting factor. The size of the antennas compensates to some

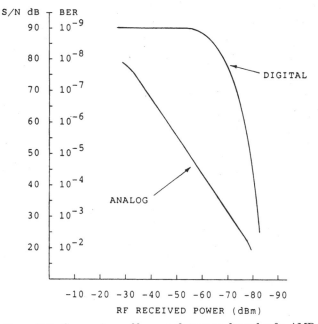

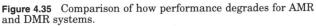

Figure 4.35 Comparison of how performance degrades for AMR and DMR systems.

extent, but there is a practical limit to the size of an antenna. The larger the antenna, the more robust and therefore expensive the tower must be to hold the antenna in position. Also, the beamwidth decreases with increasing antenna diameter, so maintaining alignment of the transmit and receive antennas can be a problem if the antennas are too large. In practice, 2- or 3-m-diameter antenna dishes are usually used, with 4 m being the exception. The receive power level P_r must be substantially higher than the minimum receivable or threshold power level P_{th}; otherwise during atmospheric fading conditions the signal will be lost. This leads to the term *fade margin*, which is the difference between the threshold level and the operational level:

$$\text{Fade margin} = P_r - P_{th} \tag{4.20}$$

The fade margin is a composite parameter, which is primarily associated with dispersive fading due to atmospheric effects, but also contains the components of fading attributable to interference and thermal noise.

The fade margin of each hop in a route is determined by the required overall availability. This is a statistical problem, due to the statistical nature of fading. Also, P_{th} is fixed by the minimum acceptable BER, which is related to availability. P_{th} is usually considered to be the received power level at which the hop becomes unavailable. The unavailability caused by fading obviously improves as the fade margin is increased. Mathematically, using Rayleigh-type single-frequency fading, the probability that fading will exceed the fade margin (*FM*) in the worst month can be approximated by:

$$P_F = 7 \times 10^{-7}(c\, f^B\, d^C\, 10^{-FM/10}) \tag{4.21}$$

where c = terrain and climate factor (4 for over the sea or coastal areas; 1 for medium rough terrain, temperate climate, noncoastal areas; 1/4 for mountainous terrain and dry climate areas)

f = carrier frequency (GHz) and $0.85 \leq B \leq 1.5$ (usually about 1)

d = hop length (km) and $2.0 \leq C \leq 3.5$ (usually about 3)

Note: The values of B and C depend on geographical location, and $p_F\% = 100\% \times P_F$.

More rigorous statistical analyses have recently been applied to fade margin calculations, but the above equation still provides a reasonable approximation. A fade margin of 30 dB results in an availability of about 99.9 percent. Ideally, in severe fading areas, hops are designed to have a comfortable fade margin of at least 40 dB. The microwave power gains and losses in a microwave hop are shown graphically in Fig. 4.36. This figure gives the typical orders of magnitude of power levels at various points in a system. For example, a 30-dBm (1-W) transmitter output has a small waveguide run loss followed by an antenna gain. The free-space loss between the antennas is the most significant loss, and the received signal power level has some uncertainty because of atmospheric effects. The receive antenna has gain, and the interconnecting waveguide and connectors have loss; this produces the resulting signal power level that must be well above the threshold level ready for processing.

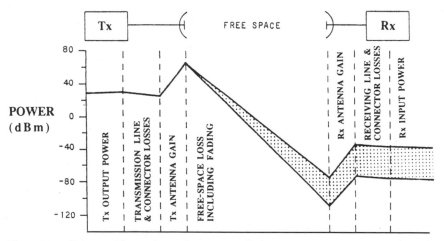

Figure 4.36 Gains and losses in a microwave radio hop.

4.8.2 Passive repeaters

When a microwave hop is required in a place that has some unavoidable physical obstacles, a passive repeater can sometimes solve the problem. For example, suppose a satellite earth station is built outside a major city, and the large satellite dish is situated in a hollow, surrounded by small hills about 150 m higher than the earth station. This is frequently a realistic situation, because the earth station dish is then shielded from the city noise. A microwave link is required to access the earth station, which might be some 40 km away from the city. It would be expensive to place a repeater station on the hilltop and the hop from earth station to hilltop might be less than 2 km (only a mile or so). A passive repeater is effectively a "mirror" placed on the hilltop to reflect the microwave beam down to (and up from) the earth station. There must be a clear line-of-sight path between the passive repeater and the other two end points. Another instance where a passive repeater is often used is when a mountain peak has to be surmounted. It might be so inaccessible that power cannot be provided for an active repeater, or even if a solar power source is used, it might be too inaccessible for maintenance purposes. Whereas helicopter access for installation of a passive repeater might be practical, it is usually excessively expensive to use a helicopter for maintenance.

There are several possible configurations for passive repeaters. First, two parabolic antennas could be placed back to back with a length of waveguide from one feed horn to the other (Fig. 4.37a). Each antenna is aligned with its respective hop destination antenna. The second type is known as the "billboard" metal reflector that deflects the microwave beam through an angle, as in Fig. 4.37b. Provided the angle is less than about 130°, only one reflector is necessary. If the two paths are almost in line (i.e., less than about 50° between the two paths), a double passive reflector repeater is used (Fig. 4.37c).

For the case of a single reflector, the distance between the reflector and the terminal parabolic antenna is important in the path loss calculations. The reflector is said to be in the *near field* or the *far field*. The far field is defined as the distance from the antenna at which the spherical wave varies from a plane wave (over a given area) by less than $\lambda/16$. The near field is consequently the distance from the antenna to that point and is calculated to be

$$r = \frac{2D^2}{\lambda} \tag{4.22}$$

where D = the diameter of a circular reflector or length of each side of a square reflector
λ = the wavelength in the same units as D

As an example, the extremity of the near field at 6 GHz for $D = 6$ m is r, which is about 1.4 km.

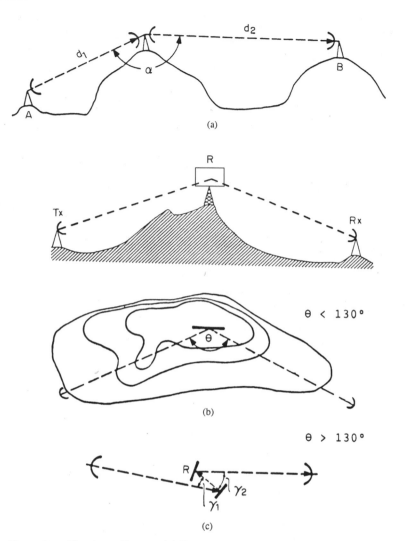

Figure 4.37   Passive reflectors. (*a*) Passive repeater with two parabolic reflec-tors. (*b*) Passive repeater with one plane reflector. (*c*) Passive repeater with two plane reflectors at one location.

The gain of a passive reflector G_p is that of two back-to-back parabolic antennas:

$$G_p = 10 \log_{10} \left(\frac{4\pi a^2 e}{\lambda^2} \right)^2 \quad \text{dB} \tag{4.23}$$

where the efficiency is much higher than a parabaloid (i.e., approximately 95 percent) and the effective area a^2 is related to the actual area of the passive reflector by

$$a^2 = D^2 \sin\left(\frac{\theta}{2}\right) \tag{4.24}$$

where D^2 = actual area of the passive reflector
       θ = deflection angle

If the passive reflector is in the far field, the calculation of path loss is straightforward, but for the near field case the gains of the antennas and passive reflectors interact with each other, producing a complex path loss calculation. Figure 4.38 compares the path losses for all combinations of near field, far field, and single or double passive reflectors. For the single passive, far field case (Fig. 4.38b), the path loss is simply the sum of the gains of the antennas and the passive minus the free-space loss of each section. However, as Fig. 4.38c indicates, for the single passive, near field case, the free-space loss of the short section is not included. Instead, a factor α_n is used, which can be a gain or loss depending upon the size of the passive, the distance between the passive and the antenna, and the frequency. Figure 4.39 is a graph that is used to evaluate α_n, which depends on the frequency and the physical

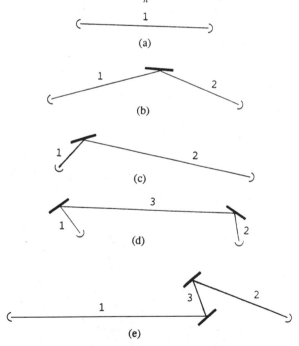

Figure 4.38 Path loss comparison for passive reflectors.
(a) Path with no passive; $L = L_1 - G_t - G_r$.
(b) Single passive, far field; $L = L_1 + L_2 - G_t - G_p - G_r$.
(c) Single passive, near field; $L = L_2 - G_t - G_r - \alpha_n$.
(d) Double passive, far field; $L = L_1 + L_2 + L_3 - G_t - G_{pa} - G_{pb} - G_r$.
(e) Double passive, close coupled; $L = L_1 + L_2 - G_t - G_{pa} - \alpha_c - G_r$.

dimensions of the system. Similarly, the path loss for the double passive in the far field of both antennas is an algebraic sum of the gains of the antennas, the passives, and the section free-space losses. Whenever the two passives are closely coupled, the term α_c replaces the term $(G_p - L_3)$, where α_c is found from the graph in Fig. 4.40.

Passive reflectors are mainly used at 6 GHz and above because they are more efficient as the frequency increases, because the passive gain factor appears twice, whereas the free-space loss appears only once. The gain of a passive reflector increases with its size. A maximum size of 12 m $\times$ 18 m is used up to 12 GHz, and this maximum size gradually decreases as the operating frequency increases. This is because the 3-dB beamwidth decreases as the area of the passive increases, and it should not be allowed to fall below 1°. Even for a 1° beamwidth, twisting movements of the tower holding the passive must not be more than $\frac{1}{4}$°, because this angle appears as $\frac{1}{2}$° for the deflecting angle.

The single passive, near field case can also be used in the "periscope antenna" mode, as shown in Fig. 4.41. Here, the antenna is located at ground level and is pointed directly upward toward the passive reflector mounted at the top of the tower. This system has the multiple benefits of reducing waveguide runs

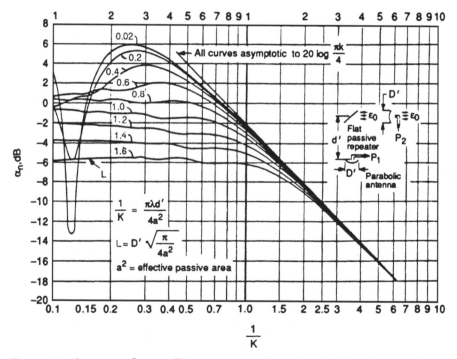

Figure 4.39 Antenna-reflector efficiency curves. (*Reproduced with permission from McGraw-Hill. Panter, P. F., Communication Systems Design, McGraw-Hill, 1972.*)

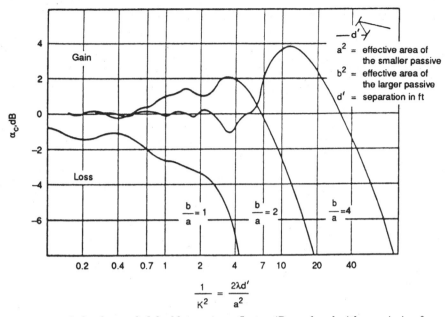

Figure 4.40 A closely coupled double-passive reflector. (*Reproduced with permission from McGraw-Hill. Panter, P. F., Communication Systems Design, McGraw-Hill, 1972.*)

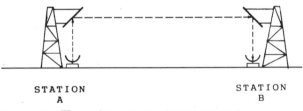

STATION
A

STATION
B

Figure 4.41 The periscope antenna arrangement.

and, with appropriate choice of dimensions, has a net gain greater than that of the parabolic antenna alone. Numerous mathematical analyses have been made for all possible combinations of dimensions including various curvatures for the reflecting surface. Despite the benefits, potential intermodulation noise problems have limited their widespread use. Also, their use is usually confined to the 6- to 11-GHz frequency range.

4.8.3   Noise

Interference.   Each receiver in a digital radio-relay network is exposed to a number of interference signals, which can degrade the transmission quality. The main sources of interference are:

- Intrasystem interference
  - Noise
  - Imperfections
  - Echo

- Interchannel interference
  - Adjacent channel
  - Cochannel cross-polarization
  - Transmitter and receiver
  - Spurious emission

- Interhop interference
  - Front-to-back
  - Overreach

- Extra-system interference
  - Satellite systems
  - Radar
  - Other radio systems

Intrasystem interference.   This type of interference is generated within a radio channel by thermal receiver noise, system imperfections, and echo distortions. Good system design ensures that imperfections do not introduce significant degradation. However, echo distortion caused by reflections from buildings or terrain and due to double reflections within the RF path (antenna, feeder) cannot be neglected in higher-order QAM systems. Echo delay causes interference, which increases as the delay increases. Ground reflections cause echo delays in the region of 0.1 to 1 ns. Reflections near antennas or from distant buildings are usually greater than 1 ns and can cause severe echo interference. Transversal equalizers (as described later) help to improve the situation. Echoes caused by double reflection in the waveguide feeder cause very long delays of 100 ns or more. The only countermeasure against this problem is to ensure that it does not occur. In other words, good return loss (VSWR) is essential.

Interchannel interference.   Interchannel interference between microwave transmission frequency bands is described by the diagram in Fig. 4.42. Adjacent channel interference can be either:

1. Cross-polar.
2. Copolar.

Cochannel interference can only be cross-polar.

For adjacent channels, the copolar interference can be suppressed by filtering, and adjacent channel cross-polarization interference is not usually a problem with today's antennas.

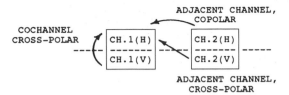

Figure 4.42   Interchannel interference.

Figure 4.43 is a graph of the cochannel interference for various S/N values. The noise introduced by one channel into the other is due to inadequate cross-polarization discrimination of the antennas, and the S/N values are effectively the cross-polarization discrimination values for channels operating at the same output power level.

This type of interference can be more serious than adjacent channel interference, and the value of the S/N should be better than 25 dB and preferably at least 30 dB. For example, at a received power level of -70 dBm, an S/N = ∞ would give a BER equal to about 10^{-10}. At S/N = 30 dB the BER would degrade to about 2×10^{-8} and, if the S/N were only 20 dB, the BER would be worse than 10^{-5}.

The trend toward cochannel operation, to enhance spectral utilization efficiency, has promoted the widespread use of cross-polarization interference cancellers, as described in Sec. 5.2.2.

Interhop interference.   This type of interference can occur because of front-to-back or nodal interference from adjacent hops and by *overreach* interference. The signal-to-interference ratio (S/I) is determined by the angular discrimination of antennas and can decrease during fading. Careful route and frequency planning is necessary to keep the degradation smaller than 1 dB. Some examples of the above intrasystem, interchannel, and interhop sources of interference are shown in Fig. 4.44.

Extra-system interference.   This interference can be caused by other digital or analog channels using the same RF band or by out-of-band emissions from other radio systems (e.g., radar). It is important to remember that the nature of digital radio is such that the spectrum of the transmitted band is completely full. In other words, the energy is spread evenly across the band. This is different from the analog radio signal, which has its energy concentrated in the middle of the band with a smaller portion of the energy in the sidebands. This means the *analog* radio is susceptible to interference from the *digital* radio. This is an adjacent channel type of interference, which can be a problem for high-capacity systems (i.e., if an 1800-channel analog radio is adjacent to a 51.84-Mb/s digital radio).

When planning new microwave links, a very important consideration is the frequency coordination to ensure there is no interference with existing or future proposed frequencies.

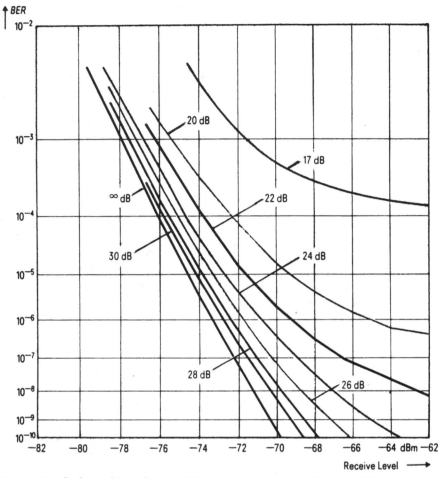

Figure 4.43 Cochannel interference. (*Reproduced with permission from Siemens Telcom Report: Special "Radio Communication," vol. 10, 1987, p. 95, Fig. 10.*)

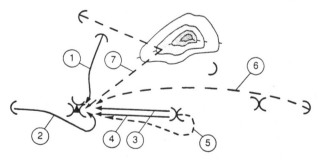

Figure 4.44 Various sources of interference. 1 = cochannel or adjacent channel signal from a different hop direction; 2 = opposite hop front-to-back reception; 3 = adjacent channel (same hop); 4 = cross-polarization (same hop); 5 = front-to-back radiation; 6 = overreach; 7 = terrain reflections.

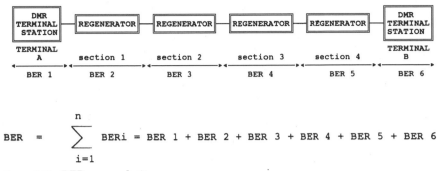

$$BER = \sum_{i=1}^{n} BERi = BER\ 1 + BER\ 2 + BER\ 3 + BER\ 4 + BER\ 5 + BER\ 6$$

Figure 4.45 BER accumulation.

DMR noise calculations. The quantity of major interest for determining the quality of the digital link has, until recently, been the *BER*. The minimum required BER values recommended in ITU-T Recommendation G.821 have been superseded by ITU-T Recommendation G.826. Quality of service is becoming increasingly important and, while the term BER might refuse to die, a value of 10^{-5} might replace 10^{-3} as the failure criterion.

As always stated in DMR texts, one of the main advantages of digital over analog radio is the fact that thermal and intermodulation noise accumulation is eliminated by the baseband regenerative process at repeaters.

For speech communication, a value of 1×10^{-6} used to be considered adequate for good quality performance. When the value was worse than 1×10^{-6}, the link was considered to be *degraded*, and maintenance initiated to improve the BER. After 10 s at a value of 1×10^{-3}, the link was considered to be *unavailable* (i.e., failed). For data transmission it used to be that the BER should be better than 1×10^{-7}. Even this goal was not always easy to achieve because a DMR link consists of several sections, and the BER of the link is the sum of the BERs of each section, as illustrated in Fig. 4.45. This meant that in order to have a satisfactory link to support data traffic, each section should have a BER better than 10^{-9}, or even 10^{-11}. Again, these figures are becoming even more stringent with the latest standards.

Regardless of how digital signal quality is measured, a more fundamental term is often used to describe the performance of a DMR—*the energy per bit per noise density ratio* E_b/N_o:

$$E_b = \frac{\text{carrier peak power}}{\text{bit rate}} = \frac{C}{B_r} \tag{4.25}$$

$$N_o = \frac{\text{noise power in } B_{eq}}{\text{equivalent Rx 3-db noise bandwidth}} = \frac{N}{B_{eq}} \tag{4.26}$$

$$= -204 \text{ dBW} + NF_{dB}$$

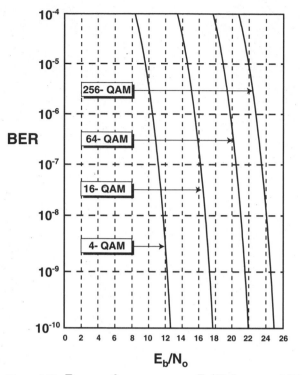

Figure 4.46 Error performance versus E_b/N_o for several QAM values.

for a perfect receiver at room temperature, 290 K, where NF_{dB} is the receiver noise figure, and N_o is the noise in a 1-Hz bandwidth. So,

$$\frac{E_b}{N_o} = \left(\frac{C}{N}\right)\frac{B_{eq}}{B_r} \qquad (4.27)$$

Calculations have been made to relate BER to E_b/N_o, and graphs have been plotted for various modulation schemes. The graphs for several QAM schemes are plotted in Fig. 4.46, which shows that as the QAM level increases, the E_b/N_o required to achieve a particular BER increases.

So, for a required minimum BER, the E_b/N_o value can be evaluated, which can be related to parameters of the path by

$$\frac{E_b}{N_o} = P_r - (-204 + \text{NF}_{\text{dB}}) - 10 \log B_r \qquad (4.28)$$

where the modulated carrier level is considered to be the received signal level in this case. Substituting for P_r [Eq. (4.18)],

$$\frac{E_b}{N_o} = P_t - \alpha + 204 - \text{NF}_{\text{dB}} - 10 \log B_r \qquad (4.29)$$

Using this equation, the hop can then be designed to provide an adequate fade margin.

Design example.   Consider a 155-Mb/s (STS-3) DMR operating at 6 GHz with 64-QAM over a 45-km, medium rough terrain hop, with 3-m antennas on 30-m towers. Establish the fade margin and therefore determine if any system specifications need to be modified.

The theoretical E_b/N_o for a BER of 1×10^{-9} is about 22 dB. The practical value could be up to 5 dB higher than this value. A 1-W (0-dBW) transmit power is relatively cheaply available, with a high linearity. The noise figure of a typical receiver in the 6-GHz band is about 4 dB, so, from Eq. (4.29),

$$27 = 0 - \alpha + 204 - 4 - 81.9$$

$$\alpha = 91.1 \text{ dB}$$

because, from Eq. (4.19),

$$\alpha = \alpha_{fs} + \alpha_b + \alpha_f - G_t - G_r$$

Free-space loss = 141 dB

Branching network loss = $1.4 \times 2 = 2.8$ dB

Suppose for adequate obstacle clearances, two towers each 30 m high are needed. Then

$$\text{Waveguide feeder loss} = 1.5 \times 2 = 3 \text{ dB}$$

$$G_t = G_r = 44 \text{ dB}$$

$$\alpha = + 141 + 2.8 + 3 - 44 - 44$$

Therefore,

$$\alpha = 58.8 \text{ dB}$$

Comparing this figure with 91.1 dB, there is a safety factor, or fade margin, of 32.3 dB.

Common carrier availability of 99.999 percent for a *path* has always been a difficult target to reach. This is a very simple number to work with, compared to the more complex way in which the network is now subdivided into its national and international sections and quality objectives based on distances. If a DMR *hop* has an availability of 99.9999 percent, the outage for that hop is 32 s per year, and that is after protection switching and other countermeasures improve the situation. The question here is, Can 99.9999 percent be achieved by diversity measures alone?

The outcome of using that figure will now be assessed. Because the outage percentage has to be divided equally between the two transmission directions, for one channel $P_F = 0.00005$ percent.

Taking fading into account, the required nondiversity fade margin can be calculated from Eq. (4.21); that is,

$$P_F\% = 7 \times 10^{-5} c f^B d^C 10^{-FM/10}$$

Because $f = 6$ GHz, $c = 1$, $B = 1$, $C = 3$, and $d = 45$ km,

$$P_F\% = 0.0227\%$$

The system in this configuration is not good enough to meet the initial requirement of 0.00005 percent under these fading conditions. A diversity system would be one solution. Space diversity would improve the outage time as denoted by Eq. (4.16) by a factor of:

$$I_s \approx \frac{1.2 \times 10^{-3} \eta s^2 f 10^{(F - V)/10}}{d}$$

A 10-m separation would be the recognized minimum of about 200 wavelengths at 6 GHz. A 15-m separation could be used only if the lower antenna in these circumstances satisfies the path profile requirements of maintaining adequate line-of-sight path clearance of any obstacles. If $s = 10$ m, $f = 6$ GHz, $V = 0$, $d = 45$ km, $\eta = 1$, the outage improvement with space diversity would be:

$$I_s = 26.94$$

$$P_{F(SD)}\% = \frac{0.0227}{26.94} = 0.00084\%$$

Increasing the antenna spacing to 15 m from 10 m would increase I_s to 60.6, giving:

$$P_{F(SD)}\% = 0.00037\%$$

Because this figure is still too high, the inclusion of frequency diversity would provide the necessary improvement in performance to overcome fading. From Eq. (4.17), the frequency diversity improvement factor is approximated by:

$$I_f \approx \frac{80 \Delta f 10^{F/10}}{f^2 d}$$

If the frequency separation $\Delta f = 120$ MHz (2 percent)

$$I_f = 9.98$$

$$P_{f(SD + FD)} = \frac{0.00084}{9.98} = 0.000038\%$$

This configuration would just meet the 99.9999 percent requirement.

During heavy rainfall, there would be no multipath fading in this system. In such circumstances the space diversity (or frequency diversity if it were included) would not provide any performance improvement. The entire 32.3-dB fade margin initially calculated would then be available to combat the rain attenuation. At 6 GHz a cloudburst (100-mm/h rainfall) would cause an attenuation of 1 dB/km. To consume the whole 32.3-dB fade margin, a cloudburst would have to cover 32.3 km of the 45-km hop. The statistical likelihood of this happening is extremely remote. Rainfall of such intensity usually occurs over only a few (< 10) kilometers, although extreme weather conditions such as hurricanes or cyclones can exceed this figure.

In this example, space diversity alone would not be a sufficient system improvement, and frequency diversity would be necessary as an additional countermeasure. For some telecommunications organizations, frequency diversity is included as a policy on backbone routes to provide protection against equipment failure.

It must be emphasized that the above calculation is an oversimplification that does not take into account all the countermeasure characteristics of the radio. However, it does provide a starting point to show how various parameters associated with diversity can affect the availability. In early DMR designs, the unavailability caused by fading was so alarming that its future viability was in question. Since then, several countermeasures in addition to diversity have been designed and proven to be very effective for most paths (except for the most severe propagation conditions). They include protection switching, IF adaptive equalizers, transversal equalizers, and FEC, all of which will be discussed in detail in Chap. 5. These countermeasures are a welcome technological development because they are generally much cheaper than the expensive space and frequency diversity equipment. They have enabled the combined space and frequency diversity techniques to be reserved for the long hops, in areas with difficult propagation conditions.

Chapter 5

Digital Microwave Radio Systems and Measurements

Significant changes in DMR technology have taken place during the 1990s. Despite the loss of ground to optical fiber transmission systems, there is still a need for DMR systems. However, unless terrain conditions preclude optical fiber, DMR is no longer used for new countrywide backbone systems, and designers have had to move toward SONET/SDH system bit rates for compatibility reasons. Radio clearly has bandwidth limitations, but STS-1 can now occupy 20 MHz and STS-3 (STM-1) 40 MHz. DMR is now finding new applications for closing SONET/SDH rings, particularly in the low-bit-rate portions of the network, or the access network.

Because there are still many DMR systems (and will be beyond 2000) operating on PDH bit rates, both types of technology will be addressed. The BER parameter is so entrenched in the minds of most engineers and technicians when discussing performance that it will be used here, together with some references to the newer block-error-based standard, *the errored second*.

5.1 System Protection

As discussed in Chap. 4, space and frequency diversity methods are implemented to enhance the performance (increase the availability) of microwave radio systems. The manner in which diversity is incorporated into the system design will now be considered. In addition, there are further measures that can be taken to provide an even higher availability, such as IF adaptive equalizers and baseband adaptive transversal equalizers. These will be described in detail.

5.1.1 Diversity protection switching

In a space diversity protected system, the received signals are usually combined at the IF. In a frequency diversity system, the RF received signals are either combined or the stronger of the two signals is used. There is usually a main channel and a standby protection channel. This arrangement is called a *1-for-1 protection* frequency diversity system. A switching technique is used to ensure that the stronger signal is accepted and the weaker signal is unused. This technique is often used in an *M*-for-*N* protection switching arrangement, where *M* protection channels are used to protect *N* information channels. For example, one protection channel might be used for protecting seven information channels. Obviously, 1-for-1 offers a much better level of protection than 1-for-7, often written 7 + 1.

5.1.2 Hot-standby protection

A hot-standby protection system is a fully redundant radio configuration operating in a *power ON* mode, ready for switching into operation in the event of a failure. *The frequency diversity system is one form of hot-standby protection.* The system illustrated in Fig. 5.1 is a hot-standby configuration used in a space diversity system. If a transmitter failure occurs, the RF switch disconnects the failed transmitter and connects the standby transmitter to the antenna. Unfortunately, this type of switching causes a brief disruption of the digital *bit stream*, resulting in an *error burst*. Because the equipment is built for a high degree of reliability, this type of switchover very seldom occurs, and therefore the long-term BER degradation due to the switchover is negligible. As in the frequency diversity system, the switching at the receiver end is done at the baseband by comparing the regenerated bit streams.

5.1.3 Combining techniques

Protection switching or signal combining can be done at the RF, the IF, or in the baseband. Post-detection switching (at the baseband) provides the best reliability, because the receive switch is the last component of the diversity system. In general, space diversity systems combine at the IF level, and frequency diversity systems combine in the baseband (even though the decision to switch over

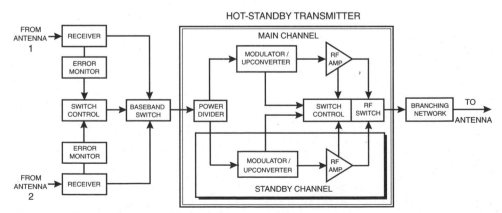

Figure 5.1 Hot-standby/space diversity DMR repeater.

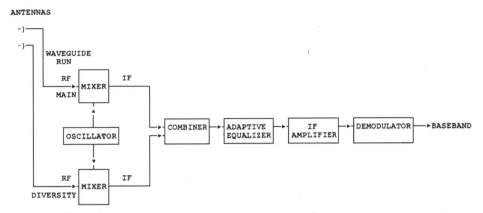

Figure 5.2   IF combining.

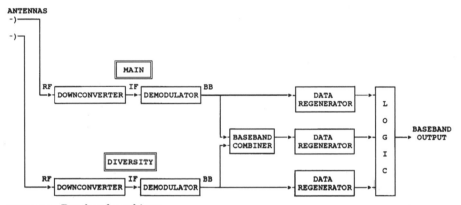

Figure 5.3   Baseband combiner.

might be based on the RF signal level). Figure 5.2 shows a typical IF combined signal. Note the presence of the *adaptive equalizer*. This is a very important component that will be discussed in detail later. In this system, the two IF-modulated carriers are dynamically delay equalized prior to combining. The combiner adds these IF carriers on a voltage basis to improve the C/N. If the C/N ratio is the same on both channels, the combined C/N could be up to 3 dB higher than that of the individual channels. IF carriers must be combined in phase because a small phase error can cause a high BER. For example, for 4-PSK systems, as the phase error becomes worse than 10° the C/N degrades by about 2 dB. Higher levels of modulation (QAM, etc.) are even more susceptible to phase errors. An example of a baseband combiner is shown in Fig. 5.3. The signals from the *main* and *diversity* channels are down-converted and then demodulated to the baseband before combining. The advantage of this type of combiner is that it does not require such careful delay equalization as does the IF or RF combiner. In addition, this combiner can be set for hitless selective switching if the C/N for the individual channels differs by a predetermined threshold level (e.g., 10 dB).

5.1.4 IF adaptive equalizers

It has been found that during multipath fading, the shape of the IF passband changes drastically. Some frequencies are attenuated more severely than others. As a result of this drop in level at selected frequencies, severe distortion occurs in the baseband, which can cause excessive errors and link failure. An IF adaptive equalizer is required to minimize this problem. In order to fully understand this very important equipment component, it is necessary to review the propagation conditions existing between the two antennas of a DMR hop. Figure 5.4 illustrates multiple path propagation between the transmitting and receiving antennas of a microwave hop. In addition to the direct path, atmospheric conditions can produce other paths above or below the direct path. Because the two paths differ in length, the time taken for each signal to propagate between the transmitter and receiver will clearly be different. This time difference can easily be calculated as follows:

Path difference: $d = D_2 - D_1 = \dfrac{2h_1 h_2}{D}$

For example, assume the values $h_1 = 60$ m, $h_2 = 500$ m, and $D = 50$ km. Therefore,

$$d = 1.2 \text{ m}$$

The time difference between the two paths is

$$\tau = \frac{\text{path distance}}{\text{velocity of propagation}}$$

$$= \frac{1.2}{3 \times 10^8}$$

$$= 4 \text{ ns}$$

This is equivalent to a frequency $\Delta f = 1/\tau = 250$ MHz.

If a graph is plotted of the vector-added signals from the two paths against the inverse propagation time difference, the resulting curve is called a *cycloid* curve. This curve (as shown in Fig. 5.5) has a maximum value when the two signals are in phase. This combined signal level has a 6-dB increase over the individual direct signal. However, when the two signals are in phase opposition, the resulting power level is theoretically zero. Fortunately, because of irregularities at the point of reflection, the combined level will never be zero for a significant length of time. Nevertheless, 40-dB fades are not uncommon when severe atmospheric conditions exist. In the above example, the in-phase and phase opposition points occur every 250 MHz. All remaining points on the cycloid curve will have a phase difference between the two waves ranging between 0° and 180°. If the IF band to be demodulated is, for example, 36 MHz

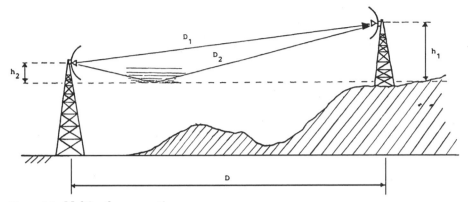

Figure 5.4 Multipath propagation.

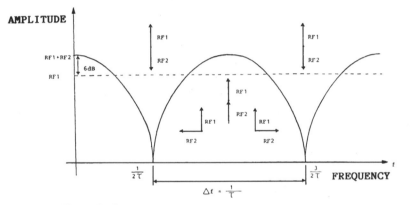

Figure 5.5 The cycloid curve.

(for a 40-MHz channel), only a portion of the cycloid curve is under consideration. This means that the IF passband will have a shape of part of the cycloid curve. Optimum demodulation conditions occur whenever the IF band to be demodulated falls in the middle of the cycloid curve (Fig. 5.6a). Conversely, demodulation is most adversely affected when the demodulation IF band falls on or around the notch (Fig. 5.6b). The notch can even be in the middle of the IF band. The situation is made more difficult to correct because of the *dynamic* nature of fading. This means that the IF band to be demodulated moves about the cycloid curve in an unpredictable manner. A *dynamic* or *adaptive* IF equalizer is consequently necessary to correct for the unflat IF passband response caused by multipath fading.

Figure 5.7 shows the actual IF passband for three cases of multipath fading. Figure 5.7a and b show the multipath fading causing a null at the IF band edges. Figure 5.7c shows the multipath distortion affecting the center of the IF passband. In each of these cases, the inclusion of the adaptive equalizer

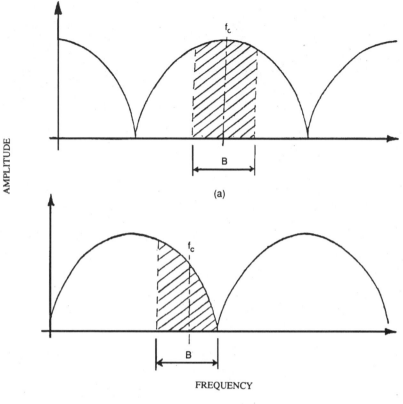

Figure 5.6 IF band location on the cycloid curve.

allows restoration of the IF passband to an almost flat condition. It also maintains a flat group delay. This eliminates (or minimizes) the existence of errors that would otherwise be caused by the multipath fading. *The word "adaptive" is used to indicate that the equalizer has the flexibility to change its character to suit the disturbance.* The equalizer operation is based upon the principle of creating an IF curve complementary to the curve of the channel affected by multipath fading. It is usually composed of two parts:

1. Slope equalization
2. Notch equalization

In a simple equalizer design, a bandwidth meter detects the amplitude distortions present in the spectrum by measuring the signal power in the proximity of three selected frequencies (e.g., 57, 70, and 83 MHz for a 70-MHz IF or 127, 140, and 153 MHz in the case of a 140-MHz IF). The process illustrated in Fig. 5.8 allows the introduction of slope or notch equalization, as necessary, to account for the multipath fading. The diagrams in Fig. 5.7a and b show how the slope equalizer restores the attenuation at the edge of the passband to a relatively flat response. Figure 5.7c shows how the notch equalizer removes the null in the passband. The improvement in the BER due to IF adaptive equalization is illustrated in Fig. 5.9 for a 6-GHz, 90-Mb/s, 8-PSK DMR. The graph shows two values of fade depth (i.e., 23.8 dB and 25.5 dB). The fade is due to 6-GHz multipath signals destructively interfering with each other at the worst possible phase of 180°. For the 23.8-dB fade, without the equalizer, a −70-dBm received signal is degraded to a BER of 1×10^{-3}. For many specifications this is a failure condition. With the equalizer, the BER improves to 2×10^{-4}. At higher received signal levels the improvement is even more noticeable. The 25.5-dB fade at a received signal level of −66-dBm improves the BER from worse than 1×10^{-3} to better than 1×10^{-7}. In each example, the presence of the equalizer almost completely restores the BER to the value obtained with no multipath fade present.

Multipath fading causes dispersive delay distortion, which can cause crosstalk between the I and Q signals in a QAM radio system. The IF adaptive slope equalizer can improve the situation to some extent, but additional equalization is often required.

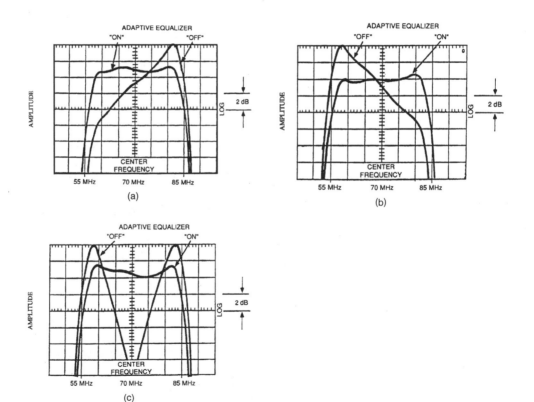

Figure 5.7 The effect of multipath fading on the IF passband. (a) IF spectrum with multipath null located at $f_0 − 20$ MHz; (b) IF spectrum with multipath null located at $f_0 + 20$ MHz; (c) IF spectrum with in-band multipath distortion.

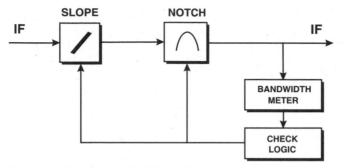

Figure 5.8 Slope and notch IF adaptive equalization.

5.1.5 Baseband adaptive transversal equalizers

The operation of the transversal equalizer can best be explained by looking at the impulse response of a distorted channel. Figure 5.10 compares an ideal channel with a distorted one. The impulse response of the ideal channel has *equal zero crossing at the symbol interval.* However, the impulse response of the distorted channel has some positive or negative amplitude at the points where there should be zero crossings. In other words, the *ringing effect* has some delay distortion included. The transversal equalizer effectively forces the zero crossings to occur at the points where they should occur.

Although this is a simple problem to describe, the circuit required to solve this problem is rather complicated. Because this is a very important aspect of DMR, it will be discussed in more detail and simplified as much as possible. It is probably fair to say that to a large extent the success of DMR can be attributed to the design and incorporation of these equalizers to counteract the potentially devastating effects of multipath fading. The objective of the baseband equalizer is to minimize the intersymbol interference caused by adverse propagation conditions. As shown in Fig. 5.11, under optimal conditions the zero crossing instants of the "tails" of a PAM signal coincide perfectly with the sampling instants of adjacent PAM information. In this situation, it is easy to see that a cancellation of the tails produces zero intersymbol interference. During multipath fading conditions, the adjacent responses do not have tails that cross the time axis at exactly equal intervals (see Fig. 5.12). The summation of the tails, called precursors and postcursors, can result in intersymbol interference if their amplitudes are sufficiently large, as in the case of severe multipath fading. These pre- and postcursors proceed to infinity with decreasing amplitude. For a simplified analysis, it is sufficient to consider only the first postcursor and the first precursor. Figure 5.13 shows how the postcursor of the preceding PAM signal (X_{k-1}) and the precursor of the proceeding PAM signal (X_{k+1}) increase the value of the transmitted PAM signal (X_k). Mathematically,

$$X_k = \hat{X}_k + V_a + V_b$$

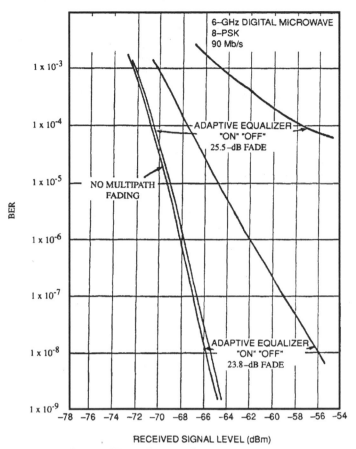

Figure 5.9 Adaptive IF equalizer performance.

where $\hat{X}_k$ = the amplitude of the transmitted PAM signal
V_a = the postcursor interference amplitude
V_b = the precursor interference amplitude

that is,

$$V_a = aX_{k-1}$$

$$V_b = bX_{k+1}$$

where a and b are coefficients whose values lie between -1 and $+1$ and direct-ly indicate the interference present. Therefore,

$$X_k = \hat{X}_k + aX_{k-1} + bX_{k+1} \tag{5.1}$$

The simplified circuit block diagram in Fig. 5.14 allows the interfering com-ponents to be subtracted from the X_k signal, resulting in X_{0k}. This equalized

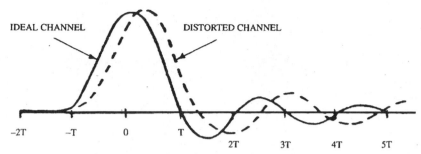

Figure 5.10 Impulse response of an ideal channel and a distorted channel.

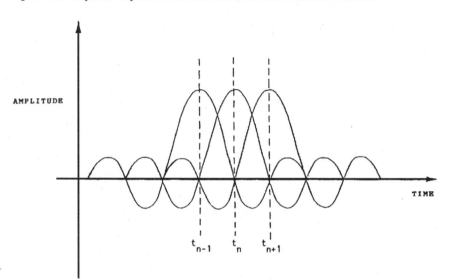

Figure 5.11 Optimal conditions for providing zero ISI.

signal is theoretically the same as the transmitted signal. X_{0k} is passed through a decision circuit to restore the binary information and then a D/A converter produces the original transmitted signal $\hat{X}_k$. That is,

$$X_{0k} = X_k - a\hat{X}_{k-1} - bX_{k+1}$$

as $a << 1$ then $a\hat{X}_{k-1} \approx aX_{k-1}$

so $X_{0k} \approx X_k - aX_{k-1} - bX_{k+1}$

Substituting for $X_k \rightarrow X_{0k} \approx \hat{X}_k$. The coefficients a and b must next be established.

Postcursor equalization (to determine a). The difference between the equalized PAM signal and the transmitted PAM signal can be defined as the error ε_k, where

$$\varepsilon_k = X_{0k} - \hat{X}_k \tag{5.2}$$

Considering an unequalized signal that for simplicity has only postcursor interference, Eq. (5.1) becomes

$$X_{0k} = X_k - a\hat{X}_{k-1}$$

It can be seen that if X_k increases in amplitude because of postcursor interference, X_{0k} will correspondingly increase. From Eq. (5.2), if X_{0k} is greater than X_k,

$$\varepsilon_k > 0 \rightarrow \text{logic 1}$$

In order to counteract the increase in X_k it is necessary for

$$a\hat{X}_{k-1} > 0$$

from which it follows that

$$\text{If } \hat{X}_{k-1} > 0 \text{ (logic 1) then } a > 0 \text{ (logic 1)}$$

$$\text{If } \hat{X}_{k-1} < 0 \text{ (logic 0) then } a < 0 \text{ (logic 0)}$$

Similarly, a reduction in X_k caused by postcursor interference decreases X_{0k}, and from Eq. (5.2)

$$X_{0k} < \hat{X}_k \text{ and } \varepsilon_k < 0 \quad \text{(logic 0)}$$

and this is counteracted by

$$a\hat{X}_{k-1} < 0$$

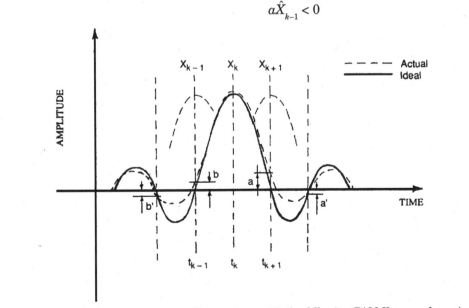

Figure 5.12 Distorted PAM X_k interferes with the following PAM X_{k+1} and previous PAM X_{k-1}
Note: Voltages (a, a') are postcursor interferences and (b, b') are precursor interferences.

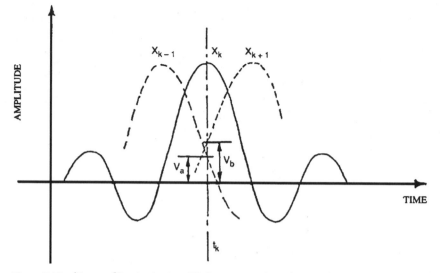

Figure 5.13 At sampling instant t_k, V_a (postcursor interference by PAM X_{k-1}) and V_b (precursor interference by PAM X_{k+1}) are to be added to the value of the transmitted PAM X_k.

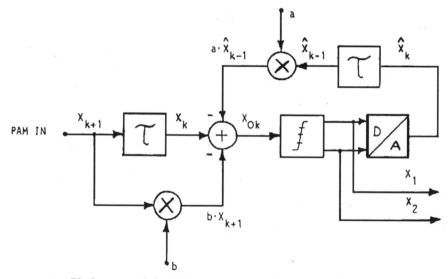

Figure 5.14 Blocks τ retard the PAM signal by one decision period to make PAM X_{k-1} available at the same time instant as PAM X_k.

Therefore,

$$\hat{X}_{k-1} > 0 \quad \text{(logic 1)} \quad \text{consequently } a < 0 \quad \text{(logic 0)}$$

$$\hat{X}_{k-1} < 0 \quad \text{(logic 0)} \quad \text{consequently } a > 0 \quad \text{(logic 1)}$$

and a can therefore be assigned a value depending upon the values of ε_k and $\hat{X}_{k-1}$.

From the above results, the following summary indicates that the sign of a is the EXCLUSIVE NOR of the sign of ε_k and the sign of $\hat{X}_{k-1}$:

Sign of ε_k	Sign of $\hat{X}_{k-1}$	Sign of a
1	1	1
1	0	0
0	1	0
0	0	1

The analog values of a are found simply by integrating the digital values of the sign of a. Figure 5.15 shows a simplified diagram of the circuit that equalizes postcursor interference.

Precursor equalization (to determine b). An analysis similar to the above precursor equalization would result in b being evaluated by passing the sign of ε_k and the sign of X_{k+1} through an EXCLUSIVE OR gate. However, X_{k+1} is not available because it has not yet passed through the decision circuit. This problem is solved by shifting the time reference by one decision instant (period). Therefore, as in Fig. 5.16, the input is now X_{k+1}, the sign of ε_{k-1} is used instead of the sign of ε_k, and the sign of X_k is used instead of the sign of X_{k+1}. This is acceptable because the signal equalization takes place over a time interval corresponding to thousands of bits (depending upon the RC time

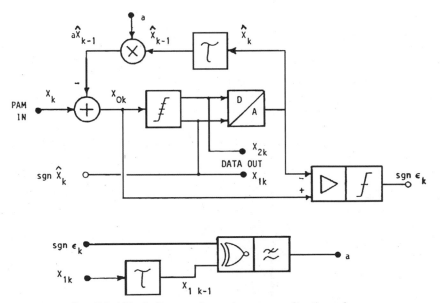

Figure 5.15 Simplified block diagram for postcursor equalization only.

constant of the integrator), so the precursor distortion of X_{k+1} on X_k is approximately the same as X_k on X_{k-1}.

So far, equalization has been considered only for the PAM X signal. For 16-QAM, there is also a PAM Y component, such that the two components are subsequently modulated by carrier signals phase shifted by 90°. The Y component is equalized by a circuit identical to that of Fig. 5.16.

Unfortunately, it is necessary to consider interactions between X and Y. This is because multipath fading causes the transmitted spectrum to be affected in an asymmetrical manner with respect to the center of the band. Equation 5.1 therefore becomes

$$X_k = \hat{X}_k + aX_{k-1} + bX_{k+1} + cY_{k-1} + dY_{k+1}$$

where cY_{k-1} = the postcursor interference generated by the PAM Y_{k-1} signal
dY_{k+1} = the precursor interference generated by the PAM Y_{k+1} signal

It is sufficient to perform cross-correlations between εX_k and Y_{k-1} and between εX_{k-1} and Y_k as indicated in Fig. 5.17.

As one can appreciate, this simplified circuit is starting to become complex, and present-day baseband equalizers are very sophisticated. The base-

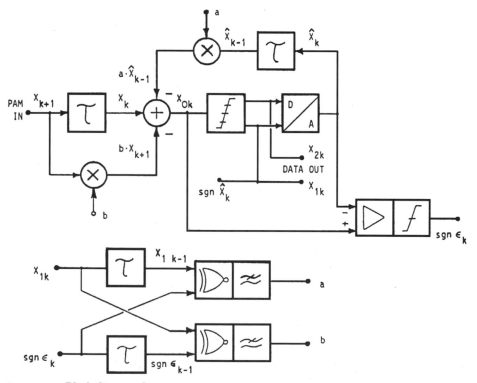

Figure 5.16 Block diagram for postcursor and precursor equalization.

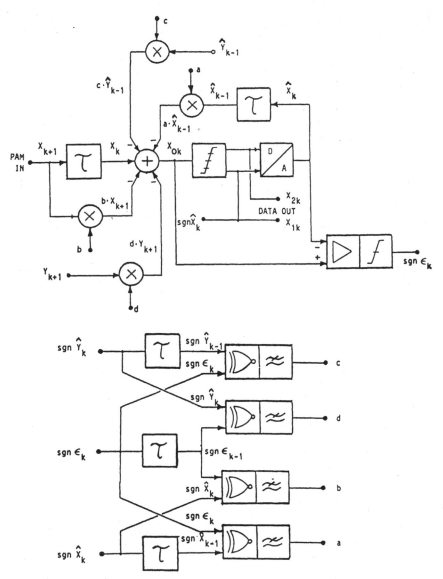

Figure 5.17 Complete equalizer block diagram.

band transversal equalizer together with the IF slope equalizer provides most hops with the necessary protection to withstand very deep multipath fades. Usually, the only air interfaces that still have a problem even after incorporating both types of equalizer are the long hops *over water*, in hot, humid climates.

5.2   155-Mb/s DMR with 64-QAM

Perhaps the most widely used DMR in the 1980s was the long-haul, medium-to high-capacity radio at 90 and 140 Mb/s, used for backbone links between major metropolitan areas. The move to higher capacity for trunk routes has favored optical fiber over DMR wherever terrain allows. In order to compete, DMR has had to become more spectrally efficient. The congestion within the 4- to 6-GHz frequency bands also led to the need for greater bandwidth efficiency. So, 64-QAM is now widespread, and 256-QAM is moving into the field, with TCM gaining in popularity.

As technology has progressed and high-quality, high-performance microwave components have become available in the frequency range above 20 GHz, a significant interest in using DMRs in the local network has evolved. This has been in the form of rural area spur hops from a backbone, SDH ring, or short hops within a city between company premises or from a large company to the exchange. In each case the transmission distance is short, usually in the range of 1 to 15 km. Incorporation of DMR into SONET/SDH upgrades is another important application.

The 140-Mb/s 16-QAM is still the PDH workhorse system that is used extensively around the world. In North America $1 \times$ DS3 or $3 \times$ DS3 are the equivalent systems. Each manufacturer has its own special design refinements but all follow a similar basic theme. In North America, SONET/SDH is being rapidly deployed at the STS-1 level and in Europe and many other countries at the STM-1 level. A typical 52 or 155-Mb/s 64-QAM SDH DMR is illustrated in Fig. 5.18. This figure illustrates a frequency diversity system where there is only one protection channel available for up to seven bearers (7 + 1). This is clearly more economical than 1 + 1 protection, but in the unlikely event of two bearers simultaneously falling below the critical BER, one of the bearers is lost. For the 7 + 1 case, the protection switch splits the signal of each bearer so that the signal is sent to the transmitter and also to a microprocessor controlled switch. Prior to each modulator there is a section overhead (SOH) insertion module that allows the addition of TMN information to the 52 or 155-Mb/s bit frame. The receivers and demodulators (including SDH extraction units) are also shown in the diagram. In the event of a failure, the microprocessor connects the faulty bearer to the standby channel and at the receiver redirects the standby to the appropriate faulty channel while disconnecting the faulty signal. The switching control signal uses a 64-kb/s signal on a dedicated SOH byte. Switching is activated in the extreme cases of no signal being received or frame loss.

There is a gray area where switching is not necessary, but further deterioration could warrant switching. The decision point is not easy to determine, because it is tied into the overall availability objective. The BER is still a parameter that is monitored, although SES is becoming a more recognized measure of performance. The *fast BER*, or early warning as it is often called, can be from the output of the Viterbi decoder, which can provide a BER having two

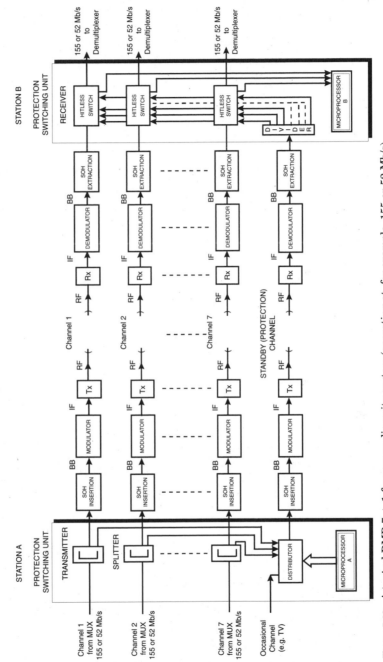

Figure 5.18 A typical DMR 7 + 1 frequency diversity system (operating at, for example, 155 or 52 Mb/s).

threshold values of, say, 10^{-8} and 10^{-12}. Parity bits still give an indication of the BER $= 10^{-3}$, although this is extremely inadequate for the level of service expected today. If radio is to compete with optical fiber on a performance basis, this fact must be recognized by service providers. If there is not an instantaneous failure of a channel as the SES or BER rapidly degrades (over a period of a few microseconds), at a predetermined BER of say 1×10^{-8}, the microprocessor ensures a hitless switchover operation. An instantaneous failure (in the nanosecond time frame) will only allow switching with an error burst. Dropped calls might even result.

Figure 5.19 shows a typical digital regenerative repeater. Down conversion and demodulation to baseband is the usual procedure. The service channels can be dropped or inserted at this point, as can individual voice channels, provided an add/drop multiplexer procedure is included in the repeater, which is usually the case today. This diagram also shows space diversity in one direction, indicating that two receivers and a combiner are now necessary for each half of the space diversity system, which means that two transmitters and four receivers are required.

5.2.1 DMR transmit path

The higher-capacity, higher-order modulation scheme DMRs tend to have the most complex system and component design. For this reason, the 155-Mb/s DMR components will be studied in more detail.

The diagram of Fig. 5.20 shows the building blocks of a typical 155-Mb/s DMR transmitter. The baseband signal inputs to the transmitter include the

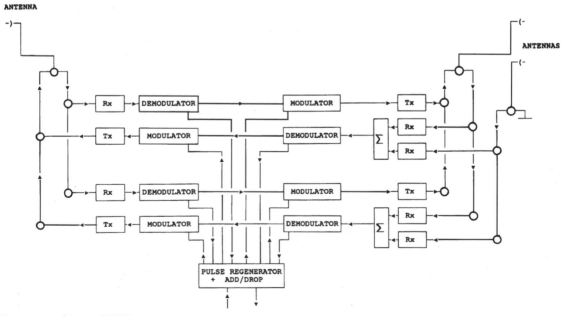

Figure 5.19 A typical DMR regenerative repeater.

multiplexed voice, data, and/or TV channels, together with service channel information. The 155-Mb/s bit stream has the CMI line code. After clock recovery, this is converted into an NRZ signal in the transmitter. The digital service channels are mixed into this bit stream in the SOH insert module, and the combined signal is scrambled to remove line spectra from the signal (line spectra cause interference in DMRs). The signal is then serial-to-parallel converted and differentially encoded. This is an encoding that allows easy demodulation in the receiver without needing to transmit absolute phase information. After QAM modulating, the signal is then filtered and amplified. Trellis-coded modulation is becoming increasingly used instead of just QAM. The IF signal is then up-converted to RF, linearized, amplified, and passed through the branching network to the antenna for transmission.

Each of the following major components of the transmitter will be discussed in detail:

1. CMI to NRZ converter

2. Scrambler

3. Serial-to-parallel converter in the modulator

4. Differential encoder

5. IF amplifier

6. RF section mixer, local oscillator, amplifier with linearizer and automatic transmit power control, branching network, waveguide run, and antenna.

The modulator is treated in detail in Chapter 3.

CMI to NRZ decoding. Electrical pulses from the MUX equipment are presented to the 155-Mb/s radio in the CMI form. The preferred code for dealing with information in the radio is the NRZ code. The simple circuit in Fig. 5.21 converts the incoming CMI signal to NRZ. The CMI signal is inverted and then passed to an EXCLUSIVE OR gate together with the $\frac{1}{2}$-bit delayed CMI signal. The output of the EXCLUSIVE OR gate is the input to a D flip-flop. The output of the D flip-flop is the NRZ required signal. The timing diagram in Fig. 5.22 shows the progress of the signal through the circuit. The final NRZ signal is $\frac{1}{2}$-bit delayed from the original CMI input.

Scrambling. In recent years, the scrambler circuit has been used in DMRs even in small-capacity systems (2 Mb/s). Its function is to transform repeated sequences of bits, such as long strings of zeros, into *pseudorandom* sequences. Pseudorandom, as discussed earlier, means almost (but not 100 percent) random. If a carrier is modulated with a random digital signal, the spectrum characteristics in the first lobe would be of the type $(\sin x)/x$. This would uniformly distribute the energy over the transmitted spectrum. If, instead, a periodic digital signal modulates the carrier, there will be concentrations of

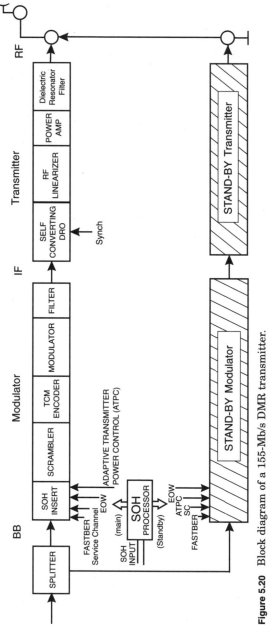

Figure 5.20 Block diagram of a 155-Mb/s DMR transmitter.

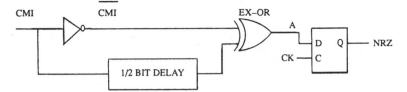

Figure 5.21 A typical CMI-to-NRZ converter.

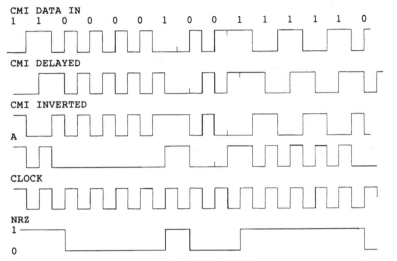

Figure 5.22 Timing diagram for the CMI-to-NRZ converter.

energy in the spectrum. These concentrations are known as *line spectra*. The lines appear at discrete frequencies depending upon the digital words, which keep repeating. This situation must be avoided because:

1. RF energy concentrations will increase the possibility of producing interferences in the adjacent RF channels.

2. Permanent lines in the spectrum can create serious problems in reception. The receiver VCO can lock onto a line in the spectrum instead of the incoming carrier, causing a loss of signal information.

3. In high-capacity DMRs an adaptive equalizer circuit is incorporated at the IF stage to restore the signal that has been impaired by selective fading. This circuit has been developed on the basis that a pseudorandom signal would be transmitted.

The scrambler/descrambler pair does have a disadvantage. Any errors that occur because of noise or interference during transmission (between the scrambler and descrambler) will be *multiplied* by the descrambler, thereby slightly worsening the S/N (typically by about 0.5 dB).

The principle of the scrambler can be described by observing Fig. 5.23. The truth table shows that the output of the second EXCLUSIVE OR gate, at y, is exactly the same as the input to the first EXCLUSIVE OR gate at a. This means that the output y is totally *independent* of the input c. So, provided c is the same in both halves of the circuit, the circuit can be split so that the *scrambler* is the left half of the circuit and the *descrambler* is the right half. The input c is made pseudorandom by using shift registers.

A simple three-flip-flop shift register pseudorandom circuit in Fig. 5.24 shows how the shift registers together with the EXCLUSIVE OR gate create an output that repeats itself after $2^n - 1$ words. This circuit when used in the configuration of Fig. 5.25 results in a three-bit scrambler circuit. If the periodicity of the a input is 2 (i.e., 01010101...), the repetition occurs after

$$(2^n - 1) \times \text{input periodicity} = (2^3 - 1) \times 2$$

$$= 14 \text{ words}$$

Figure 5.25 illustrates the circuit and truth table.

Scramblers used in present-day DMR equipment have more than three flip-flop shift registers. To ensure equipment compatibility from different manufacturers, the ITU-T has specified a scrambler circuit in Annex B of Recommendation G.954. A circuit similar to the ITU-T scrambler, shown in Fig. 5.26a, is made up of a shift register that uses 10 D flip-flops with feedback via EXCLUSIVE OR gates. It can be seen from the chart of Fig. 5.26b that the configuration of the output bits (*data*

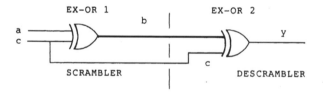

$$b = \bar{a}c + a\bar{c}$$
$$y = a$$

	a	c	b	y
TRUTH TABLE	0	0	0	0
	0	1	1	0
	1	0	1	1
	1	1	0	1

Figure 5.23   The principle of the scrambler.

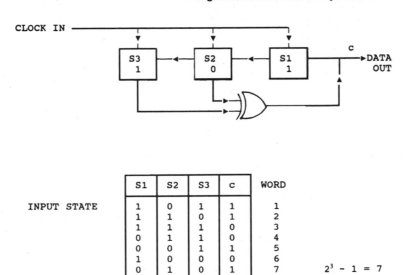

	S1	S2	S3	c	WORD
INPUT STATE	1	0	1	1	1
	1	1	0	1	2
	1	1	1	0	3
	0	1	1	0	4
	0	0	1	1	5
	1	0	0	0	6
	0	1	0	1	7
	1	0	1	1	
	1	1	0		

$2^3 - 1 = 7$
◄——— REPETITION HERE

Figure 5.24 Formation of the pseudorandom signal.

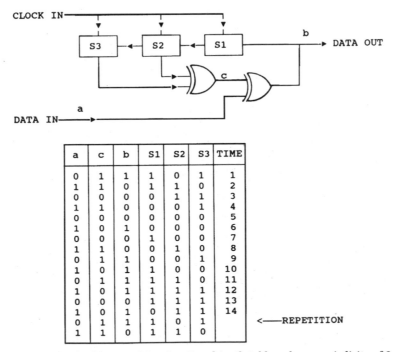

a	c	b	S1	S2	S3	TIME
0	1	1	1	0	1	1
1	1	0	1	1	0	2
0	0	0	0	1	1	3
1	1	0	0	0	1	4
0	0	0	0	0	0	5
1	0	1	0	0	0	6
0	0	0	1	0	0	7
1	1	0	0	1	0	8
0	1	1	0	0	1	9
1	0	1	1	0	0	10
0	1	1	1	1	0	11
1	0	1	1	1	1	12
0	0	0	1	1	1	13
1	0	1	0	1	1	14
0	1	1	1	0	1	
1	1	0	1	1	0	

◄——REPETITION

Figure 5.25 A 3-bit scrambler circuit and truth table; a has a periodicity of 2.

out) is not dependent only on the incoming bits (*data in*) but also on the Q-outputs of the flip-flops. Therefore, when switching on the equipment, the Q-outputs from the 10 stages of the shift register can assume any of the 2^{10} (1024) possible bit combinations. So, before any transmission, the shift register must be filled with 10 *real* bits. In other words, the first 10 bits cannot be recovered.

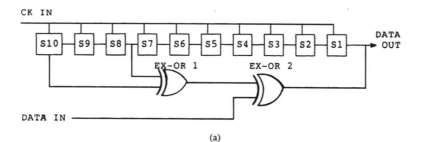

(a)

TIME	DATA IN	BIT OUT EX–OR 1	DATA OUT	REGISTER OUTPUTS									
				S1	S2	S3	S4	S5	S6	S7	S8	S9	S10
T0				1	1	0	0	1	0	0	0	1	1
T1	1	1	0	0	1	1	0	0	1	0	0	0	1
T2	1	1	0	0	0	1	1	0	0	1	0	0	0
T3	1	1	0	0	0	0	1	1	0	0	1	0	0
T4	0	0	0	0	0	0	0	1	1	0	0	1	0
T5	0	0	0	0	0	0	0	0	1	1	0	0	1
T6	1	0	1	1	0	0	0	0	0	1	1	0	0
T7	1	1	0	0	1	0	0	0	0	0	1	1	0
T8	0	0	0	0	0	1	0	0	0	0	0	1	1
T9	1	1	0	0	0	0	1	0	0	0	0	0	1
T10	0	1	1	1	0	0	0	1	0	0	0	0	0
T11	1	0	1	1	1	0	0	0	1	0	0	0	0
T12	1	0	1	1	1	1	0	0	0	1	0	0	0
T13	0	1	1	1	1	1	1	0	0	0	1	0	0
T14	0	0	0	0	1	1	1	1	0	0	0	1	0
T15	1	0	1	1	0	1	1	1	1	0	0	0	1
T16	0	1	1	1	1	0	1	1	1	1	0	0	0
T17	1	1	0	0	1	1	0	1	1	1	1	0	0
T18	1	1	0	0	0	1	1	0	1	1	1	1	0
T19	1	1	0	0	0	0	1	1	0	1	1	1	1
T20	1	0	1	1	0	0	0	1	1	0	1	1	1
T21	1	1	0	0	1	0	0	0	1	1	0	1	1
T22	1	0	1	1	0	1	0	0	0	1	1	0	1
T23	1	0	1	1	1	0	1	0	0	0	1	1	0
T24	1	0	1	1	1	1	0	1	0	0	0	1	1
T25	0	1	1	1	1	1	1	0	1	0	0	0	1
T26	1	1	0	0	1	1	1	1	0	1	0	0	0
T27	0	1	1	1	0	1	1	1	1	0	1	0	0
T28	0	0	0	0	1	0	1	1	1	1	0	1	0
T29	0	1	1	1	0	1	0	1	1	1	1	0	1
T30	0	0	0	0	1	0	1	0	1	1	1	1	0
T31	1	1	0	0	0	1	0	1	0	1	1	1	1
T32	1	0	1	1	0	0	1	0	1	0	1	1	1
T33	0	1	1	1	1	0	0	1	0	1	0	1	1
T34	1	0	1	1	1	1	0	0	1	0	1	0	1
T35	1	1	0	0	1	1	1	0	0	1	0	1	0
T36	0	1	1	1	0	1	1	1	0	0	1	0	1
T37	0	1	1	1	1	0	1	1	1	0	0	1	0

(b)

Figure 5.26 (*a*) Scrambler circuit; (*b*) scrambler timing chart.

Figure 5.27 Serial-to-parallel converter block diagram.

Serial-to-parallel conversion. The data information from the multiplexer is always transmitted in serial form. During modulation, the information must be processed in the parallel form. A serial-to-parallel conversion is therefore necessary. From Chap. 3 (Fig. 3.13), a 16-QAM signal requires the serial bit stream to be split into four parallel streams, 64-QAM requires six parallel streams, and 256-QAM requires eight parallel streams. The simplest serial-to-parallel converter transforms one serial bit stream into two parallel bit streams as shown in the example of Fig. 5.27. Fig. 5.28a shows a typical logic circuit for 64-QAM, one serial to six parallel streams conversion, together with its timing diagram in Fig. 5.28b. The $\overline{Q}$-output of FF1 is used as the input to FF2. This means that the incoming NRZ signal is delayed, inverted, and then presented to FF2. As the $\overline{Q}$-output from FF2 is passed on to FF8, the signal input to the D flip-flop FF8 is the same as the incoming NRZ signal except that it is now delayed by *two* clock periods. By a similar process, the output from FF3, which is the input to FF9, is the same as the incoming NRZ signal except that it is delayed by *three* clock periods. Also, the $\overline{Q}$-output from FF4, which is the input to FF10, is the same as the incoming NRZ signal except that it is delayed by *four* clock periods, etc., down to FF12. FF7 to FF12 are all clocked by a signal that is one-sixth of the rate of the clock input to FF1 to FF6. This enables the output of FF7 to be constructed from every sixth bit in the original NRZ bit stream. Similarly, FF8 is composed of every adjacent sixth bit; likewise for FF9 to FF12. The resulting outputs from the buffer amplifiers are parallel bit streams, thereby fulfilling the initial objective of separating the continuous input bit stream into six parallel output streams.

Differential encoding. It was previously stated that PSK and QAM types of modulation have the double sideband suppressed carrier (DSB-SC) characteristic. In other words, the carrier is not present in the received signal. A problem therefore arises. At the receiver, how does one establish the carrier phase as sent from the transmitter? This situation is further complicated because the phase delay introduced by the transmission medium is not constant because of atmospheric effects. To overcome this, the carrier is modulated with information by changing the phase of one state, depending on the previous state. At the receiving end, the phase in the present interval is compared to the phase in the previous interval. The signal received in the previous interval is delayed for one signal interval and is used as a reference to demodulate the signal in the next interval. This is differential encoding and decoding. *It allows detection without needing to know the absolute phase of the signal*

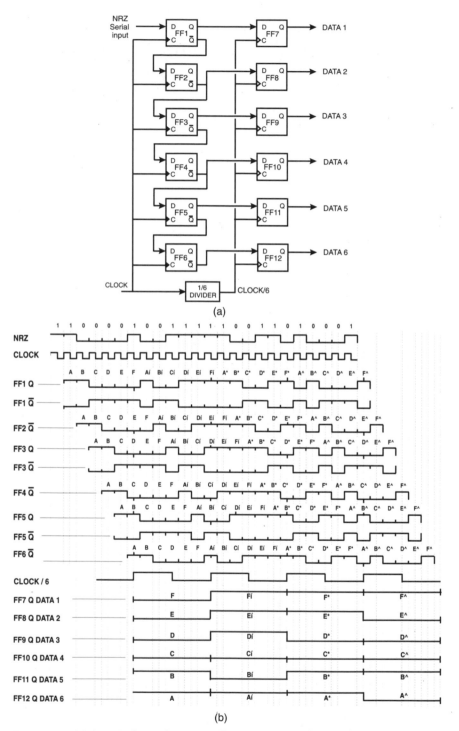

Figure 5.28 (*a*) A typical serial-to-6 parallel converter circuit; (*b*) serial-to-6 parallel conversion timing diagram.

The differential encoder forms a part of the modulator. The simplified block diagram of the 64-QAM with TCM modulator shown in Fig. 5.29 is an extension of the modulators described in Chap. 3. An incoming 155-Mb/s bit stream is split into *six* bit streams by a serial-to-parallel converter. Three of the bit streams are passed through one TCM encoder, and the other three bit streams are passed through another TCM encoder. Four outputs from each encoder are then mapped and filtered to form the *I* and *Q* signals. An LO then mixes with each *I* and *Q* signal. Because of the 90° phase shift in one side of the circuit, the resulting output is a modulated signal that has 128 states (i.e., 64-QAM-TCM, also written 128-TCM).

The differential encoder is placed between the serial-to-parallel converter and the low-pass filters. It incorporates a feedback mechanism that allows the phase associated with each pair of bits to be fixed depending upon the previous pair of bits. Figure 5.30*a* shows a differential encoder. Notice it operates on only four of the six bit streams. In the case of a 16-QAM modulator the two bit-stream pairs would be mapped into the combination of all possible binary word outputs. X1, X2, Y1, Y2, will represent the 16 possible states as shown in Fig. 5.30*b*.

The following is just one of several possible ways to differentially encode a bit stream. The input bits A1 and B1 are used to define in which of the four quadrants the output will lie. The input bits A2 and B2 will define the amplitude and therefore the exact location of the state within the quadrant.

First, observe the bits A1 and B1 in Table 5.1. The four possible combinations of these bits represent the phase with respect to the previous state. The outputs X1 and Y1 are related to the input A1 and B1 by the following equation, which includes the feedback bits H and K:

$$\phi X1Y1 = \phi HK + \phi A1B1$$

The outputs X2 and Y2 are related to the inputs A2 and B2 as in Table 5.2.

TABLE 5.1

A1	B1	$\Delta\phi$
1	1	0°
0	1	+90°
1	0	−90°
0	0	180°
H, X1	K, Y1	

TABLE 5.2

X1	Y1	Quad.	X2	Y2
0	0	3d	$\overline{A2}$	$\overline{B2}$
1	0	4th	A2	$\overline{B2}$
0	1	2d	$\overline{A2}$	B2
1	1	1st	A2	B2

NOTE: X1 and Y1 provide *quadrant* information and X2 and Y2 represent *level* (position in the quadrant) information.

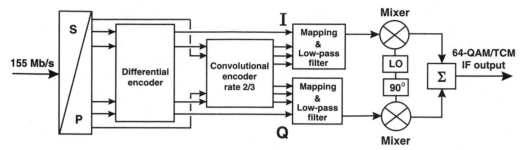

Figure 5.29   Block diagram of a 64-QAM modulator, with TCM.

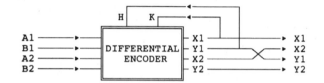

(a)

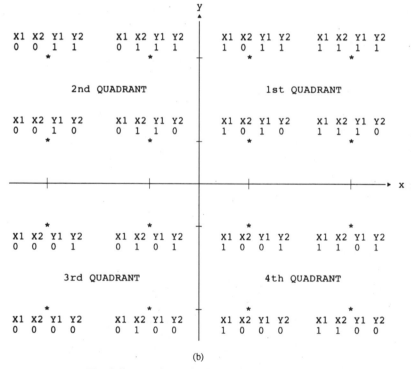

(b)

Figure 5.30   (a) The differential encoder block diagram; (b) differentially encoded states for 16-QAM.

As an example, suppose the following bit sequence is to be transmitted:

A1 B1 A2 B2

0 1 1 1   0101 0100 1011 0000 1101 0101 0100 0001 1001

The inputs and outputs are summarized in Table. 5.3.

Notice that H and K are the feedback bits, which are the same as the previous X1 and Y2 bits. 64-QAM could use a third feedback bit, say, L additional $\Delta\phi$ values of $\pm 45°$, $\pm 135°$, extra input bits C1, C2 and output bits Z1, Z2. Normally, however, manufacturers use just two out of the three input bits for the 64-QAM differential encoder because this can provide adequate receiver phase information. Consequently, the differential encoder circuit would be exactly the same for all levels of QAM at 16-QAM or above. The inputs and outputs follow the equation

$$\phi X1Y1 = \phi HK + \phi A1B1$$

For example, consider the first 4 input bits 0111. In this case, A1 = 0 and B1 = 1, so from Table 5.1 the phase change with respect to the previous state is $+90°$. Similarly, A2 = 1 and B2 = 1, so the phase is $0°$. Because the previous X1 and Y1 bits were both H = 0 and K = 0, the phase is $180°$.

Substituting this information into the equation

$$X1Y1 = 180° + 90°$$

$$= 270° \text{ or } -90°$$

From Table 5.3, X1 = 1 and Y1 = 0, which corresponds to a phase of $-90°$.

The output bits X1, X2, Y1, and Y2 are then filtered. Each pair (X1,X2 and Y1,Y2) represents a four-level PAM signal, having analog voltages +3, −3, +1,

TABLE 5.3 Transmission of Differentially Encoded Bits

Transmit side: $\phi X1Y1 = \phi HK + \phi A1B1$

Bit to be transmitted				Previous		Differentially encoded bit			
Quadrant		Level		X1, Y1 bit		Quadrant		Level	
A1	B1	A2	B2	H	K	X1	Y1	X2	Y2
0	1	1	1	0	0	1	0	1	0
0	1	0	1	1	0	1	1	0	1
0	1	0	0	1	1	0	1	1	0
1	0	1	1	0	1	1	1	1	1
0	0	0	0	1	1	0	0	1	1
1	1	0	1	0	0	0	0	1	0
0	1	0	1	0	0	1	0	0	0
0	1	0	0	1	0	1	1	0	0
0	0	0	1	1	1	0	0	1	0
1	0	0	1	0	0	0	1	1	1

−1. This signal is subsequently sent to the PSK section of the modulator for completion of the QAM process.

IF amplifier (and RF amplifier predistorter). The transmitter IF amplifier is usually very simple compared to the receiver amplifier. Its function is to supply the mixer with the modulated baseband signal at a frequency of typically 70, 140, or 240 MHz. The RF predistorter has been an important component in the DMR system. Its function was to purposely insert a nonlinearity into the IF signal that was the complement of the nonlinearity caused by the RF power amplifier. Ideally, the RF power amplifier should have an output power versus input power curve that is a perfect straight line (i.e., linear). Unfortunately, the harmonic content of the RF amplifier output power produces a graph that is slightly curved. This can result in serious degradation of the system performance if not corrected. Until recently nonlinearity compensation was rather difficult to accomplish at the microwave RF, so it was performed at the lower IF, which was considerably simpler to achieve. Nowadays RF linearizers are used as described in the next section.

RF components. The RF section is typically composed of the following components:

1. Mixer
2. Local oscillator
3. RF amplifier
4. Power control
5. Branching network
6. Waveguide run
7. Antenna(s)

These components are fabricated using a variety of "media." For example, in order to minimize losses and group delay distortion, filters are still made using waveguide cavities. The miniaturized stripline or microstrip integrated circuit form is inadequate because of higher losses. Similarly, the transmission line from the output amplifier to the antenna must have as low a loss as possible to ensure a low noise figure and to maximize system gain. Waveguides are therefore necessary for this purpose. However, the mixer, local oscillator, and amplifiers are all now made in the microwave integrated circuit (MIC) form. In fact, the RF amplifiers are currently at the peak of microwave technology since they have evolved from the hybrid technology to the *monolithic* technology. Monolithic MICs are smaller, more reliable, cheaper, and have better performance than the previous hybrid or discrete component structures.

Mixers. The mixer is part of the upconverter. It is usually made using hybrid MIC technology. This means the circuit is etched on a ceramic (Al_2O_3) substrate, and the diode devices are very small chips that are bonded into place on the substrate. Because of the miniaturization of these circuits, it is not possible to repair them in the field. They must be returned to the manufacturer if a fault occurs.

Local oscillators. The local oscillator, which is the other main component of the upconverter, is also fabricated in the miniaturized MIC construction. Local oscillators are usually designed using a GaAs FET transistor oscillator that is frequency stabilized by a dielectric resonator. The dielectric resonator makes the output frequency relatively insensitive to ambient temperature changes. The complete circuit is constructed in the microstrip circuit form and is small, cheap, and reliable.

The dielectric resonator oscillator (DRO) has recently replaced many upconverters. Traditionally, a local oscillator and IF signal feed into a mixer to produce a signal at $f_{LO} + f_{IF}$ for the upconverted output. The DRO, however, has self-converting qualities that provide upconversion with low phase noise and high stability (± 20 ppm).

RF amplifiers. The two main requirements for the RF power amplifier in a high-capacity DMR system are

1. High output power
2. Good linearity

Solid-state amplifiers are used almost exclusively instead of their older "tube" counterparts, the TWT and the klystron amplifier. It is the improvement in reliability that makes the solid-state amplifier so attractive. GaAs FET amplifiers can be made with output power values of approximately 10 W at the 1-dB compression point, but the emphasis is now on better linearity rather than higher power. Enhanced semiconductor device performance has improved receiver sensitivity, which allows a similar improvement in system gain to be achieved. In order to operate the GaAs FET amplifier at an adequate linearity, an output power *back-off* of several decibels is necessary. Until recently, to compensate for nonlinearity, adaptive *predistorters* were introduced, as already mentioned. GaAs FET power amplifiers with a 1-dB compression point of 10 W can reliably produce output power levels for 64-QAM in the 1-W range without predistortion, and in the 2-W range with predistortion.

Power control. Two advances in power control have been implemented over recent years: (1) adaptive transmitter power control and (2) RF linearization. While they fulfill different functions, they both form Tx-Rx-Tx feedback loops and both affect the RF solid state power amplifier.

Adaptive transmitter power control. This circuit is used to maintain the lowest power output that during normal operation provides adequate error performance. During fading conditions, the transmit power level is increased by up to 20 dB. It can also reduce power when constructive interference, often called upfading, increases the received signal level. As shown in Fig. 5.31, the received signal level value is fed back to the transmitting station via the SOH which is fed into the modulator at the receiving end. When this information is extracted from the transmit side demodulator, it is fed into the transmit power amplifier to increase or decrease the output power as required.

RF linearization. At the higher levels of QAM it is imperative that linearity be treated very seriously to obtain good error performance. The RF amplifier must operate well within its linear region and so minimize AM/AM distortion.

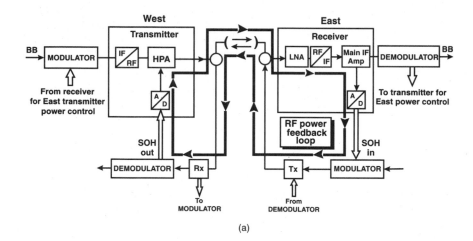

(a)

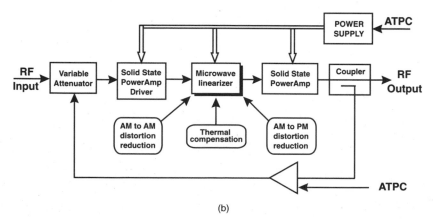

(b)

Figure 5.31 (*a*) Adaptive transmit power control feedback loop; (*b*) RF linearizer and power control.

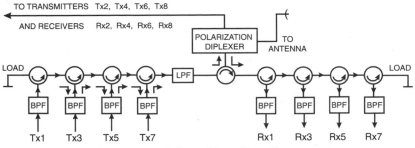

TO TRANSMITTERS Tx2, Tx4, Tx6, Tx8

AND RECEIVERS   Rx2, Rx4, Rx6, Rx8

Figure 5.32 Schematic diagram of channel branching filters.

Biasing the active devices (GaAs FETs) at low levels is the key. This means accepting a lower gain per stage, but monolithic microwave amplifier devices are inexpensive these days, and extra amplifier stages are easy to incorporate into the design. The biasing aspect of linearization is effectively a large variable back-off facility. In this respect it is part of the adaptive transmit power control loop and less back-off occurs during fading; thus we choose the lesser of two evils. AM/PM conversion is also unavoidable in amplifiers and this can be minimized by a feedback loop around the amplifier stages with a phase-shifting device such as a varactor in the loop (see Fig. 5.31b).

Branching networks. A branching network is necessary to allow several transmitters and receivers to operate on a single common antenna. An example of a branching network is shown in Fig. 5.32. Channel-branching filters are used to combine the various signal paths in the transmission section without mutual interference and to separate them again in the receiver. In modern DMR systems multiple-resonator bandpass filters are interconnected via circulators. Normally, the orthogonally polarized channel groups (1, 3, 5, 7 and 2, 4, 6, 8 in Fig. 5.32) are mutually displaced in frequency by one-half of the channel branching filter spacing in order to achieve adequate adjacent channel displacement (interleaved pattern). In digital systems, the orthogonally polarized channel groups can, however, also be operated on the same radio frequencies (cochannel pattern), thereby improving bandwidth utilization. The attenuation of the filters must be kept to a minimum. The metal cavity filters typically have a center band loss of 0.8 dB and the stripline circulators 0.1 dB.

Waveguide run. The most important aspect of the waveguide run is that it should have the lowest possible loss. This must be achieved with a minimum of reflected power. In the past, a rigid rectangular waveguide has been commonly used, with oxygen-free, high-conductivity copper being the preferred material. In the 6-GHz band, WR137 has a loss of approximately 6.7 dB per 100 m. Although a circular waveguide has lower loss than a rectangular waveguide, it has significant *moding* problems. The semiflexible elliptical waveguide has become very popular over recent years. It is very attractive because it is

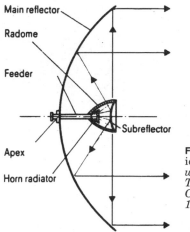

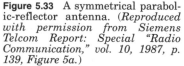

Figure 5.33 A symmetrical parabolic-reflector antenna. (*Reproduced with permission from Siemens Telcom Report: Special "Radio Communication," vol. 10, 1987, p. 139, Figure 5a.*)

supplied in continuous lengths on a drum. This enables easy installation. Bending is accomplished without the need for transitions, and its attenuation is even better than that of a standard rectangular waveguide. At 6 GHz, EW-59 has a loss of about 5.75 dB per 100 m. When correctly installed, it has excellent VSWR performance. However, it can be deformed easily, and must be handled with great care; even a small deformation (dent) can cause a mismatch that results in severe echo distortion noise.

Antenna(s). During the transition from AMR to DMR systems, more stringent requirements were demanded of antenna properties. The most important features in the design of today's antennas are:

- Higher sidelobe attenuation in all areas of the azimuth radiation pattern outside the main lobe

- Substantially improved cross-polarization properties

- Considerably increased bandwidth, allowing dual-band antennas that can be operated simultaneously in two frequency bands with two planes of polarization in each band

The construction of a typical present-day antenna is shown in Fig. 5.33. The electrical data for the parabolic-reflector type of antenna is shown in Table 5.4. When the multiband use of antennas is required, the shell type of construction is necessary. This is illustrated in Fig. 5.34.

5.2.2 DMR receive path

The diagram of Fig. 5.35 shows the building blocks of a typical 155-Mb/s DMR receiver. First, the incoming RF signal from the antenna and wave-

TABLE 5.4   Electrical Data for Microwave Parabolic Antennas

Frequency band (GHz)	Diameter (m)	Gain (dB)	3-dB beamwidth at midband (degrees)	Front-to-back ratio (dB)	Cross-polarization discrimination (dB)
3.580 – 4.200	3.0	39.6	1.83	76	36
	3.7	41.2	1.48	76	36
5.925 – 6.425	2.0	39.7	1.72	76	40
	3.0	43.6	1.15	85	40
	3.7	44.8	0.93	87	40
6.425 – 7.125	2.0	40.5	1.57	80	40
	3.0	44.2	1.05	85	40
	3.7	45.4	0.85	87	40
7.725 – 8.500	2.0	41.9	1.31	78	40
	3.0	45.7	0.88	85	40
	3.7	46.4	0.71	87	40
10.700 – 11.700	2.0	45.0	0.95	80	40
	3.0	48.8	0.63	90	40

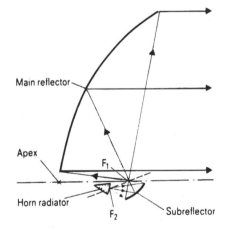

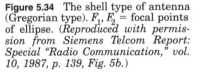

Figure 5.34 The shell type of antenna (Gregorian type). F_1, F_2 = focal points of ellipse. (*Reproduced with permission from Siemens Telcom Report: Special "Radio Communication," vol. 10, 1987, p. 139, Fig. 5b.*)

guide run is downconverted to the IF. After preamplification, the signal level is maintained as flat as possible across the IF band by the adaptive equalizer. It is then filtered, amplified, and delay equalized. The AGC circuit is incorporated at this point. Next is the demodulation process. This is followed by regeneration and timing extraction, after which the signal is differentially decoded and then parallel-to-serial converted. The signal is then descrambled, after which the service channels are extracted. Finally, the NRZ signal is converted into CMI for transmission to the demultiplexer for recovering the voice, data, or TV channels. The RF section has already been discussed, which leaves the following processes to be described:

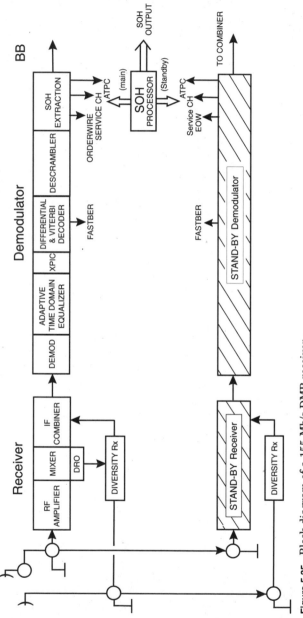

Figure 5.35 Block diagram of a 155-Mb/s DMR receiver.

1. IF in-phase combiner (for space diversity)

2. Demodulator

These circuit functions are basically the inverse of those discussed in the transmit path.

IF in-phase combiner (for space diversity). For a space diversity DMR, combining at the IF is usually the preferred technique. The purpose of the IF combiner is to combine the IF signals sent by the main mixer and the diversity mixer to offset any signal-level degradation caused by multipath fading. In-phase combining for the best case of equal levels and noise out of phase produces a 6-dB power level improvement and a 3-dB S/N improvement. Figure 5.36 illustrates one possible design for an in-phase IF combiner.

Two PIN diode attenuators, one on the main IF signal path and one on the diversity IF signal path, are driven by the AGC comparator circuit. The AGC comparator circuit continuously compares the RF signals received by the two mixers and drives the PIN diode attenuators as follows. There is no benefit in combining two signals when one of them has a high level of distortion. Therefore, should the difference between the received RF signals exceed, say, 10 dB, the attenuation introduced into the lower level IF signal is proportionately increased to ensure the combined signal is weighted in favor of the higher-quality signal. For example, a 14-dB difference between the IF signals might be designed to cause a 20-dB difference between the IF levels at the combiner input.

The phase shifter is the key component necessary for successful operation of the in-phase IF combiner. This exists only in the diversity mixer path of the space diversity receiver. It has an IF input amplifier that has two outputs, phase shifted by 180°. Each output feeds a coupler, so the four coupler outputs supply IF signals with relative phases of 0°, 90°, and 180°, 270° which are sent to four variable attenuators driven by a microprocessor, according to the following criteria:

■ At every instant, two of the four attenuators insert a large attenuation into the IF signal so that its contribution to the combiner is almost zero.

■ The two remaining attenuators are regulated to produce two output signals with amplitudes such that the resultant (vector sum) has the required phase.

Therefore, by selecting the two signals to be combined and by modifying the amplitudes of the selected signals, a resulting *constant amplitude* IF signal is obtained, and the phase can continuously shift between 0° and 360°. This

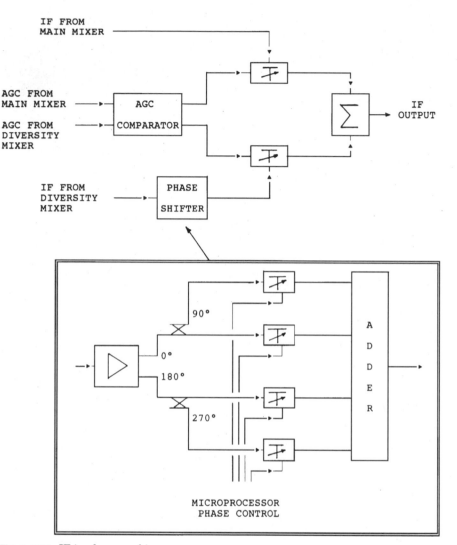

Figure 5.36 IF in-phase combiner.

allows the incoming main and diversity signals which have traveled different path lengths over the microwave hop to be combined on an in-phase basis. In addition, if one incoming signal is too low and is therefore excessively noisy, the combined output signal is weighted in favor of the better *noise quality* RF input signal.

An additional component can be included to detect IF-band amplitude distortion. Again, the signals to be combined are weighted, this time in favor of the lower dispersion signal. Sophisticated algorithms now control the combining process to minimize the power and dispersion effects of multipath fading. Maximum benefit of space diversity therefore results from this microprocessor-controlled scheme.

Demodulator. The demodulator primarily converts the 64-QAM IF signal to the baseband bit stream at 155 Mb/s. In addition, it contains the following circuits:

- Adaptive time domain equalizer [baseband transversal equalizer (Sec. 5.1.5)]
- Cross-polarization interference canceller (XPIC)
- Differential and Viterbi decoder
- Parallel-to-serial conversion
- Descrambler
- SOH extraction (for adaptive transmit power control and other service channel functions)
- NRZ to CMI encoding

Cross-polarization interference canceller. The XPIC in Fig. 5.37 reduces horizontal polarization interference on the vertically polarized signal, and vertical polarized interference on the horizontally polarized signal. This is accomplished by using a multi-tap (e.g., 11 taps) transversal filter. The principle is to generate a correction signal having an identical frequency response to the interfering signal which is used to subtract and therefore nullify the interference. This is done as part of the demodulation circuitry so that a correlation circuit in the modulator is used to feed the taps in the transversal filter. This whole process is very dynamic in that the interference changes rapidly, and the feedback circuit that adapts to nullify that interference must be even faster.

This device has become invaluable in doubling the capacity of DMRs, by allowing cochannel operation. The cross-polarization improvement factor is about 20 dB and this is added to the cross-polarization discrimination already achieved by the antenna.

Differential decoding. In the differential decoding process, the two PAM signals obtained in the demodulator are fed into a logic circuit that extracts bits X1, X2, Y1, and Y2, containing the differentially encoded information. From

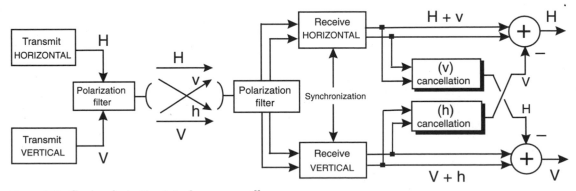

Figure 5.37 Cross-polarization interference canceller.

these bits, the originally transmitted bits A1, B1, A2, and B2 are recovered. The input and output bits are shown in Fig. 5.38 for the receive-side demodulator, using the same transmitted bit sequence example as in the *differential encoding* section. It is a useful exercise to convince oneself that the received bit sequence is correct by applying the equation and table in this figure.

Parallel-to-serial conversion. This process is accomplished by a simple circuit using flip-flops and gates. The six incoming parallel bit streams have one bit sequentially transferred from each stream to the output. This is assisted by a clock whose rate is six times that of each of the individual streams. The resulting serial output has a bit rate six times that of each input parallel stream.

Descrambling. The circuit of Fig. 5.39 is an example of a descrambler. Its operation can be observed by analyzing the timing sequence as shown in the timing chart (Fig. 5.39b). In order to be able to recover data from time T1, it is necessary to have the same bit sequence in the descrambler in T0 as when starting the scrambling process. A comparison of the descrambler and scrambler timing charts shows the only difference to be that the *data in* and *data out* bits are reversed. In other words the output from the descrambler is exactly the same as the input to the scrambler. This fulfills the initial objective.

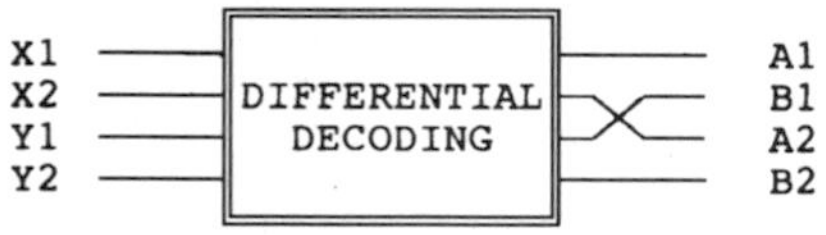

X1	Y1	QUAD	A2	B2
0	0	3rd	$\overline{X2}$	$\overline{Y2}$
1	0	4th	X2	$\overline{Y2}$
0	1	2nd	$\overline{X2}$	Y2
1	1	1st	X2	Y2

Diagram: X1, X2, Y1, Y2 → DIFFERENTIAL DECODING → A1, B1, A2, B2

Rx SIDE:	$\phi A1B1$	$=$	$\phi X1Y1$	$-$	ϕHK				
BITS RECEIVED				PREVIOUS X1,Y1		RECEIVED BIT			
X1	X2	Y1	Y2	H	K	A1	B1	A2	B2
1	1	0	0	0	0	0	1	1	1
1	0	1	1	1	0	0	1	0	1
0	1	1	0	1	1	0	1	0	0
1	1	1	1	0	1	1	0	1	1
0	1	0	1	1	1	0	0	0	0
0	1	0	0	0	0	1	1	0	1
1	0	0	0	0	0	0	1	0	1
1	0	1	0	1	0	0	1	0	0
0	1	0	0	1	1	0	0	0	1
0	1	1	1	0	0	1	0	0	1

SEQUENCE TRANSMITTED:
A1 B1 A2 B2
0 1 1 1   0101 0100 1011 0000 1101 0101 0100 0001 1001

Figure 5.38 Differential decoding.

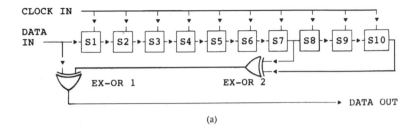

(a)

TIME	DATA OUT	BIT OUT EX-OR 2	DATA IN	S1	S2	S3	S4	S5	S6	S7	S8	S9	S10
								REGISTER OUTPUTS					
T0				1	1	0	0	1	0	0	0	1	1
T1	1	1	0	0	1	1	0	0	1	0	0	0	1
T2	1	1	0	0	0	1	1	0	0	1	0	0	0
T3	1	1	0	0	0	0	1	1	0	0	1	0	0
T4	0	0	0	0	0	0	0	1	1	0	0	1	0
T5	0	0	0	0	0	0	0	0	1	1	0	0	1
T6	1	0	1	1	0	0	0	0	0	1	1	0	0
T7	1	1	0	0	1	0	0	0	0	0	1	1	0
T8	0	0	0	0	0	1	0	0	0	0	0	1	1
T9	1	1	0	0	0	0	1	0	0	0	0	0	1
T10	0	1	1	1	0	0	0	1	0	0	0	0	0
T11	1	0	1	1	1	0	0	0	1	0	0	0	0
T12	1	0	1	1	1	1	0	0	0	1	0	0	0
T13	0	1	1	1	1	1	1	0	0	0	1	0	0
T14	0	0	0	0	1	1	1	1	0	0	0	1	0
T15	1	0	1	1	0	1	1	1	1	0	0	0	1
T16	0	1	1	1	1	0	1	1	1	1	0	0	0
T17	1	1	0	0	1	1	0	1	1	1	1	0	0
T18	1	1	0	0	0	1	1	0	1	1	1	1	0
T19	1	1	0	0	0	0	1	1	0	1	1	1	1
T20	1	0	1	1	0	0	0	1	1	0	1	1	1
T21	1	1	0	0	1	0	0	0	1	1	0	1	1
T22	1	0	1	1	0	1	0	0	0	1	1	0	1
T23	1	0	1	1	1	0	1	0	0	0	1	1	0
T24	1	0	1	1	1	1	0	1	0	0	0	1	1
T25	0	1	1	1	1	1	1	0	1	0	0	0	1
T26	1	1	0	0	1	1	1	1	0	1	0	0	0
T27	0	1	1	1	0	1	1	1	1	0	1	0	0
T28	0	0	0	0	1	0	1	1	1	1	0	1	0
T29	0	1	1	1	0	1	0	1	1	1	1	0	1
T30	0	0	0	0	1	0	1	0	1	1	1	1	0
T31	1	1	0	0	0	1	0	1	0	1	1	1	1
T32	1	0	1	1	0	0	1	0	1	0	1	1	1
T33	0	1	1	1	1	0	0	1	0	1	0	1	1
T34	1	0	1	1	1	1	0	0	1	0	1	0	1
T35	1	1	0	0	1	1	1	0	0	1	0	1	0
T36	0	1	1	1	0	1	1	1	0	0	1	0	1
T37	0	1	1	1	1	0	1	1	1	0	0	1	0

(b)

Figure 5.39 (a) An example of a descrambler circuit; (b) the descrambler timing chart.

NRZ to CMI encoding. The NRZ waveform is suitable for manipulating data within the radio equipment, but it cannot be transmitted in this form to or from the multiplexer. The reason is that *the NRZ signal spectrum contains a dc component.* When long sequences of 1s occur, dc wander produces errors caused by 1s being detected as 0s. Figure 5.40 illustrates how dc wander causes uncertainty

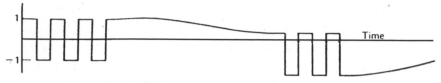

Figure 5.40 DC wander of an NRZ signal. (*From Bellamy, J., Digital Telephony. © 1982. Reprinted by permission of John Wiley & Sons, Inc.*)

in the position of the pulses relative to the dc level. This problem exists not only for long strings of 1s or 0s but also whenever there is an imbalance in the number of 1s and 0s. Consequently, the final step before transmitting the pulses to the multiplexer for demultiplexing is to convert the NRZ signal back to CMI in the case of a 155-Mb/s signal, or B3ZS in the case of 45 Mb/s or less. The circuit in Fig. 5.41a converts NRZ to CMI. The timing diagram of Fig. 5.41b shows the steps in the conversion process. FF1 delays the input data stream by one clock period. The stream is passed on to FF2 via an EXCLUSIVE OR gate whose second input is derived from a feedback loop around FF2. The inverted output of FF2 is fed to FF4, which delays the signal by one clock period before entering the upper NOR gate. The second input to the upper NOR gate is the inverted output of FF1 delayed one clock period by FF3. The output of the NOR gate at point A is the CMI 1s, which are part of the required CMI output. The CMI 0s are formed at point B by feeding into the lower NOR gate (1) the clock signal together with (2) the original data stream, delayed two clock periods by the double inversion of FF1 and FF3. Finally, the CMI 1s and 0s are summed by the OR gate to produce the complete CMI output at C.

5.2.3   DMR and the telecommunications management network

This has become a very involved and sophisticated subject now that networks are fully computerized and real-time QoS surveillance is possible. The intention here is to explain how the latest SONET/SDH systems provide the equipment status information for the overall telecommunications management network (TMN). This was always known as the *service channel* in PDH systems and this name is often retained in SONET/SDH systems.

Service channel methods. Service channel transmission is required for the efficient maintenance, performance monitoring, and control of digital transmission systems. This includes control, supervisory, and order-wire signal transmission. In order to have conversations between maintenance personnel at different locations, it is necessary to have a system in which it is easy to combine the main subscriber traffic with the service channel signal.

Furthermore, it is essential to continue maintenance conversations in the event of a failure of the main traffic signal. In PDH systems, there are several techniques for transmitting the service channel information together with subscriber traffic. These methods involve introducing extra bits into the main bit stream, usually done at the modulator.

SONET/SDH service functions. The newer SONET/SDH systems use the considerable overhead allotted for service channel use within the STS or STM frame. Remember, for the 155-Mb/s frame (Sec. 2.5) there is a path overhead

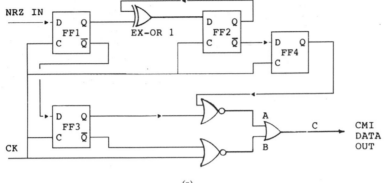

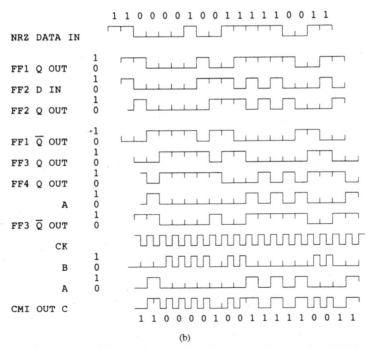

Figure 5.41 (*a*) CMI encoder circuit; (*b*) timing diagram for NRZ-to-CMI converter.

(POH) and a section overhead (SOH). The SOH is split into *regenerator* and *multiplexer* sections (Fig. 5.42). Note that each byte contains information at 64 kb/s, so a large amount of detail can be carried. Rows 1 to 3 (27 bytes) are for the regenerator and rows 5 to 9 (45 bytes) are for the multiplexer. While some bytes have been allocated to media-specific functions, there is latitude to use these bytes for other service channel activities until there is international consensus concerning their use. Some improved performance monitoring functions provide for in-service evaluation of quality of service. As indicated in Fig. 5.42, the protection switching control signals (K1, K2) each use 1 byte of 64 kb/s, and adaptive transmit power control and fast BER share one byte on a $3/4$ to $1/4$ ratio. The early warning, fast BER values of 10^{-12} and 10^{-8} are used as part of the protection-switching decision process. Notice there are bytes allocated for omnibus and express voice order-wire channels. TMN functions have a generous allocation of 3 bytes D1 to D3 (192 kb/s) for regenerator and 9 bytes, D4 to D12 (576 kb/s) for multiplexer equipment.

5.2.4 DMR with higher modulation levels

The 256-QAM DMRs have many similarities to the 64-QAM system. Theory dictates that the S/N for the 256-QAM system must be about 5 dB better than

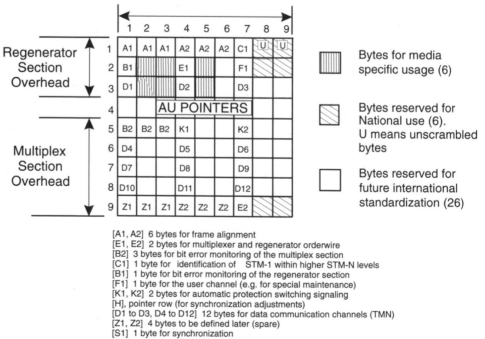

[A1, A2] 6 bytes for frame alignment
[E1, E2] 2 bytes for multiplexer and regenerator orderwire
[B2] 3 bytes for bit error monitoring of the multiplex section
[C1] 1 byte for identification of   STM-1 within higher STM-N levels
[B1] 1 byte for bit error monitoring of the regenerator section
[F1] 1 byte for the user channel (e.g. for special maintenance)
[K1, K2] 2 bytes for automatic protection switching signaling
[H], pointer row (for synchronization adjustments)
[D1 to D3, D4 to D12] 12 bytes for data communication channels (TMN)
[Z1, Z2] 4 bytes to be defined later (spare)
[S1] 1 byte for synchronization
[M1] 1 byte for section far-end block error reporting
6 bytes reserved for national usage
26 bytes reserved for future international standardization

Figure 5.42 SOH for DMR.

the 64-QAM system for a BER of 10^{-9}. This translates into requiring the components of the 256-QAM system to have a higher linearity.

The FEC technique is a very satisfactory way of improving the C/N without having to place very tight tolerances on the linearity of the components. In other words, the designer accepts that "dribble" errors will occur because of the C/N value being inadequate; then an error-correcting device is incorporated into the system to reduce these errors, thereby effectively improving the C/N. In a 256-QAM transmitter, the FEC encoder could be placed between the differential encoder and the modulator. At the receive side, the FEC decoder could be placed between the demodulator and the differential decoder.

The higher-level QAM systems should be designed for high linearity even though an FEC codec accounts for some degree of nonlinearity. The nonlinearities in the microwave portions of QAM radios are primarily the AM-to-AM and AM-to-phase-modulation (PM) conversions. AM-to-AM conversion can be defined as a device nonlinearity in which the gain of the device is dependent upon the input power. AM-to-PM conversion can be defined as a device nonlinearity in which the *transfer phase* of the device is dependent upon the input power.

Upconverters must have minimal AM-to-AM and AM-to-PM distortion. If a mixer is used, it must be operated at output levels well below the 1-dB compression point. The LO leakage is consequently large relative to the desired sideband signal. Additional filtering following the upconverter is therefore necessary.

The RF power amplifier is the main culprit in causing nonlinearity problems. A predistorter helps to correct nonlinearity created by the amplifier. But, in addition, power reduction (back-off) for 256-QAM radios is of the order 8 to 9 dB, giving an output power in the region of 20 to 30 dBm (about 100 mW to 1 W). The power reduction increases to 11 to 14 dB if no predistorter is used. Power amplifiers operating at 10-dB back-off have a poor dc-to-microwave conversion efficiency, typically less than 5 percent.

The LOs used in the up- and downconverters can be almost identical designs. For space diversity receivers it is convenient to use the same LO for each downconverter, with the power being split and fed to the respective mixers. The frequency variation and phase noise must be minimized for the LO. This can be done by temperature controlling a dielectric resonator FET oscillator (DRO) with a thermoelectric heater/cooler. This type of oscillator approaches the stability of a crystal oscillator with the advantages of lower cost and smaller size.

If cochannel operation is used, cross-polarization interference cancellers are essential for the higher-QAM systems. Using custom-built integrated circuit technology, these components are now inexpensive.

5.2.5 Multicarrier transmission

The introduction of this long-known technique into DMR systems has provided another countermeasure against multipath fading. Because multipath fading is highly frequency selective, when fading occurs in a multicarrier system some signals will be less affected than others. As Fig. 5.43 describes, the IF band is severely degraded in terms of amplitude dispersion during fading. A permissible 5-dB value across a single-carrier band translates to a total of 30 dB across the six IF bands of a multicarrier system.

5.3 Low-Capacity DMR

There is a large demand for low-capacity radio relay systems spanning short distances in the range of 2 to 20 km. The applications are mainly for rural area spur routes from a backbone link, or intracity hops between large company premises and exchanges. There are two different approaches that can be used to satisfy this objective: (1) low-microwave-frequency bands in the 1.5- to 2.5-GHz range, or (2) high-microwave-frequency bands in the 14- to 38-GHz range.

In the past, the low-frequency approach has been the most widely used method. Recent technological advancements at the higher-frequency microwave bands have increased the interest in this style of link. Despite the rain attenuation problems, the size and cost of these systems are becoming very attractive to network planners.

Because the gain of a parabolic antenna is proportional to $\log f^2$, in the Ku-band or Ka-band (about 12 to 26 GHz) the size of antennas becomes manageably small. For example, at 18 GHz a 1-m-diameter antenna has a gain of about 42 dB. Technology has recently improved to the point where high-quality microwave components are now readily available at 18 GHz and above. When these factors are combined with the fact that many large cities have heavily congested 2-, 4-, 6-, 12-, and 14-GHz frequency bands, the obvious next step is to move to higher operating frequencies.

As stated previously, the major drawback of the higher-frequency bands is the high rain attenuation. A rainfall analysis must be done to be aware of the margin available under the worst possible rainfall conditions. Rain is placed in the category of flat fading because, although the rain attenuation does increase with frequency, the variation over a relatively narrow radio band is

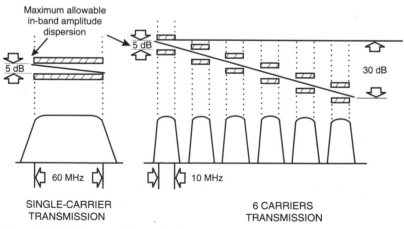

Figure 5.43 Multicarrier transmission.

very small. When the raindrop size approaches the wavelength of operation, the attenuation increases significantly. ITU-R Report 563 gives information concerning rainfall for various climatic zones that are characterized by a rain intensity or point rain rate (in mm/h) that is exceeded for 0.01 percent of the time. This is called $R_{0.01\%}$. Many studies have been pursued, and some observers have made rainfall measurements over periods of time up to 10 years or more. There is clearly merit in using their zone information when contemplating a new design, but the most valuable information is the rainfall information at the specific location being considered. Although guidelines can be followed and some characteristics apply to all locations, no two paths are the same. For example, raindrops become flattened as they fall, making the size of the horizontal dimension greater than the vertical dimension. This favors a vertically polarized propagation to minimize rainfall attenuation irrespective of location. However, the size and direction of a rain cell are very significant in determining rainfall attenuation. A tropical thunder cell will release rain well in excess of 100 mm/h over a distance of only 1 or 2 km for half an hour or less, whereas a temperate climate cloud overcast might release rain at about 20 mm/h, but over many kilometers for several hours. Hurricanes and cyclones are perhaps the least predictable because they release enormous amounts of rainfall for 24 h or more, but they might pass over a particular tropical location only once every 10 years. The prevailing wind can also play an important role in rain attenuation because if the rain cells tend to pass *across* a path, they will cause less disruption than if they proceed *along* a path. A nearby lake or mountain range can affect the rainfall pattern significantly.

Because of the enormous number of variables concerned, it is evident that the subject of how rain affects a radio link is not a very exact science. When all the data have been taken and manipulated to predict the maximum path length, the following rules of thumb seem to apply to the 18-GHz band. For an unavailability between 0.001 and 0.01 percent per year, radio hop lengths up to 20 km for an $R_{0.01\%}$ of 30 mm/h are allowed and up to 7 km for an $R_{0.01\%}$ of 60 mm/h. The picture looks rather bleak for tropical regions. Depending upon specific local conditions, the maximum hop length might be as low as 5 km or less in the 18-GHz band. This is rather unsatisfactory from a planner's point of view because some tropical regions have periods of several months where the rainfall is very low or even zero, during which time a 20-km hop would function very well. Unfortunately, during the rainy season such a link would be unavailable for several minutes almost every day.

5.3.1 RF channel arrangement

ITU-R Recommendation 595 indicates the channel arrangements for 51.84- and 155-Mb/s systems operating in the 17.7- to 19.7-GHz band. There are a total of 16 channels (8 pairs) for 155 Mb/s with a channel spacing of 110 MHz. For 51.84-Mb/s transmission, the band is subdivided to allow 70 channels (35 pairs) with a channel spacing of 27.5 MHz. This is similar to the 13- and 15-GHz bands which have a channel spacing of 28 MHz for 51.84-Mb/s transmission. The channel arrangements for 1.544 and 6.312 Mb/s can be decided at the discretion of each individual administration. A typical acceptable subdivision would provide a channel spacing of between 7 and 15 MHz for 6.3-Mb/s transmission, and between 2.5 and 5.5 MHz for 1.544-Mb/s transmission. Figure 5.44 shows the channel arrangement options for the North American usage of the 17.7- to 19.7-GHz band.

5.3.2 Modulation

4-PSK is usually the preferred modulation technique for low-capacity systems up to 51.84 Mb/s. In order to prevent interchannel interference by a 4-PSK modulator, spectrum limiting must be provided by RF or channel-branching filters. This is achieved at the expense of increased attenuation and complexity.

A significant cost saving and efficiency improvement can be attained by using the technique of directly modulating the RF band. The benefit is particularly appreciated at the higher RF frequencies, for example in the 18-GHz band. For 1.544- and 6.312-Mb/s systems (low capacity), the filter requirements are too stringent to allow 4-PSK direct modulation in the 18-GHz band. Direct FSK of the RF oscillator is a good solution for producing a low-cost, band-limited signal spectrum of acceptable efficiency. It is not even necessary to use a baseband filter with 4-FSK, because the modulated signal occupies a narrow enough bandwidth without filtering. Furthermore, the 4-FSK modulation has a relatively high immunity to gain nonlinearities. In addition, incoherent demodulation can be used in the receiver. This involves the use of a frequency discriminator, which is simpler and cheaper than the PSK demodulator because the oscillator specifications are more relaxed for 4-FSK. The S/N for incoherent demodulation is a little higher than for coherent modulation, but this is compensated by the elimination of the modulator attenuation.

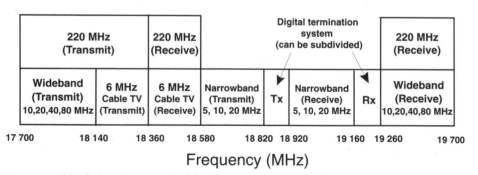

Figure 5.44 North American usage of the 17.7- to 19.7-GHz band.

5.3.3 Transmitter and receiver

Figure 5.45 shows the block diagram of a typical 18-GHz band transmitter and receiver. This system employs a transmitter that has 4-FSK direct modulation of the RF oscillator and incoherent demodulation in the receiver.

The B3ZS baseband signal is equalized, regenerated, and then converted into two parallel NRZ bit streams at half the incoming bit rate. The signals are then dejitterized, scrambled, and converted to four amplitude levels. The low-pass filter then limits the baseband spectrum, which subsequently modulates an RF oscillator to produce the 4-FSK output. The RF oscillator is voltage controllable and can be fabricated using a Gunn, HBT, or FET device, giving an output in the region of 100 mW (20 dBm). A PLL can be included for improved frequency stability.

At the receiving end, the RF signal is filtered and double downconverted via 790 MHz to the 70-MHz IF. The double downconversion allows the use of low-loss, relatively wideband RF filters. A low-noise amplifier is used between the two downconverters. The 70-MHz local oscillator can be either a surface-acoustic-wave, resonator-stabilized transistor oscillator, or a traditional crystal multiplied oscillator. The IF signal then passes through a bandpass filter for image frequency rejection and onto a limiter to provide a constant level signal to the frequency discriminator. The output from the discriminator is the baseband signal that passes through a low-pass filter prior to pulse recovery in the timing recovery and regenerator circuit. Also in the regenerator, the four-level, band-limited baseband signal is converted into two parallel signals, which are then descrambled. Finally, the two streams are combined and an NRZ-to-B3ZS conversion completes the receiver process.

Perhaps the most attractive feature of this system is the compactness of the design. The 0.6-m parabolic antenna can be mounted on a pole, together with the transmitter and receiver, which are placed in a weatherproof housing. This arrangement has the advantage of a very short waveguide run of 1 m or less. The cables into the weatherproof housing carry the baseband, service channel, power supply, and lightning protection. No expensive towers are necessary, and these systems can be installed very rapidly.

5.4 Performance and Measurements

DMR testing involves both in-service and out-of-service performance measurements in the RF, IF, and baseband sections of the system. It used to be that all tests were out-of-service tests. As microprocessors were introduced in the late 1980s and early 1990s, in-service measurements became possible. Now that SONET/SDH is installed, an array of in-service measurements can be made. First, out-of-service tests will be considered and this can apply to initial line-up after installation, or periodic maintenance tests. The usual additional features such as protection switching, service channel operation, supply voltages, and alarm conditions are also checked. The measurements in this chapter

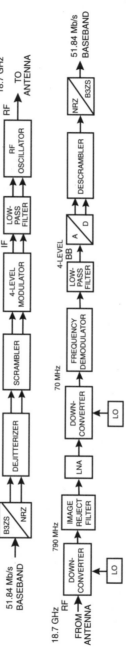

Figure 5.45 Schematic diagram of a typical 18-GHz band DMR transceiver.

refer to a typical 6-GHz band, 16-QAM DMR. In the RF section, the following tests are usually made:

1. Transmitter RF output power level
2. Transmitter RF output frequency
3. Transmitter local oscillator frequency
4. Transmitter local oscillator power level
5. Receiver local oscillator frequency
6. Receiver local oscillator power level
7. Receiver RF input power (and AGC curve)
8. Transmitter distortion level
9. Waveguide return loss and pressurization

In the IF section, the following tests are usually made:

1. IF input power level (to the mixer)
2. IF frequency and bandwidth
3. IF input power level (to the demodulator)
4. IF output spectrum
5. IF-IF frequency response and group delay
6. Demodulator eye pattern
7. Absolute delay equalization (on space diversity receivers)

At the baseband, the following measurements are usually made on the received bit stream before demultiplexing:

1. Error analysis
2. Jitter analysis
3. Constellation analysis

5.4.1 RF section out-of-service tests

Power and frequency measurements. RF power level and frequency measurements are made on the transmitter and receiver simply by using an RF power meter and frequency counter at the Tx output and Rx input. Usually, an RF coupler is provided for this purpose, with a coupling factor of typically 30 dB. The transmitter output power level is mostly in the range of 0.1 to 10 W (20 to 40 dBm) so, in addition to the coupler, an attenuator is sometimes required to reduce the level to ensure that no damage occurs to the power meter or frequency counter.

The LO in the up- and downconverters is usually less than 0 dBm, so excessive power is of no concern here. For the LO frequency measurements, very accurate values are required, because the frequencies are fixed to a very high tolerance. For example, a frequency might have a tolerance of ±30 ppm, which at 6 GHz is only ±0.18 MHz (or ±180 kHz).

In the receiver, the so-called AGC curve is plotted. This requires measurement of the IF output level as the input RF power level is varied from about −25 dBm down to the threshold level. The IF output level should remain constant until about −70 dBm or so, after which an abrupt drop in level is observed as the threshold level is approached.

Transmitter distortion level. The RF sections of *analog* and *digital* radios primarily differ in the required linearity. As already stated, the digital radio transmitter must be very linear compared to an analog radio. This is necessary because, for example, the 16-QAM has both amplitude and phase components. If the transmitter amplifiers are not extremely linear, AM-to-AM and AM-to-PM conversion can cause serious intersymbol interference and subsequent errors. The higher-order QAM systems such as 64- and 256-QAM require even higher levels of linearity. There are two tests that can be used to check the transmitter distortion level; their descriptions follow.

Using a digital signal. The pattern generator is used to send a CMI signal with a pseudorandom bit sequence (PRBS) of $2^{23} - 1$ periodicity to the modulator input. A spectrum analyzer then monitors the RF output from the transmitter amplifier. Figure 5.46 shows the test setup and expected response. The first sidelobe should be attenuated by a certain minimum value depending on the specifications of individual administrations. As a guideline, for good performance of a 155-Mb/s DMR, the first sidelobe should be at least 37 dB down at ±25 MHz from f_0. If the transmitter is out of specification, a linearization adjustment would be necessary to correct the problem.

Using the two-tone method. Two frequency generators are used to send IF signals at 130 and 150 MHz to the transmitter upconverter (as in the test setup of Fig. 5.47). The spectrum analyzer is connected to the output of the transmitter RF amplifier. The spectrum analyzer displays the two tones, together with the intermodulation products. The level of the intermodulation products is a direct indication of the linearity of the transmitter. Typically, these intermods should be at least 45 dB down from the two input tones. As in the previous method, if the transmitter is out of specification, adjustments can be made while observing the spectrum analyzer display.

Waveguide return loss and pressurization. On completion of installation, the return loss of the waveguide and antenna system is a very important measurement. If the return loss is inadequate, *echo distortion* results in a group delay problem. The return loss should ideally be better than 26 dB.

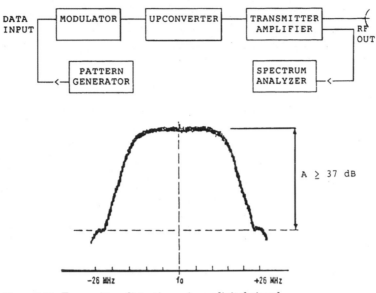

Figure 5.46 Transmitter distortion using a digital signal.

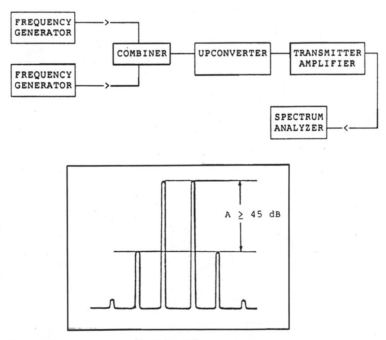

Figure 5.47 Transmitter distortion using two tones.

When the waveguide and antenna section has been commissioned, the waveguide must be maintained in a pressurized condition. This is necessary because otherwise moisture would condense in the waveguide and degrade the return loss. It is therefore an important part of the maintenance process to closely monitor the pressure of the gas in the waveguide. Usually, the pressurization mechanism is connected to an alarm so that a drop in pressure below the acceptable value alerts the technician to provide the necessary maintenance.

5.4.2   IF section out-of-service tests

Power and frequency measurements. These measurements simply involve the use of a power meter and frequency counter, as in the RF power and frequency measurements. In this case, the power levels are relatively low, so there is no danger of damaging the instruments with excessive input power.

IF output spectrum. A PRBS pattern $(2^{23} - 1)$ is sent to the modulator input in the NRZ plus clock format at a level of 1 V$_{pp}$/75 Ω. The spectrum analyzer is connected to the modulator output to display the output spectrum. It is recommended that the output should comply with the mask shown in Fig. 5.48.

IF-IF frequency response and group delay. The IF amplitude should be as flat as possible over the bandwidth of operation (e.g., < 0.5 dB at 140 ±20 MHz). The IF amplitude versus frequency response can be measured using the microwave link analyzer (MLA) test setup in Fig.5.49a. With the same setup, the group delay can be displayed on the MLA receiver. The group delay should

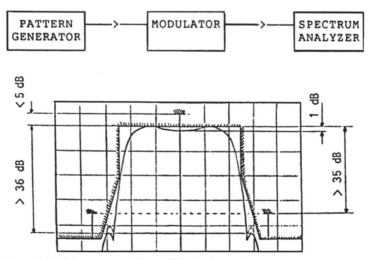

Figure 5.48   Measurement of the IF output spectrum.

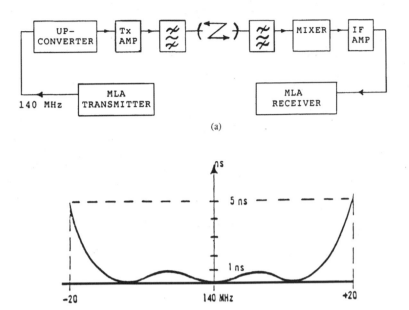

Figure 5.49 (*a*) IF-IF frequency response and group delay test setup; (*b*) group delay characteristics.

typically be within the specifications suggested in Fig. 5.49*b* (i.e., < 1 ns at 140 ±18 MHz, and < 4 ns at 140 ±20 MHz).

A group delay that is out of specification by 2 or 3 ns can be corrected by adjusting the IF amplifier module. If it is out by more than 3 ns, a microwave filter might need retuning.

Absolute delay equalization. In a space diversity system, it is necessary to ensure that the signals coming from each antenna reach the two receivers simultaneously. Otherwise, *in-phase combining* is not possible. The path lengths of the two signals must therefore be equalized prior to combining the two signals. This is done during installation, usually by inserting an appropriate length of coaxial cable in one of the two paths at the output from the downconverter (IF) point. This is not a maintenance test, but if any rearrangement or repositioning of the equipment is made during its lifetime, the delay equalization must be reestablished.

5.4.3 Baseband tests

Error analysis. On PDH radio equipment, the BER is one of the most frequent measurements a technician or engineer has to perform on a DMR system. This is the measurement that defines transmission quality and acceptability. In other words, if the BER is too high, the system is declared failed or unavailable. As stated previously, a link will be affected by several types of distortion (e.g., thermal noise, RF interference, group delay, phase jitter, etc.). The result of these combined distortions is observed as the reception of "false bits," or errors, at the receiving end. It is important to know the rate at which the errors occur, because if they are excessive, the communication link is disrupted. If the errors exceed a critical value, all calls are dropped. This situation is more severe compared to the analog system, where excessive noise impairs the audibility of the speaker but does not drop the call. Fortunately, good digital equipment design ensures that link disruption is a rare occurrence.

The latest international error performance objectives are stated in ITU-T Recommendation G.826. This is summarized in Chap. 4. Over a long period of time (e.g., a year) a link is designed to provide availability as high as possible; 100 percent availability is clearly not possible. The term *seriously errored seconds* is now an important parameter in performance evaluation in order to make the distinction between the occurrence of random errors and error bursts. BER measurements are still made on the DMR station PDH equipment, but the new SDH equipment is enforcing tighter BER specifications. For example, nowadays, BER is used to give an early warning of degradation. Depending upon the manufacturer, the SDH equipment measurements are used to provide switching criteria if the error ratio becomes worse than, say, 1 $\times$ 10^{-8} or 1 $\times$ 10^{-12}. This is a degraded state and might have an alarm indication. This is in comparison with a PDH measurement, where a degraded BER alarm would be activated if the error ratio drops below, say, 1 $\times$ 10^{-6} or 1 $\times$ 10^{-5}, depending on the required specifications. Some PDH equipment manufacturers provide the facility for observing the actual error ratio in addition to the alarm requirements. This is very useful because complete digital links or portions of a link can then be evaluated by using the loopback mode. If the error ratio cannot be read on an LED display at the DMR station, the measurement can be made by using a test set to introduce a pseudorandom digital bit stream at one terminal station, looping back at the desired receiving terminal station, and feeding the returned bit stream into the test set to evaluate the BER.

For PDH equipment performance evaluation, it is very useful to relate the BER to the S/N or the C/N. DMRs have a BER that increases rapidly as the receiver threshold is approached (Fig. 5.50). In fact, the receiver threshold for a PDH DMR used to be and often still is defined as the RF input power that degrades the BER to 1 $\times$ 10^{-3}. In QAM systems, because of the amplitude and phase characteristics, it is not possible to measure the carrier peak power C. However, the mean carrier (signal) power S can easily be measured at the demodulator input. For 16-QAM, S is approximately 2.5 dB lower than C. The received power P_r is related to the normalized S/N as follows:

$$P_r = kTB + \text{NF} + (\text{S/N})_n \quad \text{dBm}$$

where k = Boltzmann's constant = 1.38×10^{-23}
T = absolute ambient temperature (about 298 K)
B = equivalent noise bandwidth
NF = receiver system noise figure

So, for a 1-MHz bandwidth,

$$kTB = -114 \text{ dBm}$$

$$P_r = -114 + 10 \log B_s + \text{NF} + (\text{S/N})_n$$

where B_s = the symbol rate (MHz)
= bit rate/number of bits forming a symbol
= B_r / n

For example, for a 155-Mb/s system $n = 4$, so $B_s \cong 40$ MHz. Also, if NF = 4,

$$P_r = -114 + 10 \log 40 + 4 + (\text{S/N})_n$$

$$P_r = -90.0 + (\text{S/N})_n$$

From this equation, the received power can be calculated if the normalized S/N is known, or vice versa. (*Note:* The normalized S/N $[(\text{S/N})_n] = (\text{S/N}) \cdot (\text{Rx 3-dB}$ noise bandwidth)/B_s.)

BER measurement. The method in Fig. 5.51, which uses an RF variable attenuator inserted at the input to the receiver, is the most suitable for training

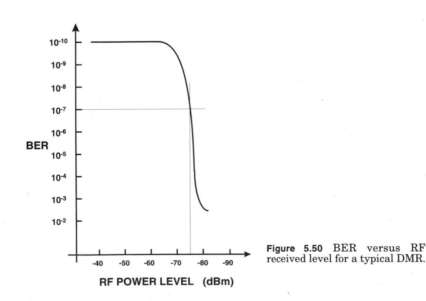

Figure 5.50 BER versus RF received level for a typical DMR.

purposes. However, in the field, it is probably more convenient to use a noise generator to degrade the S/N. The attenuator is adjusted to provide the required BER value, after which the received input power to the radio is measured. Following a sequence of such measurements, a graph can be plotted of P_r against BER. The curve is then compared with the manufacturer's installation curve or factory-recommended curve to see if any degradation has occurred. Corrective maintenance is necessary if the system has deteriorated significantly.

Test equipment for block-based error measurements is gradually being introduced. The errored second ratio, seriously errored second ratio, and background block error ratio, etc., subsequently provide a greater depth of error analysis. Improved quality-of-service requirements now demand careful control of all sources of errors. The introduction of SDH equipment has been instrumental in upgrading DMR performance. The SDH service channel measurements allow automatic in-service monitoring of the above parameters as part of the TMN surveillance.

Demodulator eye pattern. The eye pattern check is a useful, quick, *in-service, qualitative* method of observing the digital signal transmission quality, particularly useful for PDH equipment. In the absence of a BER instrument, an experienced engineer or technician can quickly evaluate the eye pattern and decide whether or not the quality is adequate. A constellation pattern check is another qualitative check that provides considerably more information regarding the source of equipment faults. Nevertheless, the eye pattern has some benefit. The eye pattern is displayed on an oscilloscope by synchronizing the oscilloscope with the demodulator clock and connecting the X and Y axes of the oscilloscope to the two eye pattern test points (I and Q) provided in the demodulator. The eye patterns of Fig. 5.52 show the deterioration of the eye as the BER degrades. Figure 5.52a is the *normal* condition (BER < 10^{-10}). When the BER = 10^{-6}, the eye opening starts to become less clearly defined. By the time the BER has increased to 10^{-3} (the PDH failure condition), the eye is very badly distorted. Incidentally, the number of eye openings depends on the level of modulation. The example shows three pairs of openings for 16-QAM. A 64-QAM signal would have seven pairs of openings.

Jitter analysis. Jitter is defined as the short-term variations of the significant instants (meaning the pulse rise or fall points) of a digital signal from their

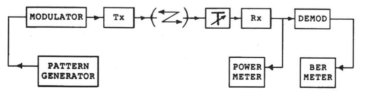

Figure 5.51 BER measurement.

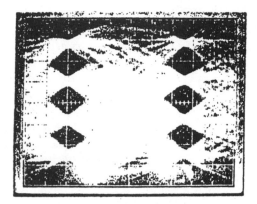

(a)

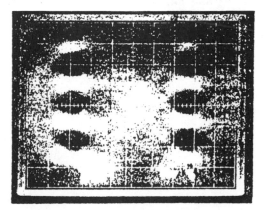

(b)

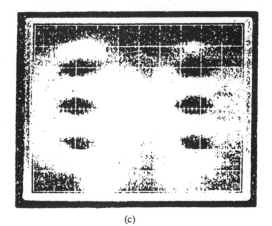

(c)

Figure 5.52 Eye pattern for various BER values.
(a) BER almost zero; (b) BER = 1×10^{-6}; (c) BER
= 1×10^{-3}.

ideal positions in time. Figure 5.53 illustrates the definition of jitter more clearly. The main jitter-creating culprits are regenerators and multiplexers. In regenerators, the timing extraction circuits are the sources of jitter and, because a transmission link has many regenerators, jitter accumulates as the signal progresses along the link. The PDH type of multiplexer uses the justification technique as described in detail in Chap. 2. Jitter is generated in the justification process and also because of the waiting time between the justification decision threshold crossing and the justification time slot. Furthermore, the newer SDH multiplexers are not immune to jitter generation. The pointer processing technique used in these multiplexers introduces jitter into the payloads they carry.

The extent of jitter acceptance depends on the hierarchical interface. That simply means the output bit rate. The ITU-T has defined maximum permissible jitter values in Recommendations G.823, G.824, and G.825. Table 5.5 shows the values for the 1.544-, 6.312-, 32.064-, 44.737-, and 97.727-Mb/s North American interfaces; the 2.048-, 8.448-, 34.368-, and 139.264-Mb/s European interfaces; and the STM-1, -4, and -16 global SDH interfaces. There are three categories of jitter measurement:

- *Maximum output jitter.* This is the value observed when an input bit stream of known jitter content is applied.

- *Maximum tolerable input jitter.* This is tested by superimposing jitter onto an input bit stream and determining the point at which errors start to occur.

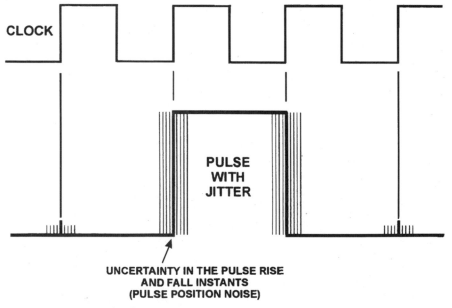

Figure 5.53 An example of jitter.

TABLE 5.5   Maximum Permissible Output Jitter

Bit rate (Mb/s)	1/(bit rate) = unit interval (UI_{pp}) (ns)	Network limit B_1 (UI_{pp})	B_2 (UI_{pp})	Measurement filter bandwidth — Bandpass filter having lower cutoff frequency f_1 and f_3 and an upper cutoff frequency f_4		
				f_1 (Hz)	f_3 (kHz)	f_4 (kHz)
			European (PDH)			
0.064[*]	15600	0.25	0.05	20	3	20
2.048	488	1.5	0.2	20	18	100
8.448	118	1.5	0.2	20	3	400
34.368	29.1	1.5	0.15	100	10	800
139.264	7.8	1.5	0.075	200	10	3500
			North American (PDH)			
1.544	648	5.0	0.1	10	8	40
6.312	158	3.0	0.1	10	3	60
32.064	31.1	2.0	0.1	10	8	400
44.736	22.3	5.0	0.1	10	30	400
97.728	10.2	1.0	0.05	10	240	1000
			Global (SDH)			
155.52	6.43	1.5	0.15	500	65	1300
622.08	1.61	1.5	0.15	1000	250	5000
2488.32	0.40	1.5	0.15	5000	†	20,000

[*] For the codirectional interface only.
† A value of 1 MHz has been suggested.

- *Jitter transfer function.* This is a measure of how the jitter is attenuated (or amplified) by passing through the system. This measurement is necessary to assess any jitter accumulation within a link.

Jitter can usually be measured with the same transmission analyzer test setup used for the BER measurements.

ITU-T Recommendations G.824 and G.825 have been established to control jitter and "wander" within digital networks. They also provide equipment designers with guidelines to ensure equipment performance reaches certain minimum standards of jitter and wander. Table 5.6 gives these Recommendations' jitter and wander tolerance of equipment input ports for North American and European PDH and global SDH networks.

The North American equivalent jitter specifications are similar to those of the ITU-T and are found in Bell Technical References 43501 and 43806. They concern the jitter tolerance, jitter transfer, and jitter generation requirements for both the North American and European hierarchies.

Higher-level QAM signals are extremely susceptible to degradation in the C/N caused by carrier and timing phase jitter. Figure 5.54 shows the graph of how the C/N degrades as the carrier phase jitter increases for several levels

TABLE 5.6   Input Port Jitter and Wander Tolerance

Bit Rate (Mb/s)	Jitter amplitude peak-to-peak			Frequency					Test signal
	A_0† (UI)	A_1 (UI)	A_2 (UI)	f_0 (Hz)	f_1 (Hz)	f_2 (Hz)	f_3 (kHz)	f_4 (kHz)	
European (PDH)									
0.064*	1.15 (18 μs)	0.25	0.05	1.2×10^{-5}	20	600	3	20	$2^{11} - 1$
2.048	36.9 (18 μs)	1.5	0.2	1.2×10^{-5}	20	2400	18	100	$2^{15} - 1$
8.448	152.5 (18 μs)	1.5	0.2	1.2×10^{-5}	20	400	3	400	$2^{15} - 1$
34.368	618.6 (18 μs)	1.5	0.15	Under study	100	1000	10	800	$2^{23} - 1$
139.264	2506.6 (18 μs)	1.5	0.075	Under study	200	500	10	3500	$2^{23} - 1$
North American (PDH)									
1.544	27.8 (18 μs)	5.0	0.1	1.2×10^{-5}	10	120	6	40	$2^{20} - 1$
6.312	113.9 (18 μs)	5.0	0.1	1.2×10^{-5}	10	50	2.5	60	$2^{20} - 1$
32.064	578.8 (18 μs)	2.0	0.1	1.2×10^{-5}	10	400	8	400	$2^{20} - 1$
44.736	807.2 (18 μs)	5.0	0.1	1.2×10^{-5}	10	600	30	400	$2^{20} - 1$
97.728	1764.7 (18 μs)	2.0	0.1	1.2×10^{-5}	10	12000	240	1000	$2^{23} - 1$

Global (SDH)

Bit rate (Mb/s)	Peak-to-peak amplitude (unit interval)					Frequency (Hz)									
	A_0 (18 μs)	A_1 (2 μs)	A_2 (0.25 μs)	A_3	A_4	f_0	f_{12}	f_{11}	f_{10}	f_9	f_8	f_1	f_2	f_3	f_4
155.52 (STM-1)	2800	311	39	1.5	0.15	12μ	178μ	1.6m	15.6m	0.125	19.3	500	6.5k	65k	1.3M
622.08 (STM-4)	11200	1244	156	1.5	0.15	12μ	178μ	1.6m	15.6m	0.125	9.65	1000	25k	250k	5M
2488.32 (STM-16)	44790	4977	622	1.5	0.15	12μ	178μ	1.6m	15.6m	0.125	12.1	5000		Under study ‡	20M

* For the codirectional interface only.

† The value for A_0 (18 μs) represents a relative phase deviation between the incoming signal and the internal local timing signal derived from the reference clock. This value A_0 corresponds to an absolute value of 21 μs at the input to a node (i.e., equipment input port) and assumes a maximum wander of the transmission link between the two nodes of 11 μs. The difference of 3 μs allows for the long-term phase deviation in the national reference clock [Recommendation G.811, Sec. 3(c)].

‡ A value of 1 MHz has been suggested for the f_3 frequency; f_2 can be derived from f_3 (using slope of -20 dB/decade).

of QAM and a BER equal to 10^{-6}. The 256-QAM system has several decibels of power penalty even when the carrier jitter is only 0.5° to 1°. Such systems require very tight jitter control. Similarly, Fig. 5.55 illustrates the C/N degradation as the timing phase jitter increases. Again, the higher the QAM level, the higher the power penalty for a given BER. For the 16-QAM case, the curves are drawn for several values of filter roll-off factor. It is interesting to see that the less steep roll-off value of $\alpha = 0.7$ has a better jitter characteristic than the more expensive, steeper $\alpha = 0.3$ roll-off filter. Remember, if α is too large, the spectral emission mask specification might be exceeded.

Constellation analysis. The constellation display provides a very useful technique for *in-service* evaluation of a DMR system and is a diagnostic tool for fault location. It is a simple measurement to perform, requiring the connection of the I, Q, and symbol-timing clock monitor points in the demodulator to a constellation display unit. Whereas the eye diagram indicates a problem only by eye closure, the constellation can often identify the source of a problem by the type of irregularity observed in the pattern. Figure 5.56 illustrates the constellation pattern and the eye diagram for a 16-QAM radio under *normal operating conditions*. Other higher levels of QAM are also applicable, but the 16 states are more manageable for illustration purposes. The 16 states of the constellation are correctly placed on the rectangular grid, and the size of each dot representing each state indicates low thermal noise and low intersymbol

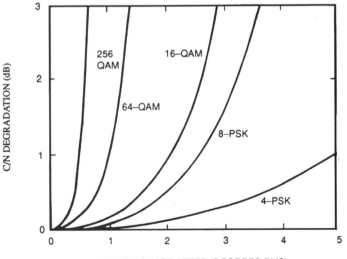

CARRIER PHASE JITTER (DEGREES RMS)

Figure 5.54 Graph of C/N degradation against carrier phase jitter. (*Reproduced with permission from IEEE © 1986. Noguchi, T., Y. Daido, and J. A. Nossek, Modulation Techniques for Microwave Digital Radio, IEEE Communications Magazine, vol. 24, no. 11, September 1986, pp. 21-30.*)

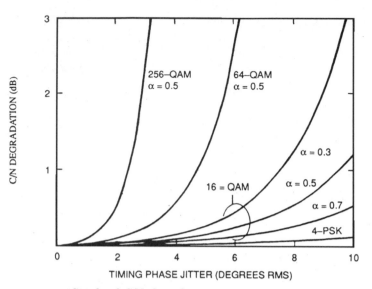

Figure 5.55 Graph of C/N degradation against timing phase jitter. *(Reproduced with permission from IEEE © 1986, Noguchi, T., Y. Daido, and J. A. Nossek, Modulation Techniques for Microwave Digital Radio, IEEE Communications Magazine, vol. 24, no. 11, September 1986, pp. 21-30.)*

interference. The corresponding eye diagram has a closure of approximately 10 percent with negligible angular displacement.

If the 16-QAM radio develops a fault, the constellation display, in many instances, can quickly identify the source of the problem. The following faults will be discussed:

- Degradation of C/N
- No input signal
- Sinusoidal interference
- Modem impairments
  - I and Q carriers not exactly 90°
  - Phase-lock error in the demodulator
  - Amplitude imbalance
  - Demodulator out of lock
- Phase jitter
- Transmitter linearity
- Amplitude slope

Figure 5.57 shows how the displays are modified when a radio has a *degraded C/N* (20 dB). The constellation states are enlarged and the eye closure is more pronounced (about 20 percent). The measurement of a higher BER would also confirm the degradation. In this case, the constellation pattern and the eye diagram provide the same information, or diagnostic result. For the case of no input signal at all, the constellation display produces a large cluster of random points (Fig. 5.58). The eye diagram would show complete eye closure for this condition without indicating the reason for the radio failure. Figure 5.59 shows a radical difference between the two displays. The small circles, or "doughnuts," generated for the constellation states are characteristic of a radio affected by a *sinusoidal interfering tone* such as the strong carrier component of an FM radio. Note that the deterioration of the eye closure provides degraded BER information only, without indicating the reason for the degradation.

Several modem impairments can be observed on the constellation diagram. First, Fig. 5.60 shows the display that would be observed when the I and Q carriers are not exactly 90° to each other. The states form the outline of a parallelogram. Figure 5.61 shows the result of a *phase-lock error* in the demodulator carrier-recovery loop. The constellation states are rotated about their normal positions, which is characteristic of the phase-lock problem.

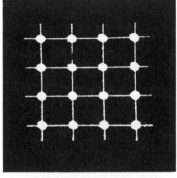
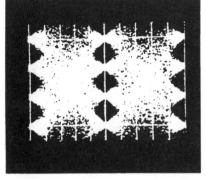

(a)   (b)

Figure 5.56 Constellation plot for a 16-QAM radio (normal operation). (*a*) Constellation plot; (*b*) eye diagram.

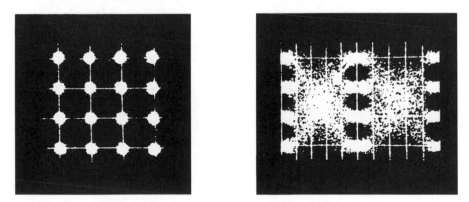

Figure 5.57 Constellation plot for a 16-QAM radio (degraded C/N = 20 dB).

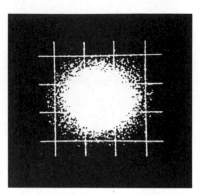

Figure 5.58 Constellation plot for a 16-QAM radio (no signal).

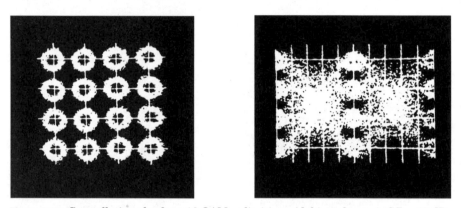

Figure 5.59 Constellation plot for a 16-QAM radio (sinusoidal interference). C/I = 15 dB.

Furthermore, the clockwise rotation indicates a negative angle of lock error (in this case, −5.6°). Once again, the eye closure only indicates BER degradation (noise) and does not specify the source of the problem.

Another modem impairment observable on the constellation display is the *amplitude imbalance* shown in Fig. 5.62. Here the amplitude levels of the quadrature carrier are not set correctly, which might be the result of a non-linear modulator or an inaccurate D/A converter. When the demodulator is out of lock, circles, as illustrated in Fig. 5.63, indicate the total lack of phase coherence, culminating in total radio failure. Although the eye diagram would show eye closure, it would not indicate the location of the fault in the equipment.

Phase jitter on a carrier, local oscillator, or recovered carrier causes a circular spreading of the constellation states as in Fig. 5.64. Again, note that the eye diagram does not indicate the specific source of this degradation.

Transmitter power amplifier nonlinearities can be detected on the constellation plot. In this respect, Fig. 5.65 shows phase distortion (AM-to-PM) by the twisted and rotated states, and amplitude distortion (AM-to-AM) is indicated by compression of states at the corners of the plot. As one can appreciate, it is very useful to observe this constellation diagram when adjusting predistorters. As the transmitter linearity gradually improves during the adjustment, the constellation diagram clearly indicates when a satisfactory performance is achieved.

Finally, the *IF amplitude slope* conditions that occur during multipath fading can be viewed. The 45° oval shape of the states in Fig. 5.66 indicates crosstalk between the I and Q signals causing an asymmetrical distortion. Here is one instance where the eye diagram does indicate the fault because in this case the eye opening is shifted to the right of the position for normal operation. From the above constellation diagram observations, it is clear that pattern aberrations provide very useful information for in-service diagnosis of QAM digital microwave radio faults.

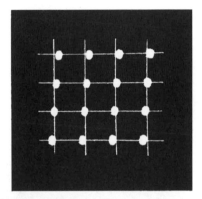

Figure 5.60 Constellation plot for a 16-QAM radio (quad without lock).

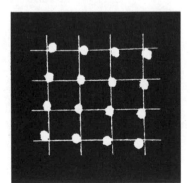

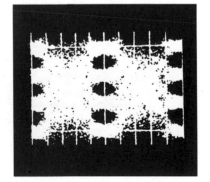

Figure 5.61 Constellation plot for a 16-QAM radio (– 5.6° carrier-lock error).

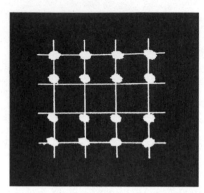

Figure 5.62 Constellation plot for a 16-QAM radio (amplitude imbalance).

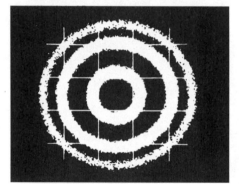

Figure 5.63 Constellation plot for a 16-QAM radio (carrier-recovery loop out of lock).

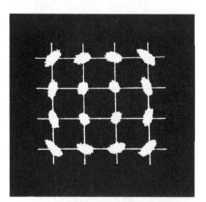

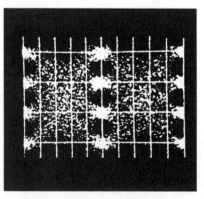

Figure 5.64 Constellation plot for a 16-QAM radio (phase jitter).

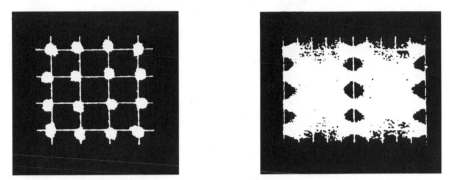

Figure 5.65 Constellation plot for a 16-QAM radio (amplitude nonlinearities).

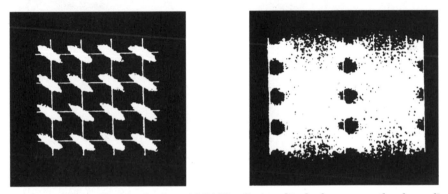

Figure 5.66 Constellation plot for a 16-QAM radio (amplitude slope across the channel).

Satellite Communications

6.1 Satellite Communications Fundamentals

There are three categories of communications satellite:

1. Fixed service
2. Broadcast service
3. Mobile service

The fixed service type is associated with geostationary satellite orbit (GSO) systems for international and sometimes domestic traffic. Very-small-aperture terminal (VSAT) systems are fixed services that use 1-m or so diameter antennas, and provide the important platform for two-way data communications.

Broadcast services provide distribution of TV to inexpensive receive-only terminals using antennas as small as about a half meter (18 in) in diameter.

Mobile services initially used GSO systems for applications such as maritime or news gathering. The new LEO and MEO systems provide telephony and limited data to individual pocket phones.

Before embarking on an exploration of this fascinating technology, it is important to appreciate the market forces driving the current developments, along with their advantages, limitations, and applications. The need for satellite communications arose foremost from the need for global international communications, in an era before optical fiber systems existed, when distances between continents were not easily bridged by other terrestrial systems. Subsequently it was realized that domestic satellite systems could also be an attractive option at the national level. Indonesia, a country composed of thousands of islands, springs to mind as an example of cost-effective interconnection by satellite. There are now more than 50 countries using satellites for domestic (national) telecommunications.

The broadcast nature of satellite communications is a unique feature that will probably prolong its survival even in the presence of intense competition from optical fiber networks. However, the extent of its future use will depend on the cost of satellite services. The severe bandwidth limitation of satellite systems compared to optical fiber systems does not bode well for the long-term prospects of broadband satellite communications, especially when fiber becomes extensively installed to the home. However, the satellite communications fraternity is proving to be remarkably resilient and resourceful. Some of the innovations presently being implemented promise a new lease on life for satellite services. Two key areas of interest are:

1. The role satellite systems will play in constructing global mobile networks or personal communications networks (PCNs) or universal mobile telecommunications systems (UMTSs). Low- and medium-earth-orbit (LEO and MEO) satellite constellations could play a major part if the economics are right.

2. Digital electronics innovations that improve bandwidth efficiency and also allow SDH, ATM, and bandwidth-on-demand over satellite.

Unless there is a major breakthrough in reducing the cost per kilogram of satellite payload, the room for price maneuvering is somewhat limited. Nevertheless, satellite technology is evolving, and the objective here is to present the technical details of present and possible future systems, indicating how the marketplace has dictated the direction of technology and how new techniques (both digital and analog) are cost-effectively matched to present-day telecommunication requirements.

6.1.1 Satellite positioning

Placing satellites in orbit is a precarious business at best. Many satellites have been lost at various stages of the launch and orbit placing. LEO constellations require literally hundreds of satellites to be placed in orbit for initial start-up and then more for old-age replacement. The LEO is much easier to attain than the geostationary satellite orbit (GSO), but it will be interesting to see the overall success rate 20 years on.

The most widely used method for GSO positioning has been to attach the satellite to a rocket and launch at (or close to) the equator in an easterly direction, taking advantage of the earth's rotation to gain velocity. Most of the rocket fuel is consumed in placing the satellite in an initial "parking orbit" about 200 km above the earth. The next phase is to fire the final expendable stage of the rocket to place the satellite in an elliptical "transfer orbit," with the perigee (closest point to the earth) the same as the parking orbit and the apogee (farthest point from the earth) at the required geostationary altitude of about 36,000 km above the equator. When the apogee is reached, the apogee motor, which is a nonexpendable, fixed part of the satellite, is fired to place the satellite into a circular GSO. Once in the right GSO, small position adjustments are made to ensure the correct plane of orbit and longitude position.

Position corrections are necessary throughout the satellite's life to keep it at the correct point in space. There are gravitational forces acting on the satellite from the sun and moon, in addition to the gravity of the earth, which tend to cause the satellite to move in a figure-eight as seen from the earth. From time to time it is necessary to fire a small rocket to correct the position of the satellite. This "station-keeping" activity is the satellite's main lifetime-determining factor, because eventually the fuel carried to reposition the satellite is consumed, and the satellite then drifts until it is burnt up on reentry into the earth's atmosphere.

6.1.2 Frequency allocation

The world has been split up into three regions as indicated on the ITU map of Fig. 6.1. This is accompanied by the uplink and downlink frequency allocations for fixed satellite communications as in Table 6.1. The main bands of interest are usually referred to by their popular names, which were allocated more than 50 years ago. They are summarized as follows:

Frequency range (GHz)		
Uplink	Downlink	Frequency band
6	4	C-BAND
14	12	Ku-BAND
30	20	Ka-BAND

The traditional satellite bands have used the radio window of 1 to10 GHz, which is relatively free from ionospheric absorption and rain attenuation. The graph in Fig. 6.2 highlights the severe attenuation encountered by rain at frequencies above 10 GHz. Satellite communications at the Ka-band (30/20 GHz) frequencies have recently started to gain more attention as the C- and Ku-bands have become congested. At the uplink frequency of 30 GHz, an attenuation of about 15 dB is experienced for rain falling at 25 mm/h. In tropical regions this would be a serious problem.

6.1.3 Polarization

In addition to the horizontally and vertically polarized waves for frequency reuse, circularly polarized waves are also encountered in satellite communications. The reason is clear if one visualizes the situation for the satellite link. In space, "this end up" has little meaning. On earth, vertical is relative to a radial line emanating outward from the center of the earth. Keeping the satellite in a position so that the plane of polarization of a linearly polarized transmitted wave remains fixed during its journey to earth involves considerable technical effort, especially because the ionosphere can rotate the plane of polarization to some extent. The polarization discrimination for signals transmitted on vertical and horizontal polarizations on the same frequency is just as important for satellite systems as it is for terrestrial microwave radio systems. Atmospheric effects such as raindrops also reduce the polarization discrimination of satellite communication signals. Although circular polarization does not eliminate cross-polarization, it does provide better performance over long-distance satellite links.

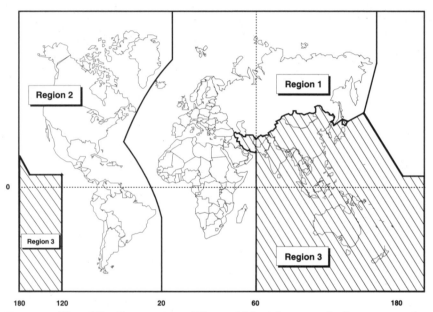

Figure 6.1 Map of the three regions of the world for telecommunication purposes (as designated by ITU).

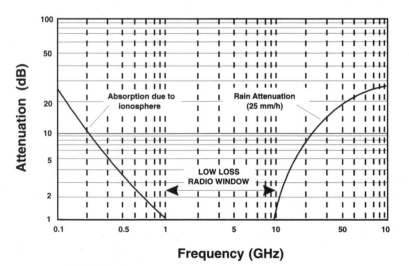

Figure 6.2 Radio window.

TABLE 6.1   Frequency Allocation for Fixed Satellite Communication Services

Frequency range, GHz	Up link			Down link		
	R1	R2	R3	R1	R2	R3
2.500 - 2.535					▓	▓
2.535 - 2.655					▓	
2.655 - 2.690		▓			▓	
3.400 - 4.200				▓	▓	▓
4.500 - 4.800					▓	▓
5.725 - 5.850	▓					
5.850 - 7.075	▓	▓	▓			
7.250 - 7.750				▓	▓	▓
7.900 - 8.400	▓	▓	▓			
10.700 - 11.700	▓	▓	▓	▓		
11.700 - 12.200				▓	▓	
12.200 - 12.300					▓	
12.300 - 12.500					▓	▓
12.500 - 12.700	▓				▓	▓
12.700 - 12.750	▓	▓				▓
12.750 - 13.250	▓	▓	▓			
14.000 - 14.500	▓	▓	▓			
14.500 - 14.800	▓	▓	▓			
17.300 - 17.700	▓	▓	▓			
17.700 - 18.100	▓	▓	▓			
18.100 - 21.200				▓	▓	▓
22.550 - 23.550 *	▓	▓	▓	▓	▓	▓
27.000 - 27.500		▓	▓			
27.500 - 31.000	▓	▓	▓			
32.000 - 33.000 *	▓	▓	▓	▓	▓	▓
37.500 - 40.500				▓	▓	▓
42.500 - 43.500	▓	▓	▓			
50.400 - 51.400	▓	▓	▓			

Note: (1) R1, R2, R3, are the geographical regions designated by the ITU.
       (2) * means intersatellite services common to all three regions.

It is easy to rotate the electric field vector to resemble the helix shown in Fig. 6.3, so that a receiver does not have to be oriented to a specific position in order to allow reception of the polarized wave as would be the case for linear polarization. A circularly polarized wave is generated by combining two linearly polarized waves with polarizations 90° to each other. If the electric field rotates clockwise as the wave propagates away from an observer, it is a *right-hand circularly polarized wave* (RHCP), and if it rotates counterclockwise, it is a *left-hand circularly polarized wave* (LHCP). The rate of rotation is equal to the frequency of the carrier. The polarization state of a wave can be described by the electric field's maximum and minimum amplitudes. The *voltage axial ratio* (VAR) = E_{max}/E_{min} is the term normally used to describe the quality of a circularly polarized wave. Ideally, VAR = 1 for a circularly polarized wave and, as the degree of ellipticity increases, so does the VAR.

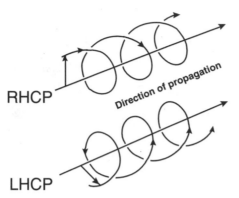

RHCP

LHCP

Direction of propagation

Figure 6.3 Circular polarization.

It is not necessary to adjust the direction of the receiving horn to receive an incoming circularly polarized wave containing RHCP and LHCP signals. It does, however, complicate the feeder hardware design, which must separate the two before any signal processing can be done.

6.1.4 Antennas

There are several types of antenna used in satellite communication systems. Those used in the earth station and the satellite are critical to the performance of the whole system and therefore great care is taken in their design and construction. Ideally, antennas should have high efficiency, low noise, and good sidelobe and polarization characteristics.

In the early days of satellite systems design, the high free-space loss could be overcome only by huge earth station parabolic antennas of 30 m in aperture (diameter). Fortunately, during the late 1980s, receiver noise temperature improved with advancements in semiconductor technology. This allowed the antennas to be reduced to 15 m in diameter. On the other side of the equation, satellites were designed to have high gain, with regional and spot beam coverage of specified areas. This allowed much smaller earth station aperture antennas of about 2 m or less to receive good signals, leading to the term *very-small-aperture terminals* (VSATs).

Figure 6.4 illustrates the different types of antenna feed. The focal feed used in many terrestrial microwave radio systems is also used for receive-only satellite earth terminals. The main problem with this type of antenna is that it has poor sidelobe characteristics. The cassegrain type has the feed horn at the base of the main reflector and illuminates the dish via a subreflector. The sidelobe characteristics are improved by this technique, so the cassegrain is widely used for large-scale earth stations, especially for international traffic. In both the focal-feed and the cassegrain designs there is an unavoidable obstruction of the main beam of energy caused by the feed horn or the subreflector. While this is not serious, it does reduce the efficiency.

An offset parabolic antenna can overcome the obstructing hardware from the boresight signal path. Low sidelobes are experienced with this construction and, although it has a slightly asymmetric beam, this antenna style can be used for small-scale, high-performance antennas. An offset Gregorian style, which has an elliptical subreflector, has extremely low sidelobes and is suitable for high-performance earth stations of all sizes. An offset cassegrain has a hyperbolic subreflector with similar characteristics.

Antennas designed for the satellite are mainly offset parabolic types for spot beams, and horn types for global beams.

Gain. The gain is the most important parameter in determining the scale of an earth station. From Eq. (4.1) in Chap. 4, the antenna gain for a conservative efficiency value of 0.6 (60 percent) $G = 18.18 + 20 \log D + 20 \log f$, where f is the frequency in gigahertz and D is the diameter in meters.

Because the frequency is also very important in determining the size of earth station antennas, Table 6.2 has been constructed to display some typical values of antenna gain for different parabolic antenna diameters at four satellite frequency bands.

Notice that the D^2 relationship gives a 6-dB gain increase by doubling the diameter. Also, for a given diameter, at 14 GHz the gain is 7.4 dB higher than at 6 GHz, and at 30 GHz it is 14 dB higher. A 2-m antenna at 30 GHz has the same gain as a 10-m antenna at 6 GHz, which would translate into a significant installation-cost savings. The question is "Does this increase in antenna gain compensate for higher atmospheric losses in the higher frequency bands?" There is no simple answer, because overall system gain must be considered. In general, however, areas with low rainfall benefit most from the higher-frequency bands, whereas tropical areas can experience high attenuation fade problems during the rainy season.

Beamwidth. While there is no size or weight constraint for earth stations, the satellite antenna weight and therefore size must be minimized. The satellite antenna gain is, to a large extent, fixed by the area of earth coverage required. For example, a full earth coverage antenna, usually called a global beam, has a specific beamwidth easily calculated by geometry. Remember from Chap. 4 that the antenna half-power beamwidth, in degrees, is given by the simple

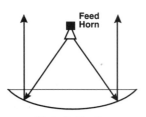

**Parabolic Antenna
(Focal Feed)**

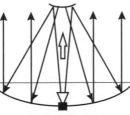

Cassegrain Antenna

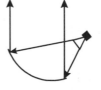

**Offset Parabolic
Antenna**

**Offset Gregorian
Antenna**

Figure 6.4 Types of antenna feed.

TABLE 6.2   Antenna Gain for Various Apertures at the Major Satellite Bands*

Diameter, m	Antenna gain, dB						
	2 GHz	4 GHz	6 GHz	11 GHz	14 GHz	20 GHz	30 GHz
1	24.2	30.22	33.7	39.00	41.1	44.19	47.7
2	30.2	36.23	39.7	45.02	47.1	50.21	53.7
4	36.2	42.25	45.7	51.04	53.1	56.23	59.7
6	39.7	45.77	49.3	54.56	56.7	59.75	63.3
8	42.2	48.27	51.7	57.06	59.1	62.25	65.7
10	44.2	50.21	53.7	59.00	61.1	64.19	67.7
15	47.7	53.73	57.2	62.52	64.6	67.71	71.2

* Efficiency = 60%.

relationship $\theta = 21/Df$. The angle for full earth coverage is found to be about 17°; therefore the diameter of a global parabolic satellite antenna is fixed at a particular frequency. For example, at 4, 12, and 20 GHz, $D = 0.309, 0.103,$ and 0.062 m, respectively.

An antenna with a larger value of gain will have a narrower beamwidth and therefore will cover a smaller area, or *footprint*, as it is known. The term *spot beam* is used to describe a high-gain satellite antenna that covers a relatively small land area. While this might be a problem in limiting the number of customers that can be provided with service, there is a major benefit for service providers in the spot beam zone. They experience an increase in power flux density, observed as an increase in the signal strength by their earth station receivers.

Noise.   A satellite system antenna might receive noise from several sources:

1. Thermal noise created by waves passing through atmospheric gases and water (worse at higher frequencies).

2. Thermal noise produced by the earth and entering the satellite antenna on the uplink path, and on the satellite-to-earth downlink path entering the antenna by the sidelobes (worse at higher frequencies and lower elevation angles).

3. On the downlink path the sun can cause severe noise at certain times of the year when it is directly behind the satellite. This effect is worst for three days (about 5 min each day) around the vernal and autumnal equinoxes on March 21 and September 21, respectively.

4. Noise from deep space (several Kelvin, almost negligible).

5. Noise caused by the feed system loss (about 7 K per 0.1-dB loss).

For an earth station low-loss, low-sidelobe antenna, the typical antenna noise temperature is 15 to 20 K (for elevation angles greater than 30°). These noise contributions are explored in more detail in the calculations of Sec. 6.4.2.

Interference. Interference by signals from adjacent satellites must be taken into account in both geostationary and nongeostationary satellite systems. The unwanted signals enter the antennas via the sidelobes.

Interference entering an earth station antenna from a terrestrial microwave link must be minimized. This is done mainly by locating major earth station installations such as Standard A stations for international traffic outside major cities. Where possible, they are shielded from cities by a convenient intervening hill.

6.1.5 Digital satellite communication techniques

Perhaps the most important developments in signal processing for satellite communications technology are those that improve bandwidth efficiency, particularly speech-encoding algorithms and modulation techniques. Chapters 2 and 3 discussed various techniques used for digital processing in transmission systems. Most of those techniques are applicable to satellite communications, and some are worth special mention.

Low-rate encoding (LRE). This is another expression for bandwidth compression to enhance the efficiency of bandwidth utilization. Because bandwidth is in short supply in satellite systems, it is essential to use sophisticated compression techniques to pack as many channels as possible into the available bandwidth. The voice channel bit rate is reduced to the minimum possible by today's technology, while maintaining good subjective quality.

The 64-kb/s PCM channel is reduced to 32 kb/s by ADPCM as designated by ITU-T Recommendation G.726. Further reduction to 16 kb/s (ITU-T Recommendation G.728) by *lower-delay code-excited linear predictor* (LD-CELP) and 8 kb/s (ITU-T Recommendation G.729) by *code-excited linear predictor* (CELP) is already producing an eightfold improvement in voice channel capacity. The difficulty has always been to reduce voice channel bit rate without any perceptible echo caused by delay time. The problem is to ensure that the cumulative delay on a GSO circuit does not overstep the maximum delay criterion, which is ideally 300 ms and never more than 400 ms.

Digital speech interpolation (DSI). Further compression can be achieved by voice activation. This is called DSI in satellite communications terminology, and results in a gain factor of about 2.5. In other words, between two and three voice channels can be packed into each satellite channel. This figure depends on the number of available channels and the speech activity of each channel. This technique is effectively a concentrator mechanism. The main difficulty with DSI is to ensure that the number of occasions when there are more talkers than available channels is kept to a very small fraction of the total call time. In a concentrator, when all available channels are in use, a busy signal will be sent to a customer trying to set up a call. DSI works on calls already set up so the first part of a speech phrase is lost when the system has no spare channels to allocate to a new speech burst. The percentage of lost speech time is called *freeze-out* or *clipping*. If enough channels are made available so that freeze-out is only about 0.5 percent, the clipping is unnoticeable to the customer.

It must be emphasised that DSI can be used only on speech channels and not with data traffic. The same ruling applies to channels that have voice compression (LRE channels). Nonspeech traffic should bypass the DSI and LRE equipment; otherwise data can be lost. Channels reserved for data are designated as digitally noninterpolated channels (DNIs). If the number of DNI assigned channels exceeds about 10 percent of the total, the DSI gain would have to be reduced to ensure freeze-out does not increase to an unacceptable level.

Digital channel multiplication equipment (DCME). This is the process of packing more voice channels into a satellite communications link by a combination of LRE and DSI. ITU-T Recommendations G.763 through G.766 define DCME for synchronous systems, and Recommendations G.764 and G.765 define DCME for plesiochronous systems. INTELSAT VIII, for example, offers a total capacity of 22,500 two-way telephone circuits, which can be increased to a maximum of 112,500 by using DCME. Note that this factor-of-5 improvement in capacity does not mean a fivefold improvement in the bandwidth efficiency (defined in bits per second per hertz). It is achieved by the improvement in voice A/D and D/A conversion efficiency, together with the elimination of idle channel time.

Intermediate data rate (IDR). The bulk of INTELSAT traffic is now carried by IDR and also by INTELSAT Business Services (IBS). To overcome the bandwidth-limited mode (see Sec. 6.4.4) existing in some IDR-carrying satellite transponders, bandwidth efficiency needs to be improved by higher modulation techniques. 4-PSK modulation with convolutional encoding and Viterbi decoding has been the standard for satellite communications IDR/IBS for many years now. Concatenation with an outer Reed-Solomon (RS) codec is a means of improving BER performance at the expense of some increase in transmission delay. The bandwidth efficiency of 4-PSK modulation with rate $\frac{3}{4}$ Viterbi plus RS coding is only 1.15 b/s/Hz. The move to 8-PSK modulation with rate $\frac{2}{3}$ trellis plus RS coding to give high-quality service improves the efficiency to 1.54 b/s/Hz. This change is now necessary, but to make it a less financially painful transition for Standard A earth station operators, software-driven switching between the 4-PSK and 8-PSK is a highly desirable feature of the next generation of modem equipment.

The BER versus E_b/N_0 performance improves as the code interleaver/deinterleaver depth increases. Unfortunately, the overall delay gets longer as the interleaver performance improves, and satellite circuits cannot afford to include much extra delay in addition to the unavoidably long path delay. Overall transmission delay for a 2-Mb/s carrier should be kept below 10 ms, which is achievable with 8-PSK.

Satellite links that incorporate scrambling and FEC experience errors that occur in clusters or short bursts, and which can be called *error events*. They occur randomly (they are Poisson-distributed). The subsequent block error rate is the same as if caused by randomly occurring bit errors.

6.1.6 Multiple access techniques

Fundamental to any transmission link is the decision to use either analog or digital technology. Clearly, digital technology is used wherever possible, but where radio transmission is concerned, the carrier waves are always analog in nature. Nowadays, a digital signal is (in most cases) superimposed on the analog carrier. Frequency-modulated systems are still in service in many parts of the world, especially for TV transmission, but the focus here will be on digital techniques.

A satellite transponder can be shared primarily in three ways, each defined by a multiple access technique:

1. Frequency division multiple access (FDMA)
2. Time division multiple access (TDMA)
3. Code division multiple access (CDMA)

The satellite bandwidth allocation (about 500 MHz at C-band) can be split up into numerous single voice channels by a multiplexed signal containing many voice band channels, or by a digital bit stream containing a combination of voice and variable bit rate data. These options lead to the terms frequency division multiple access (FDMA) and time division multiple access (TDMA) as illustrated in Chap. 3 (Fig. 3.43).

Frequency division multiple access. FDMA sounds like an analog technique, perhaps because it contains the FDM which is indeed the analog frequency division multiplexing technique. Before the digital revolution, all satellite systems used FDM signals which were frequency modulated onto a carrier within the FDMA bandwidth available. Nowadays, FDMA uses digital transmission packaging. FDMA describes the way in which the information passes through the transponder. There can be many carriers, and the bandwidth used by each carrier is a measure of the number of voice or data channels transmitted. At one extreme there is the single channel per carrier (SCPC), or a carrier might contain many channels in a time division multiplexed (TDM) bit stream.

In FDMA systems, intermodulation products created in the satellite transponder by the many carriers necessitate a reduction of the output amplifier output power to ensure that it operates in the linear region, well below its saturation value. This "back-off" results in a reduction of transmitted power and consequently the total number of channels that can be transmitted.

Single channel per carrier (SCPC) systems previously contained a 64-kb/s PCM voice or data channel, superimposed on a carrier by 4-PSK modulation, using a transponder bandwidth of about 38 kHz. With a carrier spacing of about 45 kHz, a 36-MHz transponder could therefore carry about 800 channels of traffic.

The dramatic improvement in digital compression techniques is reducing the voice channel bit rate down to 1 kb/s or so. The minimum subjective quality level is the main point of discussion these days. Even at a voice bit rate of 16 kb/s, which is relatively high by today's capability, this equates to 3200 channels per transponder.

This figure can be improved by more than a factor of 2 by carrier voice activation. During the gaps in speech, a carrier is not transmitted, making space in the transponder for another carrier that has been voice activated.

Time division multiplexing.   The TDM bit stream is also transmitted as a 4-PSK modulated carrier. The bit stream can be a PDH or SDH bit rate. INTELSAT was very successful during the 1990s with IDR transmission, which uses a TDM-PCM-PSK combination. The modulation is 4-PSK with $\frac{3}{4}$-rate FEC coding and soft decision decoding. For thin routes, 2.048-Mb/s (30-channel) PCM is the basic building block, but rates up to 44.736 Mb/s are acceptable. When used with DCME, these carriers offer a sizeable traffic flow at a cost that is generally less than other methods of transmission. While the voice channel information is all contained in a TDM digital format, it is important to realize that it is then packaged for delivery through a satellite transponder in an FDMA scheme and not TDMA.

Time division multiple access.   Satellite TDMA follows a similar style to PCM in that information from many different sources is time interleaved. A frame structure is constructed in which each earth station places its information. Careful synchronization is necessary to ensure that information bursts from different earth stations do not collide with each other. Just like PCM, there is a maximum number of slots available and so a maximum number of earth stations that can participate. As usual, these numbers depend on the overall TDMA bit rate and finally on the satellite transponder bandwidth available.

A fourth type of access is evolving for fast packet transmission in the form of the asynchronous transfer mode (ATM). This is essentially a TDMA technique that is made more efficient by filling the gaps that would normally occur in conversations with information from other sources, be it voice, video, or data.

Code division multiple access.   The initial excitement of the early 1990s at the theoretically large capacity improvement of CDMA systems over TDMA has, to some extent, subsided; this is because the deployment of CDMA systems has only partially delivered on that expectation. While CDMA systems have found

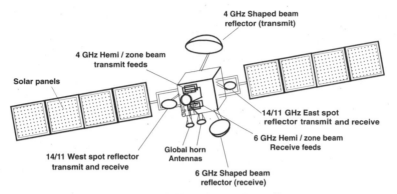

4 GHz Shaped beam
reflector (transmit)

4 GHz Hemi / zone beam
transmit feeds

Solar panels

14/11 GHz East spot
reflector transmit and receive

6 GHz Hemi / zone beam
Receive feeds

14/11 West spot reflector
transmit and receive

Global horn
Antennas

6 GHz Shaped beam
reflector (receive)

Figure 6.5   Artist's impression of INTELSAT VIII satellite structure.

relatively little penetration into GSO systems, several LEO and MEO systems are committed to CDMA. This is not surprising, because some major cellular and PCS operators have designed CDMA into their next generation of mobile systems, and the LEO and MEO systems are primarily intended for mobile users.

6.2   Geostationary Satellite Communications

GSO satellites orbit in the same rotational direction as the earth at an altitude of about 36,000 km above the equator as calculated by simple Newtonian mechanics. Interestingly, that altitude is the same regardless of the mass of the satellite. Because of its high altitude, a GSO satellite is visible from about 42 percent of the earth's surface. A satellite that orbits at an altitude less than 36,000 km above the equator has to move at a higher velocity to remain in orbit which means the satellite moves relative to the earth. For the orbiting altitude of 36,000 km the velocity is 3.1 km/s, and for a 780-km altitude (LEO) the velocity is 7.5 km/s. Clearly, any nonequatorial orbit regardless of altitude will cause the satellite to move relative to the earth; that is, the satellite will be in a nongeostationary orbit.

6.2.1   Satellite parameters

The geostationary satellite is composed of several subsystems, and a view of INTELSAT VIII is shown in Fig. 6.5. It has:

- *An antenna subsystem* for receiving and transmitting signals.
- *Transponders*, which receive signals, amplify them, translate their frequency, and retransmit them back to earth. In some cases demodulation to baseband, switching, rerouting, and remodulation functions are also included.
- A *power generating and conditioning subsystem* to provide power for the satellite and to supply all of the different voltages and currents the electronic circuits require.

- *A command and telemetry subsystem* for transmitting data about the satellite to earth and receiving commands from earth.

- *A thrust subsystem* for adjusting the satellite orbital position and attitude (orientation in space relative to the earth).

- *A stabilization subsystem* to keep the satellite antennas pointing toward earth.

Antenna subsystem. New types of antennas are being introduced with each new generation of satellite. In addition to parabolic reflectors, metallic lens antennas and dielectric lens antennas are being used. Array antennas are also used. These antennas are made up of several individual dipoles, helical elements, or open-ended waveguides. Many of these radiating elements are combined to form the desired pattern. Beam shaping with array antennas is much easier than with other types, and they are less prone to interference than other antennas.

Sometimes it is necessary to concentrate the transmitted energy into a narrow beam and direct it to a much smaller area than over the full earth. This is known as a spot beam antenna. Figure 6.6 shows the shaping of the beam to cover specific land areas. This area is often called a "footprint." Spot beam antennas clearly have a higher gain and therefore a larger aperture than global coverage antennas. A large, single-antenna dish can produce many spot beams. Many small feed horns are positioned so that their signals are reflected into narrow beams by the dish. Multiple spot beams have the following advantages:

- Frequencies can be reused several times in different locations on earth.

- The number of transponders can increase, because of frequency reuse.

- Spot beam signals can be switched from one satellite antenna to another.

Transponders. A transponder is the equipment in the satellite that receives, amplifies, frequency translates, and retransmits the RF signals back to earth. Figure 6.7 shows the frequency plan for the transponders in the INTELSAT VIII satellite in which there are several transponders for each of the different antennas. In the C-band, the antenna arrangements provide a sixfold frequency reuse using two spatially isolated hemispherical beams and four zonal beams. In addition, there are two C-band global beams. At the Ku-band there are spot beams on vertical and horizontal polarizations.

Power supply subsystem. The satellite is powered by silicon solar cells. Cylindrical satellites such as INTELSAT VI have their bodies covered with silicon solar cells. The cells are fixed to the whole cylindrical surface. Because

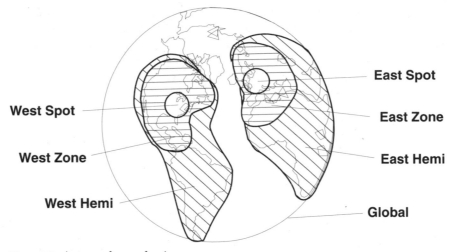

Figure 6.6 Antenna beam shaping.

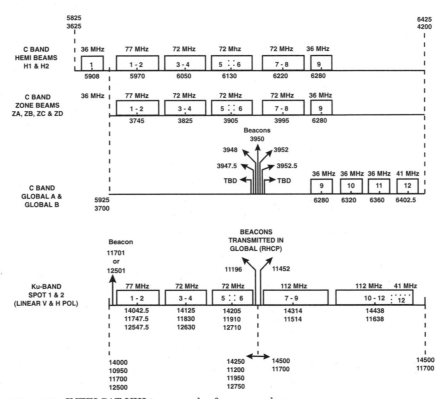

Figure 6.7 INTELSAT VIII transponder frequency plan.

the satellite is spinning on its axis, at any time only half the cells are exposed to sunlight. Also, most of the cells are at an angle, which grossly reduces the amount of light energy that can be converted to electrical energy.

A more efficient arrangement is to use flat solar panels often call "sails." INTELSATs VII and VIII have this configuration. If solar panels are used for power generation, the satellite must be three-axis stabilized. This is achieved by using momentum wheels. The solar panels must also be directed toward the sun, and this is done by using a sun sensor and stepping motor.

Eclipses are an unavoidable nuisance for geostationary satellite communications. The situation would be worse than it is now if the earth's rotational axis were at right angles to the plane of orbit around the sun, because in that case there would be eclipses every day when the satellite is in the earth's shadow. In reality, the situation is worst at each of the two equinoxes, at which time a satellite is in shadow for 69 min. On days to either side of the equinoxes the time spent in shadow gets progressively smaller, but it takes 21 days on either side to be completely eclipse-free. That means that a total of 84 days per year are affected. If batteries were not used on the satellites to keep the system alive during those periods when the sun does not provide power via the solar panels, the system would be unavailable for service.

Batteries are a limiting element in the life of the satellite. Their length of service depends on factors such as the number of charging and discharging cycles, and depth of discharge. Because batteries are so heavy, their quantity (and therefore the total charge capacity) must be minimized, but a maximum allowable depth of discharge (typically 70 percent) defines the capacity. INTELSATs I to IV used nickel-cadmium cell batteries, whereas nickel hydrogen ones were used in the V to VIII series.

Command and telemetry. A satellite contains instruments that relay information concerning the condition of the subsystems on board. Special earth stations monitor this information and they are equipped to:

- Spin-up or spin-down the satellite (INTELSAT VI)
- Perform orbit and course corrections
- Switch transponders into and out of service
- Steer spot beam antennas
- Activate feeds to shape the beam
- Control the charging of storage batteries

Thrust and stabilization. A satellite in orbit must always have its antennas pointing toward the earth in order to maintain continuous communication. For this reason, the satellite must be stabilized. A satellite can rotate about its three axes without moving from its orbital position. The three motions are referred to as yaw, pitch, and roll. There are three methods of stabilizing a satellite:

- By spinning a *cylindrical* satellite about its own axis
- By gravity stabilization using the motion of a fluid, such as mercury
- By employing gyroscopes or flywheels on board the satellite

If a satellite is spin stabilized (as is INTELSAT VI), the antenna assembly must be despun, so that it always faces the earth. A servo mechanism controls the speed at which the antennas are despun relative to the satellite body. Also, satellites tend to wobble, like a spinning top. This motion, called *nutation*, is minimized by nutation dampers. INTELSAT satellites have earth and sun sensors that position a satellite and maintain that position throughout its life.

The satellite carries a comprehensive thrust system because from time to time the orbit needs to be corrected, the orbital velocity needs to be increased, and the spin velocity needs to be increased. Hydrazine gas is the propulsion fuel that is used to provide the thrust to make the adjustment. The release of gas from the respective thrusters is controlled by solenoid valves, operated by commands from ground telemetry, tracking, and command stations. The apogee motor gives the final orbit velocity.

6.2.2 The INTELSAT series

There are now so many satellites in geostationary orbit belonging to a variety of international and domestic organizations and providing a wide range of services that the space within that orbit is considered crowded. This might seem hard to believe, because the physical distance between satellites is large (about 1300 to 2000 km), but it is the angle as seen from the earth between satellites that must be preserved to ensure that there is no interference between adjacent satellites. A 3° sector was originally allocated to each satellite. That figure defined the maximum number of satellite locations to be 120. Satellite beam shaping to reduce interference can now enhance that number to about 180 if a 2° sector is used.

The formation of the international cooperative organization INTELSAT in 1964 paved the way for the sophistication of satellite systems available today. There are other international organizations but none has stood the test of time as well as the very successful INTELSAT systems, which provide voice, data, and video services. Because of INTELSAT's dominance of the field of satellite communications, a brief history of its satellites will now follow, to show how early satellite systems have evolved into present-day configurations. INTELSATs VII and VIII are now in service and will be until at least 2003 and 2012, respectively.

INTELSAT I. Known as Early Bird, this was launched in 1965 and was the first commercial communications satellite operating over the Atlantic Ocean. It was a cylindrical, spin-stabilized satellite and had two transponders, each with about 30 MHz of bandwidth. The RF transmit power was about 4 W, and the EIRP was 12 to 14 dBW, indicating the antenna was an omnidirectional type. The *equivalent isotropic radiated power* (EIRP) is the product of power × gain.

About 600 solar cells provided 45 W of power for all the circuitry. It had a capacity of either 240 voice telephone circuits, or one good-quality television transmission. Its 18 months of designed lifetime extended in real life to more than three years.

INTELSAT II. In 1967, three INTELSAT II satellites were put into operation. The first was placed over the Pacific Ocean to extend coverage to the Pacific region. The second was placed over the Atlantic Ocean to increase capacity for the Atlantic region, and the third was also placed over the Pacific Ocean and acted as a spare. INTELSAT II was very similar to Early Bird in construction, but with improved channel capacity. With satellites over the Atlantic and Pacific, two thirds of the world's area was covered by communication satellites. INTELSAT II was designed to have a lifetime of three years.

INTELSAT III. During 1969, three INTELSAT III satellites were launched over the Atlantic, Pacific, and Indian Oceans. This configuration saw the first global satellite coverage as initially proposed by Arthur Clarke in 1945. Figure 6.8 depicts the three regions and the coverage of each satellite placed over the three oceans. Several failures hampered the overall success of INTELSAT III. Three failures in the launch phase and two failures while in orbit meant that only three satellites out of eight were in service for the full design lifetime of five years. The two in-orbit failures were due to bearing seizure between the satellite body and the despun platform. Each INTELSAT III satellite had two

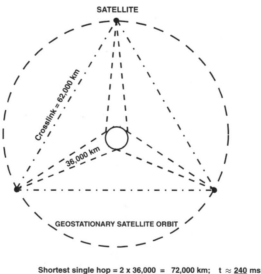

Shortest single hop = 2 x 36,000 = 72,000 km;   t ≈ <u>240</u> ms

Shortest double hop = 154,000 km;   t ≈ <u>513</u> ms

Shortest single hop + crosslink = 72,000 + 62,000 = 134,000 km;   t ≈ <u>447</u> ms

Figure 6.8 Global satellite configuration.

transponders of 225-MHz bandwidth, and a capacity of 1200 telephone channels, or 700 telephone channels plus one TV channel.

INTELSAT III introduced some new features, such as the despun antenna. Unfortunately, this design causes the line-of-sight path between earth and satellite antennas to be lost unless the satellite antenna is fixed at the end of the cylindrical satellite and pointing along the axis of spin, as in INTELSATs I and II. In those satellites, the omnidirectional antenna radiation patterns overcame the spin problem, but the low gain caused a high percentage of radiated power to be lost into space instead of being directed to earth. INTELSAT III was also the first in the series to use a highly directional antenna. It had a beamwidth of 19°, which is just 2° more than the 17° requirement for global coverage. The 2° margin was to allow for pointing error. The global antenna had a conical shape with a flare angle of 6°, and pointing accuracy was maintained by controlling the despin motors with infrared sensors that detected the earth's horizon.

INTELSAT IV and IV-A. During the period from 1971 to 1975, eight INTELSAT IV satellites were launched, of which seven reached orbit and provided service. The 1500-telephony-channel capacity of INTELSAT III was achieved by using most of the 500 MHz of the available, ITU-allocated bandwidth. Increasing the INTELSAT IV capacity to 4000 telephony channels plus two TV channels without increasing the bandwidth required some additional frequency reuse ingenuity. INTELSAT IV satellites each had twelve transponders of 36-MHz bandwidth, with a 4-MHz guard band between adjacent transponders. One transponder was used for SCPC transmission as described later.

The satellites had global coverage antennas and two 4.5° beamwidth spot beam antennas that were steerable from earth. One of the two spot beams was directed to the east and the other to the west of an ocean region. The objective of the spot beam transmissions was to provide more capacity for the heavier traffic-generating routes, the spot beam capacity being approximately double the global beam capacity. Capacity was also increased by frequency reuse achieved by transmitting two signals at the same frequency, but with different polarizations. Circular polarization provides a higher cross-polarization discrimination for 6/4-GHz-band satellite links than linear (horizontal/vertical) polarization. Left-hand circular polarization (LHCP) was used for the downlink, and right-hand circular polarization (RHCP) for the uplink.

By the early 1970s, the growth of international satellite traffic was outstripping the increase in satellite capacity. In 1972, the INTELSAT IV design was modified to increase the capacity to 6000 telephone and two TV channels, and the first INTELSAT IV-A satellite was launched in 1975.

The capacity increase was achieved by frequency reuse that employed spatial separation. In addition to using global beam antennas and higher-gain spot beams, hemisphere (or hemi) directional antennas were designed for the INTELSAT IV-A satellites. These hemis were directional antennas with offset parabola reflectors fed by an array of waveguide horns. The shape of the beam could be modified by selecting the appropriate horns for transmission.

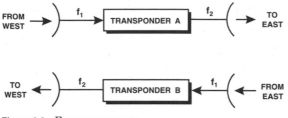

Figure 6.9   Frequency reuse.

Figure 6.9 shows how the two transponders operating on the same 36-MHz frequency band at 6 GHz have signals coming in from opposite sides of the globe via hemi antennas. After amplification and downconversion to 4 GHz they are retransmitted via hemi antennas to their destinations. The different shaped beams are illustrated in Fig. 6.6.

Capacity increase could easily be obtained by adding transponders at the 14/11-GHz band. Unfortunately, the technology was not sufficiently advanced in the mid-1970s, and high-quality components were not available for that frequency band until about 1980. SCPC systems were first introduced on INTELSAT IV-A to provide more flexibility, especially for low-traffic (thin) routes.

INTELSAT V.   New technology employed on INTELSAT V included the use of the 14/11-GHz band. Linear polarization was used in this band because it provides better cross-polarization discrimination at this frequency than does circular polarization. It also used three-axis body stabilization for the first time, instead of spin stabilization. The solar arrays producing 1228 W (after 10 years in orbit) were on flat panels instead of around the cylindrical body, and a momentum wheel within the body of the satellite provided gyroscopic stabilization. TDMA was first introduced for use with INTELSAT V.

INTELSAT V-A uses the spatial and polarization frequency reuse techniques to enhance bandwidth utilization. Onboard switch interconnection between global, hemi, zone, and spot beams resulted in a very flexible system that had a total channel bandwidth of 2250 MHz achieved from the available bandwidth of 912 MHz.

INTELSAT VI.   Satellite-switched TDMA was introduced for the first time with INTELSAT VI. The total available transponder bandwidth is 3330 MHz by using C- and Ku-bands, orthogonal polarization, and spatial isolation. The 6/4-GHz frequencies are reused six times, and the 14/11-GHz frequencies are reused twice. DCME started to have widespread use for more cost-effective voice traffic.

INTELSAT VII.   The INTELSAT VII/VII-A series of satellites is a high-capacity and versatile fleet of spacecraft deployed for global service. Design of the series focused on the special requirements of the Pacific Ocean region, but the satellites have unique features facilitating operation in other ocean regions as well. This series uses digital operation, is optimized for operation with smaller-aperture earth stations, and has efficient use of capacity.

Eight of the nine VII/VII-A spacecraft were launched successfully. The capacity provided by this series is impressive. The INTELSAT VII design has 26 C-band and 10 Ku-band transponders that allow 18,000 telephone calls and three color TV broadcasts simultaneously, or up to 90,000 telephone circuits using DCME. INTELSAT VII-A has 26 C-band and 14 Ku-band transponders, providing 22,500 telephone calls and three color TV broadcasts simultaneously, or up to 112,500 telephone circuits using DCME. VII-A also has independently steerable C-band spot beams for better traffic-handling capability. This satellite upgrades the IDR BER performance from 10^{-7} to 10^{-10} under clear sky conditions.

A new feature introduced by the INTELSAT VII/VII-A satellites is the ability to reconfigure the satellite's coverage and to meet changing traffic patterns and service requirements on a real-time basis. Three high-powered Ku-band spot beams can be independently steered so that 10 transponders can be assigned individually as demand requires.

The satellite lifetime is 13 to 18 years, depending on the launch vehicle, and launches took place from October 1993 to March 1996.

INTELSAT VIII. The INTELSAT VIII series, with six launches between February 1997 and July 1998, has improved C-band coverage and service capacity for public switched telephony and INTELSAT Business Services, provided better quality for video services, and encouraged new international VSAT applications. These satellites have sixfold C-band frequency reuse, twofold frequency reuse of expanded C-band capacity, and the highest C-band power level ever for an INTELSAT satellite.

INTELSAT VIII satellites include two independently steerable Ku-band spot beams that can be pointed anywhere on the earth's surface visible from the satellite. There is also interconnected operation between C- and Ku-bands. *Satellite news gathering* (SNG) service is expanded by the capability to connect spot beams to global beams.

Each satellite has a total of 44 transponders, of which 38 are C-band and 6 are Ku-band. This provides a total capacity of 22,500 two-way telephone circuits and three TV channels. If DCME is used, this increases the total capacity to 112,500 two-way telephone circuits. The satellite system lifetime is 13 to 18 years, depending on the launch vehicle.

Some of the main features of the INTELSAT satellites are compared and contrasted in Table 6.3.

6.3 Very Small Aperture Terminals

The technical improvements made in the EIRP from satellites and receiver noise figures of satellites and earth stations in the late 1980s and early 1990s inspired designers to create earth stations with smaller and smaller antennas,

TABLE 6.3 Comparison of INTELSATs' I to VIII Characteristics

	Retired from service								In service now (or soon)		
	I	II	III	IV	IV-A	V	V-A	VI	VII	VII-A	VIII
Launch date of first in series	1965	1966	1968	1971	1975	1980	1985	1989	1994	1995	1997
Design life, years	1.5	3	5	7	7	7	7 – 12	10 – 15	13 – 18	13 – 18	13 – 18
Capacity (2-way telephone circuits or TV channels)	240 or 1 TV	240 or 1 TV	1200 or 4 TV	4000	6,000 and 2 TV	12,000 and 2 TV	15,000	33,000 and 3 TV	18,000 and 3 TV; or up to 90,000 using DCME	22,500 and 3 TV; or up to 112,500 using DCME	22,500 and 3 TV; or up to 112,500 using DCME
Number of transponders	2	1	2	12	20	27 (21 C-band, 6 Ku-band)	38 (32 C-band, 6 Ku-band)	46 (36 C-band, 10 Ku-band)	36 (26 C-band, 10 Ku-band)	40 (26 C-band, 14 Ku-band)	44 (38 C-band, 6 Ku-band)
Power generated by solar array, watts	45	45	300	460	595	1200	1475	2100	4000	5000	5100
Stabilized	Spin	Spin	Spin	Spin	Spin	3-axis	3-axis	Spin	3-axis	3-axis	3-axis
Frequency bands, GHz	6/4	6/4	6/4	6/4	6/4	6/4, 14/11	6/4, 14/11, 12	6/4, 14/11	6/4, 14/11, 12	6/4, 14/11, 12	6/4, 14/11, 12
Bandwidth after reuse, MHz	50	125	450	480	800	2200	2400	3200	2390	3160	2550
Mass in orbit, kg	39	45	300	720	795	970	970	1800	1437	1823	1587

resulting in earth stations called *very small aperture terminals* (VSATs). There is no universal definition of VSATs, but in general they are low-cost earth stations equipped with antennas typically 1 to 2.4 m in diameter and a low transmit power of about 1 to 10 W. They usually operate in the 6/4- and 14/11-GHz bands. Generally, VSATs are digital communication networks, operated by private groups where the remote VSATs are installed at the user premises.

VSATs often compete directly with terrestrial packet-switched data networks. VSATs have the usual satellite advantages of distance independence and broadcast capability, together with rapid deployment and reconfiguration to meet varying traffic demands. Transmission bit rates are usually 64 or 128 kb/s, but can be higher. They have found widespread applications. Operators boast that they can establish operating VSAT links in less than one month. In the early days of VSATs, their primary purpose was data transmission. Because of the ease and speed with which VSATs can be installed, they have recently become attractive for very remote location connectivity, particularly for thin-route data and telephone traffic to developing countries.

Figure 6.10 shows the widely used star configuration, where a central hub is used for data distribution and return information from remote locations. One-way applications include broadcasting data from the hub to all remotes, providing news releases, real-time stock market figures, pricing and inventory details, weather bulletins, computer program downloading, and digitally compressed

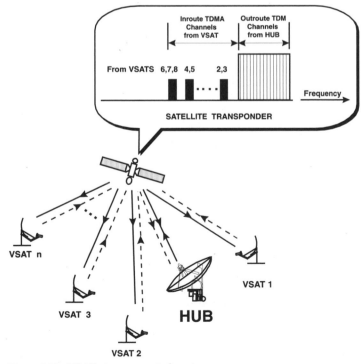

Figure 6.10 VSAT star-connected system.

video at 1.544 Mb/s or less. For example, the VSAT has brought about a blossoming TV-broadcast-to-the-home industry. Two-way applications include financial, banking, videoconferencing, credit card verification services, reservations for airlines, hotels, etc., and e-mail. A VSAT star network can be viewed as a geographically dispersed LAN.

When designing a large VSAT system, some of the important factors to consider are:

- User's application
- Network configuration
- Multiple access technique
- Transmission characteristics (power, modulation type, coding rate, etc.)
- Network management

There are many possible network configurations and perhaps the most fundamental considerations are, How many terminals are needed, and how are they to be interconnected? A star configuration is the simplest and least expensive, but a mesh or combination might be required. If one hub is broadcasting to more than 100 remote terminals, should the access technique be TDMA or FDMA? How would the network differ for telephony applications instead of data? A brief attempt will be made to answer some of these questions.

The hub of a star network has a relatively large antenna aperture of 6 to 9 m. Hub outroute broadcasts use TDM-multiplexed information placed on either a TDMA or FDMA carrier. The inroute traffic from VSATs to the hub also uses FDMA or TDMA, but the traffic is either continuous or occasional. For the continuous case, SCPC or narrowband TDMA is used, and for occasional or bursty traffic DAMA or random assignment (Aloha) is used. CDMA can be used in either situation.

Network configuration.   The choices for the network topology are point-to-point, star, mesh, or star/mesh combination. Star topologies are the most common and are good for operating in the broadcast mode. Star topologies are not effective for remote-to-remote voice connections because of the unavoidable double satellite hop.

Mesh networks provide direct interconnections between VSAT remotes. A hub is not necessary in this mode of operation, although a central earth station is needed for call assignment and setup. Mesh networks are ideal for voice communications because there is only a single hop between earth stations. The major drawback of mesh networks has been the larger EIRP needed from each VSAT. In the star configuration, the hub acts as a repeater to boost the signal with its larger dish and power output. This problem is being solved by modern satellites having larger EIRP, multiple hopping spot beams, and onboard processing and switching. This mesh network is almost the same as a star network with a "hub in the sky."

6.3.1   Multiple access

FDMA.   FDMA is well known for its uncomplicated earth station equipment and FDMA/SCPC is most suitable for thin routes. Its modularity allows circuit increases by incorporating channel units at the hub and remote stations. They can be used for point-to-point or point-to-multipoint networks. Data information transmission can be from 64 Kb/s to 2 or even 8 Mb/s. INTELSAT Business Services (IBS) is a typical example. A channel unit has selectable modem levels of 2-PSK or 4-PSK and 8- or 32-kb/s encoding rates. Voice activation and FEC further contribute to efficiency.

FDMA/SCPC.   This access method does not require any multiplexing. It is used for point-to-point, point-to-multipoint, and mesh networks. It is the VSAT equivalent of the conventional leased line, delivering up to about 2 Mb/s of bandwidth to individual VSATs. Satellite channels are either preassigned or demand assigned. Preassigned multiple access (PAMA) dedicates channels to specific earth stations regardless of the network call activity. SCPC/DAMA systems are simple and cost-effective for small networks with less than four or five sites and several channels per site. *Demand-assigned multiple access* (DAMA) is a more efficient way of using the limited frequency resource. In SCPC/DAMA systems, users from different earth stations share a common pool of channels. For each call a request is sent and if a channel is available, it is assigned on demand. The DAMA system is more complex and the earth station equipment is more expensive, but the recurring space segment costs are lower. This is a type of concentrator mechanism, and traffic requirements need to be carefully studied; otherwise blocking can reduce the system effectiveness. DAMA is suitable for many remotes when only a few channels are required for each remote. If the traffic is too light, the additional cost of the DAMA control equipment negates the reduction in satellite charges. The hub station controls the DAMA system by a common Aloha signaling channel. Remember, the Aloha system allows random contention (first come, first served) until the traffic becomes relatively heavy, at which time it changes to a reservation mode.

SCPC VSAT networks are well suited to thin-route, rural telephony, and can even be the primary communication method for some developing countries. SCPC can accommodate voice or data traffic, whereas TDMA is best suited to data. Because SCPC is in direct competition with leased lines, it is not surprising that costs are similar, whereas TDMA services are comparatively cheaper.

FDMA/MCPC.   Multiple channels per carrier (MCPC), as its name implies, is another FDMA technique in which each carrier contains several channels. Again, star networks with thin routes find MCPC to be a good alternative in some situations. Voice, data, or fax channels are time-multiplexed into one or several preassigned signals and then sent via a modem for transmission.

Using speech coding to allow 16 kb/s for each voice call, four calls can be multiplexed into a 64-kb/s signal for one carrier. Data channels must be preassigned because speech encoders cannot be used with data traffic. Usually, data is sent at 1.2, 4.9, 9.6, 56, or 64 kb/s, and several different-rate users can be multiplexed for one carrier. Carrier preassignment is more suitable for star or point-to-point applications where a few earth stations use up to only six channels.

A VSAT network would evolve as traffic increases, often beginning with a star network using MCPC to an SCPC/PAMA and eventually to an SCPC/DAMA. Further upgrades to a thin-route mesh network could follow. A TDMA star configuration would be a major upgrade that would be cost-effective only with more than 25 remote stations each allocated at least 15 voice circuits.

TDMA. Very efficient and of medium capacity (typically less than 40 Mb/s), TDMA is normally used in the star topology, and most current VSAT networks use this technique. TDMA is effective for broadcasting from the hub to remotes. In these systems a central hub station broadcasts multiplexed packets in a time division format to all of the remote stations. Each remote receives the time division multiplexed (TDM) bit stream and filters out information from unwanted time slots. Transmission from remotes back to the hub operates in either the TDMA or random assignment (Aloha) mode. For the TDMA mode each remote is assigned a specific time slot during which it transmits its bursts of data. In the Aloha mode any remote can transmit when it is ready to send data, and it is accepted by the hub on a "first come, first served" basis.

Star-networked earth stations that use the Aloha protocol statistically expect a certain number of collisions. Normally, if a hub correctly receives a packet, it sends an acknowledgment back to the VSAT (Fig. 6.11a). If bursts collide, the VSAT detects the collision because no acknowledgment is sent from the hub. After a random time interval the bursts are retransmitted. For "*pure Aloha*," the collisions increase as the traffic increases, to such an extent that the maximum system information throughput is, on average, only about 18 percent of the total channel capacity. Even worse, the total packet delay, including the roundtrip satellite propagation delay, increases to about 1.5 s at 15 percent throughput (Fig. 6.12).

Slotted Aloha, also known as random-access TDMA, improves the maximum throughput to about 35 percent. This system requires synchronization of the VSATs in the network. As indicated in Fig. 6.11b, packets must only be sent from the VSAT during the time slot periods. This figure shows that a collision will occur if two VSATs transmit during the same time slot. If this happens, no acknowledgment is sent back to the VSATs and both retransmit following a random delay. It is the elimination of partial collisions that improves the overall throughput of slotted Aloha. However, if the packet length is less than one time slot, the throughput will be reduced and eventually becomes the same as pure Aloha when the packets are one-half time slot.

Slotted reject Aloha overcomes the need for time synchronization by formatting the packets into a sequence of subpackets. As Fig. 6.11c illustrates, the fixed-length subpackets, each having their own header and acquisition preamble, can cause partial collisions. Only the most severe overlapping subpackets (collisions) are rejected and retransmitted. The throughput is better than pure Aloha (typically 20 to 30 percent) and gives a good performance for variable-length messages.

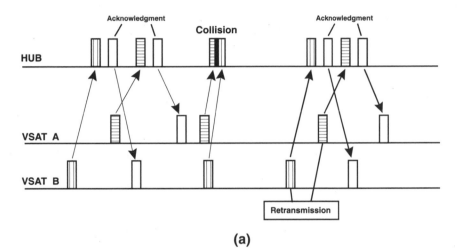

(a)

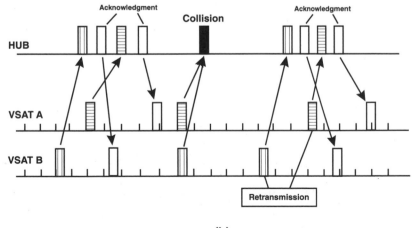

(b)

Figure 6.11 Packet transmission. *(a)* Pure Aloha; *(b)* slotted Aloha; *(c)* slotted reject Aloha; *(d)* reservation mode.

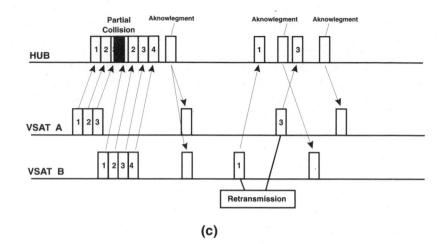

(c)

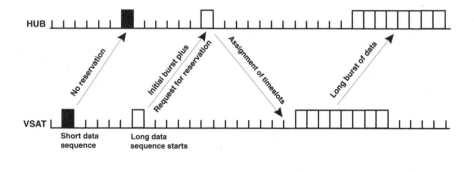

(d)

Figure 6.11 *(Continued)* Packet transmission. *(a)* Pure Aloha; *(b)* slotted Aloha; *(c)* slotted reject Aloha; *(d)* reservation mode.

Throughput can be greatly enhanced by *slot reservation* (Fig. 6.11*d*), sometimes called demand-assigned TDMA. The VSAT sends a request to the hub to transmit a certain number of packets and the hub responds with slot assignments. The VSAT then makes collision-free data packet transmission, because all other VSATs are made aware of the reservation schedule. 60 to 90 percent throughput can be achieved by this method at the expense of longer message delay (typically up to 2 s or even more) as in Fig. 6.12.

Note: Packetized voice mixed into the data messages is possible, but the quality can be degraded depending on the extent of the variable delay. Improved quality can be achieved by prioritizing the voice packet delivery.

Video for conferencing, education, medical services, etc., uses VSATs, and FDMA/SCPC (with DAMA) is usually used. Digital processing has reduced the typical bit rate to 384 kb/s or less.

Performance objectives. In addition to ITU-R Standards for performance of VSATs, individual operators have their own objectives, depending on the type of service. For example, the INTELSAT *super IBS* system has the following quality objectives, expressed in terms of the maximum percentage of total time that the BER is exceeded:

BER	Worst month	Year average
10^{-3}	0.2 %	0.04 %
10^{-6}	2 %	0.64 %
10^{-7}	10 %	4 %

The combination of TDMA, TDM, and FDMA enables thousands of interactive VSATs to operate within one satellite transponder.

CDMA. Spread-spectrum techniques are not very efficient as far as VSAT transponder power utilization and traffic capacity are concerned. Furthermore, spreading interference noise for each channel is added to the usual thermal noise, which means the satellite power (EIRP) must be increased to offset it. CDMA does, however, have the advantage of being able to operate effectively when severe interference from other radio systems is present, particularly in the congested 6/4-GHz band. For example, a C-band receive antenna can be as small as 0.6 m without suffering noticeable adjacent satellite interference degradation.

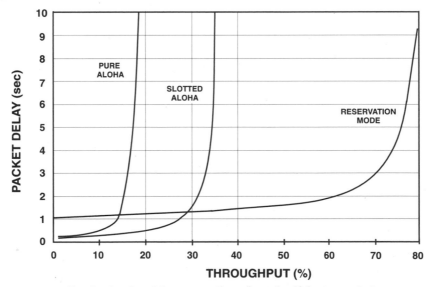

Figure 6.12 Graph of packet delay versus throughput for Aloha transmissions.

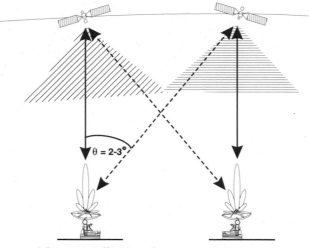

Figure 6.13 Adjacent satellite interference.

Interference. Adjacent satellite interference is a major consideration for VSATs. It appears to be a rather clever trick that two satellites, separated by only 3° of geostationary orbit, can transmit signals on the same frequency band, each of which can be received by two earth stations on the same rooftop, so that one earth station communicates with one satellite and the other satellite with the other earth station, without noticeable interference (Fig. 6.13). This is possible only if the beamwidth of each earth station antenna is narrow enough so that the adjacent satellite transmission is received only by the sidelobes of the antenna; consequently, the sidelobes must be minimized. VSATs just satisfy this criterion. Horizontal or truncated ellipsoid antennas provide better spatial discrimination than front-feed parabolic antennas. Offset parabolic antennas also have improved sidelobe characteristics because there is no blockage within the antenna aperture. Antennas smaller than about 1.8 to 2.4 m tend to have this style.

In order to ensure that adjacent satellite interference is acceptably low, for the 14-GHz band, VSAT earth stations should typically have sidelobe EIRP density values as follows:

$EIRP_{max}$ density per 40 kHz	Off-axis angle
$33 - 25 \log \phi$ dBW	$2.5° \leq \phi \leq 7°$
12 dBW	$7° \leq \phi \leq 9.2°$
$36 - 25 \log \phi$ dBW	$9.2° \leq \phi \leq 48°$
−6 dBW	$\phi > 48°$

Cost. The nonrecurring costs consist of the initial equipment and installation of a hub and as many remotes as required, plus an annual space segment cost that increases with the required amount of bandwidth. This space segment cost is levied on a preemptible or nonpreemptible lease basis. Nonpreemptible in this case means that in the event of a catastrophic satellite failure there is a much higher priority for restoration than for the preemptible type of lease. There can, however, be almost a factor-of-2 price difference.

6.3.2   Ultrasmall aperture terminals

The USAT is simply an extension of the VSAT concept and is loosely defined as earth terminals having antenna diameters of less than 1 m. Because antenna gain increases with frequency, the USAT is more appropriate for the Ku- and Ka-bands. Antenna discrimination is the main limitation to decreasing the antenna diameter to less than 1 m. Perhaps the worst interference situation is when an analog FM/TV is on the same carrier frequency (from a different satellite) as a narrowband digital signal. The carrier-to-interference ratio (C/I) is the parameter that describes antenna discrimination, and it should be at least 20 dB for satisfactory voice quality on digital channels, and preferably 23 dB. For FM-transmitted TV, 22 dB discrimination would be good and 28 dB studio quality. These figures are difficult to achieve at C-band, especially if the satellite spacing is less than 3°.

To achieve 20-dB antenna discrimination at Ku-band, an 0.8-m diameter parabolic antenna will suffice if the satellite spacing is 3°, but for 2° spacing the diameter would need to increase to 1.2 m. A 0.8-m-diameter parabolic antenna has a beamwidth of 1.8° to 2.5° in the 11- to 14-GHz band. This clearly presents a sidelobe interference problem in the direction of adjacent satellites only 2° away. The situation is still under evaluation.

A move up to the 30/20-GHz band for USAT operation would achieve adequate interference characteristics. For example, at 20 GHz the 0.8-m antenna has a beamwidth of only 1.3°. Unfortunately, rain attenuation becomes a problem in that band (typically 8 dB for 0.1 percent of the time and 20 dB for 0.01 percent of the time at 30 GHz in coastal areas). It is estimated that by the year 2010 the number of USATs and VSATs in the United States operating at 30/20 GHz will be about the same as in the 14/12-GHz band at present (about 250,000).

6.3.3   Hub in the sky

A major refinement of VSAT technology will be to move the star-connected hub from the ground to the satellite. In one move, the system is improved in many ways. Decoupling the uplink and downlink budgets from the overall link budget is perhaps the most significant. Eliminating double hops is more than a trivial improvement, because voice then has better quality. Unfortunately, this benefit might be offset by increased transmission delay, depending on the amount of onboard processing. To fulfill the role of a hub, onboard transponder processing should include demodulation, demultiplexing, baseband regeneration, switching between users, and back up to RF for downlink transmission.

To maintain timing and synchronization, an onboard atomic clock would be beneficial (if not essential) for SDH and ATM services. To improve performance benchmarks such as satellite G/T and transponder EIRP, multiple steerable spot beams received and transmitted by phased array satellite antennas are necessary. This also has the benefit of enabling frequency reuse of RF bands in different spot beams. Switching can be at RF, IF, or baseband frequency depending on the application. Regardless of the onboard interconnection technique, a major limitation is the geographical spacing between users. If spot beams cover only, say, 1 to 5 percent of the country in a VSAT network, possibly 100 beams or more are needed to interconnect population centers or even rural locations. This can be done by forming, say, 10 spot beams and hopping between specific locations in the microsecond time frame so that TDMA can operate over all 100 footprints. A 0.3° spot beam would have a very desirable antenna gain of about 50 dBi. This is satellite-switched TDMA, and it is still used on active INTELSAT VI satellites, but it does not process down to baseband, performing only RF-to-RF switching. This is zone-to-zone switching instead of user-to-user switching. Onboard baseband switching (including regeneration) enables earth station EIRPs to be reduced.

6.4 Geostationary Satellite Path/Link Budget

Many of the techniques used for terrestrial microwave radio transmission are applicable to satellite links. The satellite link can be viewed as a microwave link connecting two points on earth with a repeater at the midpoint distance of the link. The repeater in this case is the satellite. Whereas many terrestrial repeaters would be necessary to connect two points that are considerably further apart than the maximum line-of-sight distance allowed by the earth's curvature, the satellite system can do the same job with just one repeater. As for the terrestrial microwave link case, the satellite repeater can simply be an RF amplifier and frequency shifter for analog or digital transmission, or a full baseband regenerator for digital transmission. Switching and crossconnecting are additional features available to the more modern satellite designs. On each part of the link (uplink and downlink), the microwave signal has to pass through the earth's atmosphere. It is this atmospheric loss that has in the past determined the best frequency bands to use.

Figure 4.15 in Chap. 4 shows the atmospheric loss due to rain and oxygen, which poses a restriction on the high-frequency spectrum usage. Ionospheric absorption and reflection at the low-frequency end of the spectrum, particularly HF, limits the low-end usage This defines a satellite frequency transmission window as illustrated in Fig. 6.2. During the early satellite communication days of the 1960s and 1970s, when microwave technology was in its infancy, even C-band equipment components were considered difficult to manufacture. High-performance Ku-band components were simply not available until the 1980s. Today, technology has advanced to the point where the losses caused by rain and oxygen at the higher frequencies can be offset by the improvement in system gain. The prize for pushing toward higher and higher transmission frequencies is an increase in available absolute bandwidth for a given launch weight, which translates into more channels per kilogram of launch cost. Provided the satellite fabrication cost does not increase to the same extent, the cost-effectiveness of satellite communications is enhanced.

To proceed with link calculations, as is usual with any radio transmission system, the most important parameters affecting the system design are:

1. The signal to noise power ratio in the baseband channel

2. The RF carrier to noise power ratio at the input to the receiver

The signal-to-noise power ratio in the baseband channel is affected by the coding and modulation techniques used. In the older analog satellite systems, many voice channels were combined by the frequency division multiplexing (FDM) scheme to form the baseband, and then frequency modulation (FM) was used to modulate the RF carrier.

6.4.1 Performance objectives

Circuit quality was an important issue for terrestrial microwave links, and it is even more important for satellite links. Terrestrial digital microwave radio system quality requirements have become more stringent with the application of ITU-T Recommendation G.826. In its effort to improve bandwidth efficiency, satellite technology has created an error burst problem from what used to be a less serious random error problem.

The ITU-R has derived design masks based on ITU-T Recommendation G.826. To take into account the effects of error clusters (bursts) caused by FEC and scrambling, sometimes called error events, a term α is introduced, which is the average number of errored bits in a cluster. α is therefore the ratio of the BER to the error event ratio, and is a measure of "burstiness." α typically varies between 1 and 10, where 1 would be the case of no FEC or scrambling, and 10 would be for a laboratory-simulated INTELSAT IDR type of transmission having a rate of 3/4 FEC and a scrambler.

The graph (masks) in Fig. 6.14 plots the bit error probability divided by the average number of errors per burst (BEP/α) against percentage of total time (for the worst month) for several bit rates (ITU-R Rec. S.1062-1) at the output of either end of a satellite hypothetical reference digital path (HRDP). Table 6.4 shows typical performance objectives. Notice that for a STS-1 for $\alpha = 10$, the BEP of 2×10^{-9} can only be exceeded for a maximum of 10 percent of the total time for the worst month, and for STM-1 the figure is 1×10^{-9}.

Availability.   Regardless of the distance over which the satellite communicates, any satellite hop in a national or international section is allocated 35 percent of the overall end-to-end outage time. As a consequence, Table 6.5 indicates the performance objectives for a satellite hop.

The quality of a satellite link, just as in a terrestrial microwave radio link, is estimated by the carrier-to-noise ratio (C/N) and the amount of distortion in

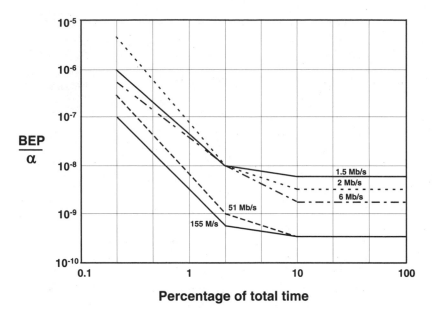

Figure 6.14 Graph of BEP/α versus percentage of the total time for the worst month.

TABLE 6.4   Satellite Systems BEP/α Performance Objectives

Bit rate, Mb/s	BEP/α for percentages of total time (for worst month)		
	0.2 %	2.0 %	10.0 %
1.5	7×10^{-6}	3×10^{-8}	5×10^{-9}
2.0	7×10^{-6}	2×10^{-8}	2×10^{-9}
6.0	8×10^{-7}	1×10^{-8}	1×10^{-9}
52	4×10^{-7}	2×10^{-9}	2×10^{-10}
155	1×10^{-7}	1×10^{-9}	1×10^{-10}

TABLE 6.5   Satellite Systems Error Performance Objectives

Bit rate	ESR	SESR	BBER
1.5 to 5	0.014	0.0007	1.05×10^{-4}
> 5 to 15	0.0175	0.0007	0.7×10^{-4}
> 15 to 55	0.0262	0.0007	0.7×10^{-4}
> 55 to 160	0.056	0.0007	0.7×10^{-4}
> 160 to 3500	Under evaluation	0.0007	0.35×10^{-4}

the received signal. In digital satellite systems the relationship between C/N and BER is most significant, whereas in analog systems the relationship between C/N and S/N is significant. In both cases the modulation scheme has a significant effect on the quality of the received signal.

6.4.2   Carrier-to-noise ratio (C/N)

Because of the large distance in the GSO satellite link, the gain within the system must increase to offset the correspondingly large free-space attenuation. This leads to the massive parabolic antennas one associates with satellite communications. The noise side of the equation is very important, because any improvement in the noise characteristics of the equipment allows a reduction in the gain requirements. Technological developments in the late 1980s made this point very noticeable when the size of the antenna for an international gateway (Standard A) was reduced from 30 m to just 15 m in aperture. A reminder of the main thermal noise sources follows, where noise is categorized as external or internal depending on whether the source is generated inside the equipment (satellite or earth station) or outside. The main interference noise sources are included to complete the picture.

	Thermal noise		Interference noise	
	External	Internal	Same route	Other system
Uplink	Earth surface	Receiver	Depolarization Adjacent channel	
Downlink	Sky Rainfall	Receiver Antenna Antenna feeder	Depolarization Adjacent channel	Terrestial microwave route

There are minimum acceptable values of C/N needed to attain the required system availability. To evaluate the complete satellite signal path the C/N must be calculated for both the uplink and the downlink, and then the two must be combined.

Carrier power.   For the downlink, the carrier signal power received by an antenna on earth from a signal transmitted by a satellite is:

$$C_D = \frac{P_s G_s G_{ES}}{\alpha_p} \quad \text{W} \tag{6.1}$$

where P_S = power transmitted by the satellite
$\quad\quad G_S$ = satellite antenna gain
$\quad\quad G_{ES}$ = earth station antenna gain

The total path loss is

$$\alpha_p = \alpha_{FS} + \alpha_A + \alpha_I + \alpha_R$$

where α_{FS} = free-space loss
α_A = atmospheric absorption
α_I = ionospheric absorption
α_R = rainfall loss

When seeking to maximize the carrier received power at the satellite or the earth station, the attenuation side of the equation should be minimized. Clearly the path loss for a geostationary satellite is fixed and cannot be reduced. Atmospheric and ionospheric losses are negligible in the satellite frequency window of about 1 to 40 GHz. Rainfall attenuation must be taken into account when operating above 10 GHz. On the gain side of the equation, there are limits to the size of the satellite antenna imposed by the launch weight. The larger the antenna, the heavier the satellite, and therefore the higher the cost of the entire system. Similarly, the power transmitted by the satellite is, to a large extent, also constrained by weight. This is one important parameter that is highly dependent on the advancement of technology. As a general trend, the output power achievable by a microwave amplifier for a given gain, size, and weight increases slightly every year. Also, the noise figure (or temperature) of amplifiers is decreasing with time as technology improves. The earth station receiver antenna gain is limited only by the economics of building large antenna structures. The 32-m-diameter antennas that were characteristic of the first three decades of satellite communications are now gradually decreasing in size as low-noise amplifier technology improves, so that less system gain is needed from the antennas. The path loss and antenna gains are both frequency-dependent. Path loss is proportional to frequency squared and antenna gain is also proportional to frequency squared, so as gain appears twice in Eq. (6.1), the received carrier power is proportional to frequency squared. This is a significant point in favor of moving toward higher operating frequencies.

Remember, the product P_SG_S is called the satellite *equivalent isotropic radiated power* (EIRP)$_{SAT}$.

Noise.   There are several sources of noise in a satellite link. There is the usual *thermal noise*, which arises in the receiver electronic equipment and antenna feeder. There is also thermal noise produced by the earth, which mostly affects the uplink because the satellite antenna is pointing toward the earth. Provided the earth-station antenna has a good front-to-back ratio and low sidelobes, negligible earth surface noise should enter the earth station antenna and receiving equipment. This ground noise is about 250 to 300 K. The earth station antenna pointing toward the satellite does pick up some thermal noise from the sun and from deep space, known as sky noise. This is

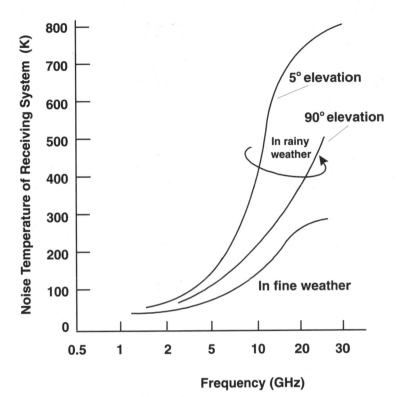

Figure 6.15 Rain attenuation.

particularly noticeable during the equinox when, on two days per year, the sun appears to move across the path between the earth station and the satellite. For several days before and after the equinox, high noise levels enter the earth station antenna for a few minutes each day. This is certainly noticeable in direct-broadcast receivers in regions where satellite TV performance is already borderline for a particular dish diameter. During these equinox periods the noise appears as "snow" on the TV picture. On other occasions, the moon passes directly behind the satellite. Although not sufficiently powerful a noise source to disrupt transmission, it does degrade performance during the pass.

Atmospheric noise entering the earth station receiver via the sidelobes might be insignificant when the weather is fine, but rain affects not only the attenuation but also the thermal noise. As Fig. 6.15 indicates, for the worst-case combination of heavy rain and low antenna elevation angle, the noise contribution becomes more significant as the frequency increases. For operation at 20 GHz, the antenna elevation angle and weather patterns become serious considerations. If operating in the tropical regions at such high frequencies, unacceptably high values of unavailability can be expected at certain times of the year.

The noise power N at the input to the earth station receiver is:

$$N = kTB \quad \text{W} \tag{6.2}$$

where k = Boltzmann's constant = 1.38×10^{-23} J · K^{-1} = -228.6 dBW
  T = the effective noise temperature of the receiver, kelvins
  B = the receiver noise bandwidth, hertz

The effective receiver noise temperature takes into account not only the equivalent noise temperature of the LNA multistages and mixer in the receiver, but also the earth's atmospheric conditions, the antenna type, angle of elevation, and the sky noise in the direction of pointing. The cassegrain type of antenna is almost universally accepted for large earth stations because of its high performance. The offset style is used for small antennas (particularly USATs) because they have no signal obstruction by the reflector and stays, which improves the sidelobe performance. There is no control over the earth's atmosphere, except that it is known that the effects of water in the atmosphere are worse as the operating frequency increases. In the early satellite days there was a significant difference in the amount of power that could be generated by a 4-GHz amplifier compared to a 6-GHz one (for a given size). Furthermore, the sky noise decreases with increasing frequency. These facts led to the use of the 6-GHz frequency band for uplink and the 4-GHz frequency band for the downlink. This is despite the fact that, as a general rule, the equipment size, weight, and satellite payload costs decrease as the frequency increases simply because the wavelength decreases with increasing frequency, and the size of many components is wavelength-dependent.

Intermodulation noise created in the satellite transponder power amplifier can be a problem for FDMA signals if the amplifier is operated in its nonlinear region near saturation. Figure 6.16 indicates how just three signals mix in a nonlinear device to produce an output spectrum having many intermodulation products. In the past this problem has necessitated operating with an amplifier output level 6 to 10 dB below the saturation level. The so-called *back-off* is worse for the older TWT amplifiers than the newer solid-state amplifiers. GSO earth stations transmit several hundred watts of RF power, and TWT/klystron amplifiers are still used, so back-off is necessary. At the

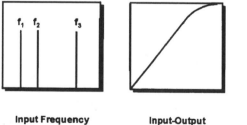
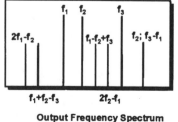

Input Frequency Spectrum   **Input-Output Characteristics**   **Output Frequency Spectrum**

Figure 6.16 Intermodulation noise.

satellite, a further power control mechanism is used. The total input power to the satellite from all participating earth stations is also controlled by enforcing a strict output power maximum from each earth station.

C/N. The important carrier-to-noise power ratio for the downlink from Eqs. 6.1 and 6.2 is

$$\left(\frac{C}{N}\right)_D = \frac{P_S G_S G_{ES}}{\alpha_p kTB}$$

$$\left(\frac{C}{N}\right)_D = \text{EIRP}_{SAT}\left(\frac{G}{T}\right)_{ES} \cdot \left(\frac{1}{\alpha_p kB}\right) \tag{6.3}$$

It has to be emphasised that C and N must be referenced or measured at the same point in the system. The term G/T is an important figure of merit for either a satellite transponder or an earth station, and is independent of any reference point.

Equation 6.3 is valid for both the uplink and downlink paths and is usually written with each parameter in decibels, as follows:

Uplink

$$[10 \log \left(\frac{C}{N}\right)]_U = [10 \log (\text{EIRP})]_{ES} + [10 \log \left(\frac{G}{T}\right)]_{SAT} - [10 \log \alpha_p]_U -$$

$$[10 \log k] - 10 \log B] \, dB$$

$$\left(\frac{C}{N}\right)_U \, dBHz = (\text{EIRP})_{ES} \, dBW + \left(\frac{G}{T}\right)_{SAT} \, dB \cdot K^{-1} - (\alpha_p)_U \, dB - k \, dBW \cdot$$

$$Hz^{-1} \cdot K^{-1} - B \, dB$$

Downlink

$$\left(\frac{C}{N}\right)_D \, dBHz = (\text{EIRP})_{SAT} \, dBW + \left(\frac{G}{T}\right)_{ES} \, dB \cdot K^{-1} - (\alpha_p)_D \, dB - k \, dBW \cdot$$

$$Hz^{-1} \cdot K^{-1} - B \, dB \tag{6.4}$$

For satellite systems that do not have onboard processing (such as digital regeneration or satellite switching), the overall round-trip C/N takes into account both the noise power for the uplink and downlink together with an interference component. They must be added arithmetically and also numerically (not in dB).

$$\frac{1}{(C/N)_{Total}} = \frac{1}{(C/N)_{Uplink}} + \frac{1}{(C/N)_{Downlink}} + \frac{1}{(C/N)_{Interference}} \tag{6.5}$$

This equation is important because it indicates that the total C/N is weighted in favor of the worse half of the link, together with an interference component. Onboard regeneration uncouples the uplink from the downlink so that the C/N is simply the sum of the uplink and downlink values (numerically added, not in dB).

6.4.3 Energy per bit to noise power density ratio, E_b/N_o

For digital satellite systems, the most important parameter is the total energy-per-bit to noise-density ratio $(E_b/N_o)_{\text{Total}}$.

$$\text{The bit energy or } E_b = \frac{C}{B_r} \qquad (6.6)$$

where C = the average wideband carrier power
B_r = the bit rate, b/s

The noise density N_o is the noise power present in a normalized 1-Hz bandwidth and is given by:

$$N_o = kT_e = \frac{N}{B} \quad \text{W} \cdot \text{Hz}^{-1} \qquad (6.7)$$

where T_e is the equivalent system noise temperature in kelvins.
From Eq. (6.7):

$$\frac{C}{N_o} = \frac{C}{kT_e} = \left(\frac{C}{N}\right)B \quad \text{dBHz} \qquad (6.8)$$

This carrier to noise-power-density ratio is also often used in satellite link budget calculations.
From Eqs. (6.6), (6.7), and (6.8):

$$\frac{E_b}{N_o} = \left(\frac{C}{B_r}\right)\left(\frac{B}{N}\right) = \left(\frac{C}{N}\right)\frac{B}{B_r} = \left(\frac{C}{N_o}\right)\frac{1}{B_r} \qquad (6.9)$$

Substituting for C/N from Eq. (6.3):

$$\frac{E_b}{N_o} = \left[\text{EIRP}\left(\frac{G}{T}\right)\frac{1}{\alpha_p kB}\right]\frac{B}{B_r}$$

$$\therefore \frac{E_b}{N_o} = (\text{EIRP} - \alpha_p) + \left(\frac{G}{T}\right) - k - B_r \quad \text{dB} \qquad (6.10)$$

Combining the uplink and downlink components and including interference:

$$\left(\frac{E_b}{N_o}\right)^{-1}_{\text{Total}} = \left(\frac{E_b}{N_o}\right)^{-1}_{\text{Downlink}} + \left(\frac{E_b}{N_o}\right)^{-1}_{\text{Uplink}} + \left(\frac{E_b}{N_o}\right)^{-1}_{\text{Interference}} \quad (6.11)$$

Interference noise power. In addition to systems noise, there is also interference noise caused by (1) adjacent satellites and (2) terrestrial microwave systems. Antenna patterns show that radiation can be measured in all directions around an antenna, and not just in the required direction. These sidelobes must be minimized for *geostationary* satellite communications antennas so that satellites can be parked as closely together as possible without causing interference. The sidelobe characteristics have been specified by INTELSAT for Standard A earth stations to be:

$$G_S = 29 - 25 \log \phi \text{ dBi} \qquad \text{for } 1° \leq \phi < 20° \qquad (6.12)$$

$$= 3.5 \text{ dBi} \qquad \text{for } 20° \leq \phi < 26.3°$$

$$= 32 - 25 \log \phi \text{ dBi} \qquad \text{for } 26.3° \leq \phi < 48°$$

$$= 10 \text{ dBi} \qquad \text{for } \phi > 48°$$

where G_S = the sidelobe gain
ϕ = the angle from boresight

Sidelobe characteristics are illustrated in Fig. 6.17 for a parabolic antenna, together with performance recommendations for antennas whose aperture-to-wavelength ratio (D/λ) is greater than 150. For a 3° angle between satellites, the physical separation is about 2000 km. The term *congested* is used to

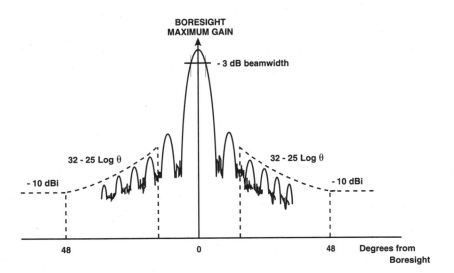

Figure 6.17 Antenna sidelobe performance.

describe this close packing of satellites, so while there are hundreds of satellites in orbit simultaneously, they are certainly not close enough to bump into each other during their daily, gravitationally induced excursions from their ideal fixed points. For VSATs the FCC specified the 1.2-m antennas for the 14/12-GHz band to be $29 - 25 \log \phi$ for off-axis angles of 1.5 to 7°.

While the satellites are spaced far enough apart to prevent interference, terrestrial microwave systems can interfere with satellite systems. Earth stations are usually located at least 20 km from a major city and conveniently located intervening hills prove effective for further shielding.

Same-route interference within a system can be depolarization interference noise or adjacent channel interference noise. Depolarization occurs in dual polarized systems and is interference due to one polarized signal affecting a channel using the other polarization. Remember, the cross-polarization discrimination, which should be at least 30 dB, degrades with rain, especially above 10 GHz. Adjacent channel noise should not be such a problem because the receiver filters can be designed with cutoff characteristics that ensure it is insignificant. This highlights the necessity for operating at the precise frequency allocated.

6.4.4 TDMA channel capacity

The satellite channel capacity or information throughput as determined by the bit rate is either *bandwith limited* or *power limited*. The bandwith limit is simply the maximum bit rate, Br_{max} that can pass through a transponder of bandwith W Hz depending on the bandwith efficiency B_e b/s/Hz, as defined by the modulation scheme used. Therefore, $Br_{max} = W \times B_e$ b/s. For QPSK, B_e is 2 but in practice would be about 20 percent less after filtering. For a 72-MHZ bandwith GSO satellite transponder, $Br_{max} = 72 \times 2 = 144$ Mb/s (= 81.6 dB).

Before reaching the bandwith limited bit rate, the maximum bit rate can be power limited by the uplink earth station EIRP or downlink satellite EIRP which must overcome the overall noise to achieve a desired BER. Until recently, the earth station has more flexibility for increasing the EIRP than the satellite because of the weight constraints imposed on satellite designs. But, in the case of small earth stations, such as VSATs, LEOs, or MEOs, the capacity is uplink *power* limited as demonstrated by the following analysis.

From Eq. (6.10), for the uplink:

$$Br_{max} = EIRP_{ES} - \alpha_p + (G/T)_{SAT} - k - E_b/N_o - \text{margin} \quad \text{dB}$$

For the 14/12-GHz band, consider the following typical VSAT values:

$$Br_{max} = 42.5 - 207 + 0.7 + 228.6 - 6 - 4$$

$$= 54.8 \text{ dB}$$

$$= 302 \text{ kb/s}$$

This is the power-limited maximum bit rate and the earth station EIRP or satellite G/T would need to be increased to improve the bandwith (capacity). For example, a 10-dB increase in earth station EIRP by increasing the antenna diameter and/or earth station output power would increase the Br_{max} to 3.02 Mb/s. However, there is a limit to the bandwith increase that can be achieved by increasing the EIRP. As indicated above, a transponder bandwith allocation and modulation type define the ultimate bandwith limitation.

6.4.5 Link budgets

The link budget differs significantly depending on the type of system (GSO, VSAT, LEO, or MEO), the type of signal passing through the satellite system (e.g., TV, telephony, or high-bit-rate data), frequency bands, transmit power levels, and antenna apertures. It is useful to appreciate the various factors affecting satellite transmission characteristics for different types of systems, so TV GSO and data VSAT systems are considered here, and LEO/MEO telephony later (Sec. 6.7).

GSO (TDMA-PSK) link budget. The first link budget considers the case of an operator broadcasting five TV channels for reception on very small Ku-band antennas. The system shown in Fig. 6.18 has five digitally compressed TV channels multiplexed into a 25-Mb/s signal. After $\frac{3}{4}$ FEC and RS coding to give a required receive E_b/N_o = 4.66 dB, the resulting 37.7-Mb/s signal QPSK modulates a carrier that is transmitted to the satellite. The objective is to calculate the earth station receive antenna aperture.

Uplink. The EIRP level is chosen to be sufficient to saturate the satellite transponder which, in this case, uses only one carrier. The satellite used has a saturation flux density value (manufacturer's statistic) of –87 dBW/m^2. This is also known as the illumination level per unit area, P dBW/m^2.

$$P = \text{EIRP} - \alpha_p + \text{G}_{1\text{m}^2}$$

G_{1m^2} is the gain of a 1 m^2 antenna with 100 percent efficiency = $20 \log f$ + 21.4 dB $\cdot$ m^2. The path loss α_p includes a rain margin of 3 dB, and pointing loss of 0.5 dB. So,

$$-87 = \text{EIRP} - 210.4 + 44.3$$

The required earth station EIRP is therefore

$$\text{EIRP} = 79.1 \text{ dBW}$$

This could be achieved, for example, with a 6-m antenna (57.3-dB gain) and a transmit power of 21.8 dBW (150 W).

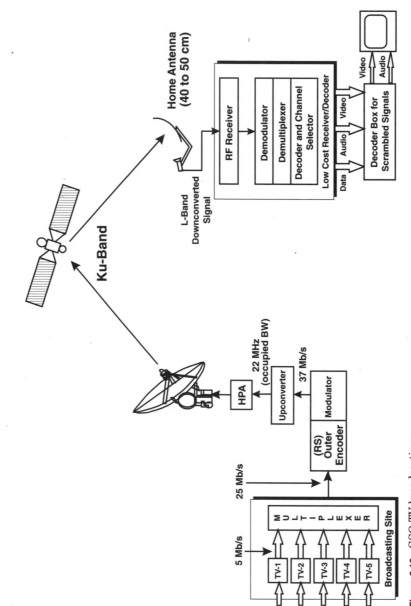

Figure 6.18 GSO TV broadcasting.

Downlink. From Eq. 6.10:

$$\left(\frac{E_b}{N_o}\right)_D = (\text{EIRP}_{\text{SAT}} - \alpha_p) + \left(\frac{G}{T}\right)_{\text{ES}} - k - B_r - \text{margin}$$

The satellite EIRP is at least 52 dBW over the region of beam coverage. The minimum required E_b/N_o is 5 dB, and the bit rate is 37.7 Mb/s. A 6-dB rain margin is included, as well as 4 dB for beam edge received signal strength reduction.

$$5 = 52 - 212.3 + \left(\frac{G}{T}\right)_{\text{ES}} + 228.6 - 75.8 - 4$$

$$\left(\frac{G}{T}\right)_{\text{ES}} = 16.5 \text{ dB}$$

For an LNA noise temperature of 35 K, a receive antenna diameter of about 0.4 to 0.5 m (40 to 50 cm) would be adequate for this application. The link budget presents the above calculation in tabular form in Table 6.6.

VSAT (star) link budget. VSAT star topology link budgets have two sections: (1) outroute and (2) inroute. The outroute is from the hub via the satellite to the VSAT, and the inroute is the return journey from the VSAT to the hub. The figures in the link budget presented in Table 6.7 are for a Galaxy IV satellite star system with a 6.1-m hub antenna transmitting 256 kb/s outroute and a 1-m-diameter VSAT antenna transmitting 256 kb/s inroute. The information rates are only 128 kb/s, and the FEC code rate of 0.5 increases the carrier rate to 256 kb/s.

Notice how the overall C/N for the outroute is dominated by the weaker downlink compared to the inroute, which has the weaker uplink. This is mainly the effect of the larger 6.1-m hub antenna and 80 W HPA output, compared to the 1-m VSAT antenna, which has a transmit power of only 1 W.

The system is dimensioned so that the aggregate number of VSATs in the network almost saturates the transponder when all are operational. The hub, on the other hand, has enough RF power to saturate the transponder and maintain a large margin for rain fading.

6.5 GSO Earth Stations

Before the age of mobile communications, satellite earth stations were huge facilities with massive equipment and installation costs. In those days, only GSO systems were used for long-distance international telephony traffic until optical fiber started eroding their market monopoly. LEO satellite systems will no doubt change the competitive as well as the physical landscape once more.

Because the Standard A satellite earth station is a feature common to almost all countries in the world as an international telecommunications gateway, the

TABLE 6.6   Link Budget for INTELSAT TV Broadcast (14/11-GHz Band)

Uplink (14 GHz Band)		
1 Earth station transmit power	21.8	dBW
2 Earth station antenna gain	57.3	dB
3 Uplink earth station EIRP	79.1	dBW
4 Pointing loss	0.5	dB
5 Uplink path loss	206.9	dB
6 Rain attenuation (+ air absorption loss)	3.0	dB
7 Satellite received power [3–4–5–6]	−131.3	dBW
8 Satellite (G/T)	−5.2	dB/K
9 Boltzmann's constant	−228.6	dBW/Hz/K
10 C/N_o [7+8–9]	92.1	dB
11 Bandwidth (37.7 Mb/s)	75.8	dBHz
12 E_b/N_o [10-11]	16.4	dB
13 Required E_b/N_o	11.0	dB
Margin	5.4	dB
Downlink (11 GHz Band)		
1 Downlink satellite EIRP	52.0	dBW
2 Downlink path loss	205.7	dB
3 Rain attenuation	6.0	dB
4 Air absorption loss	0.6	dB
5 ES Carrier received power level (C) [1–2–3–4]	−160.3	dBW
6 E_b/N_o	5.0	dB
7 Boltzmann's constant	−228.6	dBW/Hz/K
8 Bandwidth (37.7 Mb/s)	75.8	dBHz
9 Link margin	4	dB
10 $E_b/N_o + k + Br + $ margin [6+7+8+9]	−143.8	
11 Required earth station (G/T) [10–5]	16.5	dB/K
12 LNB (+ antenna) noise temperature	35	K
13 Earth station antenna gain [11+12]	31.9	dB
Earth station antenna aperture	0.42	m

Uplink path distance (earth station to satellite) = 38000 km; downlink path distance (satellite to earth station) = 40000 km; uplink frequency = 13.9 GHz; downlink frequency = 11.5 GHz; earth station antenna diameter = 6 m.

equipment for a typical station will be described in more detail. However, this time-tested GSO workhorse is now being challenged by the availability of smaller and smaller earth stations right down to the LEO satellite pocket handset "earth station" (Sec. 6.6).

6.5.1   INTELSAT earth station Standards A to Z

INTELSAT systems are categorized in terms of their earth station antenna sizes, frequencies of operation, and services provided. A summary of these categories is shown in Table 6.8.

INTELSAT has set Standards for various grades of earth station, designated A, B, C, D, E, F, G, and Z. There are many INTELSAT documents detailing each of the earth station standards, and only a very brief comparison will be made here. The most significant characteristic for any earth station is the G/T value. Standard A earth stations carry high and medium traffic and operate in the 6/4-GHz band with a minimum G/T of 35 dBK^{-1} at a minimum elevation angle of 5°. Standard B earth stations have a minimum G/T of 31.7 dBK^{-1}. Earth stations operating in the 14/11-GHz and the 14/12-GHz bands, satisfying a minimum G/T value of 37 dBK^{-1}, are called Standard C earth stations. Standards A, B, and C are for international voice, data, and TV.

INTELSAT diversified its operation to offer service for businesses. These stations are categorized as IBS and LDTS (low-density telephony service) or

TABLE 6.7 VSAT Link Budget

Uplink (14 GHz band)	Inroute (VSAT to satellite)	Outroute (HUB to satellite)	Unit
Earth station transmit power	0	19.0	dBW
Earth station antenna gain	41.5	57.3	dB
Uplink earth station EIRP per carrier	41.5	41.8	dBW
Uplink path loss	207.4	207.4	dB
Pointing loss	0.5	0.7	dB
Air absorption loss	0.3	0.3	dB
Satellite carrier received power level (C)	−166.8	−166.6	dBW
Boltzmann's constant	−228.6	−228.6	dBW/Hz/K
Satellite G/T	0.7	5.1	dB/K
Carrier noise bandwidth (307.2 kHz)	54.9	54.9	dBHz
Carrier to noise ratio at satellite $(C/N)_{up}$	7.7	12.2	dB

Downlink (12-GHz band)	Inroute (satellite to hub)	Outroute (satellite to VSAT)	Unit
Satellite saturation EIRP	45.2	45.2	dBW
Downlink satellite EIRP per carrier	19.6	22.5	dBW
Downlink path loss	205.9	205.9	dB
Air absorption loss	0.2	0.2	dB
Earth station pointing loss	0.4	0.4	dB
Earth station carrier received power level (C)	−186.9	−184.0	dBW
Boltzmann's constant	−228.6	−228.6	dBW/Hz/K
Earth station G/T	35.9	20.5	dB/K
Carrier noise bandwidth (307.2 kHz)	54.9	54.9	dBHz
Carrier to noise ration at ES $(C/N)_{down}$	22.6	10.3	dB

Composite (Uplink and downlink)	Inroute	Outroute	Unit
Carrier to noise ratio at satellite $(C/N)_{up}$	5.9	16.7	
Carrier to noise ratio at ES $(C/N)_{down}$	183.9	10.7	
$(C/N)_{uplink\ \&\ downlink}$ (numerical value)	5.7	6.5	
$(C/N)_{uplink\ \&\ downlink}$ (in dB)	7.5	8.1	dB
$(C/I)_{adjacent\ satellite,\ cross-polarization,\ intermodulation}$	10.0	10.0	dB
$C/(N+I)_{total\ (uplink\ \&\ downlink)}$	5.6	5.9	dB

Distance from hub & VSAT to satellite = 39,500 km; frequency of uplink = 14.25 GHz; frequency of downlink = 11.95 GHz; hub antenna diameter = 6.1 m; VSAT antenna diameter = 1 m; hub power output = 80 W (= 19.03 dBW); VSAT power output = 1 W (= 0 dBW); carrier information rate = 128 kb/s; FEC coding rate = 0.5; carrier transmission rate = 256 kb/s.

TABLE 6.8   Characteristics of INTELSAT Earth Station Standards A to Z

Earth station standard	Antenna size, m	G/T, dBK^{-1}	Services	Frequency band(s), GHz
A	15 – 18	35	All services: international voice, data, and TV, including IBS and IDR	6/4
B	10 – 13	31.7	All except TDMA	6/4
C	11 – 13	37	International voice, data and TV, including IBS and IDR	14/11 and 14/12
D1	4.5 – 6	22.7	Vista	6/4
D2	11	31.7	Vista	6/4
E1	3.5 – 4.5	25	IBS	14/11 and 14/12
E2	5.5 – 6.5	29	IDR	14/11 and 14/12
E3	8 – 10	34	IBS and IDR	14/11 and 14/12
F1	4.5 – 5	22.7	IBS	6/4
F2	7 – 8	27	IBS and IDR	6/4
F3	9 – 10	29	International voice and data, including IBS and IDR	6/4
G	All sizes	—	International lease services	6/4 or 14/11 and 14/12
Z	All sizes	—	Domestic lease services	6/4 or 14/11 and 14/12

IDR. Earth stations operating these categories of service were originally called D, E, and F earth stations. However, Standards A, B, and C now also carry these services. A Standard D earth station is operated for the LDTS in the 6/4-GHz band and falls into one of two categories. A Standard D1 earth station carries one circuit, but the circuit capacity can be expanded if the performance requirements are met. A Standard D2 earth station carries heavier traffic. These are both comparatively light traffic stations and therefore use single-channel-per-carrier with companded frequency modulation (SCPC/CFM). They can therefore be directly linked between themselves. Standard E and F earth stations offer three types of INTELSAT business service. These are a closed network, an open network, and a type which can be interconnected with a public terrestrial digital data network. The Standard E earth station operates at 14/11 GHz or 14/12 GHz. It is broadly divided into E1, E2, and E3 earth stations based on three different G/Ts, according to how close or far the business facility is located. The Standard F earth station operates on the 6/4-GHz band and is further classified into F1, F2, and F3 earth stations according to the G/T values. The only approved modulation method for E and F earth stations is coherent quadrature phase-shift-keying (QPSK, i.e. 4-PSK) and low-bit-rate QPSK/TDMA/FDMA, TDM/QPSK/FDMA or QPSK/FDMA systems for special networks. Standard E and F earth stations can be directly linked among themselves.

Earth stations that access leased space segments in the 6/4-, 14/11-, and 14/12-GHz frequency bands and provide domestic communication services, satisfying certain performance characteristics set by INTELSAT, are called Standard Z earth stations.

6.5.2 INTELSAT Standard A earth stations

Look angle. As with all GSO earth stations, one of the main problems to overcome is correct orientation of the antenna to receive signals from the satellite and, once located, to maintain that alignment condition. This is a high-precision exercise. Two angles provide the necessary information for pointing the antenna and they can be calculated quite accurately (Fig. 6.19). They are:

1. The elevation angle, which is the angle from the earth's horizontal to the satellite

2. The azimuth angle, which is the compass angle between the satellite and due north

The elevation angle becomes smaller for earth stations further from the equator. If the angle is very small, the radio beam has to pass through a large

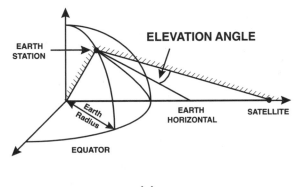

(a)

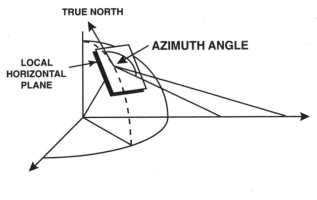

(b)

Figure 6.19 Look angle. *(a)* Elevation; *(b)* azimuth.

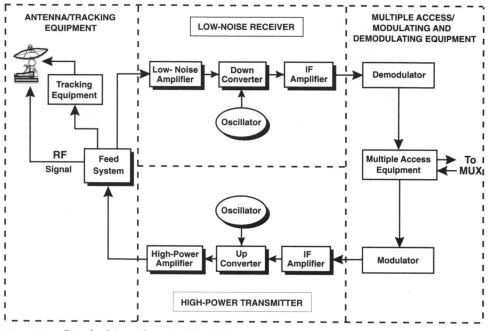

Figure 6.20   Standard A earth station block diagram.

earth's atmosphere distance, during which it is severely affected by noise and atmospheric absorption. For this reason, most satellite systems specify a 5° elevation angle as the minimum practical angle for operating an earth station. With reference to the block diagram in Fig. 6.20, there are five main subsystems for a Standard A earth station:

1. Antenna subsystem

2. LNA receiver subsystem

3. HPA transmit subsystem

4. Ground communication equipment subsystem

Antenna subsystem.   There are two types of antenna mount:

1. Kingpost or pedestal

2. Wheel and track

Standard A antenna designs use beam waveguides (Fig. 6.21) that have reflecting surfaces so that any elevation or azimuth motion of the antenna does not affect the signal as it passes between the feed horn, which is fixed near ground level, and the subreflector on the cassegrain antenna. It acts as a kind of periscope to guide the beam through the antenna rotation axes. The entire system, with the exception of the feed horn, rotates with the main reflector around the vertical azimuth axis, while just one of the beam waveguide reflectors rotates in the elevation axis. Surprisingly, the cross-polarization performance is relatively insensitive to mechanical movements of the reflectors.

Satellites are controlled from special earth stations generally referred to as telemetry, tracking and command (TTC) stations. In addition to maintaining the correct position of the satellite by performing orbit and course corrections, they control the charging of its batteries, steer the spot beam antennas, and activate feed horns to shape the zone or hemi beams. They also switch on or off transponders and switch in backup equipment, in the event of electronic failure or changes in traffic patterns. The control frequencies are usually in the VHF or S band but can be in the band allocated for the main satellite telecommunications mission (e.g., 6/4 GHz). The antenna used during the satellite launch phase is omnidirectional. However, once the satellite is stabilized and located in orbit ready for operation, a narrow-beam horn antenna is used for telemetry and control.

Antenna drive system.   There are duplicated drivers for both the azimuth and elevation angles. There are two sets of large dc motors on each axis in a counter

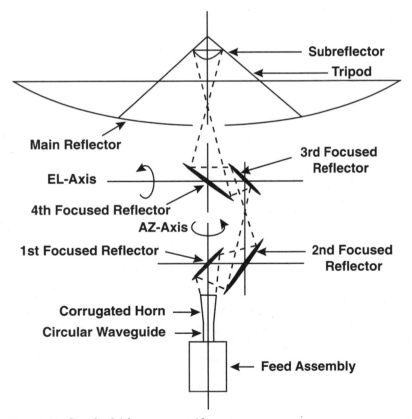

Figure 6.21   Standard A beam waveguide.

torque arrangement. The drive power is supplied by thyristor power amplifiers, with feedback to control the system and motor torque.

Principles of tracking modes used in earth stations. The satellite makes a small figure eight about its geostationary position because of higher space harmonics of the earth's gravitation, and gravitational "tidal" forces from the sun and moon. From a viewpoint on earth, the satellite continually moves from its fixed position by a small amount. INTELSAT specifications ensure that the satellite is kept at its fixed position relative to earth to within a daily tolerance of, for example, $\pm 0.1°$ north-south and $\pm 0.1°$ east-west. Any movement within these limits must be tracked by the antenna; otherwise communication will be lost. Earth station antennas have automatic tracking to cope with the satellite motion. There are generally two modes of tracking operation:

1. Monopulse (single-horn, higher-mode detection)

2. Step-track

The monopulse is a radar technique and, although it was used extensively in the early days of satellite communications, it is rarely used nowadays.

Step-track. The most commonly used method is the *automatic step-track* technique with a manual control option for rare occasions such as when snow needs to be cleared away, or serious maintenance is required. The antenna is simply moved step by step to maintain a direction of maximum received signal strength from a signal, called the beacon signal, sent from the satellite specifically for tracking purposes.

The signal level coming from the beacon is monitored by the earth station. The antenna is then moved (stepped) a small amount, say, vertically upward, and the beacon signal level measured again. If the signal has increased, the antenna is stepped again by the same amount in the same direction. If the signal level has decreased, it has been moved too far and it is stepped back in the opposite direction. Next, the process is done in the left and right directions until the maximum signal level is found. The process is repeated every so often, for example every 30 s, so the antenna is moved very frequently to keep it pointing in the best direction.

Antenna feed system. Dual circular polarization is used for C-band frequency reuse, which requires good cross-polar isolation of the circularly polarized waves. For example, a hemispherical beam would have one polarization, while a zonal beam within the hemispherical beam area would have the other polarization.

In order to operate dual polarization frequency reuse, the feed must be equipped with two types of devices:

1. Orthomode junction (OMJ)

2. Orthomode transducer (OMT)

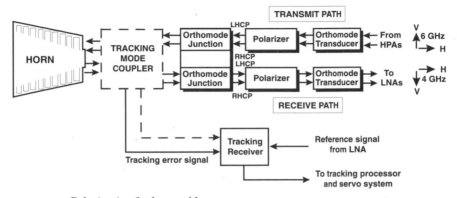

Figure 6.22   Polarization feed assembly.

Figure 6.22 shows the block diagram view of an orthogonal polarization feed assembly, which in reality is a rather unusually shaped waveguide plumbing structure. In the 6/4-GHz band, the 6-GHz transmit signal and 4-GHz received signal are routed separately. Two 6-GHz *linearly polarized* signals are combined in the transmit orthomode transducer (OMT). The output of the OMT passes through two transmit polarizers: (1) a quarter-wave differential phase shifter ($\pi/2$-polarizer) and (2) a half-wave differential phase shifter (π-polarizer), which converts it into two orthogonal *circularly polarized* signals—one LHCP, and the other RHCP. These are then sent to the orthomode junction (OMJ) for diplexing with the 4-GHz received circular polarized signal. The 4-GHz received circularly polarized signals are fed to the OMJ. These received LHCP and the RHCP signals are converted into orthogonal linearly polarized signals, using a $\pi/2$-polarizer and a π-polarizer. The receive OMT is able to discriminate the RHCP signal from the LHCP signal by a phase-conversion process.

Additionally, the tracking beacon signal is routed to the tracking receiver that, together with a reference from the low-noise amplifier, sends information to the tracking processor and servo system to move the antenna to the correct position.

LNA receive subsystem.   The low-noise amplifiers are mounted so that their orientation does not vary with the attitude of the antenna, thus eliminating the need for flexible waveguides or rotating joints. The waveguide feeder loss between the antenna and the receiver preamplifier needs to be as short as possible to minimize its attenuation. For example, an attenuation of 0.3 dB would contribute about 19 K to the system noise temperature. The low-noise amplifier receiving subsystem is designed to optimize the receive system figure of merit, G/T. The LNAs always operate in a hot-standby mode, and waveguide transfer switches are used for automatic switchover in the event of an LNA failure.

HPA transmit subsystem.   This subsystem has to be capable of supporting more than one carrier and capable of expansion. The use of air-cooled klystron HPAs (40-MHz bandwidth) avoids the proximity of high voltage and water of some systems and has significant advantages in terms of reliability and maintenance. The HPAs are usually driven by low-power TWT amplifiers. Again, the HPAs always operate in a hot-standby mode, and waveguide transfer switches are used for automatic switchover in the event of an HPA failure.

Ground communication equipment subsystem (GCE). This subsystem contains all the equipment that is specific to a particular RF carrier. It provides the interface between the baseband terrestrial multiplexing equipment and the broadband transmit/receive equipment. Depending on the specific requirements of each installation, a GCE might contain TDMA and/or FDMA, and TV equipment.

In the transmit path, the TDMA or FDMA baseband information would be 4-PSK or 8-PSK modulated and upconverted to the 4-, 12-, or 20-GHz band and combined with other signals, such as SCPC, in an RF combiner. On the receive side, an RF divider provides several signals at the 6-, 14-, or 30-GHz band, so that each path can extract specific information. This information would include the beacon signal, the SCPC band, TDMA or FDMA signal, and TV signals. After downconverting and demodulating the signals, the SCPC, TDMA, or FDMA signals are demultiplexed, and individual speech channels are passed through the echo cancellers prior to onward travel to their destinations by digital microwave radio links.

6.5.3   Time division multiple access

The most significant difference between FDMA and TDMA is that in a TDMA system all earth stations transmit on the same carrier frequency (at different times), and information from only one earth station passes through the satellite transponder at any particular time. This means intermodulation does not exist in a single-carrier TDMA transponder amplifier, so no back-off is required.

Figure 6.23 shows a simple case of just two earth stations operating into the same satellite that combines the uplink traffic bursts from each earth station into a single bit stream that is downlink transmitted to all earth stations. This simple concept is not easy to implement. All earth stations are given a specific time slot to transmit bursts of information according to the *burst time plan*. This is complicated by each earth station being a different distance from the satellite depending on its location, which causes differing transmission delay times. This means all earth stations must buffer the bit streams between the terrestrial network and the uplink/downlink space segment.

In typical TDMA style, a frame is used to define the precise sequence of transmission. As shown in Fig. 6.24, a frame starts with a burst from a reference earth station. This is followed by a burst from each earth station in a specified order, until all time slots have been completed. Each burst has a guard time of 1 μs to allow for small timing discrepancies to ensure bursts do not overlap and cause loss of information. A new frame begins immediately following the previous one,

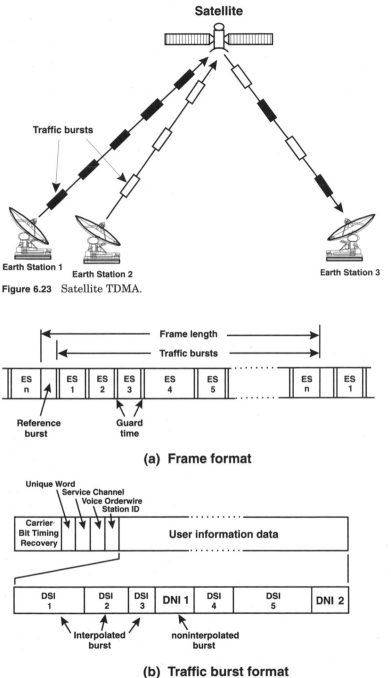

Figure 6.23 Satellite TDMA.

(a) Frame format

(b) Traffic burst format

Figure 6.24 TDMA frame.

and the process continues. The time slots will always be fully occupied even when the traffic load from earth stations is very small, or even zero. This is an inherent inefficiency of the system. A peculiar fact of TDMA is that the whole transponder bandwidth of 72 MHz is occupied by information from only one user at any point in time.

Each burst starts with a preamble that carries housekeeping information. First, a *carrier-and-bit-timing-recovery* (CBTR) sequence is transmitted, which depends on the type of modulation used. This is followed by a unique word to confirm that the burst is about to be sent. Next, service channel information, a voice order-wire group of bits, and then station identification information is transmitted.

After the preamble, the traffic burst begins. This can be a single burst of many digitally multiplexed customer voice channels, or the burst can contain subbursts, so that each subburst is sent to a specific destination or destinations. The subbursts are either interpolated (DSI) speech channels or group of bits and then noninterpolated (DNI) data channels.

The INTELSAT frame length is 2 ms duration, and the typical TDMA system bit rate is 120 Mb/s. 4-PSK carrier modulation is normally used, which has a bandwidth efficiency of 2. This would mean a 60-MHz required bandwidth, but after filtering it increases to 72 MHz as indicated in a typical satellite transponder FDMA plan. The total capacity of a 120-Mb/s TDMA depends on the mix of voice and data channels, and the extent of DCME incorporated. For example, taking the 64-kb/s channels as a baseline, the satellite channel capacity would be 120 Mb/s divided by 64 kb/s = 1875. The housekeeping bits amount to about 15 percent of the total bits, so the capacity is therefore about 1600 channels. If 10 percent of these channels (i.e., 160) are pure data channels at 64 kb/s, that leaves about 1440 for LRE and DSI conditioned channels. If the encoding is 16 kb/s (fourfold bit rate improvement) and the DSI gain is say, 2, the combined improvement factor is 8, meaning the total voice capacity is $1440 \times 8 = 11,520$, plus the 160 data channels.

A satellite has a total available bandwidth of 500 MHz in the 6/4-GHz band. Transponders typically have 40-, 80-, or 82.5-Hz frequency spacing in this band which, after allowing for filtering, permits 12×36-MHz, 6×72-MHz, or a combination of 36-, 72-, and 77-MHz transponders to be packed into the 500-MHz total bandwidth. After frequency reuse, one can see that a completely TDMA system transponder capacity is more than 100,000 voice channels.

Acquisition.   This is the process by which the bursts from earth stations fit precisely into the slots allocated to them in the TDMA frame. The distance of a satellite from an earth station can vary by about 5000 km depending on the earth station latitude and satellite longitude (approximately 36,000 to 41,000 km), causing the uplink transmission delay time to vary from about 123 ms near the equator to 139 ms at 5° latitude. The complete frame length is only 2 ms and the guard time between bursts is a miniscule 1 μs. Clearly, the transmission delay from each earth station to the satellite must be very precisely known for the information burst to arrive at the satellite at the correct position in the frame. There is an additional problem of the satellite straying 200 or 300 km per day from a fixed reference position relative to the earth. This cyclical position variation is fortunately predictable, but still needs to be taken into account in setting up the frame burst timing.

A burst time plan is essential for TDMA transmission. This is a catalog of all earth station burst time slot positions, durations, destinations, or DSI or DNI for each transponder. All earth stations must store the portion of the burst time plan relevant to their transmission options and all stations must simultaneously update the plan when changes are made by a controlling earth station.

The following techniques are used for acquisition. First, an earth station sends a short burst of the CBTR and unique word to arrive at the satellite at a time calculated to follow the reference burst by an amount precisely allocated to that earth station. If the earth station can receive its own initial short data burst after a round-trip through the satellite, it can then establish exactly where in the frame its burst has been placed. Then, a time shift can easily be made to correct the transmission to its exact allocated burst time slot. The disadvantage of this method is that satellites these days use spot beams that do not allow earth stations to receive their own transmitted bursts. In this case a second earth station is used to send error correction information back to the transmitting station. Once acquisition has been achieved, the synchronization is maintained.

Because a TDMA satellite transponder can manage only a limited number of earth stations within its 2-ms frame, transponder hopping is done to expand the service to a larger number of destinations. In other words, mid-frame, a burst can be transmitted on a different transponder to an earth station in a different TDMA group. In another case, stations in an east spot beam can transmit to stations in a west spot beam using one transponder, while the return information is sent through a different transponder. This principle is expanded to satellite-switched SS-TDMA, where the satellite has many spot beams and a complex microwave (RF) switch matrix that enables interconnection of transmit and receive signals between selected spot beams. Synchronization is even more difficult in this situation.

As mentioned in Sec. 6.3 on VSATs, there are several types of TDMA, such as TDMA-DA and TDMA-FDMA, that might also incorporate Aloha transmission depending on the application. Onboard processing and switching at the baseband adds another dimension of timing control, which should ease the synchronization difficulties. However, wideband SDH has its own special synchronization problems that are so serious that onboard atomic clocks might be necessary.

6.5.4   VSAT equipment

A typical VSAT earth station configuration is illustrated in Fig. 6.25, where the antenna in the 6/4-GHz C-band is on average about 2.4 m, and 1.2 m for Ku-band. Spread spectrum allows 1 m or less even in the C-band. Offset parabolic

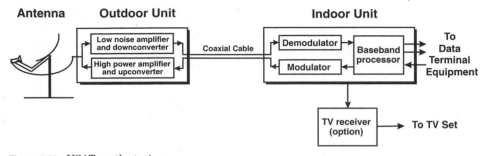

Figure 6.25 VSAT earth station.

antennas are popular because they allow wall mounting as well as rooftop or ground installation.

The outdoor unit contains the circuitry of the low-noise-block downconverter, the upconverter, and solid-state RF power amplifier in a weatherproof casing that is attached to the back of the antenna feed. The PSK modulator can also be included in this housing. So, the IF or baseband signal is carried by a 100- to 300-m coaxial cable to the indoor unit, which is placed close to the data terminal equipment (DTE). The indoor unit usually contains the modulator/demodulator, and the baseband processor that is connected to the DTE by a standard interface. An optional TV receiver can usually be fed from the indoor unit to receive a TV channel, if it is offered by the satellite signal being received.

Hub earth stations are generally much larger than VSAT remotes because of the 3.5- to 11-m antennas typically used for Ku-band, and 7- to 18-m antennas for C-band. These can be similar in appearance to the 15-m Standard A earth stations used for international traffic. Transmit power levels can vary substantially, from a few watts for a solid state HPA up to about 400 W for a TWT, depending on the system application.

6.6   Nongeostationary Satellite Systems

LEO and MEO satellite capacity is very low compared to terrestrial cellular, which should lead to a peaceful coexistence because the systems complement each other. Indeed, cellular service providers see satellite systems as an opportunity to enhance their business. The claim of global operation is a grand statement, but a closer view shows that only limited coverage is possible in urban areas and indoors because of radio path shadowing and building attenuation. A key market exists in very remote areas where the telecommunications infrastructure is sparse or nonexistent.

The most important reason for using polar orbits is to enable communication from and to any point on earth. Clearly, a single polar orbiting satellite can cover only a small area at any one time, even though it is over much of the globe during the months and years that the earth rotates beneath it. A constellation of several satellites in each of several orbits is needed for full earth coverage at all times.

The relatively low altitude of LEOs means that the path loss is considerably smaller than for geostationary orbits. Consequently, the overall system gain for LEOs favors the use of small mobile handset "earth stations" instead of the large GSO earth stations. The economic potential of a LEO system is very attractive. During the next few years the extent of the success of nongeostationary orbit systems will become clear when two or more competing systems are fully deployed. As with all telecommunications services, success or failure depends on offering the service at an attractive price and the subsequent market share acquired by that price. The sources of market share in this case have two major components: (1) customers already using a terrestrial mobile system, and (2) potential customers who presently have no access to any mobile system, or at least not at an affordable price. Capturing long-term market share from existing terrestrial mobile operators will not be easy, especially because the performance of systems such as GSM, IS-54/136, and IS-95 is hard to beat. It is recognized that a major customer base for the LEO and *medium earth orbit* (MEO) systems will be from developing countries that have large rural populations and relatively modest telecommunications infrastructures. The per minute cost and, equally important, the handset cost will determine the extent of their usage.

From the technical viewpoint, there is already a battle among LEO and MEO participants. First, the 1992 World Administrative Radio Conference (WARC-92) allocated for worldwide services the use of frequencies 2483.5 to 2500 MHz in the S-band for the downlink, and the frequencies 1610 to 1626.5 MHz in the L-band for the uplink. This is a very small bandwidth and no doubt there will be serious lobbying from both proposed and potential operators for these bandwidths to be increased. If no further bands are allocated, several LEO and MEO systems must use the same frequency bands. Band sharing poses some technical questions that are difficult to answer. While multiple satellite-based CDMA systems can coexist in the same frequency band, the same is not true for multiple TDMA systems. Four major competing consortia have emerged. Iridium and INMARSAT-P are TDMA-based systems, whereas Globalstar and Odyssey are CDMA systems. Furthermore, the coexistence of a CDMA system on the same band as a TDMA system is possible, but the percentage drop in total capacity due to interference between the signals from the two systems is still being evaluated.

A nongeostationary earth orbit satellite constellation combines with the terrestrial network to form what is often referred to as a universal mobile telecommunications system (UMTS). The UMTS network architecture is illustrated in Chap. 7, Fig. 7.37. Fixed satellite earth stations are necessary to control the traffic within a geographic area known as the *guaranteed coverage area* (GCA). This is a relatively simple task for the geostationary satellite case because the area of the earth covered by each satellite does not change. The situation becomes considerably more complex when one or more satellites are in motion above the earth. Based on the satellite orbital details, each fixed earth station has information that allows it to predict the position of all satellites and spot beams passing over its area at any time in the future. Clearly, a satellite or spot beam does not have to pass directly overhead to enable communication but, as a passing LEO satellite moves farther away from the direct overhead pass, its visibility time (time it remains above the horizon) becomes shorter and the path loss correspondingly increases, becoming closer to the worst-case value. Satellites passing directly overhead are above the horizon for the longest time, but high population concentrations enforce the use of off-overhead satellite path usage because the capacity of just one satellite is insufficient to accommodate all the traffic.

Fortunately, in most cases, the speed of ground mobility even in fast cars is small relative to that of the satellite; otherwise the systems become even more complex. Communication to aircraft customers using LEO systems requires only slightly more handoff actions. The worst-case situation is when the aircraft is moving latitudinally (east-west or west-east) and not, as one might intuitively imagine, when moving longitudinally in the opposite direction to that of the satellite (north to south or south to north). In addition, the Doppler effect is a concern for high-speed mobile units.

Low-earth-orbit satellite systems.   These satellite constellations form the basis of a new category of global cellular radio systems. There is a definite sense of elegance that makes them very attractive. Even so, the financial risks involved are significant, because there are some technical difficulties that are not easy to surmount, as well as the serious competition that will evolve when several constellations are in service. The variation in propagation attenuation caused by the varying path length as a satellite makes a pass from the horizon (worst case of, say, $8°$ elevation) to overhead (best case) is about 10 dB of free-space loss plus rain attenuation. That attenuation difference is not small, and a power control mechanism is needed to cope with the variation. The shadow problem caused by a satellite moving behind a high building during a conversation is not trivial, and can lead to dropped calls in downtown areas. Some constellation designs allow for this. For example, Iridium has a 20-dB variable satellite EIRP to cope with shadow and path loss variations.

The additional attenuation experienced by a mobile unit operating within a building is high. Mobile satellite systems are primarily designed for outside use. Direct mobile-to-satellite connections within buildings are possible, but not in high-rise locations, where reasonable quality might be obtained only on the top one or two floors. Many factors can affect the call quality; for instance, being close to a window might provide unobstructed sight of a satellite for the duration of a call. In conditions that create poor signal quality, arrangements therefore have to be made to use the terrestrial cellular or PCS network. LEO service providers are already linking up with cellular counterparts. Initially, the cost of LEO connection per-minute charges will be higher than that of terrestrial operators, so a dual-mode handset is essential with the terrestrial mode as the default setting.

The LEO has the advantage of negligible time delay between earth and the satellites compared to the geostationary satellite systems. This relatively small link distance also allows communications via a low RF power output handset.

6.6.1 The Iridium LEO system

The first low earth orbit (LEO) satellite system is being pioneered by Motorola, providing partial operational status in 1998, and full operation two or three years later. The concept has six low-earth polar-orbital planes, each containing 11 satellites, totaling 66 satellites. These form a fixed "cage" around the earth, which rotates within it. The altitude of each satellite is 780 km above the earth's surface. The original design had seven satellites per orbit, totaling 77 satellites. The atom containing 77 electrons rotating about the nucleus is called iridium, which lends its name to the system. Because cost reduction efforts have reduced the total number of satellites to 66, perhaps the system is now more akin to the atom dysprosium. The orbits are separated by 31.6° for the corotating satellite planes and 22° for the counterrotating pair (1 and 6).

LEO satellites move at an amazing 27,000 km/h, and the antenna spot beam moves at 6.6 km/s relative to the earth as the satellite approaches the horizon. The complete orbital period is only 100 min and 28 s. Each satellite appears above the horizon for only 5 to 10 min, so a sophisticated interconnection between satellites is necessary to ensure continuity for a long call during which one satellite hands over the call to another overflying satellite either in the same or an adjacent orbit. Figure 6.26 shows the system overview.

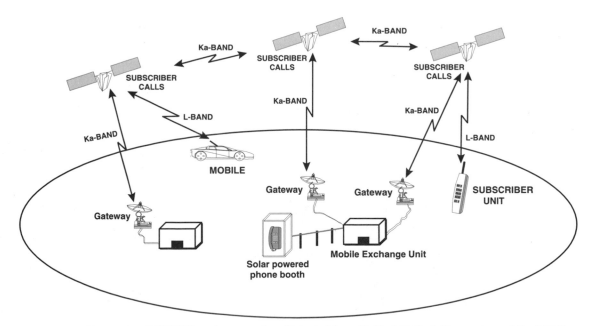

Figure 6.26 IRIDIUM system overview. (*Adapted from Hubbel, Y. C., A Comparison of the IRIDI-UM and AMPS Systems, IEEE Network, March/April 1997, pp. 52-59. © 1997 IEEE.*)

The system uses both FDMA and TDMA to make the most use of the limited spectrum available. The satellite has two modes of communication. First, a very remote or mobile user can communicate directly with a satellite moving overhead by an L-band frequency radio connection. An individual subscriber unit is a small hand-held, pocket-sized phone similar to present cellular style telephones. A fixed user accesses one of the satellites via a gateway. Intersatellite crosslink communications are at Ka-band, as are the satellite-to-ground system control links. Because the satellites orbit above the stratosphere, the path losses for the crosslinks do not include the rain attenuation losses that uplinks or downlinks suffer. The crosslinks operate in the 22.55- to 23.55-GHz frequency band at a transmission rate of 25 Mb/s. Each satellite can have four crosslinks. One is between each of the two satellites in front and behind within the same orbit, and the other two are to satellites on either side in adjacent orbits.

The large number of satellites necessitates a large number of launches, and the finite lifetime of each satellite will require almost a permanent, ongoing production and replacement of satellites. These systems are obviously very expensive. The charges, however, need to be competitive with other call charges, with a premium for the unique feature of *global roaming*. Iridium targets business travelers as its major customer base.

Iridium satellites all use onboard baseband processing to receive, route, and switch traffic between any of the satellites within the constellation. All satellites in adjacent orbits rotate in the same direction except for one pair of adjacent orbits, approximating a hexagonal cell structure rotating about the earth. The cell overlap increases toward the poles, and some beams are switched off to reduce interference and maintain effective frequency reuse.

Each satellite has 48 circular spot beams generated by phased-array antennas, which project a total earth coverage footprint of 4600 km in diameter. For earth traffic each satellite has three phased-array panel antennas oriented at 120° to each other. Each antenna forms 16 spot beams (Fig. 6.27), totaling 48 beams per satellite; there are consequently $48 \times 66 = 3168$ beams for the global constellation. Because the satellites converge toward the poles, only about 2150 spot beams need to be on for global coverage. The rest are turned off, which also conserves power.

Frequency reuse is on a 12-beam pattern basis. Only one satellite is accessed by a user at a given time and so no satellite diversity is used. The 12-dB of shadowing margin allows some building penetration. A satellite variable antenna gain from 7 to 15 dBi equalizes the signal into the satellite to account for variable path loss.

The level of onboard processing in Iridium is higher than all other previous satellite systems. Switching is done at the individual voice band level, so each satellite has demodulation-demultiplexing-switching-multiplexing-modulation capability. Satellite-to-earth-station gateway and system control links use Ka-band (19.4 to 19.6 GHz downlink; 29.1 to 29.3 GHz uplink) and earth stations use 6.25-Mb/s backhaul links for terrestrial interconnects. The overall throughput of each satellite is about 100 Mb/s.

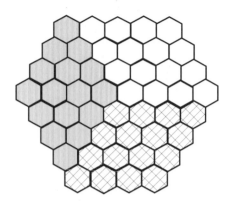

Figure 6.27 Iridium antenna beams pattern.

The subscriber links use the 1616- to 1626.5-MHz frequency band (10.5 MHz) for both up- and downlinks. The 12-frequency reuse (FDMA) is enhanced by each frequency containing a 90-ms TDMA frame that provides four 50-kb/s user channels. Voice channels use 4.8 kb/s, and data 2.4 kb/s. By using voice activation, each beam contains 80 channels, so there are $12 \times 80 = 960$ channels per 12-beam frequency reuse cluster. Because there are 2150 beams globally, the total channel capacity is $80 \times 2150 = 172,000$ channels available globally.

If one considers the global surface to be about 70 percent water, and perhaps another 15 to 20 percent dense forest, desert, or mountains, where population is also very sparse, it would appear that some spot beams could be underutilized.

6.6.2 The Globalstar LEO system

Globalstar is a LEO 48 satellite constellation with six satellites in each of eight 1414-km altitude orbits, inclined at 52° to the equator, covering areas up to ±80° latitude by two satellites simultaneously. It uses simple, traditional, "bent-pipe" satellite transponders. Each mobile unit uplink call is downlinked to one of a number of earth stations, with no intersatellite crosslinks in the system. Each satellite footprint is 7600 km in diameter, and at least one earth station is within every footprint at any time. Satellite diversity reduces shadowing loss problems. CDMA assists this diversity with soft handoff (see Sec. 7.8.4). Each satellite has 16 beams that are elongated as illustrated in Fig. 6.28, and every beam reuses the same frequencies. This beam shaping is designed to reduce the number of handoffs (compared to circular beams) during a long call. The mobile uplinks and downlinks use the L-band and S-band, respectively. Each band is FDMA divided into thirteen 1.25-MHz CDMA channels and used in all cells. The system is compatible with IS-95 terrestrial CDMA technology and a dual-mode terminal (handset) can be used. As with the terrestrial systems, voice coding is at rates that are variable from 1.2 to 9.6 kb/s. Each signal is direct-sequence spread over 1.25 MHz, and a 2-W transmit power is used only when speaking.

A major difference between Globalstar and Iridium is that Globalstar ground stations will be owned by a country's local service provider, which might be

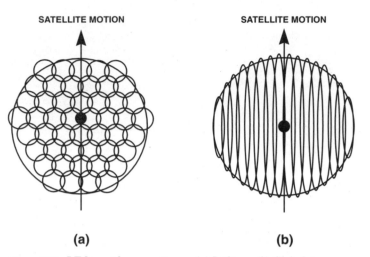

(a)   **(b)**

Figure 6.28   LEO spot beam patterns. (*a*) Iridium; (*b*) Globalstar.

government- or privately-owned. That can be more attractive to small, developing countries, which might prefer the local service provider to maintain control. Globalstar is targeting the rural-area populations.

It is recognized that the capacity of a CDMA system is power-limited on the downlink (from satellite to mobile) and interference-limited on the uplink (from mobile to satellite). The interference is due to two main sources: (1) code noise (internal), and (2) other systems sharing the same frequency bands (external).

6.6.3   The Teledesic LEO system

This proposed system is perhaps the grandest of them all. In its initial form, it envisages 840 LEO satellites, using Ka-band earth-satellite transmission, with onboard processing. The intention is to provide fixed earth terminals with data services ranging from 16 kb/s to 2 Mb/s using 25-cm aperture antennas.

There might be a need for such a system, but the huge number of satellites involved poses some serious questions concerning the practical viability of such a scheme. A problem that immediately comes to mind is the satellite launch rate. To achieve full satellite deployment could take several years. At an average rate of one deployment per day, it would take 2.3 years. Perhaps a scaled-down version with about 200 satellites would be more economically and logistically viable.

6.6.4   The MEO Odyssey satellite system

This satellite system is a medium earth orbit (MEO) system that has some interesting comparisons with the LEO system. The MEO system has some technical compromises between the GSO and LEO characteristics.

Twelve satellites orbit at an altitude of 10,355 km, providing global coverage as indicated in Fig. 6.29*a*. There are three circular orbits inclined at 55° to the equator and four satellites in each orbit. The deployment and operational costs should therefore be less than for the LEO network. Odyssey will provide voice, data, paging, and messaging services to customers worldwide. Inevitably, there are some trade-offs. The MEO satellite-to-earth link distance is larger than that of the LEO, which translates into higher transmit power needed from the satellite and mobile unit to overcome the additional free-space loss. As a consequence, the customer might have to accept either a larger handset or less talk time between battery recharges.

Satellite transponders are of the "bent pipe" design, with no onboard baseband processing. Each satellite has a multibeam antenna pattern that divides its assigned coverage area into a set of overlapping cells so that two or three satellites are always visible from any point on earth. The antennas might be pointed off vertical to track the specific desired coverage areas. This key feature of the Odyssey system allows multiple satellite coverage over high-population-density land areas while simultaneously maintaining single-satellite coverage of other areas including ocean regions.

Odyssey does not use satellite diversity. Each spacecraft has three gimbal-mounted Ka-band antennas, independently pointing toward earth for control links. Each satellite has 37 spot beams for mobile up- and downlinks. The MEO system enjoys a higher elevation angle than the LEO from the point of view of a customer on earth, which reduces shadowing losses. The 6-h orbital period means handovers are unnecessary for the MEO system. Satellite crosslinks are not needed, thereby considerably reducing system complexity. The MEO link margin should generally offer a high availability of service. The Odyssey plan has six satellites in orbit at the end of 1999 and the full 12 at the end of 2000.

The power limitation in handsets used in MEO systems and the need for synchronization between terminals does not favor the choice of TDMA operation. Odyssey uses CDMA, with mobile uplinks in the L-band and downlinks in S-band. Three channels, each of 4.833 MHz, are used for DSSS. Each 4.833-MHz signal is BPSK modulated. Mobile terminal transmit power is in the range 0.5 to 5.0 W.

6.6.5   The INMARSAT-P (I-CO) MEO system

INMARSAT has embarked on a project that will also be a MEO satellite system. This project, called I-CO, was previously known as INMARSAT-P and will operate 10 satellites in MEO and two spares in two separate MEO planes.

This 10-satellite constellation operates 5 satellites in each of two 45° inclined orbits (Fig. 6.29*b*). At present, the system design is still evolving; currently its high capacity derives from 163 beams to and from each satellite to earth, with each beam steerable so as to remain geographically fixed during traffic activity. Because each beam contains 28 channels, the capacity is 4564 simultaneous calls per satellite.

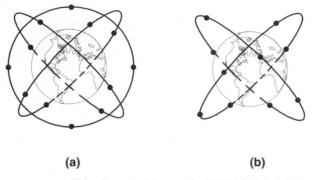

(a)   **(b)**

Figure 6.29   MEO orbits. (a) Odyssey; (b) INMARSAT-P (ICO).

Satellite diversity is used to enhance availability by reducing the effects of shadowing. Mobile terminal up- and downlinks are in the IMT-2000 2-GHz bands. Although "bent pipe" transponders route calls from mobiles to their nearest earth stations as do Odyssey and Globalstar, frequencies are reused similar to Iridium (geographically fixed and in this case a four-cell reuse).

6.6.6 Brief comparison of LEO and MEO systems

Table 6.9 compares some of the technical specifications of the present four major contenders in the race to capture nongeostationary satellite service market share. Iridium and Globalstar are LEO satellites, and Odyssey and INMARSAT-P are MEO satellites. Iridium is, for practical assessment, a polar-orbit system, whereas the other three use inclined orbits. The Globalstar LEO satellites have a higher altitude than Iridium, and consequently they have about 5 dB more path loss. The two MEO satellites are designed to operate at almost the same altitude, and their path loss is about 25 dB higher than the Iridium LEO system. INMARSAT has 10 satellites, two orbits each containing five satellites. Odyssey has 12 satellites, three orbits each containing four satellites. The difference is more significant for the two LEO constellations, with Iridium having 66 satellites compared to Globalstar's 48. At the higher altitude, Globalstar satellites have a larger duration of visibility (about 16 min compared to 12 for Iridium for a given latitude), so they can provide earth coverage with fewer satellites. Iridium has a link shadow margin of 12 dB compared to several dB less for the other three systems. The transmission delay is insignificant for LEO satellites, and the minimum 34.5-ms delay for MEO satellites is not high enough to cause any problems, that is, unless the MEO hop is combined with other mobile service hops containing very high levels of digital signal processing. In that case, there is a combined delay up around the 300-ms region. INMARSAT has the highest number of spot beams at 163, which gives more flexibility for traffic management than systems with fewer spot beams. The high-gain spot beams also help to compensate for the higher path loss of the MEO satellites.

TABLE 6.9 Comparison of LEO and MEO System Parameters

Uplink parameters (Earth-to-space)	LEO		MEO	
	Iridium	Globalstar	Odyssey	INMARSAT (ICO)
Orbit				
Altitude, km	780	1414	10354	10355
Satellite separation, degrees	32.7	60	90	72
Number of satellites	66	48	12	10
Orbital planes	6	8	3	2
Inclination angle, degrees	86	52	50	45
Orbital period, min	100	113	359.5	358.9
Satellite max. visibility time, min	11.1	16.4	94.5	115.6
Frequency bands, GHz				
mobile-to-satellite, service				
Uplink frequency, GHz	1.616 – 1.6265	1.610 – 1.6265	1.610 – 1.6265	1.98 – 2.01
Downlink frequency, GHz	1.616 – 1.6265	2.4835 – 2.50	2.4835 – 2.50	2.17 – 2.20
earth station-to-satellite, feeder				
Uplink frequency, GHz	27.5 – 30	5.09 – 5.25	29.1 – 29.4	5
Downlink frequency, GHz	18.8 – 20.2	6.875 – 7.055	19.3 – 19.6	7
Intersatellite crosslink, GHz	23.18 – 23.38	None	None	None
Polarization				
Feeder link	RHCP	RH/LHCP	RHCP	CP
Service link	RHCP	LHCP	LHCP	RHCP
Satellite characteristics				
Number of spot beams per satellite, for the service links	48	16	61	163
Frequency reuse, cells per cluster	12	1	3	4
Beam size, km²	1.8×10^5 to 7×10^5	6.3×10^5 to 2.3×10^6	9.7×10^5 (6.3°)	5×10^5 to 2×10^6
Average beam sidelobes, dB	–20	–15	–20	–20 (peak)
Total satellite output power, dBW	31.5	30	37.9	34
Average gain / beam, dBi	17 to 25	N/a	24 to 28	30
Satellite mass, kg	700	450	2207	1925
Link characteristics				
Nominal user EIRP, dBW	–4 to + 6 (peak)	–3	–5.8 to –11	–1 (avge) +7 (peak)
Satellite G/T, dB/K	–3 to –10	–17	–1.4 to 1.8	2.0
Unshadowed user EIRP, dBW	7 to 15	N/a	20.6	*
Shadowed user EIRP, dBW	19 to 27	0 to 5	24.6	*
EIRP / CDMA channel, dBW	N/a	0 to 16	*	N/a
Handset G/T, dB/K	–23	–23	–22.2 to –24	–24
Free space loss [up / down], dB	154 / 154	160 / 163	177 / 181	179 / 179
Minimum elevation angle, degrees	8.3	10	20	10
Transmission parameters				
Voice telephony, kb/s	2.4 / 4.8	2.4 / 4.8 / 9.6	4.8	4.8
Data for BER < 10^{-3}, kb/s	2.4	7.2	2.4 – 9.6	2.4
Modulation	QPSK	QPSK	QPSK	QPSK
Coding	FEC	FEC	FEC	FEC
Access scheme	FDMA/TDMA	FDMA/CDMA	FDMA/CDMA	FDMA/TDMA
Duplex scheme	TDD	FDD	FDD	FDD

TABLE 6.9   Comparison of LEO and MEO System Parameters (*Cont.*)

UPLINK parameters (Earth-to-space)	LEO		MEO	
	Iridium	Globalstar	Odyssey	INMARSAT (ICO)
Frame length, ms	90	N/a	N/a	40
Burst rate, kb/s	50	N/a	N/a	36
Chip rate, Mchips/s	N/a	1.2288	~2.0	N/a
Voice activity factor	0.4	0.4	0.5	0.4
Modulation bandwidth, MHz	0.0315	1.2	2.5	0.025
Required E_b/N_o, dB	6.1	4.8	4.0	2.5

* means still under consideration.
N/a means not applicable.

6.7   Nongeostationary Satellite Path/Link Budget

The small size requirement characteristic of handset mobile communications poses some severe technical problems for a satellite-based global system. The path length is considerably more favorable than in the very high altitude GSO systems, but there are also some additional complications that were not present in GSOs. As more satellite constellations are deployed, interference between users within the same constellation and from other constellations reduces the system capacity from the theoretical maximum values. The following sections place all this into perspective by looking at the link budget numbers for LEO and MEO systems.

6.7.1   LEOs

For LEOs the path length (and therefore free-space loss) is much shorter than that of GSOs. The LEOs also operate at lower frequencies than GSOs, which further reduces the free-space loss. To provide some numbers, the GSO has a 6-GHz free-space loss of about 200 dB compared to the LEO, which has a 1.6265-GHz free-space loss of about 154.5 and 159.6 dB for altitudes of 780 and 1414 km, respectively. This 40- to 45-dB difference is mainly absorbed by the lower mobile handset EIRP. The handset must have a simple omnidirectional antenna and a transmitter output power as low as possible.

The link equations for the LEO take the same form as for GSOs, with a few exceptions. Three additional losses are:

1. Variable path loss with satellite position

2. Shadow loss from obstructions

3. Building loss for indoor use

These losses now appear in the link equations and were not present for GSOs. The LEO variable path loss is due to the path distance varying as the satellite moves from the horizon to its maximum height in the sky and down to the horizon again. The worst-case low horizon loss is about 10 dB higher than directly overhead. The 8° elevation angle path distance is about three times the satellite orbiting altitude. Shadow losses are due to buildings, trees, or other objects temporarily or permanently blocking the direct line-of-sight path during a call. This is caused either by a satellite moving overhead and passing behind an obstacle while the mobile caller is stationary, or by the mobile caller moving into a shadow. Building losses are caused by concrete, wood, or even worse, metal obstructing the line-of-sight path. Obviously, this will be much worse for handsets used in the lower stories of high-rise buildings.

Equation 6.10 for the E_b/N_o ratio now has some modifications from the GSO figures. The $(E_b/N_o)_{total}$ required to produce a specific quality value does not change, and in fact determines the other parameters that are needed to achieve a particular $(E_b/N_o)_{total}$. The variables of concern are therefore the EIRP and G/T values for the earth station and satellite, the path loss and bandwidth.

Uplink.

$$[10 \log \left(\frac{E_b}{N_o}\right)]_U = [10 \log (\text{EIRP})]_{ES} - [10 \log \alpha_p]_U + [10 \log \left(\frac{G}{T}\right)]_{SAT}$$
$$- [10 \log k] - [10 \log B_r] \text{ dB}$$

The LEO uplink budget based on Eq. 6.10 for a typical TDMA LEO satellite system is presented in Table 6.10.

EIRP$_{ES}$. For the uplink, the earth station EIRP is much less than for GSOs, because the handset antenna gain must be 0 dBi. The amount of power a handset can generate is also severely limited; otherwise the weight of the battery becomes unacceptable. A handset mobile transmitter average power output of 10 mW would be ideal to be consistent with GSM handsets, but a value between 300 and 400 mW is more realistic. An EIRP$_{ES}$ value in the region of −5 dBW (average) can be expected for LEOs.

(G/T)$_{SAT}$. The satellite G/T has a typical spot beam value of −3 dB/K per carrier (user).

Downlink

$$[10 \log \left(\frac{E_b}{N_o}\right)]_D = [10 \log (\text{EIRP})]_{SAT} - [10 \log \alpha_p]_D + [10 \log \left(\frac{G}{T}\right)]_{ES}$$
$$- [10 \log k] - [10 \log B_r] \text{ dB}$$

The LEO downlink budget is also shown in Table 6.10.

TABLE 6.10   LEO Link Budget

	Uplink	
1	Earth station (handset) transmit power	−5.0 dBW
2	Earth station (handset) antenna gain	0.0 dB
3	Uplink EIRP [1+2]	−5.0 dBW
4	Uplink path loss (at zenith)	154.5 dB
5	Variation in path loss	8.0 dB
6	Air absorption loss	0.2 dB
7	Satellite received power [3−4−5−6]	−167.7 dBW
8	Satellite G/T per carrier (user)	−3.0 dB/K
9	Boltzmann's constant	−228.6 dBW/Hz/K
10	C/N_o [7+8−9]	57.9 dB
11	Bit rate B_r (50 kb/s)	47.0 dB
12	E_b/N_o [10−11]	10.9 dB
13	Required E_b/N_o	6.1 dB
	Margin [12−13]	4.8 dB

	Downlink	
1	Satellite EIRP per beam	22.0 dBW
2	Downlink path loss	154.5 dB
3	Variation in path loss	8 dB
4	Air absorption loss	0.2 dB
5	Handset received power [1−2−3−4]	−140.7 dBW
6	Handset G/T	−23.0 dB/K
7	Boltzmann's constant	−228.6 dBW/Hz/K
8	C/N_o [5+6−7]	64.9 dB
9	Bit rate B_r (50 kb/s)	47.0 dB
10	E_b/N_o [8−9]	17.9 dB
11	Required E_b/N_o	6.1 dB
	Margin [10−11]	11.8 dB

Uplink frequency = 1.6265 GHz; downlink frequency = 1.625 GHz; path length (distance from handset to satellite) = 780 km.

EIRP$_{SAT}$. The satellite EIRP for a LEO spot beam is lower than for a GSO spot beam system. For a given spot beam area of coverage on earth, the antenna focusing power (gain) has to be greater the higher the orbiting altitude of the GSO satellite. Whereas a GSO satellite needs a 15-dB gain antenna for full earth coverage, such an antenna on a LEO satellite would produce a spot beam of about 2×10^5 km^2 in area, or 500 km in diameter. LEO spot beams are usually even more focused, and have a gain of about 17 to 25 dBi per beam, with gain variation to maintain a near-constant input to the satellite receiver. The total available satellite power is split equally between its spot beams, so the satellite EIRP per spot beam is relatively low. However, it is still about 22 dBW, which is more than adequate to offset the 155-dB or so L-band free-space loss.

$(G/T)_{ES}$. The earth station (handset) gain has to be 0 dBi so that reception is possible regardless of the orientation of the user's handset. Therefore, the earth station G/T is typically –23 dB/K for each user, which is clearly the weak link in the chain.

Comments. The E_b/N_o required to give a satisfactory QoS is in the order of 6 dB. This figure is smaller as the amount of handset and onboard processing increases.

The 11 MHz of bandwidth (within 2.48 to 2.50 GHz) planned for band sharing by Globalstar and Odyssey was considered acceptable because (1) CDMA is used and (2) one is a LEO and the other is a MEO. However, some calculations indicate a reduction in the total available capacity of such band-sharing systems. Because the LEO and MEO systems are recent innovations, a few adjustments can no doubt be expected in the early days of operation.

6.7.2   MEOs

Table 6.11 is the link budget for a typical MEO satellite system. There are several similarities between the LEO and MEO systems. The handset EIRP is higher than the LEO. The higher altitude has a free-space loss of close to 180 dB at 2 GHz. The MEO satellite beams might have a similar gain to the LEO, but because there are fewer satellites, more beams are necessary for economically attractive capacity. The available onboard RF power generated is limited and therefore satellite EIRP per beam might not be as high as desirable. Using more solar panels is only part of the solution.

The additional free-space loss of about 20 to 25 dB compared to LEOs must be offset by a combination of satellite G/T, and user and satellite EIRP. The other parameter in the equations is the E_b/N_o. By increasing the handset and satellite signal processing at the expense of latency, a value as low as 2.5 dB can be accepted.

The MEO does not have the variable altitude overpass problem of the LEO system, or not to the same extent. A satellite passing directly overhead (zenith) would be above the horizon for 1.5 to 2 h. It does have the same building loss problem indoors, but shadow loss might be less as the satellites appear higher in the sky. The margin values calculated here might not be high enough, and increased satellite or handset EIRP might be necessary for margin improvement.

6.8   Satellite TV Systems

TV signals transmitted over satellite using analog techniques (FM) are limited to one, or possibly two, TV channels per transponder. The remarkable advances in video compression techniques over the past few years have now enabled five to ten digital TV signals to be transmitted, depending on the level of quality required. Digital video signals for home use are typically 2 to 8 Mb/s depending on the program content. Obviously a football game contains more rapid motion than a talk show. Only a few years ago data rates of 30 to 100 Mb/s were required to achieve similar subjective quality.

TABLE 6.11   MEO Link Budget

	Uplink (2.0 GHz)	
1	Earth station (handset) transmit power	−1.0 dBW
2	Earth station (handset) antenna gain	0.0 dB
3	Uplink EIRP [1+2]	−1.0 dBW
4	Uplink path loss	178.7 dB
5	Air absorption loss	0.2 dB
6	Satellite received power [3−4−5]	−179.9 dBW
7	Satellite G/T per user	2.0 dB/K
8	Boltzmann's constant	−228.6 dBW/Hz/K
9	C/N_o [6+7−8]	50.7 dB
10	Bit rate B$_r$ [36 kb/s]	45.6 dB
11	E_b/N_o [9−10]	5.1 dB
12	Required E_b/N_o	2.5 dB
	Margin [11−12]	2.6 dB

	Downlink (2.2 GHz)	
1	Satellite EIRP per beam	24.2 dBW
2	Downlink path loss	179.6 dB
3	Air absorption loss	0.2 dB
4	Handset received power [1−2−3]	−155.6 dBW
5	Handset G/T	−22.2 dB/K
6	Boltzmann's constant	−228.6 dBW/Hz/K
7	C/N_o [4+5−6]	50.8 dB
8	Bit rate B$_r$ (36 kb/s)	45.6 dB
9	$E_b N_o$ [7−8]	5.3 dB
10	Required E_b/N_o	2.5 dB
	Margin [9−10]	2.8 dB

Path length (distance from handset to satellite) = 10355 km.

Direct-to-the-home video has certainly benefited from the advances made in satellite EIRPs, earth station G/Ts, and concatenated coding, but the most significant advances have been made by video compression. In 1990 it was difficult to place one video signal on a 36-MHz transponder. By 1996, five to ten video channels could be accommodated in a 36-MHz transponder.

6.8.1   Satellite TV broadcasting

Broadcasting TV via satellite is a huge business. At the receiving end, individuals have their own dishes and receivers. Hotels put a dish on the roof and then distribute the channels to every room in the hotel. Cable companies can use the broadcast from several satellites and package the large number of channels for distribution along their coaxial cables.

For individual users receiving satellite to the home, the equipment is very simple and getting cheaper every year. All that is needed is an antenna, low-noise block downconverter (LNB), satellite receiver, and TV set (see Fig. 6.30a). The dish size, as usual, depends on the frequency band, satellite EIRP, LNB

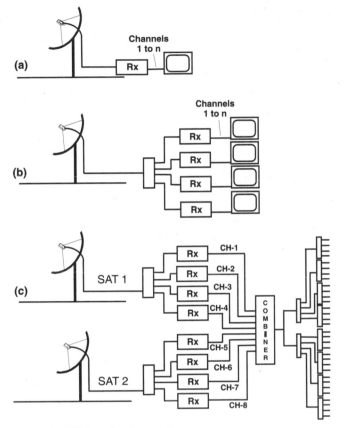

Figure 6.30 TV broadcast reception.

noise temperature, and the latitude of the receiver location. The satellite EIRP is designed so that, for example, the C-band dishes are typically 2 to 2.5 m in diameter in the tropical zone while at high latitudes, say 35° or more, larger diameters of 3 to 3.5 m are needed. The noise temperature of the LNB for the 2-m dishes at about 30° latitude must be about 25 K, which is achievable with a cheap, mass-produced HEMT amplifier within the LNB. A 35-K LNB would cause noise that would appear on the TV picture as "snow," and a 45-K LNB would be almost unwatchable.

Ku-band satellite systems are designed for use with very small antenna sizes, typically 0.5 to 1 m in diameter.

For a small number of users (say four) sharing the same C-band dish and LNB, a 4-way splitter will allow the downconverted IF signal (about 950 to 1050 MHz) to feed four satellite receivers and four televisions, so that each receiver is tuned to a specific channel on each TV and each user can tune to a chosen satellite program by tuning the receiver to a specific program frequency (see Fig. 6.30*b*).

If the desired programs are not available from one satellite, two dishes are needed together with two satellite receivers for an individual user. Widespread distribution of many channels would require more complexity as shown in Fig. 6.30c. For example, four channels from each satellite would be received using two four-way splitters and eight receivers. These would all be colocated in a cabinet and the outputs from the receivers all combined into a single amplified signal, then split into as many coaxial cables as required. In the case of a large hotel this could be hundreds of rooms.

In some places, C-band reception from two satellites has been achieved using one dish and two LNBs. One LNB is fixed at the focus of the parabaloid, and the other is fixed off-axis to receive a glancing-angle signal. The extra aperture blockage reduces the antenna efficiency. This setup, although not recommended, can give reasonable reception if the satellites have orbits that are neither too far apart nor so close that the LNBs would need to overlap.

6.9   Future Developments

There are several challenging obstacles to be overcome for future satellite communication systems and perhaps none is more pressing than the improvement in satellite information throughput and packaging. It is not just a case of providing more voice channels for the available bandwidth, but also providing dynamic bandwidth allocation or bandwidth-on-demand. Closely related to this is the incorporation of multimedia services, including the simultaneous transmission of real-time (voice and video) and non-real-time (data) information. To elaborate, the transmission of SDH packages of bit streams at the STM-1 rate and above will be essential. Also, ATM via satellite will be necessary as an alternative medium to optical fiber.

The severe bandwidth restriction problem of satellite communication systems compared to optical fiber systems will probably never be solved with the present bandwidth allocations. Significant advances have been made over the past decade in the areas of speech and video compression, DSI, modulation techniques, and multiple access techniques. There are limits to the bandwidth efficiency that can be achieved while maintaining adequate quality of service. Eventually, unless more spectrum is made available for satellite transponders in the existing operating bands, the only option is to operate at higher and higher frequency bands. Technology is now well developed at the Ku-band and rapidly maturing at the Ka-band (20/30-GHz band), but the laws of physics are not in the designers' favor at higher frequencies, where water and gases in the earth's atmosphere cause severe attenuation.

The attractive broadcast feature of satellite communications will guarantee a niche for its future survival, particularly in the well-tested area of video to the home. For nonbroadcast applications, LEO and MEO should give voice communication via satellite a new lease on life. However, the future for serious data transmission (multigigabits) via satellite is more questionable.

Bearing these statements in mind, it is interesting to examine some of the technological developments still in the laboratory phase, or just in the process of implementation.

6.9.1 Onboard switching

The more sophisticated antenna beam arrangements on the latest generations of GSO satellites improve the EIRP from the satellite to the small spot beam areas and also allow more frequency reuse. The disadvantage of spot beams is that it becomes increasingly difficult to interconnect all the earth stations. Since INTELSAT VI, it has been possible to switch from an uplink hemi or zone beam to one of several downlinks, mid-TDMA frame. This satellite switched or SS-TDMA is done at the RF and not the baseband level.

Onboard switching has been the subject of much debate in recent years, especially concerning the level and complexity that should be used. The idea of a geostationary satellite becoming a user-to-user switching center like a CO is not realistic for thousands of users, but a nongeostationary satellite does use individual voiceband switching. For the geostationary scenario, ideally there would be a satellite spot beam directed to every household, which, now and in the foreseeable future, is not technically feasible. Furthermore, GSO transceiver capability would have to be on the customer premises. Presently, the only options for geostationary satellites are for the satellite to switch incoming information among several downlink beams and to carry out performance-improving signal processing. SS-TDMA is already functioning well for improved routing efficiency. The other question is, Should there be onboard regeneration of the GSO uplink signal? The fact that individual channel drop/insert at the satellite has little or no meaning unless there are customers in space, leads to the conclusion that regeneration would be a purely signal-improving technique. Onboard regeneration separates the uplink and downlink C/N components so that the total C/N value is simply the sum of the up- and downlink values. The downlink budget is further improved by using FEC. While there is usually merit in improving quality, in the case of regeneration it would be at the expense of additional weight (and therefore cost) and potential reliability problems. VSAT systems for a relatively small number of users (up to about 100) can benefit from onboard switching, although the geographical distribution of the users places a constraint on the total number of users.

LEO systems also certainly benefit from a CO style of onboard switch. System flexibility is necessary to enable a mobile user in a remote part of the world to access another mobile user in a remote location. Onboard processing routes the call through the network of satellites by intersatellite links, and the downlink can be direct to the destination mobile handset.

Reliability has always been a major concern in satellite systems, and increased onboard processing makes it an even more critical factor. It means equipment redundancy is essential to prevent system lifetimes from being too

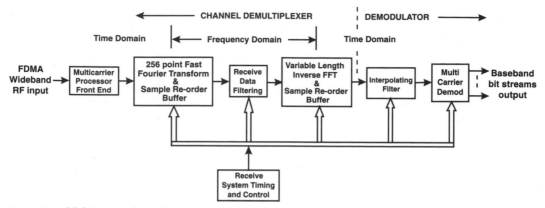

Figure 6.31 Multicarrier demultiplexer and demodulator.

short. Space-shuttle-style maintenance is possible for satellite repairs in space, but the cost is immense.

Onboard regeneration requires demultiplexing and demodulation of multiple uplink carriers. The multicarrier demultiplexer/demodulator (MCDD), which can simultaneously demodulate many channels, is a recent innovation considered to be essential for future VSAT satellites. An MCDD takes a block of spectrum (at microwave frequencies) and performs an A-to-D conversion similar to the sampling and quantization of PCM. The number of carriers contained in the block is irrelevant, because they are individually demodulated digitally. Separate conventional demodulators would be unsatisfactory for a number of reasons such as size, weight, and power consumption. Figure 6.31 shows a typical MCDD. The digitized signal is demultiplexed by performing a discrete Fourier transform (DFT), resulting in a frequency domain equivalent signal. The signal is multiplied by individual frequency coefficients, which is a form of filtering process to isolate channels (i.e., demultiplexing). An inverse Fourier transform on each carrier then converts the demultiplexed signals back into the time domain. Individual call data bit streams are then formed by demodulating the QPSK signals into 64-kb/s signals in the case of SCPC, or time division multiplexed signals at 2 Mb/s or higher. In fact, the above processes have been known for many years. It is only recently that the speed and level of integration of electronic circuits have become sufficient to cope with the complexity of the components within an MCDD.

6.9.2   Ka-band transmission

There is a general trend to operation at higher-frequency bands. There are two reasons for this: (1) the lower bands are already crowded, and acquiring new allocations is becoming increasingly difficult; and (2) for a given percentage bandwidth, the higher frequency bands have a wider absolute bandwidth. The satellite Ka-band has 2500 MHz of available bandwidth, compared to 1000 MHz at Ku-band and only 500 at C-band. Transmission at the Ka-band having a 30-GHz uplink and 20-GHz downlink is technically more demanding on the hardware design. The higher rain attenuation at Ka-band must also be compensated for by dynamic transmitter output power control.

There are several Ka-band satellite systems already operational, for example, the U.S. ACTS, Japanese CS-SAT, and German DFS-Kopernicus. Intersatellite links operate above the atmosphere, so their "clear skies" performance does not suffer the severe rain attenuation of the earth-satellite links. Availability and performance objectives for the Ka-band are still evolving. In this category, to mention a few, there are VSATs and USATs, satellite news gathering (SNG), teleconferencing, and distance learning.

Some economic aspects of Ka-band satellite operation are very encouraging. The circuit capacity or information throughput is several times that of lower frequency satellite systems on a comparable weight basis. This is achieved by several factors, such as employing reusable frequency spot beams with a TDMA hopping network, and reduced component size and therefore less weight at these higher frequencies. That is in addition to the factor of 5 bandwidth advantage compared to C-band.

To give a measure of the size reduction advantage at Ka-band, the gain of a 5-m uplink antenna at Ka-band is the same as an 11-m uplink antenna at Ku-band. Unfortunately, at present one of the main techniques proposed for overcoming the fading at Ka-band is to use frequency diversity. This is a simple matter for a terrestrial microwave radio link, but duplicating the satellite earth station antenna and RF equipment is very expensive.

6.9.3   Intersatellite links

Because all satellites operate above the atmosphere, frequencies at Ka-band or even much higher can be used for intersatellite transmission, including optical frequencies. Large bandwidths (typically 1000 MHz or more) are available for intersatellite communications. This is more bandwidth than is needed for present applications, but historically such situations do not last very long.

GSO satellite systems can benefit slightly from intersatellite links to reduce time delay on links that would otherwise need a double hop. For a system of three equally spaced GSOs, the intersatellite link is a total path distance of about $(2 \times 40,000) + 62,350 = 142,350$ km for high latitude earth stations, compared to a double hop distance of about $(2 \times 40,000) + (2 \times 40,000) = 160,000$ km. The difference in time delay is not very significant; it is 474 ms compared to 533 ms. For six equally spaced satellites in GSO, the path distance improves to $(2 \times 40,000) + 40,000 = 120,000$ km, with a delay of about 400 ms.

On the other hand, compared to GSOs, LEOs already have a small delay, less than 3 ms. Because they must be able to communicate with their nearest neighbor satellites for call routing, intersatellite links are essential for LEO systems.

6.9.4   Wideband transmission through satellites

Although the total available bandwidth of a transponder is relatively small compared to optical fiber technology, the bandwidth efficiency of the existing satellite transponders continues to be enhanced. The transmission of a 155.52-Mb/s SDH signal is now possible over a 72-MHz satellite transponder. This does not offer any more digital voice channels than the older analog technology, but it does allow the flexibility of offering variable bandwidth or "bandwidth-on-demand," which is necessary to compete with services offered by terrestrial technology (optical fiber).

The maximum bit rate that can be transmitted over a 72-MHz transponder is limited by (1) out-of-band emission requirements, and (2) channel distortions increasing with bandwidth. Out-of-band emission is usually the dominant factor, because INTELSAT specifies the transmit spectrum emission to be 26 dB down outside the user's bandwidth of 80 MHz (transponder spacing). Taking filtering into account, the maximum usable bandwidth is about 64 MHz, and even this reduced bandwidth necessitates the use of steeper roll-off filters than in previously established satellite services. The required bandwidth efficiency is therefore 2.43 b/s/Hz (155.52 b/s divided by 64 MHz), which can be achieved by using a modulation scheme such as 8-PSK or 16-PSK. Incidentally, multilevel QAM is considered unsuitable because of the nonlinearities experienced in satellite channels, but that position could change in the future as technology improves. Although 16-PSK modems are potentially 1 dB more power-efficient than 8-PSK, at 2.5 b/s/Hz they are more susceptible to channel and modem impairments. 8-PSK with trellis coding is a suitable solution. The code rate is established, for example, by allowing a bandwidth expansion of 8.5 percent for the (235,255) RS outer code, meaning the inner code should be at least $2.43 \times 1.085/3 = 0.879$. A rate of $8/9 = 0.888$ would therefore be suitable for trellis coding.

6.9.5   ATM bandwidth-on-demand over satellite links

The packet-based transmission and networking medium known as ATM is usually associated with optical fiber technology and STM-n bit rates. ATM, as described in Chap. 10, is the broadband ISDN standard. Although this scheme has already been adopted, it will be well beyond the year 2000 before it becomes a widespread physical reality. In the meantime, many ATM islands are growing and they need to be interconnected. Eventually this will be done by optical fiber but, in the interim, satellite is a medium that can fulfill the role.

Because ATM incorporates voice, video, and data into one transmission medium it has some improved efficiency characteristics compared to TDM systems. Also, the attractive ATM feature of bandwidth-on-demand has made satellite operators seriously consider ATM over satellite links. ATM is intended to have a minimum bit rate of STM-1 (i.e., STS-3 or 155.52 Mb/s), but DS3 (45 Mb/s) or STS-1 (51.8 Mb/s) are options for ATM through satellite links. The INTELSAT IDR allows up to 45 Mb/s, but the performance of the IDR link in terms of BER, errored seconds, etc., as recommended by ITU-T G.826 is not easily achievable. The channel coding necessary in satellite links causes bursts of errors. This is a problem because while ATM can correct single ran-

TABLE 6.12   ATM Typical BER Objectives

Applications	Bit Rate	BER
Existing video phone	64 kb/s	$< 10^{-6}$
Future video phone	2 Mb/s	$< 3 \times 10^{-10}$
Video conferencing	5 Mb/s	$< 1 \times 10^{-10}$
TV distribution	20 to 50 Mb/s	$< 3 \times 10^{-12}$
MPEG1	1.5 Mb/s	$< 4 \times 10^{-10}$
MPEG2	10 Mb/s	$< 6 \times 10^{-11}$

dom errors, it cannot correct error bursts, resulting in cells being discarded by the network nodes. Objectives for ATM performance are quite stringent. Table 6.12 shows typical objectives figures for ATM networks. For satellite IDR links to achieve these BER values, tighter link specifications are needed, or more robust coding schemes such as Reed-Solomon codecs with adequate interleavers.

In the never-ending struggle to remain competitive with optical fiber systems, satellite systems must be able to offer ATM services. Unfortunately, the provision of limited ATM services *at low bit rates* of 155 Mb/s or less will only accentuate the bandwidth inadequacy of the satellite transmission medium compared to that of optical fiber.

Mobile Radio Communications

7.1 Introduction

From the customer's point of view, it is mobile systems that are perhaps the most exciting telecommunications development since the invention of the telephone. The developments in optical fiber technology might sound very impressive and the statistics involved are mind-boggling, but the average subscriber does not usually appreciate the full extent of the benefits at a personal level. The developments in this field are taking place mainly behind the scenes, and are not really tangible service improvements to the subscriber. The pocket telephone, on the other hand, is a revelation that is center-stage and whose benefits can be instantly appreciated by anyone who purchases one of these devices. The cellular telephone industry has experienced explosive growth over recent years. It is an area of telecommunications that has benefited not only the developed world, but also many developing countries. From the service provider's point of view, cellular systems are very fast to install compared to extending new cables to customer premises. When a leading telecommunications company forecast in 1996 that by the year 2000 there would be 350 to 400 million mobile radio units worldwide, there was widespread disbelief. So far, all forecasts of this nature have turned out to be conservative.

Despite the many advantages of being in contact with business associates or friends at all hours of the day, no matter where one might happen to be within a city, the pocket telephone does have its drawbacks. The antisocial aspects of receiving calls in places such as quiet restaurants or theaters, or during conferences, have prompted many establishments to require customers to turn off their pocket phones.

Cellular mobile telephone systems are not easy to classify. They could be considered as part of the local loop because they extend out to the subscriber handset, or because of the long distances that can be bridged between a fixed subscriber and a mobile subscriber (or mobile-to-mobile subscribers), they could be called long-haul circuits. The technology incorporates some of the most advanced radio transmission techniques. In addition, the call processing requires high-level digital switching techniques to locate the mobile subscriber and to set up and maintain calls while the mobile subscriber is in transit.

The portable telephone was made possible by the miniaturization resulting from VLSI electronic circuitry. Even today there are still some technological problems to solve, such as increasing the time between recharging the batteries of portable telephones. More serious is the fact that high-mobility cellular telephones and low-mobility radio telephones have some severe technical incompatibilities. These take two forms. First, the systems developed in different parts of the world (e.g., Europe, North America, and Japan) do not yet even have the same operating frequencies, and the system designs vary quite considerably. However, ANSI and the ITU are currently coordinating recommendations with all major parties to enable equipment compatibility for the next generation of cellular mobile radio systems. Second, one can say there are three major categories of cellular radio systems: (1) analog FM, (2) narrowband TDMA, and (3) CDMA. The analog systems that have been around for a number of years are giving way to digital technology. Narrowband TDMA is currently seeing widespread deployment, and CDMA started deployment in 1996. Significant equipment incompatibilities are encountered when moving from one system to another or trying to incorporate two or three of these into the same network. Many of these problems stem from the difference in power outputs. High-mobility cellular radios transmit at relatively high power levels in the region of 1 to 10 W, whereas the latest low-mobility portable units transmit at relatively low power levels of 1 to 10 mW. While this is fine for the customer, it makes coexistence of the two systems a network planner's nightmare. *High tier* is a term often associated with high-mobility systems, whereas *low tier* is associated with low-mobility systems.

In summary, cellular telephony is the culmination of several technologies which have progressed in parallel over the past two decades. In fact, the progress has been so rapid that the standards bodies have had difficulty organizing meetings fast enough to determine standards that are consistent with the new technology.

7.2   Frequency Allocations

As usual, in the absence of global standards, mobile radio system designs in different parts of the world have evolved using frequencies in the bands made available by their national frequency coordinators. Inevitably, the frequencies chosen are different from one country to another. This is not a problem in a large country like the United States unless mobile telephone users want to use their handsets internationally, but in Europe it would render the mobile telephone useless to many users who travel across the borders of several countries to conduct their normal daily business. In the 1970s the Federal

Communications Commission (FCC) in the United States allocated the frequencies shown in Table 7.1 for land-based mobile radio telephones. The accompanying diagram (Fig. 7.1) illustrates how the allocated frequency spectrum is used for typical FDMA, frequency division duplexed (FDD) systems.

The 800-MHz band has been allocated for the use of cellular mobile radio, and the first systems used analog modulation. In the not-too-distant future, this band will be completely digital, although at present it contains both analog and digital radio. Full-duplex operation is made possible by using 25-MHz uplink and 25-MHz downlink carriers, separated by 45 MHz; 869 to 894 MHz is the base station transmit frequency band (downlink), and 824 to 849 MHz is the mobile station transmit frequency band (uplink). The carrier spacing is 30 kHz. (Note: A *mobile station* is also known as a *user's terminal* in North America.)

Personal communications services or *network* (PCS or PCN) are terms now associated with microcellular technology (see Sec. 7.10) which specifically uses pocket-sized telephones. The trend is to use higher frequencies than previous mobile systems, notably the 1.7- to 2.3-GHz band. Although most of this band is already allocated, the FCC is under considerable pressure to allocate this band exclusively, or at least up to 200 to 300 MHz of it, for PCNs.

In Europe, the frequencies in Table 7.1 have been adopted for land-based mobile radio telephones. Digital modulation systems operate in the 935- to 960-MHz (base transmit) and 890- to 915-MHz (mobile transmit) bands.

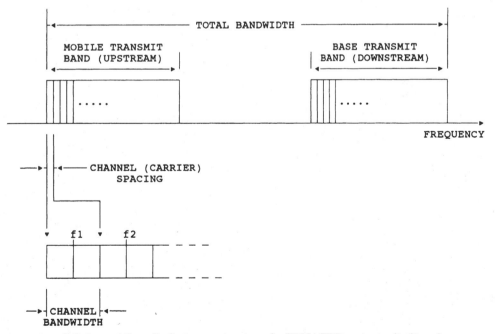

Figure 7.1 Cellular mobile radio frequency structure for FDMA/FDD systems. *f*1, *f*2, ... *fn* are carrier frequencies.

TABLE 7.1 Comparison of Cellular Radio and Cordless Specifications

Standard	High-Tier Cellular Systems				Low-Tier Cordless Systems		
	GSM	DCS 1800	IS-54	IS-95	DECT	PWT-E (licensed band)	PACS (licensed band)
Frequency band (MHz)	Europe	UK	USA	USA	Europe	US	USA
Uplink (mobile transmit frequency)	890-915	1710-1785	824-849	824-849	1880-1900	1850-1910	1850-1910
Downlink (base transmit frequency)	935-960	1805-1880	869-894	869-894		1930-1990	1930-1990
Multiple access	TDMA + FDMA		TDMA+FDMA	CDMA+FDMA	TDMA+FDMA		TDMA+FDMA
Duplex mode	FDD		FDD	FDD	TDD		FDD (licensed) TDD (unlicensed)
Duplex spacing (MHz)	45	95	45	45	-	-	80
Carrier spacing (kHz)	200	200	30	1250	1728	1000	300
No. of radio carriers in frequency band	(25/0.2)-1 = 124	(75/0.2)-1 = 374	832	10	10	15	50
No. of channels per carrier	8 (full rate) 16 (½ rate)		3 (full rate) 6 (½ rate)	Variable	12	12	8 / 16 / 32
Total traffic channels per cluster (before frequency reuse)	124 x 8 =992 (full rate) 124 x 16 =1984 (½ rate)	374 x 8 =2992 (full rate) 374 x 16 =5984 (½ rate)	832 x 3 =2496 (full rate) 832 x 6 =4992 (½ rate)	variable (approx. 1280 per CELL maximum)	120	180	400 / 800 / 1600
Modulation	GMSK	GMSK	π/4 DQPSK	QPSK BPSK	GFSK	π/4 QPSK	π/4 QPSK
Carrier bit rate (kb/s)	270.8	270.8	48.6	1288	1152	1152	384
Speech coder	RPE-LTP		VSELP	QCELP	ADPCM	ADPCM	ADPCM
Net bit rate (kb/s)	13 (full rate) 6.5 (½ rate)		7.95	(variable rate: 8, 4, 2, 1)	32	32	32 / 16 / 8
Channel coder for speech channels	½ rate convolutional + CRC		½ rate convol. + CRC	½ (down) 1/3 (up) convol. + CRC			CRC
Gross bit rate (kb/s) [speech + channel coding]	22.8 (full rate) 11.4 (½ rate)		13 (full rate) 6.5 (½ rate)	var. rate 19.2, 9.6, 4.8, 2.4	32	32	32 / 16 / 8
Frame size (ms)	4.6		40	20	10	10	2.5
Handset (Mobile Station) Transmit power (W)	Peak 20 8 5 2 / Avge. 2.5 1 0.625 0.25	Peak 1 0.25 / Avge. 0.125 0.031	Peak 9 4.8 1.8 / Avge. 3 1.6 1.6	0.66	Peak 0.25 / Avge. 0.01	Peak 0.09	Peak 0.2 / Avge. 0.025
Handset power control	Y	Y	Y	Y	N	N	N
Base station power control	Y	Y	Y	Y	N	N	N
Operational C/I (dB)	9		16	6	12	12	>11
Equalizer	needed	needed	needed	Rake receiver	option	option	option
Handover	Y	Y	Y	Soft handoff	Y	Y	Y

354

7.3 Cellular Structures and Planning

Cells. In a mobile telecommunications system the mobile handset beams into a nearby radio receiver called the base station. As the mobile user moves further away from the base station the received signal level decreases until communication is too noisy and eventually, if extra measures are not taken, the call would be dropped. This area around the base is called the *cell*, and multiple base stations form what is now known as a *cellular radio structure*.

Cellular systems use a hexagonal honeycomb structure of cells, and the base station at the center of each cell connects into the public telephone network (Fig. 7.2). The hexagonal cell pattern arises from the best method of covering a given area, remembering that radio coverage is ideally radial in nature. Figure 7.3 illustrates three possible methods of covering a particular area.

A quick observation of Table 7.2 reveals that the area of coverage of a hexagon is twice that of a triangle, with a square midway between the two. The area of overlap is calculated for a completely surrounded cell (i.e., by six cells for the hexagon, four for the square, and three for the triangle). The hexagon has a small overlap compared to the triangle. To cover an area of three hexagonal cells, or $7.8r^2$, would require six triangular or four square cells. The obvious conclusion from this simple analysis is that the regular hexagon is the most advantageous and therefore widely used structure, with the triangle suitable only in difficult propagation areas that require deep overlapping of radio zones.

Initially, the limited available power transmitted by the mobile radio telephone set determined the cell size. As the user moves (*roams*) between cells during a journey, the communication with the base station of the departing cell ceases and communication with the base station of the entering cell commences. This process is known as *handoff* in North America and *handover* in Europe. Each adjacent base station transmits at a different frequency from its

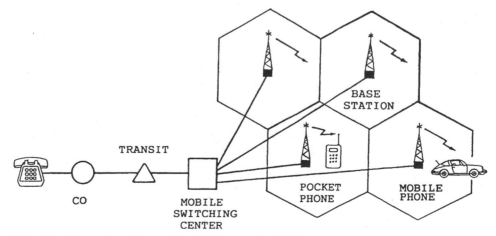

Figure 7.2 Connection of the mobile telephone to the PSTN.

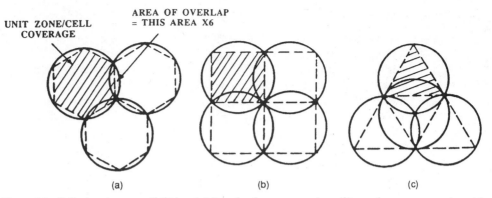

Figure 7.3 Cell structure possibilities. (*a*) Regular hexagon zoning; (*b*) regular square zoning; (*c*) regular triangle zoning.

TABLE 7.2   Cell Characteristics for the Three Main Cell Types

Cell type	Center-center distance	Unit zone coverage	Area of overlap	Width of overlap	Min. number of frequencies
Triangle	r	$\approx 1.3r^2$	$\approx 3.7r^2$	r	6
Square	$r\sqrt{2}$	$2r^2$	$\approx 2.3r^2$	$0.59r$	4
Hexagon	$r\sqrt{3}$	$\approx 2.6r^2$	$\approx 1.1r^2$	$0.27r$	3

neighbor. The handoff is accomplished when the received signal from the base station is low enough to exceed a predetermined threshold. At the border between two cells the subscriber is under the influence of two or even three base stations, and the link could pass back and forth between base stations as the moving subscriber receiver experiences a fluctuating field strength depending on the nature of the immediate environment, such as being surrounded by tall buildings. Some further intricacies of handoff are described in Sec. 7.8.

In the real world, there are no true circular cells. The signal strength contour for each cell does not produce the pleasing precise pattern of Figs. 7.2 or 7.3, but can be very distorted and is usually more like the strange jigsaw puzzle shapes of Fig. 7.4. One can say that only the *nominal* cells are hexagonal. The nominal cell diameter also varies depending on the traffic density. Typically, the center of a city is the most populated area, with the suburbs gradually decreasing in population. This leads to the cells in the center having a small diameter with a gradual increase in diameter when moving outward. Variations on the nominal plan would be necessary to account for irregular field strength contours caused by buildings and irregular terrain. So, while the hexagon is the hallmark of mobile radio technology, it is only the design starting point, and the final real-life cellular structure might bear little resemblance to multiple hexagons.

Frequency reuse.   To enable the available bandwidth to be used efficiently and so increase the number of users, a frequency reuse mechanism is built into the cellular structure. In the diagram of Fig. 7.5, the cells are clustered into groups of seven, each group having the same pattern of seven base station frequencies. The distance between different base stations using the same frequency is:

$$D = R \sqrt{3N}$$

where R = the cell radius
        N = the number of cells per cluster

For the case of seven cells per cluster, D is 4.6 times the cell radius R. A considerable effort is made to check that signal strength contours for each cell within the whole area of the system are not distorted to the point where two calls could seriously interfere with each other. Note that each cell can use only one-seventh of the channels available within the system.

There is a trade-off here between interference and channel capacity. If the number of cells in the pattern were increased beyond seven, clearly the distance between the base stations of identical frequency would be greater. The

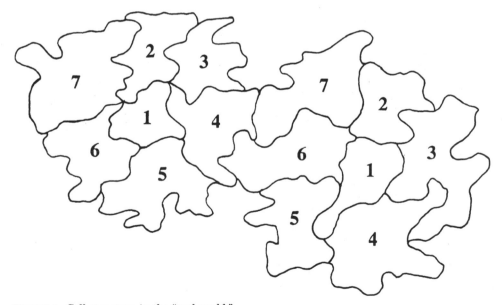

Figure 7.4   Cell structure in the "real world."

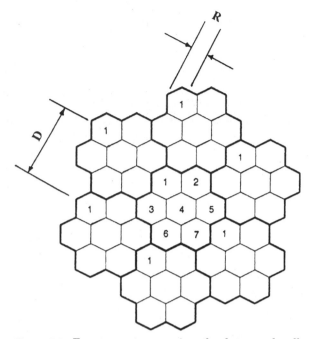

Figure 7.5 Frequency reuse using the hexagonal cell structure. $D/R = 4.6$. Note: $D/R = \sqrt{3N}$, where N is the number of cells per cluster.

cochannel interference would correspondingly decrease. Unfortunately, the system capacity would drop because the number of channels per cell is simply:

$$\frac{\text{Total number of channels per repeat pattern}}{\text{Number of cells (base stations) within the repeat pattern}}$$

For the hexagonal cell, the number of cells within a regular repeat pattern can be only 1, 3, 4, 7, 9, 12, 13, 16, 19, 21... etc. Seven is the most usual choice for cellular radio systems, as a compromise between degradation due to cochannel interference and high system-channel capacity.

A cellular structure with smaller and smaller cells is evolving for two main reasons: first, the increasingly limited available power that can be transmitted by the smaller and smaller mobile subscriber telephone sets, and second, increased capacity. Whereas the base station can have a large power output and therefore cover a radius of 50 km or more, the move to smaller, pocket-sized mobile telephones restricts the return path distance. For example, a UHF transmitted output power of 10 mW can work reliably over a distance of only about 200 to 300 m in a city center. The small cell size of less than 1-km radius is called a *microcell* structure and has the major benefit of providing orders of magnitude improvement of system capacity over the large (20-km) cell size. The increase in channel capacity is approximately inversely proportional to the square of the microcell radius. For example, reducing the cell radius from 8 km to 150 m increases the network capacity by a factor of more than 2500. In a rapidly expanding market, keeping up with the ever-increasing customer demand for access to the mobile network is a major problem. So far, the microcell structure is satisfying that demand.

Sectorization. Capacity can be enhanced even further by cell sectorization. As cell sizes decrease, the distance between base stations of identical frequency also decreases. This can be offset by careful control of the power radiated from either the base station and/or the mobile station (user terminal). That process is described later, but another way to reduce the cochannel interference coming from the six surrounding cells of a seven-cell structure is to use several directional antennas at each base station. If three antennas each cover 120°, the cell is split into three sectors, each having its own channel set (Fig. 7.6a). In a seven-cell repeat pattern, the three sectors allow 21 channel sets instead of 7. Further sectorization using six antennas (Fig. 7.6b) results in a four-cell repeat pattern (instead of seven), providing a total of 24 channel sets. Notice that greater sectorization has diminishing returns because the cell pattern becomes repeated more often (over a shorter distance) as the number of sectors increases. Sectorization might appear to be a disadvantage because there are fewer channels per sector and therefore a worsening of the trunking efficiency, which means reduced traffic capacity for a given level of blocking. In fact, sectorization results in an *enhanced* capacity because the cell sizes can be greatly reduced.

As most telecom organizations have followed the path from analog FM to digital mobile radio, there have been some formidable technical and economics problems. When the system becomes saturated for a given cell size, it becomes necessary to add new radio base stations for more channels, and consequently the cell size reduces and frequency reuse increases. This must be done without impairment of voice quality. Just as important, the infrastructure cost per subscriber must not exceed critical limits that would preclude affordability. In addition, there is the major technical problem of the coexistence of mixed analog and digital technologies during the transition from the

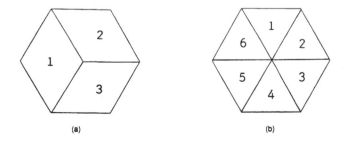

(a)   (b)

Figure 7.6 Cell sectorization. (*a*) Three sectors; (*b*) six sectors.

old analog to the new digital systems. Perhaps just as complex is the coexistence of low-power and relatively high-power mobile radio systems. The frequencies of operation must be different for the analog and digital systems; otherwise the severe cochannel interference would be intolerable.

From the planning point of view, there are some problems that differ quite considerably from the usual network planning. For example, the traditional telephone network generally expands as the population of urban areas expands. Demographic data usually provide a starting point for the network planner and projected population increases allow an estimation of future demand. The customers of a mobile telephone system live mainly in towns and consist of up to millions of people. A large percentage of homes might already possess a telephone. The question is, How many of them want to "go mobile" and have a pocket telephone? If the mobile service could be provided without any additional cost to the customer, the answer is, probably, *all of them*. The mobile network planner must be acutely aware of how much extra, on average, the subscriber is prepared to pay for the privilege of mobility. The potential demand is already known to be enormous when the price is right. Maximizing system economy while maintaining voice transmission quality is the difficult balancing act. If the requirement of data transmission is brought into the picture, there is an added dimension of difficulty. Furthermore, the newly mobile population, by the definition of mobile, is not predictably at specific locations for known lengths of time. Suppose 50,000 people decide to go to a football game, and 25 percent of them want to use their portable telephones to make a call at halftime. This obviously presents a problem. Similarly, cities are the focus of population migration during the working day, but when people return home in the evening they cause the mobile channel requirements to shift to the city suburbs.

The evolution of the vehicular or wide-area high-mobility telephone has been substantially different from the low-mobility portable telephone. In the initial start-up phase, in a new area, the high-mobility system has only a few customers and therefore only a few cell sites that are widely separated to provide a large area of coverage while minimizing the cost per customer. As the number of users increases, the density of base stations increases accordingly. The low-power low-mobility pocket telephone, in contrast, provides service to a large number of customers within a relatively densely populated area. Such areas, or islands, can then be interlinked to provide a wider area of coverage.

Cellular or cordless.   At this stage it is instructive to make the distinction between the terms *cellular* and *cordless* in relation to mobile systems. Definitions vary from text to text and in some cases they are obscure. Basically, cellular refers to land mobile radio networks generally used over wide areas, for example national or even international coverage, to be used with medium- or high-power vehicular or portable mobiles and for providing mobile access to the PSTN. The cellular network structure enables frequency reuse in nonadjacent cells. Cordless systems are characterized as simple, low-power portables usable within a short range of the base station, to provide access to a fixed public or private network. The three cordless applications originally intended were broadly (1) residential service, (2) PSTN access in high-population-density locations, and (3) wireless PABX for PSTN access in the office or business environment.

These systems are different in terms of cost, transmit power, talk and standby time, and speech quality. Nevertheless, there have been a significant convergence and even merging of some aspects of the two technologies. Originally, it was not envisioned that cordless technology would provide handoff from one cell to another, but this feature is now offered by most cordless operators. Cordless technology is also being deployed from customer premises to the cell distribution point or CO, for what is called the wireless local loop (WLL). Wide area cellular network planning now has the new LEO and MEO satellite links also factored in. Ideally, from the customers' perspective, the mobile handsets they carry should be low-cost, high-voice-quality, have the office environment access of cordless technology, and also be usable over the wide range (high mobility) offered by cellular technology. Satisfying that demand is not easy.

7.3.1 Quality of service

Ideally, cellular radio service should have the same quality as fixed telephones. While cordless systems achieve this goal, it is not so easy for cellular systems. From the traffic viewpoint, the quality can be as good as fixed services, but transmission quality is another matter. The following parameters are used to assess transmission quality:

- Signal-to-cochannel interference ratio (S/I_c)
- Carrier-to-cochannel interference ratio (C/I_c)
- Mean opinion score (MOS)
- Signal-to-noise ratio (S/N)
- [Signal + noise + distortion] to [noise + distortion] ratio (SINAD)
- Adjacent channel interference selectivity (ACS)

S/I_c is a parameter that is very subjective, but usually expected to be about 18 dB. The associated C/I_c will depend on the type of modulation, being around 7 dB for GMSK.

MOS is an ITU designation. It is a score derived from exposing numerous subjects to various laboratory-controlled speech-quality tests. The scores are rated as: 5 = excellent, 4 = good, 3 = fair, 2 = poor, 1 = very poor. For example, 32-kb/s ADPCM encoded speech rates about 4.1 on the MOS scale.

The SINAD is usually taken to be 12 dB for analog FM (25-kHz bandwidth). S/N varies from standard to standard, but 18 dB minimum is typical. Similarly, ACS is usually about 70 dB minimum.

Traffic. The crucial feature of the cellular concept is frequency reuse. The same set of channels is reused in different areas spaced apart sufficiently to keep cochannel interference to tolerable levels. Quality of service (QoS) is becoming increasingly important and two parameters are critical to the grade of service: (1) carrier-to-cochannel interference ratio (C/I_c) as above, and (2) blocking probability. Carrier-to-cochannel interference needs to be high, while blocking should be low. This situation can be achieved by a large cluster in an area that has few users and therefore low traffic and minimal blocking. In this case the grade of service is good, but the resources are used inefficiently.

Performance of cellular systems can be measured in terms of the *spectrum efficiency* (η_s), which is a measure of how efficiently space, frequency, and time are used. Note this is not the same as the *bandwidth efficiency* used in other radio technology.

$$\eta_S = \frac{N_f N_{ch} \Delta\tau}{AW} \quad \text{erlangs/m}^2\text{/Hz} \tag{7.1}$$

where N_f = the number of frequency reuses
$\quad N_{ch}$ = the number of channels
$\quad \Delta\tau$ = the fraction of time a channel is busy
$\quad A$ = the coverage area
$\quad W$ = the bandwidth available

Trunking efficiency is also of interest. This relates the number of customers per channel to the number of channels per cell for a particular blocking probability. Remember, a blocking probability of 2 percent means two calls in every hundred receive the busy signal because of congestion.

Traffic management is a very important aspect of cellular systems design and operation. Grade-of-service definition is the starting point, and 2 percent blocking during the busy hour is a good target. The definition of *busy hour* is not easy, because it can mean for the busiest cell, or for the whole system. Preferably each cell should be assessed. The conventional traffic theory erlang-B formula is usually used as a starting point to determine the number of channels per cell for a given traffic load and blocking probability. If the required grade of service cannot be accomplished with the number of channels available, the size of the cell must be reduced. Clearly, the cellular traffic situation is dynamic because of the mobility factor.

Software simulation packages can be very helpful in observing the dynamics of traffic variation with user mobility patterns. One tool that is now receiving a lot of attention is *dynamic channel allocation* to rearrange channel availability as demand increases or decreases. Computer algorithms assist in dynamic channel allocation, and some prediction of traffic demand can be factored into the overall strategy. In some types of system, capacity can be borrowed from adjacent cells. If a user on the edge of two cells gets blocked by the nearer of the two bases, a *directed retry* to the base in the adjacent cell can often be accepted.

Microcells. To satisfy the increasing traffic demands, cell sizes are being reduced to less than 1 km and in some circumstances to less than 100 m. Microcellular systems are being designed to take care of downtown users on city streets, and picocells for inside office buildings. Some of the first microcells were designed for highways and freeways. Base stations and directional antennas were placed on poles, producing cigar-shaped cells (see Fig. 7.7), and omni antennas were used at base stations on intersections or junctions.

When microcells are designed for city streets, the waveguide-like cells have little penetration through buildings, and cells extend only about 200 m from the base station. In addition to the direct line-of-sight signals between the base and mobile, there are reflections from the ground and buildings on each side of the street that interfere with the direct path. There is often a dominant interfering path, which causes fading described as Rician, which becomes a Rayleigh distribution when there is no dominant path. Macrocells' fading is usually Rayleigh, whereas microcells' fading is only occasionally Rayleigh. The consequence is that the depth of microcell fades is mostly considerably less than in macrocells. This allows microcellular radios to operate at receiver levels closer to threshold without suffering error bursts.

Handoff rates are quite high in microcells. A vehicle moving at about 100 km/h on a highway will have a handoff more than once every minute. On city streets the vehicle moves more slowly, but the cell size is small. Also, pedestrians talking as they walk create a large number of handoffs. In this case the handoffs are because of many users crossing few cell boundaries. Dead spots (no signal) inevitably occur in city streets and this problem can, to some

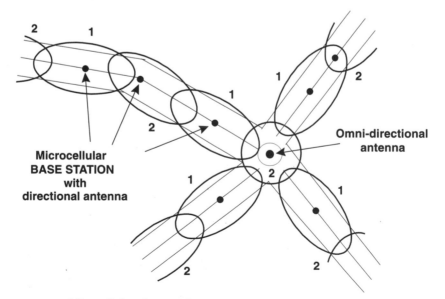

Figure 7.7 Microcellular clusters along a street.

extent, be overcome by using a macrocell within which the microcells are established. Clusters of microcells in cities can have as little as two microcells per cluster. The usual number is four, and in unusual circumstances, six or more are used. Where there is a vacant lot, unwanted signals stray into other street areas, causing cochannel interference in those areas if they are operating on the same base station frequency.

In microcells there is a gray area between cellular and cordless systems, and some cellular operators are opting to incorporate cordless technology into their cellular operations at this small scale.

Picocells. Service inside buildings is often considered to be the last frontier for cellular systems designers. In the quest for the wireless city, the different options for accessing customers within buildings are being explored. Several solutions have been proposed that incorporate building interiors into the cellular mobile radio structure. Each high-rise building could be a microcell within the cellular structure and each floor a picocell. Building coverage is complex because radio wave propagation into and within buildings, particularly up into the microwave region, suffers severe attenuation when propagating through steel-reinforced concrete. Floors and walls generally cause high attenuation, and different materials cause vastly different values of attenuation.

For example, recent measurements of the signal strength have shown the values taken on the first floor of buildings to be about 15 dB less than at the same height outside the building. The signal strength improves as one moves further up the building. Measurements have shown a typical 2.7-dB per floor improvement when moving from the first to the twelfth floor. Within buildings furniture causes losses, shadowing, and reflections depending on constituent materials. In addition, there is the problem of signals leaking in and out of windows and stairways.

Several in-building propagation techniques are being considered, such as the 18-GHz microwave frequency band (i.e., 18.820 to 18.870 and 19.160 to 19.210 GHz). Microwaves at this high frequency do not pass through steel-reinforced concrete walls, but do pass through the thin dielectric partitioning material used in an office environment. This implies frequency reuse for each office floor. The advancements in monolithic microwave integrated circuits (MMICs) have driven the cost of the technology down to levels that are attractive for in-building micro- or picocell communications. When used with "smart" sectorization antennas, these low-power microwave systems offer a good solution for the last 30 to 40 m of a PCS. They are even being considered for LANs for data transmission at 15 Mb/s using ten 10-MHz channels in an FDMA configuration.

For those who want to provide total cellularity using one handset, the technical problems are not small. As VLSI-circuit-technology-driven base stations are reduced to the size of a baseball, the costs that were prohibitive in the early 1990s become less of a concern. Low-cost base stations scattered throughout buildings like light bulbs interconnected by optical fibers is one of several possible picocell structure solutions. This picocell architecture will support the huge increase in traffic anticipated as the mobile phone becomes as ubiquitous as the wristwatch, or even a part of it.

Mobile location. For a mobile subscriber to receive a call, the subscriber's precise location must obviously be known. There are several possible ways to track the movement of a mobile user. A convenient method is to split up the whole cellular network into a number of location areas, each having its own ID number. Each base station within a particular area periodically transmits its area ID number as part of its system control information. As the mobile unit moves from the control of one base station to another, it will eventually move across to a new ID number region, and the network is updated with the new area in which it can be found. This is all done using signaling information transfer.

7.4   Propagation Problems

The following discussion is primarily related to moving vehicles but is equally applicable to portable telephone users, where the vehicle is just one of the propagation situations encountered and perhaps the most difficult to quantify. Radio propagation is one of the most fundamental parameters in mobile communications engineering. Unlike fixed point-to-point systems, there are no simple formulas that can be used to determine anticipated path loss. By the nature of the continuously varying environment of the mobile subscriber, there is a very complicated relationship between the mobile telephone received signal strength and time. The situation is not so bad if the mobile unit remains stationary for the duration of the call. If a particularly poor reception point is encountered, a short move down the road might significantly improve the reception. For a moving subscriber, as in the typical call placed from a moving car, the signal strength variation is a formidable problem that can be approached only on a statistical level. For land-based mobile communications the received signal variation is primarily the result of multipath fading caused by obstacles such as buildings (described as "clutter") or terrain irregularities.

Obstacles (clutter) can be classified into three areas, as follows:

1. *Rural areas*, in which there are wide open spaces with perhaps a few scattered trees but no buildings in the propagation path

2. *Suburban areas*, including villages where houses and scattered trees obstruct the propagation path

3. *Urban areas*, which are heavily built up with large buildings or multistory houses or apartments and greater numbers of trees

The terrain conditions can be categorized as follows:

1. *Rolling hills*, which have irregular undulations, but are not mountainous

2. *Isolated mountains*, where a single mountain or ridge is within the propagation path and nothing else interferes with the received signal

3. *Slopes*, where the up- or downslopes are at least 5 km long

7.4.1 Field strength predictions

For an ideal single-path radio transmission in free space, the received power P_r is related to the transmitted power P_t by the familiar equation

$$\frac{P_r}{P_t} = G_t G_r \left(\frac{\lambda}{4\pi d} \right)^2 \tag{7.2}$$

where G_t and G_r are the transmitter and receiver antenna gains. The term $(\lambda/4\pi d)^2$ is the free-space loss, d is the distance between the transmitter and receiver, and λ is the transmitted signal wavelength. For the mobile situation, where the propagation is over the earth plane and distances are much larger than the antenna heights or signal wavelength, this equation is modified to:

$$\frac{P_r}{P_t} = G_t G_r \left(\frac{h_1^2 h_2^2}{d^4} \right) \tag{7.3}$$

The equation is important because it indicates a power drop-off at the rate of the fourth power of distance. This relationship arises because in its derivation a ground reflected signal is considered to interfere with the direct path signal. Although this is an oversimplification, it indicates that even though the rate at which the received signal strength decreases when moving away from a base station is in one sense painfully steep, it is also helpful in reducing interference from neighboring cells in a cellular structure.

A typical plot of the received field strength measured while traveling is shown in Fig. 7.8a. When moving at any specific fixed radius from the base station between a rural area and an urban area, a 20- to 30-dB variation of field strength can be expected. The average field strength is lowest in urban areas, followed by suburban, and then rural. The rate of variation of field strength is proportional to the product of the frequency of the carrier wave and the speed of the mobile subscriber. Incidentally, for high-velocity mobile units the Doppler effect must be taken into account in the equipment design.

Figure 7.8b shows a typical relative signal level while moving away from a base station within an urban microcell. Notice the fluctuation about the path loss inverse square law due to multipath effects. At a greater distance from the base station there is a steeper signal drop more in keeping with the inverse fourth power.

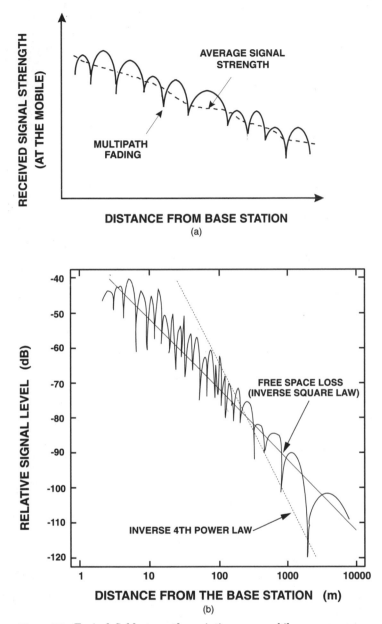

Figure 7.8 Typical field strength variation as a mobile moves away from a base. ((*a*) Within an urban microcell; (*b*) over a short distance within the microcell.

Okumura et al.,* followed by Hata,† developed a set of curves for signal prediction based on measured data. This model has become well used in the mobile radio industry, giving predicted path loss accuracy with standard deviations within 12 dB of actual measured data.

For urban areas located on relatively flat terrain, Fig. 7.9 shows a graph of the statistically predicted median attenuation relative to free space plotted against frequency at various distances from a base station that has an effective antenna height of 200 m, and a mobile antenna height of 3 m. For

---

* Okumura, Y., et al., Field Strength and Its Variability in VHF and UHF Land Mobile Radio Service, *Rev. Elec. Communication Lab.*, vol. 16, Sept. - Oct. 1968, pp. 825-873.

† Hata, M., Empirical Formula for Propagation Loss in Land Mobile Radio Service, *IEEE Trans. Veh. Technology*, vol. VT-29, no. 3, 1980, pp. 317-325. ©1980 IEEE.

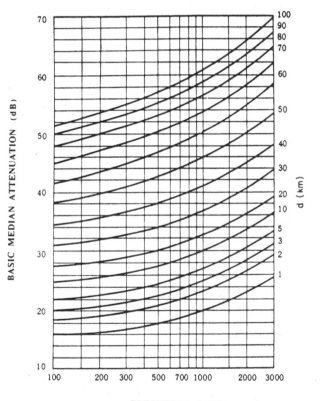

FREQUENCY f (MHz)

Figure 7.9 Graph of urban areas median attenuation against frequency for quasi-smooth terrain. Base station antenna height = 200 m; mobile station antenna height = 3 m. (*Reproduced with permission from McGraw-Hill. Lee, W. C. Y., Mobile Communications Engineering, McGraw-Hill, 1982.*)

example, at 900 MHz over a quasi-smooth terrain distance of only 1 km, the attenuation in addition to that of free space is approximately 20 dB.

In suburban areas, where the effects of obstacles are less severe than in urban areas, the field strength is generally higher. Figure 7.10 indicates the predicted correction factor that must be subtracted from the Fig. 7.9 attenuation values for suburban areas. Notice that the correction value is larger, signifying more improvement, at the higher frequencies.

Figure 7.11 shows the predicted correction factor for rural areas. The corrections to be subtracted from the Fig. 7.9 attenuation values are even greater, which indicates an even larger received field strength than in suburban and urban areas.

Clearly, the antenna height of the base station is a very important parameter in determining the field strength. Figure 7.12 shows how the estimated field strength is increased or decreased depending on whether the base station antenna height is above or below 200 m, respectively.

7.4.2 Effects of irregular terrain

For rolling hilly terrain, measurements have been made to establish the field strength correction factor, which depends on the degree of undulation of the hills. Figure 7.13 illustrates this correction factor calculated for the 900-MHz frequency band within a 10-km radius from the base station. For a peak-to-trough rolling hill of 100 m, the field strength suffers an extra loss of about 7 dB.

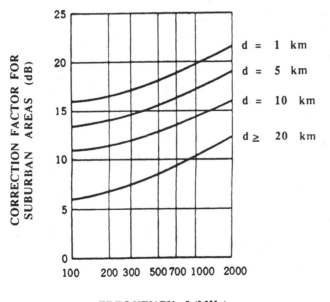

Figure 7.10 Correction factor for suburban areas.

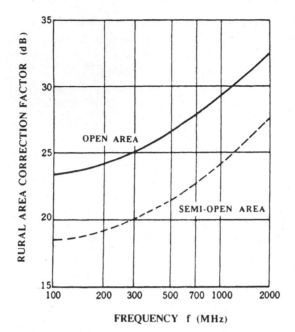

Figure 7.11 Correction factor for rural areas. *(Reproduced with permission from McGraw-Hill. Lee, W. C. Y., Mobile Communications Engineering, McGraw-Hill, 1982.)*

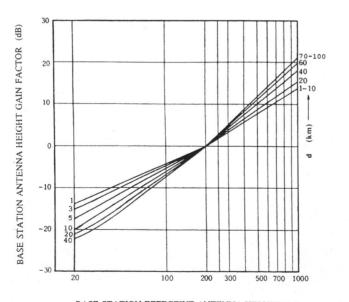

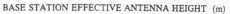

Figure 7.12 Multiplication factor for urban area antenna heights other than 200 m.

DEFINITION OF Δh

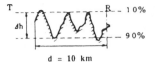

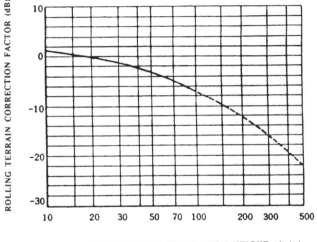

THEORETICAL TERRAIN UNDULATION HEIGHT Δh (m)
(900-MHz BAND)

Figure 7.13 Field strength correction factor for rolling hill terrain.

Operating within the vicinity of an isolated ridge or mountain, field strength measurements have been made that take into account knife-edge diffraction effects. They are very similar to the theoretical correction factors depending on the distance from an isolated ridge or mountain of nominally 200 m in height (Fig. 7.14). Notice the field strength *enhancement* on the top of the ridge, and the maximum attenuation values in the immediate shadow of the ridge.

Finally, Fig. 7.15 shows the measured and interpolated correction factors for terrain that slopes over a distance of at least 5 km. A positive slope obviously enhances the signal strength, whereas the negative slope decreases it.

There are other factors that also come into the picture—for example, tree foliage can cause as much as a 10-dB attenuation variation between the winter and summer months. Tunnels, bridges, subways, and close proximity to very high skyscrapers can produce highly unpredictable signal strength variations.

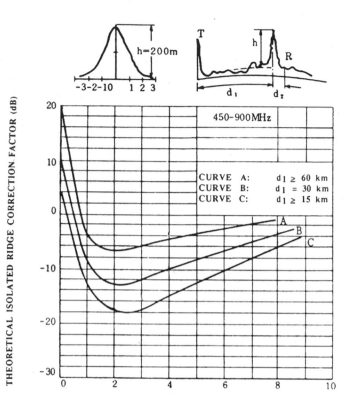

Figure 7.14 Correction factor for an isolated ridge.

The propagation evaluations are at best an approximation with a relatively high degree of uncertainty. Nevertheless, they do provide guidelines for designers deciding where to place base stations in order to provide the best signal coverage for the terrain to be covered. After rough designs are made in the office, field measurements of signal strength at many locations around a temporary base are necessary to fine-tune the configuration.

7.5 Antennas

Previously described microwave antennas have all been parabolic in nature. They were highly directional, because it was necessary to focus the energy in one direction. Conversely, mobile station radios require antennas to be omni-directional *in the horizontal plane* (looking from above), but to have very little upward radiation. This is because, at any time, the mobile unit could be at any point around the full 360° range of the base station antenna. Note that these are not isotropic antenna qualities, which would require equal radiation in *all* directions. There are several styles of antenna that can be used for this purpose. As these antennas are generally variations on the dipole antenna, a few comments on the dipole antenna are necessary. The dipole is the simplest of all antennas as far as physical construction is concerned (Fig. 7.16), but even this antenna has some rather elaborate mathematics associated with it. The radiation pattern for a short piece of metallic rod is shown in Fig. 7.16, which indicates that it is indeed nondirectional in the horizontal plane, although there is some directivity in the vertical plane. When the dipole is fed from the center of the metal rod antenna such that the total length is one-half wavelength, as one would expect, this is referred to as the half-wave dipole antenna, as in Fig. 7.17. The gain of a half-wave dipole is about 2 dB (i.e., relative to the isotropic antenna). The gain of antennas for mobile communications is often specified relative to the dipole antenna.

The impedance of antennas is important because the idea is to match the impedance of the transmitting device to that of free space so that 100 percent energy transfer takes place. If there is a bad mismatch, reflections from the antenna back into the transmitter cause severe interference. In the analogous

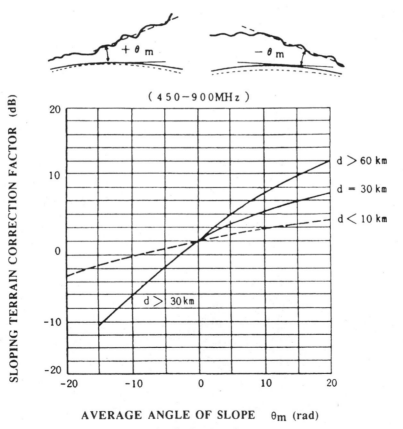

DEFINITION OF ANGLE OF SLOPE

Figure 7.15 Correction factor for sloping terrain.

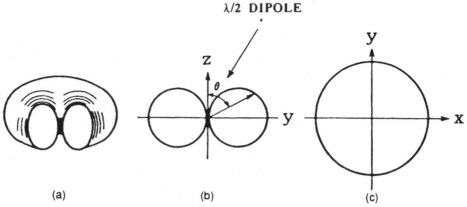

Figure 7.16 Dipole antenna radiation pattern. (*a*) Radiation pattern; (*b*) vertical plane; (*c*) horizontal plane from above.

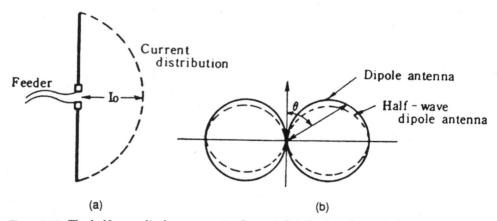

Figure 7.17 The half-wave dipole antenna. (*a*) Current distribution; (*b*) vertical pattern.

case of the TV antenna, the interference caused by mismatch is observed as "ghosting" on the picture.

7.5.1 Base station antennas

There are several types of antenna frequently used for the mobile base station, and five are illustrated in Fig. 7.18.

1. *Bent or folded dipole antenna* (Fig. 7.18*a*). This is constructed as a bent or folded conductor whose horizontal dimension is one-half wavelength.

2. *Ground plane antenna* (Fig. 7.18*b*). The coaxially fed antenna is physically convenient for many applications. The finite ground plane tends to incline the radiation pattern maximum slightly upward instead of horizontally. This is not usually desirable, and can be circumvented by several techniques for improving the ground plane.

3. *Stacked antenna* (Fig. 7.18c). A stack of several half-wave dipole antennas reduces the radiation in the vertical direction and effectively increases the omnidirectional horizontal gain. For example, a stack of four folded dipoles increases the gain by about 6 dB compared to a single dipole. Variations on the stacked antenna can be used if the gain in the horizontal direction needs to be asymmetrical to preferentially illuminate some areas of a cell that might have some geographical or building screening problems.

4. *Corner reflector antenna* (Fig. 7.18d). For the case of cell sectorization, the base station antenna must radiate only over a specific angle (e.g., 60°). For this antenna, a half-wave dipole is placed in the corner of a V-shaped wire plane reflector at 0.25- to 0.65-λ spacing from the vertex. The wires are typically spaced less than 0.1λ apart. With an angle of 90°, a typical gain of at least 8 dB relative to a half-wave dipole can be achieved. As the angle is made smaller, the directivity (gain) increases. In order to provide coverage to less accessible areas, it might be necessary to use an antenna that has a high directivity or preferred orientation of radiation. A corner reflector might not have enough gain, in which case the high-gain YAGI could be used.

5. *Zig-zag antenna* (Fig. 7.18e). Base station antennas for mobile cellular radio applications are usually required to have a beamwidth of about 120° so that three antennas can cover each of the three sectors of a cell, or 60° for six-cell sectorization. The zig-zag antenna is a suitable choice because it has a 60° beamwidth gain of more than 13 dB, and in the UHF band its dimensions are convenient (e.g., 1.8 × 0.4 m) for rapid installation. The wide horizontal beam has very low power in the vertical direction.

6. *Array antenna*. The phased array base station antenna is a very sophisticated solid state antenna that can be steered to provide gain at any desired angle within a cell sector. This beam can then be rotated around the calls in progress so that in a TDMA system each call receives additional gain for the duration of its transmitting time slot. Additional null-creating signals can be formed to neutralize multipath interfering sources. The resulting improvement in carrier-to-interference ratio can enhance the system capacity about twofold. These sophisticated "smart" antennas will no doubt be important mobile communication systems components in the future.

7.5.2 Mobile station antennas

Antennas used for car mobile radios must be omnidirectional and as small as possible and must not be adversely affected by the car body. The location of the antenna is usually on the roof, trunk (boot), or rear fender (bumper). The whip antenna of Fig. 7.19a is one possibility. This is a λ/4 vertical conductor with the body (preferably the roof) acting as a ground plane. The compact size makes this a very popular style. The center of the roof is the best location so that the car body least affects the signal. Figure 7.19b is a λ/2 coaxial antenna. Figure 7.19c is a high-gain whip antenna with a loading impedance coil used to adjust the current distribution.

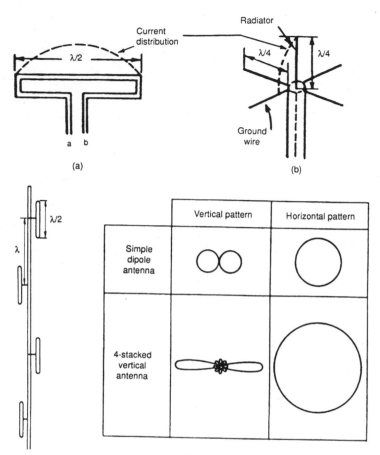

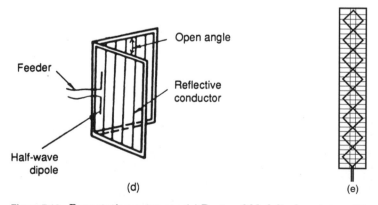

Figure 7.18 Base station antennas. (*a*) Bent or folded dipole antenna; (*b*) ground plane antenna; (*c*) stacked antenna; (*d*) corner reflector antenna; (*e*) UHF zig-zag panel antenna.

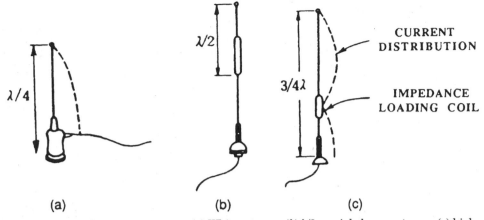

Figure 7.19 Mobile station antennas. (*a*) Whip antenna; (*b*) $\lambda/2$ coaxial sleeve antenna; (*c*) high-gain whip antenna.

Portable radio antennas are usually either short ($\lambda/4$ or less) vertical conductor whip antennas or normal-mode helical designs. There are a variety of options. Some are coaxial-fed and retractable for compactness, whereas others are detachable. Some are encased in a plastic or rubber material for ruggedness. Ideally, for maximum convenience, the antenna should not even protrude from the body of the portable radio. The antenna of a portable poses several problems. First, there is no ground plane (as is the case for automobiles), so its efficiency is reduced. Second, the user might not be pointing the antenna in the optimum direction, or, even worse, it might be held almost horizontally. In either case the antenna might not be in the best orientation for the correct polarization reception. Third, the user's head might cause disturbances by mismatching the antenna impedance. Incidentally, the long-term health effects of holding a high-power (several watts) microwave portable telephone close to the head for extended periods of time are at present unknown.

7.6 Types of Mobile Radio Systems

Historically, technology has lagged behind the design of mobile telephone systems, and it was only by 1983 that the first good-quality systems were put into operation. Since those early, low-capacity pioneering systems subscriber demand has mushroomed, despite the higher cost of calling from a mobile telephone. The different types of mobile radio systems, in terms of frequency spectrum usage, are summarized in Fig. 7.20. They differ primarily in modulation technique and carrier spacing.

Analog FM. The first-generation cellular systems in operation were analog FM radio systems that allocated a single carrier for each call. Each carrier was frequency modulated by the caller. The carriers were typically spaced at 25-kHz

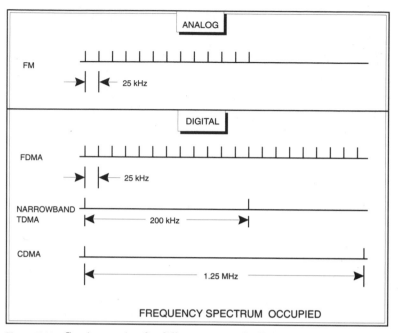

Figure 7.20 Carrier spacing for different types of cellular radio systems.

intervals (i.e., carrier bandwidth). The overall allocated bandwidth was relatively narrow, and only a few channels (typically 12) were available.

Digital FDMA. FDMA systems resemble analog FM, with the exception that the carrier is modulated by a digitally encoded speech signal. The bandwidth of each carrier is similar to the analog FM systems (typically 25 kHz).

Digital narrowband TDMA. TDMA systems operate with several customers sharing one carrier. Each user is allocated a specific time slot for transmission and reception of short bursts or packets of information. The bandwidth of each carrier is typically 200 kHz, and the total bandwidth available is in the region of 10 to 30 MHz, so many FDMA carriers each contain several customers on a TDMA shared basis. This access combination allows a reasonably large channel capacity in the region of 500 to 1000 channels, before frequency reuse.

In the United States, for example, the 824- to 849-MHz frequency band is allocated for one-way transmission from the base station to the user, and the 869- to 894-MHz band is allocated for transmission from the user to the base station. To enable two competitive systems to operate simultaneously, only half of each of these bands is available to each operator. Each system therefore has 12.5 MHz available for transmission and 12.5 MHz for reception. Each of these 12.5-MHz bands is subdivided into several carriers in an FDMA manner. Each carrier is operated in a TDMA mode having time slots for voice or data channels.

Digital wideband (spread spectrum). One form of digital wideband operation is CDMA. In these systems there is a single carrier that is modulated by the speech signals of many users. Instead of allocating each user a different time slot, each is allocated a different modulation code. Mobile users in adjacent cells all use the same frequency band. Each user contributes some interfering energy to the receivers of fellow users, the magnitude of which depends on the processing gain (see Sec. 3.3). In addition to interference from users within a given cell, there is also interference from users in adjacent cells. The distance between adjacent cell users attenuates the interference considerably more than users within the same cell. Frequency reuse is therefore unnecessary. Consequently, each cell can use the full available bandwidth (12.5 MHz) for CDMA operation.

7.7 Analog Cellular Radio

Analog cellular systems were used exclusively in the early days of mobile communications. Although they are being superseded by digital technology, a large number of systems are still in service and will probably remain in use for several years to come.

Analog FM. Analog FM cellular radio systems are relatively old technology (1980 to 1985) in this fast-paced industry. These are the first-generation cellular radio systems. However, it is informative to discuss some of their features briefly, because they provide insight into how future systems are evolving. Analog cellular radio was initially designed for vehicle-mounted operation. By 1990, already more than 50 percent of mobile radios (stations) in most networks were hand-held portables, and the demand was growing.

As far back as 1979, Bell Labs designed and installed a trial cellular mobile system called the Advanced Mobile Phone Service (AMPS). This was really the birth of cellular radio in the United States, and is still the basis of the analog systems in operation today. The AMPS system uses the hexagonal cell structure, with a base station in each cell, as illustrated in Fig. 7.21. The cells are clustered into groups of seven cells (i.e., a seven-cell repeat pattern). AMPS covers large areas with large-sized cells, and high-traffic-density areas are covered by subdividing cells. Sectorization is also used to enhance capacity. The overall control of the system is by a mobile telephone switching office (MTSO) in each metropolitan area. This digital switch connects into the regular telephone network and provides fault detection and diagnostics in addition to call processing. The mobile unit was originally installed in a car, truck, or bus. The frequencies for AMPS are 870 to 890 MHz from base to mobile, and 825 to 845 MHz from mobile to base. Each radio channel has a pair of one-way channels separated by 45 MHz. The spacing between adjacent channels is 30 kHz. The AMPS system uses FM with 12-kHz maximum deviation. FM has a convenient capture mechanism. If a receiver detects two different signals on the same frequency, it will lock onto the stronger signal and ignore the weaker, interfering signal.

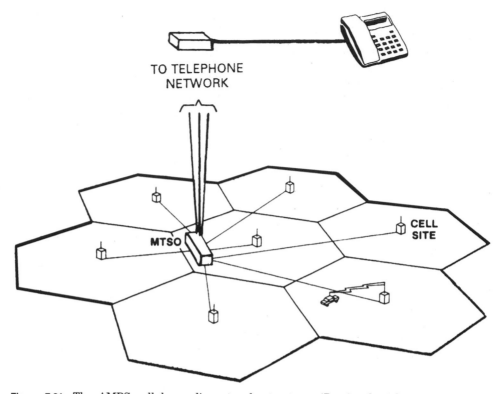

Figure 7.21 The AMPS cellular radio network structure. (*Reprinted with permission from Advanced Mobile Phone Service Using Cellular Technology, Microwave Journal, August 1983, pp. 119-126, by Ehrlich, N. Artech House, Inc., Norwood, MA, U.S.A. http://www.artech-house.com.*)

Mobile units are microprocessor controlled. The MTSO periodically monitors the carrier signal quality coming from the active mobile. If, during a call, a mobile moves to the edge of a cell boundary and crosses the boundary, the signal quality from the adjacent cell gradually becomes better than the existing service provider, so handoff is initiated. The handoff command is a "blank and burst" message sent over the voice channel to the designated cell. A brief data burst is transmitted from the base providing service to instruct the mobile microprocessor to retune the radio to a new channel (carrier). The voice connection is momentarily blanked during the period of data transmission and base station switching. This interruption is so brief it is hardly noticeable, and most customers are unaware of its occurrence.

All of the call setup is done by a separate channel. There are dedicated signaling channels that transmit information only in the form of binary data. These channels are monitored by all mobiles that do not have a call in progress. When a mobile is first switched to the ON mode or is at the end of a call, it is in the idle state. It scans the frequencies used for call setup and monitors the one providing it with the strongest signal. Each cell has its own setup channel. The mobile periodically makes a scan to see if its change of position has made the setup channel of an adjacent cell stronger than the one it is monitoring. If a customer from, for example, a home initiates a call to the mobile, the telephone number of the particular mobile station is transmitted over the setup channel of every base. This is a type of paging mechanism. The mobile identifying its own number sends a response message, and the call connection procedure starts. This is done by the MTSO, which assigns an available voice channel within the cell where the mobile is located. It then connects the home to the cell base. Another data message is sent over the setup channel to instruct the mobile to tune to the assigned voice channel. The mobile activates ringing to alert the owner to pick up the handset, thereby establishing the call. When either party hangs up, another data message is sent over the setup channel to instruct the mobile to switch off its transmitter power and revert to the idle mode. Calls initiated in the opposite direction (i.e., mobile to fixed user) are similar. The mobile customer dials a number that is stored in a register. On pressing the *Send* button, the mobile sends a message over the setup channel identifying itself and giving the number of the customer called. The MTSO then assigns an available voice channel and makes the necessary connection and ringing to the called customer. It also instructs the mobile to tune to the assigned channel, and the call is established.

The AMPS system has nationwide roaming capability. This is possible by cooperation between the service providers in different parts of the country. The AMPS system has been very successful, but its main disadvantage is that its total system capacity is inferior to the more advanced digital cellular radio systems.

7.8 Digital Cellular Radio

Digital cellular radio systems can be divided into two categories, narrowband and wideband. Narrowband systems are often considered to be the second generation of cellular radio. Although the digital narrowband TDMA systems in North America and Europe have developed along similar lines, there remain many features that are different. Because of its global success, in this text the main focus of attention for digital narrowband TDMA cellular radio will be the European-designed system called GSM, to which the U.S. system called IS-54 will be compared.

7.8.1 European GSM system

The acronym GSM originally stood for the French name Groupe Speciale Mobile, the planning organization that did much of the groundwork for the TDMA cellular system. GSM now stands for *global system for mobile communications*. A description of its features serves to highlight some of the intricacies of present-day cellular radio systems.

GSM was designed to provide good speech quality with low-cost service and low terminal cost, and to be able to support hand-held portables that have a new range of services and facilities, including *international roaming*. This last feature alone is a significant technological milestone. The system became operational in 1991 and the initial GSM Recommendations totaled approximately 6000 pages, but only the main features will be discussed here. Although voice communication from hand-held portables is the major service, the need for data communications using laptops is growing rapidly. The maximum data rate that could initially be provided by GSM was 9.6 kb/s which, although not high, was a direct data interface without the need for a modem. This is in the process of upgrading to 64 kb/s and then 115 kb/s capability.

The obvious difference between a mobile system and a fixed public switched telephone network (PSTN) is the radio section. However, some less obvious network differences appear, particularly additional interfaces, and they are highlighted in Fig. 7.22. The home location register (HLR) is where all the management data for home mobiles is stored. The visitor location register (VLR) contains selected data for visiting mobile stations, including international mobile station identity and location. The HLR and VLR can be located in the same room as the mobile services switching center, and the quantity of HLRs and VLRs depends on how extensive the network becomes. As a mobile moves into an area other than its home location, the ID number is passed on to the appropriate VLR via the nearest base station, and then on to the switching center. A temporary mobile identity and roaming number are allocated to the mobile by the network being visited. This facilitates the origination and reception of calls in that area. The HLR provides the VLR with the services that are available to a particular mobile. Even with this very simplified view of the network, one can see that there is a high level of signaling activity over both the radio link and the fixed portion of the mobile network for call processing and mobile station management. ITU-T common-channel signaling system number 7 is used for the GSM network. The mobile application part (MAP) is covered in GSM Recommendation 09.02.

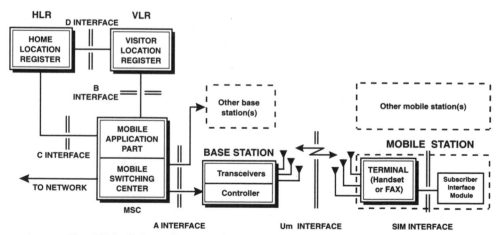

Figure 7.22 The GSM cellular radio network structure.

The radio link. These days the radio link section of mobile systems is generally referred to as the air interface. To cater to both vehicle-mounted and hand-held (pocket-sized) portable mobile stations, the peak output power ranges from 20 W (43 dBm) for vehicles and 2 to 5 W (33 to 37 dBm) for portables. The maximum receiver sensitivity of either vehicle mobile units or base stations is -104 dBm, and it is -102 dBm for pocket mobiles, giving the vehicle-mounted mobiles an operational advantage of several decibels. This is balanced by the fact that portables are intended for use within microcells in city centers or dense urban areas. The use of portables inside buildings, elevators, and subways and in other poor radio reception locations still poses some problems.

Speech coding. The initial reduction in bit rate from the 64-kb/s PCM to 32-kb/s ADPCM is inadequate for cost-effective operation of a high-capacity narrowband TDMA cellular radio network. The bit rate is reduced by various techniques such as regular pulse excitation, linear predictive encoding, and other elaborate algorithms, but the delay time introduced into the transmission increases. A 13-kb/s rate has been chosen for GSM, which causes an additional transmission delay (or *latency* as it is often called) of 20 to 50 ms depending on the manufacturer. This is sometimes called the source code rate. Other functions are also built into the codec chips, as follows:

Voice activity detection to enable discontinuous transmission, otherwise known as voice activation. This reduces the cochannel interference and therefore allows greater frequency reuse. It also saves battery power.

Comfort noise introduction so the speaker does not think that connection with the called party has been lost.

Channel coding. Channel coding is used to provide protection against error bursts and random errors. This is done by convolutional coding at a rate of half-block diagonally interleaving over eight TDMA frames. Convolutional coding is known to be more effective than binary block coding as a countermeasure against random errors. Also, when convolutional coding is combined with block interleaving, it provides a protection against error bursts that is similar to Reed-Solomon coding (see Chap. 3). Viterbi algorithm decoding is used at the receiving end.

Research into speech coding, FEC, and the speed of processors has resulted in a reduced bit rate for a given QoS. The full rate standard is 22.8 kb/s for GSM, and 13 kb/s for IS-54. These values have now been reduced to 11.4 kb/s and 6.5 kb/s, respectively, for the newer half-rate encoders. Remember, the more signal processing involved, the longer the transmission delay and also the more battery power used.

Frames. Voice or data information is transmitted over the radio by digitally processing the voice or data into a PCM multiplexing style of format, using frames and multiframes. These TDMA signals are then modulated and superimposed on RF carriers that have 200-kHz bandwidth. Each frame consists of eight time slots, one for each call, each of 576.923-μs duration, producing a 4.6154-ms frame. Figure 7.23a shows the normal burst (NB) mode. There are also bursts known as the *frequency correction burst, synchronization burst*, and *access burst,* as shown in Fig 7.23b.

Every frame clearly has considerable overhead incorporated. In each 156.25-bit time slot, two bursts of 57 bits are for data and the rest are overhead. Also, each time slot has a 26-bit training sequence. This specific bit sequence is already known to all mobile stations, and serves to improve reception quality in a multipath fading environment. The mobile station circuitry compares the known sequence with the detected sequence, and any disparity between the two allows a form of equalizer using digital filters to compensate for the errors introduced by fading. This process is invaluable for improving communication quality when the distance from base station to mobile station exceeds several hundred meters, which is typically when multipath fading can become serious. A flag bit at the beginning and end of the training sequence separates it from the two data bursts. The complete time slot begins and ends with three "tail" bits, all logical zero. Finally, there is a guard time of 8.25 bits to ensure that at any base station there is no overlap of bits arriving from several mobile stations. Because 156.25 bits are transmitted in 576.923 μs, the transmission bit rate is 270.833 kb/s, of which more than one-fourth are overhead bits.

The *traffic* frames transmitted from both the base and mobile are identical. There is a three-time slot shift between them. Because all eight mobiles within the same TDMA carrier use only one of the eight time slots, the shift ensures that sensitive mobile receivers are not switched on at the same time as their transmitters, which safeguards against damage due to power leakage. During the other seven time slots not used by each mobile, the signal levels coming from other bases in the area are monitored as part of the handoff process.

Adaptive frame alignment synchronization. This is used by GSM to equalize differences in propagation delays for mobiles at different distances from the base. The maximum cell size is 35 km, at which distance the round-trip delay time is 233.3 μs. Signals from all the mobiles into a base must be synchronized so there is no overlapping of frames. The access burst (AB) has a guard time of 68.25 bits (252 μs), which ensures no overlap, and this burst is used on first attempting a call and just after handoff. The base receives the 41 synchronized sequences and measures the signal delay which is used as a timing advance signal back to the mobile. This is a 6-bit number that instructs the mobile to advance its time base by a value between 0 to 63 bits (0 to 233 μs). This method ensures that signals from all mobiles arrive at the base in the correct time slot allocations without overlapping adjacent ones. This synchronization is precise enough to reduce the guard time for all other bursts to 8.25 bits (30.46 μs). The base monitors all mobiles' signal delays throughout their calls and, as mobility

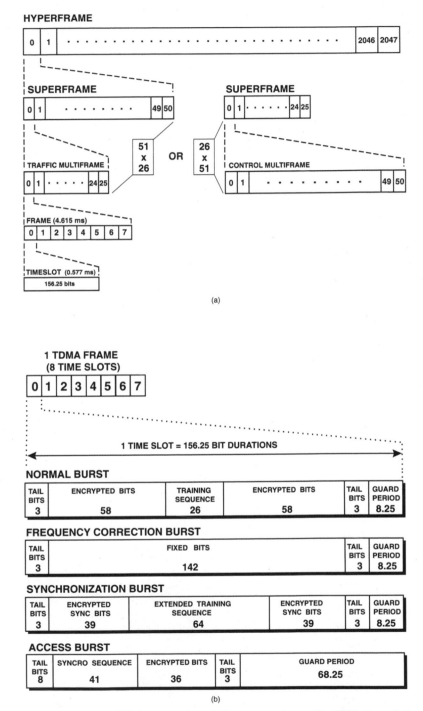

Figure 7.23 (*a*) GSM frame and multiframe structure. (*b*) GSM time slot composition.

changes the propagation delay, the base sends timing advance instructions to update the mobiles' transmit time synchronization.

Multiframes.   There are two types of multiframe, one containing 26 frames and the other containing 51 frames. The 26-frame multiframe is 120 ms in duration; it has 24 frames that carry traffic, and two frames for associated control channels (ACCH). There are two types of ACCH. One is a slow continuous stream for supervising calls and is known as the slow associated control channel (SACCH). The other operates in the burst-stealing mode and is for power control and handoff. This is known as the fast associated control channel (FACCH). If required, a traffic channel can be taken (stolen) and used as a power control and handoff time slot.

The 51-frame multiframe is 235 ms in duration and is called the control multiframe. It is different for mobile and base stations. Figure 7.24 shows that the uplink contains call setup information, but the downlink is a broadcast of information necessary for frequency correction and transmit synchronization. In the 51-frame multiframe, time slot zero is called the broadcast control channel (BCCH) and is used for downstream broadcast control to update the mobile station with the base station identity, frequency allocation, and frequency-hopping sequence information. All other time slots in the multiframe carry voice and data traffic information. This multiframe is placed on a non-frequency-hopping

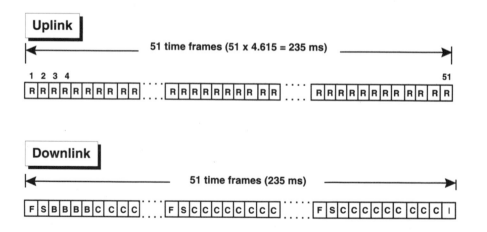

R = random access channel
C = access grant/paging channel
B = broadcast control channel
F = frequency correction channel
S = synchronization channel
I = idle frame

Figure 7.24 The GSM control multiframe.

RF carrier. Each base station is allocated several RF carriers, but only one of these includes the broadcast control channel.

The half-rate encoders allow a doubling of system capacity. The time slots, frames, etc. remain the same so it takes two frames for the 16 traffic channels to be placed on one carrier. The signaling done in the SACCH brings the traffic multiframe time slot 26 into action in addition to time slot 13 so that 16 traffic channels can be serviced. In this case the multiframe is still designated as 120 ms but now contains 2×26 frames and carries sixteen 11.4-kb/s half-rate traffic channels.

Superframes and hyperframes.   Each superframe contains 1326 TDMA frames. One type of superframe contains 26×51 multiframes and the other type contains 51×26 multiframes. The multiframe mapping is configured by the operator when optimizing the network.

One hyperframe contains 2048 superframes, which means 2,715,648 TDMA frames, and lasts for 3 hours, 28 minutes, and 52.8 seconds duration. The reason for the hyperframe is to operate the encryption algorithm on such a large number of frames (2,715,648) that security is ensured. Because a superframe contains only 1326 TDMA frames, this is insufficient for adequate encryption security.

Modulation and spectral usage.   The GSM system uses Gaussian MSK (GMSK) modulation with a modulator bandpass filter having a 3-dB bandwidth of 81.25 kHz. This is 0.3 times the bit rate, and it allows the signal baseband (270.833 kb/s) to be transmitted on a 200-kHz bandwidth RF carrier. The bandwidth efficiency is therefore $270.833/200 = 1.35$ b/s/Hz. The total spectrum available for the network is 2×25 MHz, so 124 carriers can be used, each containing eight full-rate TDMA channels. The total channel capacity is $8 \times 124 = 992$ channels per cluster (before frequency reuse). The new half-rate encoding Standard improves the channel capacity to 1984 channels per cluster. The GSM system is designed to operate at a carrier-to-interference ratio (C/I) of 10 to 12 dB. Because analog systems operate at a typical C/I value of 17 to 18 dB, there is at least a 6-dB improvement in the digital system, which translates into the use of smaller cells (greater frequency reuse) and therefore higher overall capacity.

The transmission emission spectrum for the GSM system is specified to have out-of-band radiated power in adjacent channels (adjacent channel interference) to be at least 40 dB below any observed channel. GMSK modulation assists here and takes some of the burden off the RF stage filters of the multichannel transceivers. MSK with premodulation Gaussian low-pass filtering (i.e., GMSK) is closely related to FM. In fact, it is really a type of FM. The circuitry for GMSK can take several different forms.

Power control and handoff.   Perhaps one of the most important features of any cellular radio system is handoff. The GSM system has an elaborate power control mechanism that is designed to minimize cochannel interference, and it is closely interrelated with handoff. The objective is to operate the mobile and base stations at the lowest possible power level, while maintaining good subjective service quality. During a call, the base periodically sends a signaling message to the mobile station to adjust its power level. By the very nature of TDMA, the mobile and base stations must transmit pulses of power that do not cause sideband generation that would interfere with adjacent channels. The shape of the pulses of power is therefore well controlled. The range of pulse power control is 30 dB in steps of 2 dB. This adaptive power control during calls is estimated to improve the mean system C/I values by up to 2 dB.

The decision to increase or decrease the mobile and/or the base station transmit power level(s) is made by comparing measured received signal level and quality (BER) values with predetermined threshold values. The decisions to adjust transmit power are based on *incoming* received signal values and not on values obtained at the other end of the link. The question arises: "By how much should a transmitter be increased or decreased if a threshold value is exceeded?" Also, "How often should the adjustment be updated?" GSM initially had single-step incremental changes, but software enhancements are being developed. These features affect system stability. For example, if a mobile station has its transmit power level increased because its performance has fallen below threshold, that would increase the cochannel interference between it and the mobile in the nearest cochannel cell. Subsequently, that mobile might require a transmitter power level increase to maintain satisfactory performance. In turn, it would affect the interference level in the original cell and the surrounding cochannel cells. This positive feedback mechanism would spiral all mobile transmitters up to the maximum power level and defeat the intended objective of power control. A lot of work has gone into solving this problem, even though experience has shown that such situations rarely occur. This is because of factors such as the diverse geographic distribution of users within each cell, the discontinuous transmission feature, the time distribution of call activity, and frequency-hopping sequences, etc.

Handoff decisions are based on the performance of a channel as the mobile station approaches the boundary between cells. Unfortunately, the base does not know whether the mobile is at the cell edge or whether a multipath fade is occurring, which would also necessitate a power level increase. Clearly, the power control and handoff algorithms need to be coordinated in synchronism to prevent inappropriate decisions. For example, a channel suffering deteriorating quality should not proceed with handoff if high quality could be restored by increasing the mobile transmitter power level. On the other hand, a mobile approaching the edge of one cell and about to enter another should not have its power level raised just when handoff is about to take place. Because power adjustments and handoff decisions are based on the same measurements of received signal power and quality, the algorithms have to be very sophisticated.

Each GSM *mobile* station monitors the broadcast control channel of as many as 16 surrounding *base* stations during the "deadtime" periods (no transmission or reception time slots). The mobile prioritizes up to six likely candidates for handoff and sends signal strength and quality information to the base currently providing service. The base also makes measurements of signal strength and quality for the mobile currently receiving service. The actual handoff decision is made by the network management system. In addition to factors already mentioned, there can be other reasons for handoff; for instance, it can be used to balance traffic between cells. Handoff to another cell usually requires the mobile to retune to a different carrier frequency. Another form of handoff is possible within the same cell, using the same carrier frequency but a different time slot. This can be beneficial for interference control.

The relationship between handoff and power control takes several different forms, depending on the equipment manufacturer. Two widely used algorithms are (1) the minimum acceptable performance method and (2) the power budget method. In the first method, power control takes precedence over handoff. That is to say, power increases are made while they are possible and handoff is done only when a power increase will not improve the quality of a channel that has deteriorated below the threshold value. This method tends to produce poorly defined cell boundaries so that a mobile user might be well into an adjacent cell before handoff takes place. As stated previously, this situation suffers from higher cochannel interference levels than other methods. One way to prevent late handoff is for the base to make distance measurements between itself and the mobile so that it knows how close to the cell boundary the mobile is. This technique tends to be most effective in large rural area cells. In the second method, handoff takes precedence over power control. Handoff is made to any base that can improve on the existing connection quality at an equal or lower power level.

Frequency hopping.   Frequency agility is an essential aspect of TDMA cellular mobile radio. The mobile station has to monitor the different frequencies of all surrounding base stations in order to facilitate the handoff process as the mobile user moves to another location that, from the system point of view, is in another cell. Frequency monitoring is done during the short inactive time between transmitting and receiving information. Frequency hopping, which was originally designed for military privacy, is now featured in modern cellular radio designs to enhance performance in a multipath fading environment. Because multipath fading is a narrowband phenomenon, shifting a relatively small amount in carrier frequency can place the radio link in a nonfading condition. GSM mobile stations use slow frequency hopping at 217 hops per second, that is, once every 4.6083 ms, which is almost one TDMA frame duration (4.6154 ms).

Conclusion.   As one can see, there are some mandatory specifications for the GSM system that are necessary to maintain international compatibility. However, there is considerable latitude within the system framework for circuit innovations by competing equipment manufacturers. This is essential in a system operating right at the edge of technological capability, in order to advance the state of the art.

7.8.2 Digital Cellular System (DCS-1800)

The second generation of European cellular radio systems is the DCS-1800. This is an extension of the GSM system and, as Table 7.1 indicates, it has an improved capacity over GSM. This is achieved mainly by increasing the bandwidth from the GSM 25 MHz to DCS-1800 75 MHz, by moving up to the 1.710 to 1.880-GHz band. Bandwidth efficiency is doubled by using half-rate encoding at 11.4 kb/s (instead of 22.8 kb/s for GSM). This is therefore about a sixfold increase in the number of traffic channels per cluster before frequency reuse to 5984, compared to 992 for the original GSM.

The DCS-1800 has a low transmit power level of 125 mW average or 31 mW average, compared to 2.5 W down to 250 mW for GSM. The call range is clearly lower for DCS-1800, which means more cells are needed to cover a given area in the suburban-to-rural environment. The lower transmit power translates into less battery power consumption, and therefore longer talk and standby time between battery recharges. The higher operating frequency means higher free-space loss and higher shadowing loss due to signal refraction. That in turn means more signal fading in urban areas, and a higher cost of overcoming the free-space loss in rural area coverage.

7.8.3 North American Integrated Services-136 (IS-54) system

The United States TDMA system was originally called IS-54 or digital AMPS, and has been superseded by the IS-136A upgraded standard. This system places its information on carriers spaced by 30 kHz (i.e., 30-kHz bandwidth). This is the same as the AMPS system, and allows operating companies gradually to replace the analog channels with digital channels to ease base station traffic congestion. Each digital channel carries three user signals at the full-rate coding standard compared to eight in the GSM system. The half-rate doubles the number to six. The channel transmission bit rate for digitally modulating the carrier is 48.6 kb/s. The composition of the 40-ms frame is shown in Fig. 7.25. Each frame has six time slots of 6.67-ms duration. Each time slot carries 324 bits of information, of which 260 bits are for the 13-kb/s full-rate traffic data. The other 64 bits are overhead; 28 of these are for synchronization, and they contain a specific bit sequence known by all receivers to establish frame alignment. Also, as with GSM, the known sequence acts as a training pattern to initialize an adaptive equalizer. The IS-54 system has different synchronization sequences for each of the six time slots making up the frame, thereby allowing each receiver to synchronize to its own preassigned time slots. An additional 12 bits in every time slot are for the SACCH (i.e., system control information). The digital verification color code (DVCC) is the equivalent of the supervisory audio tone used in the AMPS system. There are

256 different 8-bit color codes, which are protected by a (12,8,3) *Hamming code*. Each base station has its own preassigned color code, so any incoming interfering signals from distant cells can be ignored.

As indicated in Fig. 7.25, time slots for the mobile-to-base direction are constructed differently from the base-to-mobile direction. They essentially carry the same information but are arranged differently. Notice that the mobile-to-base direction has a 6-bit ramp time to enable its transmitter time to get up to full power, and a 6-bit guard band during which nothing is transmitted. These 12 extra bits in the base-to-mobile direction are reserved for future use.

The modulation scheme for IS-54 is $\pi/4$ differential quaternary phase shift keying (DQPSK), otherwise known as differential $\pi/4$ 4-PSK. This technique allows a bit rate of 48.6 kb/s with 30-kHz channel spacing, to give a bandwidth efficiency of 1.62 b/s/Hz. This value is 20 percent better than GSM. The major disadvantage with this type of linear modulation method is the power inefficiency, which translates into a heavier hand-held portable and, even more inconvenient, a shorter time between battery recharges.

The IS-54 speech coder uses the technique called *vector sum excited linear prediction* (VSELP) coding. This is a special type of speech coder within a large class known as *code-excited linear prediction* (CELP) coders. The speech coding rate of 7.95 kb/s achieves a reconstructed speech quality similar to that of the analog AMPS system using frequency modulation. The 7.95-kb/s signal is then passed through a channel coder that loads the bit rate up to 13 kb/s. The new half-rate coding standard reduces the overall bit rate for each call to 6.5 kb/s, and should provide comparable quality to the 13-kb/s rate. This half-rate

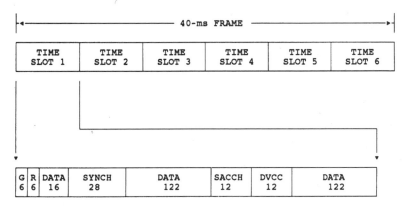

Figure 7.25 IS-54 frame structure.

gives a channel capacity six times that of analog AMPS. So, the benefit of the D-AMPS (IS-54) compared to analog AMPS is the enhanced bandwidth efficiency, which translates into more channels for a given cost. However, the need for higher-quality speech in hostile propagation environments is leading to some higher, not lower, speech coding rates. This might not be backtracking so much as a step toward future variable coding bit rates automatically adjusted by received bit-error ratio, or some other quality measurement parameter.

7.8.4   North American Integrated Services-95 system (CDMA)

Introduction.   The IS-95 standard is for CDMA cellular radio. As stated in Chap. 3, either frequency hopping (FH-CDMA) or direct sequence (DS-CDMA) provides a method of allowing multiple users to occupy the same channel (frequency band) with minimal interference. FH-CDMA is not used because the capacity would be too low for a bandwidth of 1.25 MHz. The DS-CDMA concept is reviewed in Fig. 7.26. For DS-CDMA, the system is asynchronous, meaning that the pulses from each user do not have to bear any phase relationship to one another. Present research is progressing to maximize the number of users able to operate simultaneously with an acceptable level of interference. DS-CDMA has a major drawback known as the *near-far* problem. This is defined as the differing amounts of interference contributed by different mobile stations depending on their distance from the base station. Clearly, interference from mobile stations closest to the base station is the dominant source of interference unless some transmit power control mechanism is used. Furthermore, this power control must be highly sophisticated, because the power level of multiple signals reaching the receiver must not vary by more than 1 to 2 dB if the maximum simultaneous user capability is to be achieved.

Spread-spectrum systems require the phase of the incoming signal to be determined very precisely. Phase acquisition and tracking, which collectively are called the synchronization process in spread-spectrum technology, are of paramount importance; otherwise, despreading of the desired signal is not possible (see Chap. 3).

Signals in a DS-CDMA system have different pseudorandom binary sequences that modulate a carrier and spread the spectrum of the carrier. This allows many CDMA signals to share the same frequency band. At the receiver, a correlator circuit is used to identify the signal with its specific binary sequence. The receiver despreads that signal, but not ones whose binary sequences do not match. The unspread signals contribute to system noise and the maximum number of users is reached when the aggregate noise level becomes too high. The despread signal is filtered after the correlator so that the interfering signals are reduced by the ratio of the bandwidth before and after the correlator. This signal improvement is called the processing gain and is a major factor in determining the CDMA digital cellular system capacity. Other important parameters are the E_b/N_o, efficiency of frequency reuse, number of cell sectors, and voice duty cycle.

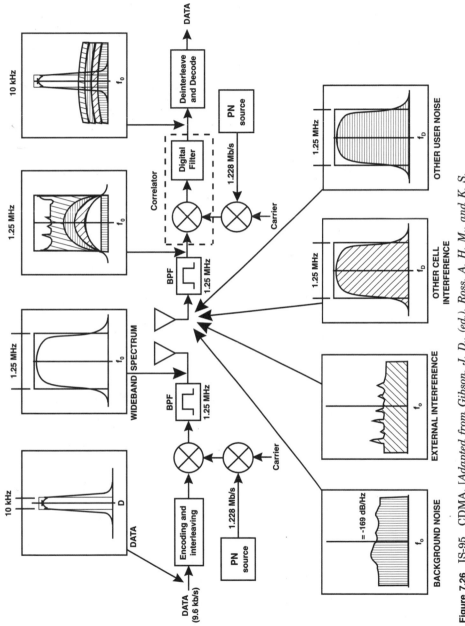

Figure 7.26 IS-95 CDMA. [*Adapted from Gibson, J. D., (ed.), Ross, A. H. M., and K. S. Gilhousen, The Mobile Communications Handbook, CRC Press, in cooperation with IEEE. © 1996 IEEE.*]

Frequency. The FCC has allocated a total bandwidth of 50 MHz for cellular services, of which 25 MHz is used for the uplink (mobile to base) and 25 MHz for the downlink (base to mobile). The FCC has divided this bandwidth into two halves so that two service providers can each operate over 12.5 MHz uplink and 12.5 MHz downlink. If CDMA is used, each operator can use ten 1.25-MHz uplink channels and ten 1.25-MHz downlink channels. By definition, this is an FDMA use of the 25-MHz bands within which each carrier contains its channels in a CDMA format. The chip rate of the pseudorandom spreading sequence for IS-95 is 1.2288 MHz. In other words, the information signal is spread over most of the 1.25-MHz transmission signal bandwidth. Until full CDMA deployment, operators can use part of the allotted bandwidth for analog FM cellular service. Guard bands are necessary in this case to ensure that other adjacent high-power analog systems do not cause interference with the CDMA.

Signal processing. IS-95 uses a proprietary 8-kb/s CELP speech coder that has variable bit rate capability. One of four bit rates (8, 4, 2, or 1 kb/s) is selected depending on input signal energy thresholds. Because the transmission bandwidth is very wide, powerful FEC can be incorporated. After channel coding, the variable gross bit rate is one of 19.2, 9.6, 4.8, or 2.4 kb/s. Convolutional coding is used with a constraint length of $k = 9$. Pseudorandom noise modulation is biphase for the uplink and quadriphase for the downlink with 64-dimension orthogonal Walsh sequences for the spread-spectrum coding.

Power control. CDMA systems must have power control of the user terminal (mobile) transmitters to overcome the near-far problem. A mobile unit near a base station must have its power reduced relative to that of a mobile far from the base; otherwise the near mobile will jam the far mobile. *Jam* is a term associated with radar in military communications, and it broadly means "to render useless." Without power control, the 1.23-MHz bandwidth (W) direct-sequence spread-spectrum modulation with a maximum of 9600 b/s data rate (R_b) will require an E_b/N_0 of at least 6 dB or higher, depending on fading conditions, etc. The ratio of the jamming interference signal to the required signal ratio is given by:

$$\frac{J}{S} = \frac{W/R_b}{E_b/N_0} = \frac{1.23 \times 10^6/9600}{4} = 32.0 \approx 15 \text{ dB}$$

J/S is the jamming margin, which is the factor by which the jamming signal can exceed the desired signal before interference is noticeable (assuming equal transmit power levels). This equation indicates that for the IS-95 system, two mobiles can exist in the same cell with one being 15 dB of path loss closer to the base than the other. Analysis shows that with no power control, CDMA systems have little or no capacity advantage over the old FM analog AMPS. The CDMA system has optimal (maximum) capacity if all the mobile signals arrive at the base station at the same power level. Therefore, *power control is central to the success of CDMA*. The problem is complex. Not only is there a path loss variation of about 80 dB from the base to the edge of the cell, but multipath fading can cause an additional 30-dB loss. Remember, multipath fading is caused by a reflected signal arriving at the receiver 180° out of phase with the direct transmitter-to-receiver signal, causing cancellation of the signal. The multipath fading conditions within a city are so complex that the signal can drop 20 to 30 dB by moving the mobile only 0.3 m. If such a situation prevails while the mobile user is traveling on the freeway, the rate of fading can be more than 100 times per second. The fading problem is further complicated by the fact that fading from base to mobile can be different than from mobile to base. This is because multipath fading is very frequency dependent, and there is a 45-MHz separation of uplink and downlink transmissions.

There are two types of power control: (1) open loop, and (2) closed loop. In the open-loop method the mobile receiver measures the strength of the signal coming from the base by using the AGC circuit on its IF amplifier. As the received signal strength decreases, the AGC voltage increases to provide a constant input to the next stage in the receiver. This voltage is also used to increase the transmitter power output. The path loss is the same for uplink and downlink paths. Unfortunately, because the fading loss is not reciprocal, a downlink received signal fade will increase the transmit power for the uplink path, which might not be fading. Consequently, the fluctuations due to fading cannot be tracked precisely by the open-loop method. The closed-loop control operates on the S/N of the signal received by the base. An estimate of the S/N made every 1.25 ms is compared to a prefixed value. If it is worse than expected, an increase power command is sent to the mobile, and if it is better, a decrease power command is sent. The closed loop mobile transmission power control voltage is added to the open loop control voltage. In this way, the open loop takes care of reciprocal path loss variations, and the closed loop is fast enough to deal with the nonreciprocal multipath fading losses.

This power control mechanism for CDMA has some other attractive features. If a cell is heavily loaded with users, a command is sent to increase the power of all mobiles. This has the effect of reducing the interference from neighboring cells and so temporarily increases the capacity of the cell. Admittedly, this is at the expense of higher interference in neighboring cells, but the initial situation recognized a lower demand in those cells. This flexibility to borrow capacity from neighboring cells is beneficial for good traffic management. The cell boundary can be effectively dynamically controlled by power variation. If a cell base lowers its transmission power level, mobiles in use at the edge of the cell might not be able to maintain contact with that base and so handoff to the adjacent cell takes place. This is the same as reducing the size of the cell until the base transmit power is increased to its nominal level. The soft handoff in conjunction with power control forms an adaptive anti-interference and congestion management mechanism. The power control also minimizes battery drain, because the mobile transmit power is kept at a low average level of about 2 mW.

Power control also allows selective variation of service quality. For example, a mobile in motion needs to operate with a higher S/N than one that is stationary. If the cell base observes a mobile signal having a high frame error rate, probably due to motion, it can increase the S/N for that mobile.

Notice that some of the power control features are not necessarily compatible with the capacity-maximizing objectives of all mobile signals arriving at the base with equal power levels.

Capacity. For CDMA the capacity tends to be related to the total interference of users in a particular cell plus the interference from users in adjacent cells. As more users make calls, there is a graceful degradation of the overall system. Frequency reuse for CDMA is simple. Each 1.25-MHz channel is reused in each cell. CDMA operates in conjunction with FDMA. Ten carriers each containing a 1.25-MHz wide channel can be used in every cell. Frequency reuse causes signals from mobiles into any base to have interference from neighboring cells using the exact same frequencies. Using the simplified hexagonal cell structure, estimates of neighboring cell interference contributions indicate the first tier to contribute 6 percent per cell, totaling 36 percent for all those adjacent neighbors. The contribution to the other outer tiers totals another 4 percent contribution. Sectorization is very effective for increasing CDMA system capacity, and cells are typically split into three sectors.

$$\text{System capacity } C = \frac{\text{jamming margin}}{1 + S} = \frac{W/R_b}{E_b/N_0\,(1 + S)} \text{ channels per cell}$$

where S is the spillover ratio (neighboring cell interference). Considering a total neighboring cell interference is 40 percent, then $S = 0.6$.

$$C = \frac{32}{1.6} = 20$$

Voice activity increases the capacity by a factor of about 2 to 2.5 up to, say, 40, and sectorization gives a further factor-of-3 improvement to 120 channels per cell. Uplink and downlink capacities are different because of the nonreciprocal nature of interfering signals entering the base and mobile units, respectively. In other words, the base is central in its cell cluster and therefore farther away from neighboring cell interference than the mobile. The downlink capacity is about 1.5 times better than the uplink.

Because 10 channels operate within the 12.5-MHz available bandwidth, the maximum capacity is therefore in the region of 1200 to 1500 channels per cell, in the absence of other enhancers such as smart antennas.

Soft handoff. This feature of CDMA is a good QoS enhancer. At or near the edge of a cell, a mobile receives incoming signals from two bases. Only when the mobile has moved far enough into the new cell to have a predetermined difference in signal levels from each base will the mobile disconnect the old cell base. This make-before-break transition process is undetectable in most handoff situations. It also eliminates the "ping-pong" effect experienced in the break-before-make style of TDMA handoff when a mobile user is meandering along paths close to cell boundaries. This soft handoff can be considered as a type of space or route diversity reception, which also enhances performance when fading is present.

Rake receiver. CDMA lends itself to an interesting receiver concept known as the rake receiver. In multipath fading environments, parallel correlator circuits receive signals incoming from different paths. Several received components are resolved at a resolution equal to the chip period and are coherently combined. In other words, instead of losing a signal by destructive interference of multipath components, two or more different path signals are received, and phase adjustment is made to provide constructive combining of these signals. This is a clever way to turn a potentially damaging phenomenon into an advantage. Figure 7.27 shows three multipath signals combined in a rake receiver.

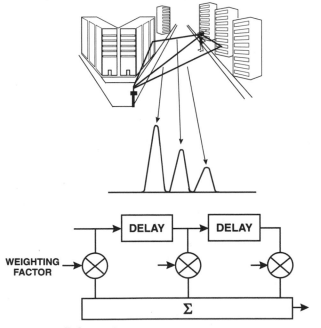

Figure 7.27 Rake receiver.

7.9   Cordless Telephone Systems

Cordless telephones were first designed to allow tetherless use in and around the home. They have now become a useful low-cost option for mobility around the workplace or in some downtown areas. They are primarily an *access* technology, in contrast to cellular radios (which form a full network).

7.9.1   Digital European cordless telecommunications

The digital European cordless telecommunications (DECT) Standard was approved by ETSI in 1992. It was intended to allow connection to PABXs to give mobility within the vicinity of a PABX, or to be used as a single cell with the base at a home or small company premises. The DECT Standard also enables users to make calls at participating locations such as airports, shopping malls, sports arenas, etc. The cost of cordless systems is lower than cellular and the data rates are higher. It must be appreciated that mobility is limited with cordless technology because the transmit power is low compared to cellular systems and is designed only for short distances. With a 10-mW transmit coverage power (250 mW peak), the line-of-sight distance is about 3 km, but more realistically, in a city environment the operating distance is only about 100 to 200 m. While a cordless system might have many cells that give the appearance of a cellular system, some of the early cordless systems did not permit a user to maintain a call while moving from one cell to another. In other words, handoff is not a feature offered by all cordless system operators. Handoff is offered by DECT operators, and this seems to be a trend that others are following.

DECT systems operate a TDMA platform using ten carriers in the 1881.792- to 1897.344-MHz band. Each carrier supports 12 channels, totaling 120 over the available 20 MHz. Figure 7.28 illustrates how high demand is met by multiple small-cell coverage depending on traffic density, which varies with urban, suburban, or rural coverage areas.

DECT differs from the uplink-downlink frequency division duplex (FDD) scheme by operating *time division duplex* (TDD) transmission, so both handset and base transmit on the same frequency at different times. Each of the 120 channels has 24 time slots; 12 for transmitting and 12 for receiving. The DECT protocol is described in ETSI documents ETS300175-1 to 5. The 10-ms frame (see Fig. 7.29) contains 24 time slots in which the first 12 are for base station to handset (downlink) channels, and the second 12 are for handset to base (uplink) channels. The duration of one time slot is 416.7 μs and 424 bits are transmitted in 368.1 μs, leaving a guard band of 48.6 μs between each time slot. The voice or data bits constitute only 320 bits per time slot, so the other 104 bits are for overhead. Because the speech coding in the DECT system is 32 kb/s ADPCM per slot (B-field), the speech quality in the *absence* of interference is toll quality. The B-field has an unprotected mode for voice and a protected mode for reliable transmission of internal control signaling and limited user data at 25.6 kb/s. There is also a 6.4 kb/s per slot A-field for control and signaling. The gross bit rate is 1.152 Mb/s. 16 frames are packed into a 160-ms multiframe.

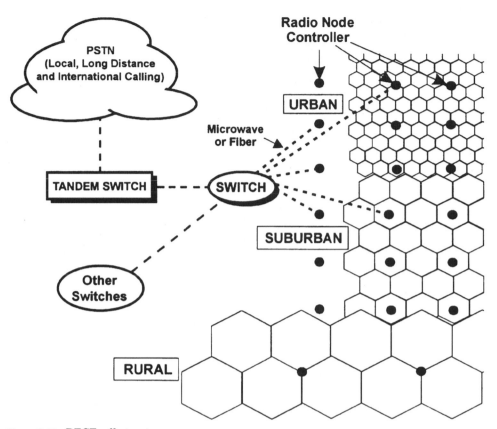

Figure 7.28 DECT cell structure.

The multiframe concept is used to allow the content of the A-field control to include paging information, identification messages for traffic bearers, and higher layer control information to be passed between the base and portable. Uplink and downlink multiframes are clearly different to accomplish the question/answer interrogation mechanism.

Because multicell configurations do not have clusters as in cellular structures, a frequency used for a channel in one cell cannot be used by its nearest neighboring cells. This results in a reduction in the practical number of channels usable by each cell from 120 to 60. High capacity within a small area is achieved by using low base and mobile station transmit power levels, and cells as small as 100-m radius or even less. DECT uses dynamic channel selection and automatic frequency planning during cell setup to select channels that have the least interference from adjacent cells.

Uplink transmission during a call occurs for only 368.1 µs (i.e., in its specified time slot) of every 10 ms, so for the rest of the time the handset performs other activities. It continually searches for a channel whose signal is of better quality. Check bits in the signaling overhead are monitored by both base and

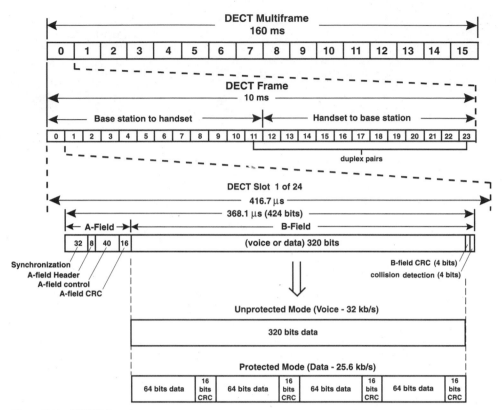

Figure 7.29 DECT framing.

handset so that reception quality can be assessed in both directions. On finding a better quality channel, a new time slot is set up while maintaining the old connection; it accesses that channel in parallel with the one providing service and seamlessly hands over.

DECT has been modified for use in North America where it is called personal wireless telecommunication (PWT) for unlicensed operation and PWT-Enhanced for licensed operation. For example, the frequency band has been changed to 1850 to 1990 MHz, and the modulation has changed to $\pi/4$-QPSK to enhance bandwidth efficiency.

7.9.2 Wireless local loop

Cordless technology is now receiving much attention with regard to the last kilometer of the connection into homes. The so-called wireless local loop (WLL), or fixed wireless, can provide service connections that are cheaper and faster than wireline (cable) connections in many areas. Obviously this application is not mobile, but the technology is very similar to, and preferably compatible with, that of cordless or mobile cellular. Some equipment manufacturers produce long-range cordless radio telephones to extend the tetherless operation beyond the home and garden to a 30-km or more radius from a base located at the home. Some cellular radio operators team up with cordless operators for the microcell portion of their cellular network. This might be an interim solution until the cost of base stations comes down to a level low enough for large-scale micro- and picocell installations (Fig. 7.30). Wireless access is a major boon to developing countries. Wireless local loop is already cheaper than wired local loop in many situations, and its rapid deployment accelerates the construction of the telecommunications infrastructure.

There is fierce competition between the low-tier radio technology suppliers for a share of this potentially vast market. Technologies based on CDMA (IS-95) and TDMA (PACS, DECT, IS-54/IS-136) are competing against each other. CDMA seems to have an edge in terms of cost-effectiveness in this field. The fixed wireless mode of operation for CDMA allows power control to be better managed than when subscribers are mobile. This translates to higher capacity per unit area. The directional antennas located at the customer premises also improve the system gain.

7.10   Personal Communications Services (or Networks)

It must be stressed that PCS or PCN is a general concept and not a specific communications system. There are several definitions of PCS. Perhaps the most appealing is the following one: *The availability of all communication services anytime, anywhere, to everybody, by a single identity number and a pocketable communication terminal.* This simple statement is loaded with technological hitches. With the potential customer base in the major U.S. urban areas alone estimated to be over 100 million, the stakes are high. The cost of PCS systems is not small, especially because they are evolving out of cellular radio systems. For example, because of the microcellular nature of

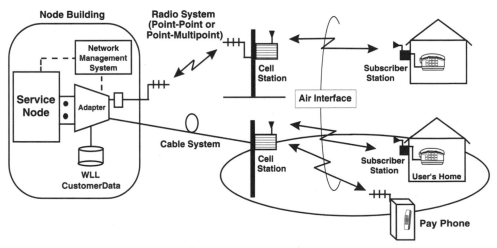

Figure 7.30   Wireless local loop.

PCSs, they will need more than five times the number of base stations of present cellular networks. Designers did not have the luxury of starting their designs from the beginning, and too much money has already been invested in cellular mobile radio to discard it.

The importance of hand-held telephones as a strong and vital market force in the telecommunications industry cannot be overemphasized. Hand-held portable radios first gained widespread recognition with the introduction of cordless extension phones, which allowed a customer to place or receive calls anywhere within the home or garden.

Because the portable phone transmitter power was only about 1 mW, the operating range was not more than a 100 m or so. The convenience of this tetherless phone facility soon prompted the idea of universal coverage for portable phones, and today's cellular radio systems go a long way toward achieving that goal.

The future promises to be very exciting in this area of telecommunications, although there are still many questions to be answered. For example, "How do you make radio contact into obscure places, such as elevators made completely from steel?" The solution might be a communication fiber or wires down the center of the steel cable operating the elevator, or to have elevators made from nonmetallic material. There are further obstacles to overcome in providing service on supersonic airplanes. In these circumstances, the Doppler effect adds another dimension of complication to the rather intricate problem of setting up cells to cover the vast areas of the Atlantic and Pacific Oceans. By the above definition of PCS, it will be some years beyond 2000 before the objective is fully realized.

PCS is primarily concerned with increased levels of mobility and high data rate communications compared to the first and second generation of cellular radio systems. This can be in the form of either *terminal mobility* or *personal mobility*. Terminal mobility means a terminal can connect to any point in the cellular network. A unique identifier is assigned to the terminal by the manufacturer, which enables validation and location information for delivery of calls. Personal mobility allows a user not only terminal mobility but also access by any terminal. This extra level of mobility requires a means of identifying the user and authenticating or validating the user's access. A unique number is a desirable option, perhaps via a portable smart card or speech recognition. A PCS system needs to keep track of terminal locations, and the terminal number will be different from the user's universal personal telecommunications (UPT) number. A user can then log on to any user terminal and once authentication is completed there is full access to the telephone network. Clearly, the main problems for the service provider are routing calls to and from the user.

Cordless, cellular, and satellite mobile systems all have their strengths and weaknesses. The dual or triple mode terminals are an interim or possibly long-term solution for providing PCS. There is a trend toward the convergence of cellular and cordless technologies. However, there remain the incompatibility aspects of transmit power (low for cordless, high for cellular) and therefore cell size, and of source coding (high bit rate for cordless, low for cellular) which greatly affects system capacity and QoS. This is being resolved by the dual handset concept.

DECT and GSM have a high level of interoperability, which forms a bridge between high-tier and low-tier cellular and cordless technologies. The dual mode GSM/DECT handset offers a synergy by combining the attributes of each to provide high QoS over a wide range from the static office environment to the high mobility of intercity vehicular communications. Switching between cordless and cellular systems is done automatically, unnoticed by the user. Figure 7.31 portrays the GSM (or DCS-1800)/DECT interworking concept. The user is contactable on a single number, regardless of the location or air interface. European regulatory bodies are moving ahead with the convergence of cellular and cordless technologies. The situation in North America is still evolving and the position of a WLL interface within the cordless and cellular platforms is still being evaluated.

If this concept is extended to triple mode handsets to include the LEO/MEO satellite selection, an intelligent terminal that can select the most cost-effective connection will provide users with global connectivity. Also, bandwidth-on-demand and QoS selection will provide further dimensions of communication services. It will probably be necessary to display a frequent charging update indication to let customers know the rate at which the bill is increasing, depending on the mode, bandwidth, and QoS in use.

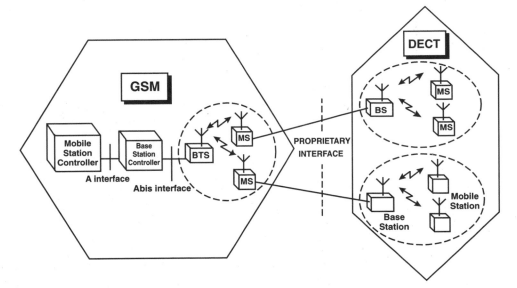

Figure 7.31 DECT/GSM.

7.11   Personal Access Communications Systems

Personal access communications systems (PACS) is a combined architecture adopting the best aspects of the Japanese PHS and the Bellcore wireless access communication system (WACS). So far, no single system achieves the goal of PCS, although the latest generation of cellular phones is moving in the right direction. PACS is an ANSI Standard (J-STD-014) that is aimed at the low-tier medium-mobility part of a PCS that can be integrated with high-tier handsets such as AMPS, GSM/PCS-1900, IS-95, and IS-54/IS-136. The PACS air interface is a TDMA/FDMA system designed for the U.S. 1.9-GHz band. It operates with frequency division duplexing and time division duplexing. It is designed for indoor picocell and outdoor vehicular use, and can support vehicular speeds up to 60 km/hr with cell handover times of 10 to 20 ms. High quality voice is assured by using ADPCM at 32 kb/s or data at rates close to 32 kb/s.

Figure 7.32 portrays the PACS functional architecture. The subscriber unit can either be portable or a fixed wireless local loop installation. The radio ports are small indoor or pole-mounted units that are powered to a distance of 3.7 km (12,000 ft) from a local exchange by 130 V on 24- or 26-gauge copper wires. 24 to 100 radio ports are connected to a single control unit in the local exchange through the P (port) interface. A radio port controller can control a maximum of 320 radio ports. The access manager takes care of the call setup for several control units, and also manages visiting users.

The PACS radio uses π/4-QPSK modulation with a 192-kbaud symbol rate (384-kb/s bit rate). Figure 7.33 shows the 2.5-ms TDMA frame that contains eight time slots. A 16-kb/s broadcast channel uses time slot 5 on both uplink and downlink. This channel is used for alerting the user to incoming calls and system information such as priority requests. Each time slot contains 120 bits,

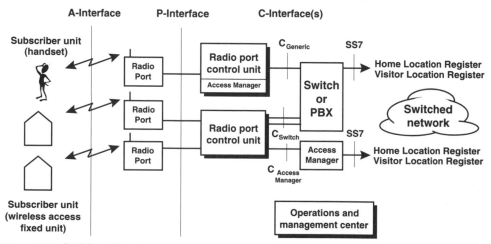

Figure 7.32   PACS architecture.

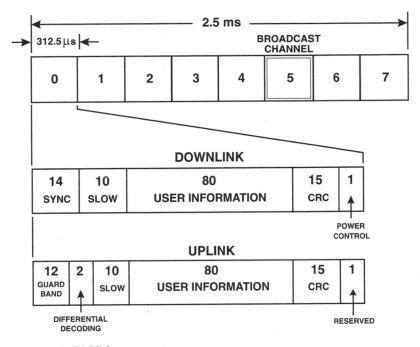

Figure 7.33 PACS frame structure.

of which 80 are payload and 40 are overhead. The uplink and downlink time slots are very similar, the only difference being the first 14 bits, which are used for synchronization on the downlink, whereas on the uplink the first 12 bits are the TDMA guard time and the other 2 bits are for differential decoding (for carrier recovery). The 10-bit slow channel carries signaling information, error indications, synchronization patterns, and user data. The 80-bit fast channel is for user information. The 15-bit CRC is for error control, and the power control bit is to optimize the handset transmit power level.

The 80 bits per time slot translates to 32 kb/s. A selection of other information rates such as 16 kb/s, 8 kb/s, or multiple 32 kb/s, can be made depending on the user's bandwidth requirements. This type of bandwidth-on-request feature is becoming essential for today's operators. For customer premises modem use, two 32-kb/s time slots can be merged into a full 64-kb/s PCM μ-law channel.

The subscriber unit uses space diversity reception, which is referred to as (1) full selection receiver diversity plus switched transmitter diversity on the uplink, and (2) preselection receiver diversity plus transmitter diversity on the downlink.

The FCC has allocated 1850 to 1910 MHz for the uplink and 1930 to 1990 MHz for the downlink in 3×15 MHz plus 3×5 MHz pairs in a frequency division duplexed operation. An additional unlicensed 1920- to 1930-MHz section is allocated to time division duplexed operation.

Delay spread. Multipath fading is related to the time difference between signals arriving at the receiver after traveling along different paths and therefore having different propagation delays. The spread (difference) in these delays is of concern to mobile systems designers. Interference occurs if the delay spread is more than about 5 to 10 percent of the symbol duration. The delay spread is clearly related to the distance between the base and subscriber. Indoors (small rooms), spreads above 100 ns are unusual. In larger, open-plan offices, auditoriums, shopping malls, etc., 300 ns or more is normal. Outside of buildings 500 ns and above is expected. For the 384-kb/s bit rate of PACS, delay spreads up to 520 ns can be accepted without diversity, and up to about 1300 ns with diversity. In comparison, the first generation of DECT can withstand up to about 100 ns of delay spread, without diversity. This is acceptable for the indoor use for which it was initially designed, or short-range WLL where there is no mobility, but outdoor use might require some specification modifications. However, there are always trade-offs in mobile systems. The short range offered by DECT gives it a capacity advantage (due to low interference) in densely populated downtown areas. Of course, the cost rises as the number of base stations increases.

Frequency assignment is done by the radio port control unit by what is called quasi-static autonomous frequency assignment (QSAFA). This simply means that after a command to turn off the portable transmitter, it scans the incoming received frequencies from several radio ports for the strongest signal. The signal strength values are communicated back to the control unit and the radio port is assigned downlink and subsequently uplink frequencies.

PACS is designed to support higher-layer circuit protocols, such as data and fax modems, and data protocols such as TCP/IP and ATM.

7.12 Mobile Data Systems

The miniaturization of personal computers has created a rapidly growing wireless data services industry. Laptop or notebook users increasingly want mobile connectivity. The bit rate required in this case is generally assessed as low when compared to wireless LAN connections. This could change with widespread Internet access. A bit rate of 9.6 kb/s might be fine for file transfer (e-mail) where small amounts of data are moved around, but Web browsing is becoming increasingly demanding on bit rate, and even a 64-kb/s rate places increased stress on the scarce bandwidth availability of wireless systems. An important difference between the mobile voice call and the data call is the level of mobility. Normally, a laptop user does not walk or drive along the street while tapping on the keyboard. While most users are stationary, others in locations such as trains or airplanes have high mobility, requiring special treatment. These can be considered to be very-high-mobility microcells. Clearly, there are many categories and all cannot yet be satisfied by one system.

7.12.1 Low data rates over radio

As a starting point, a large number of data-over-radio users can be grouped into the category of low-speed, wide coverage area. This type of connectivity can simply be a laptop with a modem/fax card connected to a mobile radio or better still with a built-in radio. The user operates the modem or fax through an analog voice channel of the telephone network. Some cellular system operators offer direct digital access so that a laptop can connect directly to the radio via the RS-232 printer or port without the need for a modem. The question is, How is data affected by the speech coding techniques used for the voice channel designed for cellular radio systems? In general they are incompatible, and the coding must be bypassed for data use. That usually means the data rate is low in comparison with modem rates.

Another option is for the mobile data service provider to have a pool of modems into which a user could connect. This could be an entry point to a packet switching network such as an X.25 service (Sec. 10.3). Such value-added services are attractive to the service provider and customer alike. Because data networks generally tend to have packet-based transmission, the marriage of wireless data with packet-switched systems is inevitable. The trend in this area is toward higher data rates and wider coverage areas. Because of bandwidth restrictions, there is a tendency toward variable bandwidth allocation. Depending on cell occupancy, a certain minimum bandwidth can be guaranteed with more bandwidth available in the less-congested cells.

Signal quality.   The use of portable (laptop) computers to access an information source in a business office using the cellular network is a rapidly growing service requirement. It is understandable that data communications over cellular radio links are susceptible to relatively high BERs. As already mentioned, impairments occur that result in the received signal being below the noise level of the system. This might be caused by (1) multipath fading, (2) shadowing of the receiver from the transmitter by obstacles such as buildings and bridges, or (3) interference from other radio sources such as automobile ignition systems. Both isolated errors or error bursts can result from the above problems. During handoff, breaks of up to about 300 ms in the continuity of data transmission can occur. Multipath fading or shadowing can produce fades of several seconds. In extreme cases, after a break of 5 s, the call is automatically dropped (i.e., cleared down).

Countermeasures such as FEC, interleaving, and ARQ are usually used to improve the BER. The Reed-Solomon class of FEC symbol block codes has been successfully incorporated into cellular systems to enhance the quality in the presence of error bursts. The relatively low-redundancy (72,68) code produces good improvements.

The short error burst problem can also be tackled by using "interleaving." Instead of transmitting the code blocks as complete blocks, the bits of each block are spread out in time so that there is an overlap of adjacent blocks. The depth of interleaving is denoted by the interval between each bit within the same code block.

Automatic requests for repeat are shown in Fig. 7.34. The blocks of data to be transmitted have a header at the beginning of each block. The complete block is protected by an error-detecting code that is usually a cyclic redundancy check. Each block is assigned a reference number, which is contained in the header, together with information on the reference number of the last correctly received block. For example, in Fig. 7.34, the second block is received incorrectly by the base station, which indicates the loss of a block in the block C header. The mobile station then repeats block 2. Bursts of errors occupying many blocks can be corrected by this technique. The obvious disadvantage of this system is the delay in transmission caused by repeating one or several blocks. Elastic stores and control of the stores are necessary. ARQ and FEC are very effective methods of reducing all types of cellular radio link errors even at received signal strengths as low as −120 dBm.

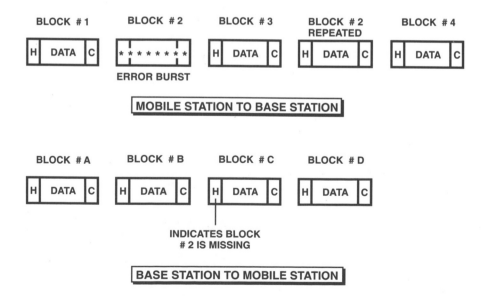

Figure 7.34 Automatic request for repeat example.

7.12.2 Wireless LANs

In some respects, the connection into a LAN by a wireless-connected laptop style of computer is easier to organize than other mobile communications. The range of mobility is very small, say, 10 to 100 m. Many applications for such connections spring to mind, such as university campuses, hospitals, warehouses, and large companies housed in several buildings, etc. The objective is to have access to the backbone LAN from one of many locations. This is like a picocell arrangement, but in this case the connect bandwidth (at 1 to 2 Mb/s) is considerably higher than regular cellular or cordless systems.

Standardization of wireless LANs had started to gain momentum by about 1996; IEEE 802.11 is the North American Standard, and the high-performance radio LAN (known as HIPERLAN) is the European (ETSI) Standard. The focus here will be on the physical and link layers of the OSI scheme.

IEEE 802.11. Initially, three physical layers were adopted: (1) FHSS and (2) DSSS radio in the 2.4-GHz industrial, scientific, and medical (ISM) band; and (3) infrared light. The three options allow for technical innovation, but cause interoperability problems. Data rates of 1 Mb/s and 2 Mb/s are specified in 802.11. For FH, there are 79 frequencies among which the signal can hop, which is slow hopping because multiple consecutive symbols are transmitted during each dwell time. Three sets of hopping patterns each composed of 26 patterns are organized so that any consecutive frequencies are separated by at least 6 MHz to avoid narrowband interference. The DSSS systems have only one spreading sequence (unlike multicoding in CDMA). The 802.11 processing gain in the United States is 10.4 dB, which allows the 83-MHz band to be segmented into 11 DS center frequencies but only three of those are usable without overlap. Note the FH systems therefore have larger capacity (more simultaneous users). Also, narrowband interference affects FH systems for only a fraction of the time, whereas DS systems are affected to a smaller extent all of the time. The transmit power of both system radios is 100 mW maximum, which translates into about 100-m indoor range.

HIPERLAN. This European wireless LAN operates at 23.529 Mb/s with a low transmit power level to provide a range in the 10- to 100-m region. This higher bit rate requires about 150 MHz of spectrum in the 5.15- to 5.30-GHz or 17.1- to 17.2-GHz allocated bands. The 5-GHz band is used initially and is divided into five channels. GMSK modulation is used to minimize adjacent channel interference. Packet error rates below 10^{-3} are recommended and a (31,26) BCH code interleaved across 16 codewords provides the necessary FEC. A block of 416 data bits is coded into 496 transmitted bits, which is quite a high level of redundancy, but at least two random errors and burst errors up to 32 bits long can be corrected. There are 47 blocks per packet and equalization is required to reduce intersymbol interference. The 17-GHz band is for higher bit rates, with the objective of 155 Mb/s for ATM.

Both IEEE 802.11 and HIPERLAN operate on a carrier sensing mechanism, with packet collision avoidance similar to that used in Ethernet style wired LANs (see Sec. 10.4). Encryption schemes are essential for all wireless LANs, and these are still in the early stages of evolution.

Mobile networking.   Networking, which is more sophisticated than a simple modem connection, is emerging for mobile communications. Internet host movement is a function that neither the Internet protocol nor the OSI network architecture anticipated. The extent of mobility and the global nature of the Internet moves the wireless LAN into the realm of the wireless WAN. The main questions to be resolved concern the details in defining a home local area connection and a foreign connection to the Internet as a mobile node. The issues to be clarified in mobile networking are more organizational and management oriented than technical. However, technical innovations such as smart antennas and high-level coded modulation will no doubt enhance the quality of mobile networking and wireless LANs in general.

7.13   Cellular Rural Area Networks

The price of base stations and mobile station equipment has decreased over the past decade because the initial research and development costs have been recouped and also because of high-volume production. The first mobile telephones were seen by many as an expensive luxury, but this viewpoint has changed as costs have fallen. An outstanding feature of the cellular system is the huge area that can be covered by a single base station. It is quite acceptable for a single, high-transmission-power base station to cover a cell radius of 30 to 45 km using an omnidirectional antenna. This is an area coverage of over 6000 km^2. The omnidirectional antenna allows a broadcast style of communication to the terminal station within a cell, or to a repeater station in the case of additional cell creation. These digital radio multiple access systems are *point-to-multipoint* in character.

A high-transmit power cellular system for rural communications is shown in Fig. 7.35. The subscriber terminal was initially designed to be a mobile station installed in a vehicle but, in this application it is, for example, installed in a village telephone call box or medical clinic. The antennas at the terminal stations are usually the high-directivity Yagi or horn type, which feed into the omnidirectional antenna at the repeater or base station. Each terminal usually serves about four to eight subscribers but, because the number of voice channels per RF carrier is up to about 60, this could be the number of subscribers served at one terminal if required. With as many as

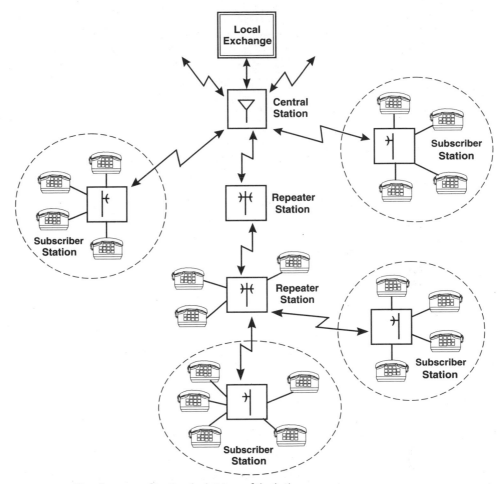

Figure 7.35 Rural area application (point to multipoint).

10 or 12 repeaters, the total coverage radius can be extended to over 600 km. The base station and repeaters operate in a microwave frequency band (typically 0.8 to 1.5 GHz or 1.5 to 2.6 GHz). The voice channels are either 64-kb/s PCM or 32-kb/s ADPCM. There are several possible modulation schemes, with 4-PSK being very popular. The choice of technology depends on the total number of subscribers, their geographical distribution, and the terrain topography.

By locating base stations at town sites along a country's backbone where multiplexers are installed, penetration deep into rural areas can be achieved. A major advantage of the cellular approach is that the system can be relocated elsewhere as needed, in the event that a small community increases in size to the point where it becomes economical to introduce fixed cable local loop using either copper or optical fiber. Other than satellite connectivity, the cellular system is the fastest method of providing telephone service to remote areas.

7.13.1 Point-to-multipoint system

Today's point-to-multipoint rural telecommunications systems have built on the early systems described above. First, they are now digital TDMA systems with many time slots (typically 24 to 96) for use throughout the repeatered cell structure. Second, each base station acts as a concentrator and not a 1:1 fixed connection, so several users can 'share' each line on a first come, first served basis.

The total number of subscribers that can be serviced by this type of system depends on the:

- Voice coding rate
- Number of time slots and therefore the bit rate
- Grade of service (blocking probability)
- Average traffic generated per subscriber

For a typical average traffic of 0.09 erl per subscriber and a 1 percent blocking rate, a concentration ratio of about 8.5 is possible. That means if, for example, a regular PCM, T1 (1.544 Mb/s) 24-channel (time slots) system is used, $24 \times 8.5 = 204$ subscribers can be serviced. If 32- or 16-kb/s voice coding is used, the number of time slots increases to 48 or 96 and the number of subscribers increases to 408 and 816, respectively.

Figure 7.36 illustrates a typical TDMA system. When a subscriber initiates a call, the base station automatically allocates an available time slot (provided they are not all busy). For just one TDMA system, only one pair of frequencies is needed, and increased capacity can be created by using multiple channel pairs in a TDMA/FDMA frequency plan.

This is another example of how mobile wireless technology is used in a fixed mode to provide a much-needed service to remote rural communities.

7.14 Future Mobile Radio Communications

The contest between TDMA and CDMA promises to be an exciting technological duel over the next decade, and it is very difficult to predict a winner. Rather than make such a bold prediction here, it is worthwhile reviewing the latest laboratory information so as to extend the visible horizon for each technology.

In Europe, third-generation systems are already being designed and standardized. These are called (1) *international mobile telecommunications* (IMT-2000), and (2) *universal mobile telecommunication system* (UMTS).

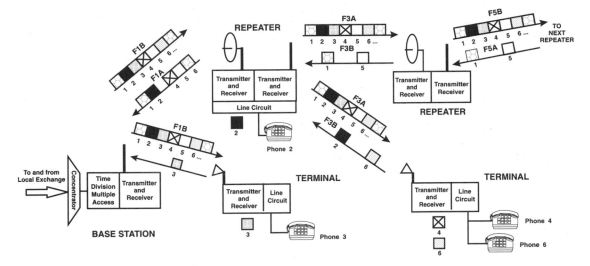

Figure 7.36 Point-to-multipoint system.

Figure 7.37 indicates the system environments, and the mobility covers all forms of transport including pedestrians, motor vehicles, boats, trains, and airplanes. The UMTS scheme, which should start to take shape in the first decade of the new century, should provide a multiband, multiservice, multifunction PCS with communication rates from 144 kb/s to 2 Mb/s with global roaming. It will support ATM for B-ISDN at rates up to 155 Mb/s utilizing the millimeter bands in the range 38 to 60 GHz for wireless LAN connect capability.

7.14.1 Comparison of CDMA and TDMA

Despite all of the capacity improvement predictions for CDMA over TDMA, practical delivery of theoretical improvement factors is proving elusive, and the progress of this technological "cat-and-mouse" chase is intriguing. System capacity has many parameters in its equation that are not easy to model or clearly define. There is, however, a clear relationship between capacity and quality of service that affects both CDMA and TDMA. As more capacity-improving signal processing is used to reduce the speech encoded bit rate of each user, the quality of service to each user drops. As the delay associated with the speech encoding and error correction (channel coding) increases, the quality drops and echo cancellation eventually becomes essential. The dc power used for extensive signal processing can offset the battery-saving benefits of lower RF transmit power. Talk time and standby time are of major concern to the customer. Unless there is a breakthrough in battery technology in the near future, there could well be a backtracking in the seemingly never-ending speech coding quest for high bandwidth efficiency.

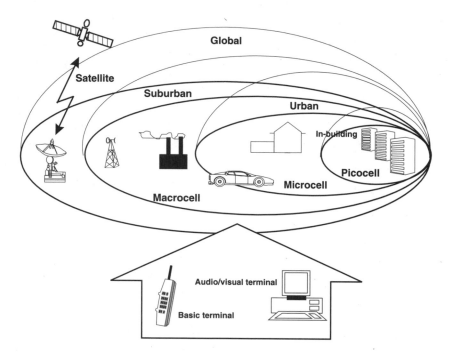

Figure 7.37 UMTS. (*Adapted from Dasilva, J. S., et al., European Third-Generation Mobile Systems, IEEE Communications Magazine, October 1996, Fig. 3, p. 70. ©1996 IEEE.*)

In other words, lower-capacity systems that give good speech quality might be worth paying for. Also of concern for QoS is voice activation. The 2- to 2.5-fold capacity improvement comes at a price. In acoustically noisy environments such as vehicles on a freeway, the system can frequently be activated unintentionally, especially if music is played.

TDMA is viewed increasingly as suboptimal for cellular systems, at least in its present form. Statistical multiplexing might improve the situation if the onset of speech in acoustically noisy environments can be detected more quickly and accurately. Statistical multiplexing used with packet reservation multiple access is a promising enhancement of TDMA. The packet dropping ratio that determines the QoS increases with more users, but at a slow enough rate to exhibit the soft or graceful degradation inherent to CDMA systems.

The complexity of CDMA, which leads to delay times as high as 200 ms, could be a major drawback in the future when QoS becomes a more serious issue. However, CDMA has some technically very pleasing attributes. When fully functional, it might well outperform TDMA or other systems.

An unbiased comparison must have preestablished ground rules, and the following factors will be used here. First, any meaningful comparison between CDMA and TDMA must be based on a unit bandwidth; 25 MHz is used here. The QoS must also be taken into account. That means capacity calculations should be based on similar voice-coding rates. Even then, a controlled MOS should be done for each competing system. The voice activation improvement factor is applicable to both CDMA and TDMA, and should not be applied to one and conveniently omitted from the other. Any sectorization factor must also be applied equally.

CDMA (for IS-95).   The capacity calculation from Sec. 7.8.4 continues as follows:

$$\text{Capacity} = \frac{W/R_b}{E_b/N_0\,(1 + S)} = \frac{1250/9.6}{4(1 + 0.6)} = 20.3 \text{ channels}$$

where W = 1.23 Mchips/s in 1.25-MHz bandwidth
R_b = 13.3 kb/s
E_b/N_0 = 6 dB = 4
S = 0.6

This figure is increased by a factor of:

- $\approx$ 2 for voice activation
- 3 for sectorization

Therefore capacity = 6×20.3 = 122 channels/cell. Because there are ten channels within the operating bandwidth of 25 MHz,

$$\text{Capacity} = 1220 \text{ channels/cell/25 MHz}$$

The initial capacity equation is based on the assumption that each base station receives equal power from all incoming mobile signals. This is an optimal situation that is not easily achievable in reality. Practical systems are limited by thermal noise in the signals from mobiles, degrading the base station S/N. In other words, there is a limit to the extent to which power control can be used to improve signal quality. Thermal noise can reduce the capacity by as much as 50 percent. Using a generous 25 percent figure, the final capacity for IS-95 CDMA is approximately 915 channels/cell/25 MHz. The bit rate is actually variable from 2.4 to 19.2 kb/s, depending on user request and traffic conditions. So, the above figure could be improved by up to a factor of 4. QoS, of course, must be weighed against the low voice coding rates.

The 1.25 MHz of bandwidth allocated to IS-95 was designed to provide compatibility with IS-54/136 AMPS. Unfortunately, this bandwidth is suboptimal for CDMA, and a wide bandwidth of 10 MHz or more would significantly improve performance.

TDMA (IS-54/136).   This system operates in the same frequency bands as IS-95, and in a 25-MHz bandwidth carriers are spaced at 30 kHz. The number of channels is therefore 25/0.03 = 832. The speech rate is 7.95 kb/s plus channel coding bit rates, totaling 13 kb/s, which is difficult to compare to IS-95 because of its variable rate. The seven-cell frequency reuse pattern means that 832 channels are available over a seven-cell cluster, so the capacity is 832/7 = 118 channels per cell, and using a sectorization of 3 to be equal with CDMA sectorization, 118 × 3 = 354 channels per cell. Voice activation is not specific to CDMA and can be used with TDMA, as satellite technology has been doing for many years with DSI. Again, using a factor of 2 to be consistent with CDMA, the final capacity is 708 channels/cell/25 MHz.

Conclusion.   The values derived here show a capacity in favor of CDMA, but not by the large factor reported in comparisons in some other texts. However, it must be remembered that these calculations are simplified, and only give approximate results. Important factors such as environmental conditions are not considered. For example, CDMA performs better than TDMA in a hostile multipath fading situation. Clearly, a direct comparison between TDMA and CDMA is not simple, and most comparisons, including this one, are inexact. This is merely a snapshot of the present time, which could change radically with future improvements in technology, such as steerable smart antennas and statistical multiplexers.

While CDMA was in its development phase, TDMA GSM systems were being rapidly deployed. It is estimated that by the year 2000, GSM will have more than 100 million customers in over 100 countries, which is a strong position in the global market. Depending on the source of information, the CDMA capacity per cell is quoted as several times that of TDMA. The exact figure is difficult to ascertain. The questions that really matter are, How does any capacity advantage translate into a resulting cost-benefit to users? and Is the quality of service any better? The answers to these questions will be known only when CDMA achieves widespread deployment.

7.14.2   Broadband wireless

CDMA has several attractive characteristics. Because system capacity is closely related to cost, it is this parameter that receives most attention. Several capacity-enhancing innovations are being proposed for TDMA systems. Adaptive is the key word. In addition to adaptive equalizers, adaptive antenna arrays, and adaptive power control, there is the very interesting option of adaptive modulation. The level of modulation dictates the spectrum utilization efficiency and therefore capacity. Ideally, 64-QAM or higher should be used, but the average carrier-to-cochannel interference power ratio (C/I_c) deteriorates at the higher levels of modulation (Fig. 7.38). Slow adaptive modulation assigns a new call to a level of modulation depending on the C/I_c value, which depends largely on the distance from the base station. Assignment might last the duration of the call or until channel reassign-

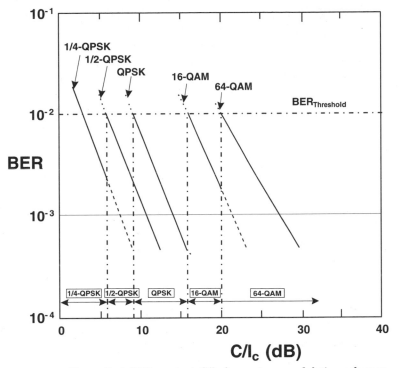

Figure 7.38 Theoretical BER against C/I_c for various modulation schemes. (*Adapted from IEEE Communications Magazine, N. Morinaga et al., Jan. 1997, p. 35. ©1997 IEEE.*)

ment. A spectral efficiency of 3.5 to 4 times the 4-PSK value can be anticipated for slow adaptive modulation. Fast adaptive modulation assigns a modulation level based on the instantaneous channel condition (C/I_c) and assigned bandwidth. Time-slot-by-time-slot modulation assignment is made based on link quality assessments embedded in each data burst. When adaptive modulation is combined with dynamic channel allocation, impressive capacity increase can be achieved.

For broadband communications, direct sequence spread spectrum would require about 2.3 GHz of bandwidth to spread a 155-Mb/s ATM signal with a processing gain of 15. Interchip interference and receiver synchronization difficulties at such high frequencies might favor other forms of CDMA such as slow frequency hopping (SFH), orthogonal FDM (OFDM), or multicarrier CDMA. Figure 7.39 shows a diagram of OFDM spread spectrum. In the transmitter the bit stream is series-to-parallel converted into n streams. Each stream is then further split into m identical parallel streams, which are interleaved for time diversity and spread onto separate orthogonal carriers for frequency diversity.

Future CDMA systems will probably be hybrid systems incorporating the advantages of several techniques. Multicarrier CDMA (MC-CDMA) systems

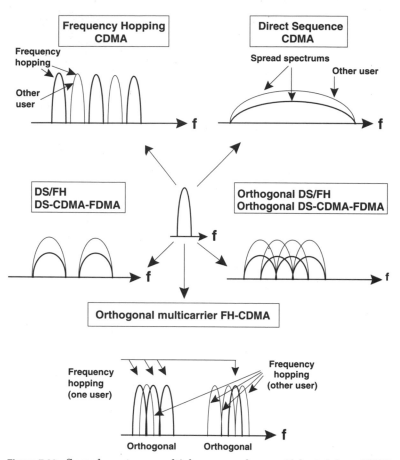

Figure 7.39 Spread-spectrum multiple access schemes. (*Adapted from IEEE Communications Magazine, N. Morinaga et al., Jan. 1997, p. 39. ©1997 IEEE.*)

are particularly attractive for broadband transmission, and OFDM minimizes the guard band necessary between each carrier. Orthogonal is defined here as when the multicarrier frequency differences are n/τ_s, where n is an integer and τ_s is the period of the data symbols. A 155-Mb/s signal could be series-to-parallel converted into, say, 16 10-Mb/s streams. The lower rate streams are more manageable and sixteen carriers would counteract frequency selective fading. The processing gain of MC-CDMA is believed to be twice that of DS-CDMA. However, the circuit complexity of MC-CDMA and its inherently inferior error performance compared to single-carrier modulated systems means it might not emerge from the laboratory for some time.

Few would dispute the technical feasibility of ATM over wireless, although more development work needs to be done. However, at the 52-Mb/s or 155-Mb/s designated ATM rate, even with extensive frequency reuse the existing allocated spectrum would be grossly inadequate for serious numbers of users.

The question is, How many users really want 155 Mb/s over wireless? Based on present trends, a much lower bit rate of, say, 2 to 10 Mb/s would appear to be more appropriate for some years to come. Whether ATM compatibility with the fixed cable environment can be assured at such rates is another question. One outcome of the potentially huge numbers (millions) of wireless data users will no doubt be the allocation of more spectrum for this application. Economic viability might depend on more bandwidth becoming available.

Introduction to Fiber Optics

8.1 Introduction

Lightwave communication was first considered more than 100 years ago. The implementation of optical communication using light waveguides was restricted to very short distances prior to 1970. Corning Glass company achieved a breakthrough in 1970 by producing a fused silica (SiO_2) fiber with a loss of approximately 20 dB/km. The development of semiconductor light sources also started to mature at about that time, allowing the feasibility of transmission over a few kilometers to be demonstrated. Since 1970, the rate of technological progress has been phenomenal, and optical fibers are now used in transoceanic service. Besides the long-distance routes, fibers are used in the inter-CO (interexchange) routes, and the subscriber loop is the final link in what will eventually be the global interconnection chain. Optical fibers are associated with high-capacity communications. A lot of attention is presently being given to optical fibers to provide a very extensive broadband ISDN. First, the characteristics of the fibers and components will be discussed.

8.2 Characteristics of Optical Fibers

The evolution of optical fibers has been extremely rapid over the past 20 years. Research and development have been directed toward reducing the signal attenuation of fibers, and also toward increasing the digital transmission rate through the fibers. Until the development of optical amplifiers, the attenuation defined the distance between regenerators, which directly affected the cost of an optical fiber route. Initial costs increase as the number of regenerators increases, so the regenerator spacing must be maximized. Note that in regenerators for optical fiber systems there is an optical-to-electrical conversion, complete regeneration of the pulse stream, retiming, and electrical back to optical conversion, ready for onward transmission. Maintenance costs also increase as the number of regenerators increases. Optical amplifiers do not entirely eliminate regenerators, but they do radically increase the distance between them. The transmission rate (bits per second) directly determines the number of channels the link can carry, so research has been aimed at maximizing this parameter. The bit rate is dependent on the linewidth of the light source, and the size and dispersion characteristics of the fiber. Dispersion causes transmitted pulses to spread and overlap as they travel along a fiber, limiting the maximum transmission rate.

Ease of coupling light into the fiber is also an important factor. This is related to the diameter of the light-carrying portion of the fiber and the characteristics of the glass waveguide. A parameter called the *numerical aperture* is associated with coupling. In summary, the three important aspects of optical fibers that must be discussed in detail are:

1. Numerical aperture

2. Attenuation

3. Dispersion

8.2.1 Numerical aperture

Figure 8.1 shows the nature of light entering a fiber waveguide. From elementary physics it is known that there is refraction of light at an air-glass interface. Similarly, there is refraction at the interface of two glass materials having different refractive indices, unless the critical angle is exceeded, in which case the light is totally internally reflected. The light waveguide is established by a glass fiber core, whose refractive index is slightly higher than the glass cladding. Light propagates along the fiber and is guided by a series of "simplified bounces" caused by internal reflection at the core-cladding interface. This is simplistically illustrated in Fig. 8.2. However, if the light enters the fiber at an angle that is greater than the "cone of acceptance angle" (ray 2 in Fig. 8.1a), instead of being internally reflected at the core-cladding interface, it is refracted and lost, so the light does not propagate along the fiber. The numerical aperture is defined as:

$$\text{NA} = \sin \theta_c = \sqrt{n_1^2 - n_2^2} \tag{8.1}$$

where θ_c = the cone of acceptance angle
  n_1, n_2 = the refractive indices of the core and cladding, respectively

The numerical aperture decreases as the diameter of the core decreases. A typical value for a 50-μm core is 0.2, and it is 0.1 for a 10-μm core fiber. Figure 8.1b gives a three-dimensional view of the cone of acceptance.

8.2.2 Attenuation

Figure 8.3 shows the relationship between silica fiber attenuation and the wavelength of light. Rayleigh scattering is the main physical loss mechanism. The first fiber to be placed in service (about 1977) operated at 0.85-μm wavelength. The attenuation at this first window, as it is often called, is about 3 dB/km.

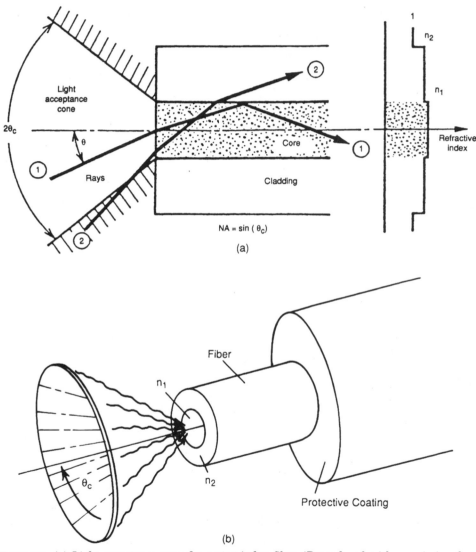

Figure 8.1 (*a*) Light-acceptance cone for a step-index fiber. (*Reproduced with permission from Yasuharu, S., Introduction to Optical Fiber Communications, ©1982, J. Wiley & Sons Inc, Fig. 2.5.*) (*b*) Numerical aperture. NA = sin θ_c = $\sqrt{n_1^2 - n_2^2}$ = $n_1 \sqrt{2\Delta}$. (*Reproduced with permission from IEEE ©1985, Keck, D. B., Fundamentals of Optical Waveguide Fibers, IEEE Communications Magazine, May 1985, p. 19, Fig. 2.*)

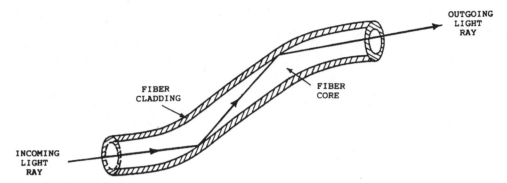

Figure 8.2 The optical fiber waveguide.

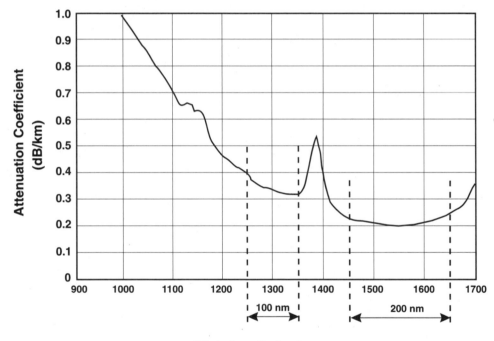

Wavelength (nm)

Figure 8.3 Graph of attentuation against wavelength for single-mode silica fiber.

Although this fiber attenuation is relatively high, it still has useful applications today, such as local area network interconnections. Since the early days, light sources have been developed for longer wavelengths; now the standard wavelengths of operation are about 1.24 to 1.34 μm for the second window, and about 1.45 to 1.65 μm for the third window. The lowest-loss silica fibers operate at 1.55 μm with a minimum attenuation of about 0.2 dB/km, and 0.35 dB/km at 1.31 μm. This 0.2 dB/km is close to the theoretical limit of about 0.16 dB/km at 1.55 μm for silica fibers.

Initial research indicated that certain chloride and halide glass materials have the potential for attenuation values of 0.001 dB/km or less at longer wavelengths (several µm). So far, such low losses have not been achievable and, furthermore, the crystalline properties of those materials would make fiber handling very difficult. However, work is still proceeding to find suitable low-loss alternative materials to the silica glass presently used. The best potential low-loss material is the crystalline KCl, which has an attenuation of 0.0001 dB/km at 6 µm, although again, the feasibility of using a crystalline material is very questionable. The attenuation values of these materials are extremely low, particularly when making a comparison with our most common type of glass (i.e., window glass, which has an attenuation of about 50,000 dB/km).

There are two types of optical fiber: multimode and single-mode. The distinction between the two types is simply the size of the fiber core. If the core diameter is made small enough, approximately the same size as the wavelength of light, only one mode propagates. Figure 8.4 shows the typical relative physical sizes of multimode and single-mode fibers; Fig. 8.4a and b are multimode fiber, whereas Fig. 8.4c is a single-mode fiber.

The 8- to 10-µm core used for single mode has a smaller aperture than the 50-µm core used for multimode operation. At 1.3 µm, both multimode and single-mode systems are in existence. The attenuation is lower for the single-mode fiber and also the maximum bit rate is higher. The 1.55-µm window of operation is used exclusively for single mode. The attenuation peaks in Fig. 8.3 are caused by the presence of impurities such as OH$^-$ ions. Above 1.55 µm, far infrared absorption causes most of the pure glass losses. Although infrared absorption is the limiting factor for the oxide glasses such as silica, other nonoxide glasses such as ZnCl$_2$ have infrared absorption peaks at much longer wavelengths: 6 to 10 µm. This is the reason for the optimism in obtaining extremely low-loss fiber at wavelengths above 2 µm in the future.

8.2.3 Dispersion

Dispersion causes optical pulses transmitted along a fiber to broaden as they progress down the fiber. If adjacent pulses are broadened to a point where they severely overlap each other, detection of the individual pulses at the receiver is uncertain (Fig. 8.5). Misinterpretation of pulses, ISI, leads to a poor error performance. So, dispersion limits the distance of transmission and the transmission bit rate. The dispersion is defined mathematically by a term which takes into account the broadening of a pulse over a distance traveled along the fiber, that is,

$$\text{Dispersion} \approx \frac{\sqrt{t_2^{\,2} - t_1^{\,2}}}{L} \quad \text{ns/km} \tag{8.2}$$

or, more precisely, at any wavelength, a pulse will be delayed by a time t_d, per unit length L, according to the equation

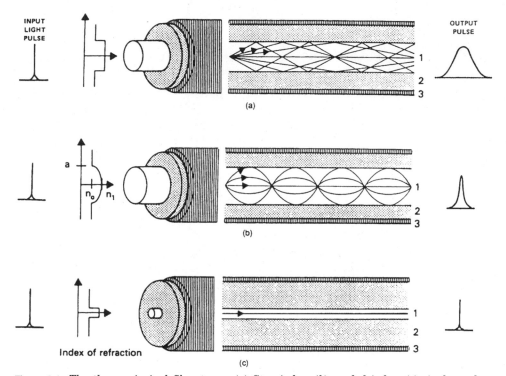

Figure 8.4 The three principal fiber types. (*a*) Step index; (*b*) graded index; (*c*) single mode. (*Reproduced with permission from ITU, Optical Fiber for Telecommunications, CCITT, Geneva 1984, Fig. 1.*)

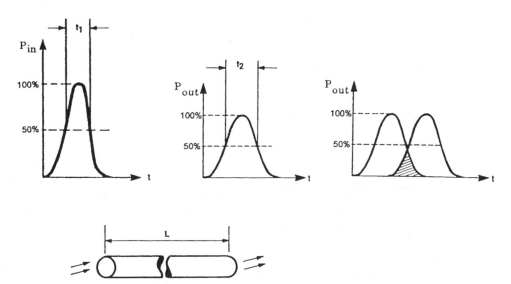

Figure 8.5 Pulse broadening due to dispersion.

$$\text{Material dispersion} = \frac{1}{L} \ \frac{dt_d}{d\lambda} \ \ \text{ps/nm} \cdot \text{km} \tag{8.3}$$

The causes of dispersion are as follows.

Spectral width of the transmitting source. The optical source does not emit an exact single frequency but is spread over a narrow band of frequencies. The laser has a narrower linewidth than the LED, and consequently is said to produce less *intramodal* or *chromatic* dispersion than the LED. If the light sources could emit only one frequency, there would be no dispersion problem. Unfortunately, that is not possible.

Characteristics of the fiber. The optical fiber causes dispersion, and the term *chromatic dispersion* is used to relate the spectral width of the source to the properties of the fiber. Chromatic dispersion is defined as the extent to which light pulses are spread for a specific source linewidth in an optical fiber due to the different group velocities of the different wavelengths within the source spectrum. The total chromatic dispersion is the sum of three components:

1. Modal dispersion

2. Material dispersion

3. Waveguide dispersion

Modal dispersion is dependent only on the fiber dimensions or, specifically, the core diameter. Single-mode fibers do not have modal dispersion. Multimode fibers suffer modal dispersion because each mode travels a different distance along a fiber and therefore has a different propagation time. Figure 8.4a shows the step-index multimode fiber. This was the first type of fiber to be produced. Because the refractive index in the core is constant, the velocity of each mode is the same, so as the distance traveled by each mode differs from one to another, so does the propagation time. Because the light in these larger core fibers is composed of several hundred different modes, a pulse becomes broader as it travels along the fiber. The graded index profile shown in Fig. 8.4b causes the light rays toward the edge of the core to travel faster than those toward the center of the core. This effectively equalizes the transit times of the different modes, so they arrive at the receiver almost in phase. The graded index fiber pulse broadening (i.e., dispersion) is significantly improved over the step-index design. For multimode fiber, modal dispersion is a major limitation to high bit-rate performance.

The single-mode fiber of Fig. 8.4c, when operating in a truly single mode, has no other interfering rays, so it does not suffer modal dispersion pulse broadening. The single-mode fiber does, however, suffer from *material and waveguide dispersion*. This is because of the frequency dependence of the refractive index (and therefore the speed of light) for the fiber material. For silica, the total dispersion drops to zero at 1.31 μm, as indicated in the graph of Fig. 8.6.

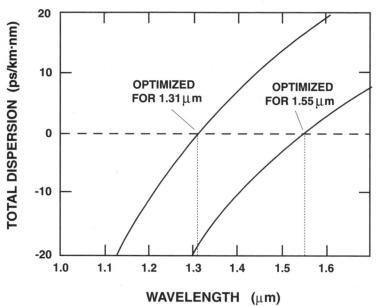

Figure 8.6 Dispersion against wavelength for single-mode fibers.

Unfortunately, at this wavelength, the attenuation is about 0.35 dB/km, which is not the minimum attenuation wavelength, so a dispersion-shifting technique has to be employed to fabricate the material for close to zero material dispersion at the lower attenuation wavelength of 1.55 μm.

Waveguide dispersion is another form of dispersion and has values of about 0.1 to 0.2 ns/km. Compared to material dispersion, these values are generally negligible in multimode fibers. However, for single-mode fibers waveguide dispersion might be significant. Figure 8.7 shows how waveguide dispersion combines with the material dispersion to shift the total chromatic dispersion curve. The zero dispersion wavelength shifts from about 1.26 to 1.30 μm. By modifying the fiber refractive index profile using precise fabrication techniques to affect the waveguide dispersion characteristic, the fibers can be designed to have the zero dispersion wavelength shifted to the lowest-loss wavelength of 1.55 μm. Manufacturing tolerances will unavoidably cause a fiber to have a small amount of dispersion even at the so-called zero value. For example, a 1.55-μm, zero dispersion shifted fiber will typically be rated at less than 3 ps/nm · km.

8.2.4 Polarization

Although single-mode fibers cure many of the problems associated with multimode fibers, they also create one. Single-mode fibers having circular symmetry about the core axis can propagate two almost identical modes at 90° to each other (a TE mode and a TM mode). The fiber is said to support *modal birefringence*. Small imperfections in the fiber's characteristics such as strain, or variation in the fiber geometry and composition, can cause energy to move from one polarization mode to the other during propagation along the fiber. This problem obviously needs to be eliminated for coherent optical fiber communication systems, and methods of treating it are discussed in Sec. 8.5.3 and Chap. 9.

8.2.5 Fiber bending

Whenever either a multimode or single-mode fiber is bent or bowed, some of the energy within the fiber core can escape into the region outside the fiber (Fig. 8.8a). This loss of energy increases as the radius of curvature of the bend decreases. If the radius of curvature approaches the fiber radius, very high losses occur, called *microbending attenuation* (Fig. 8.8b). Depending upon the circumstances, fiber bending can be either beneficial or detrimental. It is certainly necessary to fabricate the cable so that microbends are not created by impurities or defects in the fabrication process. Also, care must be taken during the installation of fibers to ensure no small-radius bends are present at the splice boxes. Figure 8.8c shows the extent of bending loss incurred for 1.3-μm and 1.55-μm fibers, depending upon the radius of curvature. This loss depends on the refractive index profile and the operating wavelength. The bending loss is very small for a 2-cm radius bend (i.e., less than 0.00001 dB/cm). As the radius of curvature becomes smaller than 1 cm, the loss increases rapidly.

In addition to the extra attenuation caused by bending, it can also be a source of unwanted surveillance of information, denying users their security. Although

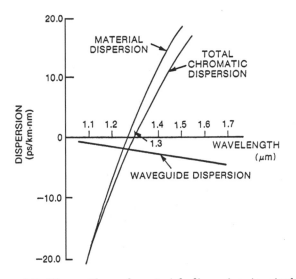

Figure 8.7 Waveguide and material dispersion in single-mode fibers. (*Reproduced with permission from IEEE ©1987, Nagel, S. R., Optical Fiber-The Expanding Medium, IEEE Communications Magazine, vol. 25, no. 4, April 1987, p. 35.*)

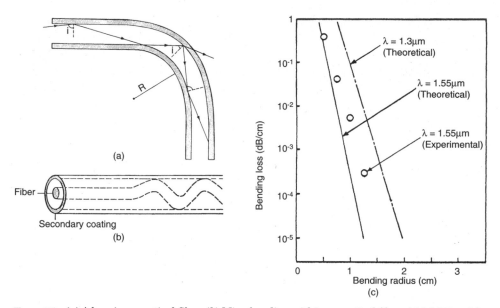

Figure 8.8 (*a*) A bow in an optical fiber. (*b*) Microbending within an optical fiber. (*c*) Additional loss caused by bending.

this could be a problem for users wanting to maintain confidential communications, it opens up an interesting means of noninvasive testing, which is particularly useful in the fiber splicing procedure, as discussed in Sec. 8.4.2.

8.3   Design of the Link

In order to assess the types of optical components and the specifications needed for those components, it is instructive to study the typical point-to-point transmission link. Clearly, the long-distance link differs significantly compared to an interexchange link, and even more radically when compared to the subscriber loop or fiber LAN. The long-distance link is primarily characterized by high channel capacity and therefore high bit rate. At the other extreme, an optical fiber from the curb to a customer's premises is relatively low bit rate. Over the past ten years major technological advances have created a sense of optimism in the information infrastructure community that fiber optics can solve the bandwidth problems which have been building up as a result of networking. Because global optical fiber interconnectivity is the objective, it is useful to concentrate first on the long-distance link. Technical solutions to shorter links are generally simpler in comparison. While networking incorporates many long-distance optical fiber links, there are some other problems peculiar to networking which are discussed later in this chapter as well as in Chap. 10.

The optical transmission link has many similarities to other types of transmission systems such as microwave systems. In its simplest form the point-to-point system has the usual transmitter and receiver with an intervening medium, which in this case is an optical fiber cable. As previously stated, one of the major objectives in optical fiber transmission is to ensure that regenerator separation is maximized in order to minimize costs. This is done by optimizing the properties of the transmitter, cable, and receiver. Ideally, the transmitter output power and the receiver sensitivity should be as high as possible. The optical fiber attenuation should be as small as possible. In the past, the fiber cable has been the major factor in determining regenerator separation. Today's optical fiber cables at the operating wavelength of 1.3 μm are readily available at 0.35 and 0.20 dB/km at 1.55 μm. In addition to attenuation, the dispersion characteristic of the cable can also limit regenerator separation. In other words, when the dispersion becomes too large, the distance is limited by intersymbol interference, causing incorrect detection of 1s and 0s. Optical amplifiers are necessary to boost the signal at intervals on a long-distance link. Dispersion needs to be "managed," to maximize the distance between regenerators.

There is one very important difference between optical amplifier repeaters and regenerators. Optical amplifiers are transparent to bit rate; it makes no difference whether the bit rate is 1 Mb/s or 100 Gb/s. On the other hand, a regenerator is designed for one specific bit rate. After the optical-to-electrical conversion, the timing and pulse reshaping require electronic circuitry that cannot be easily modified to another bit rate. Clearly, the optical amplifier repeater is amenable to future upgrades as the bit rate increases, whereas regenerators are not.

The transmitter, which uses a laser diode (LD), should have the highest possible and allowable output power. The maximum transmitter power at present is limited by semiconductor technology. Even if the present output power values increase dramatically in the near future, there is a theoretical limit to the transmitted power level because of Raman and Brillouin scattering within the fiber. However, studies have established that the nonlinearities caused by scattering can be used to advantage. Pulse compression occurs when transmitting at higher power levels (about 15 to 20 dBm), which, as described later, can dramatically improve the repeater spacing and capacity of optical systems. The early LDs had a Fabry-Perot style of construction (see Sec. 8.5.1) and an output frequency relatively broad in linewidth. These diodes have even been referred to as optical noise sources. The linewidth of the laser should be minimized in order to ensure that laser dispersion does not limit repeater separation. Laser development has resulted in the distributed feedback LD and the distributed Bragg reflector LD, which provide very narrow linewidth devices, typically 10^{-5} nm or less. Considerable impetus has been given to transmitter technology, driven by the potential improvements obtainable using coherent detection. The coherent detection mechanism, otherwise known as heterodyne or homodyne detection, requires optical sources to have extremely small linewidths. The linewidth should be 10^{-3} to 10^{-4} times the transmitted bit rate. The typical power transmitted by a laser in an optical fiber communication system is at present within the range 5 to 15 dBm.

For the receiver, the two main types of semiconductor device in use are the *avalanche photodiode* (APD) and the PIN diode followed by an FET, a *high electron mobility transistor* (HEMT) amplifier, or a *heterojunction bipolar transistor* (HBT). The development of APD and PIN diodes has proceeded in parallel and there is little to choose between the two in terms of performance up to about a 2.5-Gb/s transmission rate. The PIN FET combination is the simplest to produce (see Sec. 8.5.2 for more detail) and its high performance makes it the most widely used photodetector.

An optical transmission link usually has an objective of transmitting a certain number of channels over a specific distance. Inter-CO (interexchange) traffic can be carried without the need for repeaters (optical or regenerative). However, for long hauls, the regenerator spacing distance is extremely important for economic considerations. For example, consider the requirement of transmitting approximately 32,000 voice channels over a distance of several thousand kilometers on 1.3-µm fiber. This link could be designed by using one pair of fibers for each direction at the 9.953-Gb/s bit rate, or it could be done using four pairs of fibers for each direction at the 2.488-Gb/s bit rate. Fiber cables usually contain at least six fiber pairs, because the cost difference between two and six is not large. In addition to the extra fiber, an important difference in the cost of each of these systems is due to the different regenerator requirements. For the cheaper 2.488-Gb/s system, if regenerators are required on each pair of fibers (four regenerators) at, say, x-km intervals without the use of amplifiers, the 9.953-Gb/s system might need only one regenerator at about $x/2$-km intervals. Therefore, only about one half the quantity of regenerators would be needed in this example. Using postlaser transmitter and predetection receiver amplification, the repeater spacing would increase by the same factor for either system. If EDFAs are used on the route, the link would become dispersion-limited instead of power-limited, so regenerators would be spaced much further apart.

Another question to consider concerns the type of fiber. If standard 1.3-µm fiber is already installed, is it cost-effective to use it at the 1.5-µm wavelength? The attenuation benefit comes at a cost of poorer dispersion characteristics which might not result in a much larger regenerator spacing. 1.5-µm dispersion-shifted fiber would improve the regenerator spacing, but at the cost of installing the new fiber. The other economic factor to be considered is the price of the terminal equipment for the respective systems. At higher bit rates the equipment becomes more expensive, because more recent technology is used. Over a relatively short span (link length) the increase in the number of regenerators for the 2.488-Gb/s system might be offset by the higher price of the 9.953-GMb/s terminal equipment. A great effort is consequently being devoted to increasing both the regeneratorless span length and the transmitted bit rate. In general, the highest bit rates available are used in very-long-distance links. Optical amplifiers are gradually displacing regenerators entirely, except for ultralong links or upgrades on previously installed

1.3-μm fiber. The costs of materials and components for an optical fiber system are changing dynamically in this fast-moving industry, which makes planning difficult. For example, one year after a link design has been installed, the prices might have changed so much that a completely different design solution would be chosen at the later date. Anticipating and factoring future upgrades into present-day designs is an important aspect of optical fiber systems design engineering.

8.3.1 Power budget

The power budget is the basis of the design of an optical fiber link. To state the obvious, an optical fiber link must have

$$\text{Total gain} - \text{total losses} \geq 0$$

Therefore

$$(P_t + G_T - P_r) - (\alpha_f + \alpha_c + \alpha_s + F_m)L_T \geq 0$$

So, the total link distance

$$L_T \leq \frac{P_t + G_T - P_r}{\alpha_f + \alpha_c + \alpha_s + F_m}$$

where P_t = transmitted power
  P_r = receiver sensitivity (minimum received power)
  G_T = total optical amplifier gain (= $G_1 + G_2 + \cdots G_n$)
  α_f = fiber attenuation
  α_c = connector attenuation
  α_s = total splice losses ($\alpha_{s1} + \alpha_{s2} + \ldots \alpha_{sn}$)
  F_m = fiber margin
  L_T = total link distance (with optical amplifiers)

Gain (amplification) can only be added to this equation until dispersion or noise becomes the distance-limiting factor. The manner in which gain is included is not arbitrary. In other words, it is not always just a simple matter of placing optical amplifiers at regular intervals along the link. Optical preamplifiers are used prior to receiver inputs to improve the receiver sensitivity. They are also used before LD light is launched into the fiber. Optical amplifiers might be placed at regular intervals along the route, but gain equalization across the optical band might be periodically necessary (see Sec. 9.6.3). Some designs can benefit from remote pumping of an EDFA several tens of kilometers from the transmitter and receiver ends of a link.

The total length of a link can vary from a few meters in a local area network to thousands of kilometers in a cross-country or transoceanic trunk network. The bit rate requirement is established by the number of telephone, video, or data channels proposed. From the total length and bit rate, decisions can be made concerning the fiber characteristics and the type of transmitter and receiver to be used. As always, cost minimization is essential, but the quality (error performance) of the link must satisfy international standards.

The starting point is to establish the capacity (i.e., how many channels are needed between points A and B). This automatically defines the bit rate, which has a bearing on whether the light source is an LED or LD and whether the receiver detector is an APD or PIN device. Next, is the link a short-, medium-, or long-haul system? The answer to this question decides what type of fiber is to be used. For example, the long-haul system must definitely use single-mode and, preferably, 1.5-μm fiber, whereas a short-haul (LAN) could use silica multimode or even the cheaper plastic fiber. Again, this decision has a bearing on the choice of device for the light source and detection devices. The overall length of the link and type of fiber determines whether or not regenerators are required. For short- or medium-haul systems (interexchange routes) regenerators are not needed. For long-haul systems (trunk), regenerators might be required depending on the overall capacity and whether older 1.3-μm fiber is used or the newer 1.5-μm will be installed. All of these interrelating factors are optimized in the "power budget" calculation, which must be done for every link design.

Also, the fiber margin F_m of several decibels (depending on link length) must be factored into the analysis to account for extra splice losses in the event of future cable breaks or deterioration in the optical light source output power over the lifetime of the link.

Next is the question of whether to use wave division multiplexing to increase the total capacity. This can be designed into the link from the outset or used to upgrade a fixed number of fibers in a cable previously installed. Chapter 9 addresses these issues.

8.4 Optical Fiber Cables

8.4.1 Cable construction

The first fibers commercially available in the late 1970s operated at 0.85 μm and were grouped in cables containing relatively small quantities of fiber. By the mid-1980s, 1.55-μm dispersion-shifted fibers were being fabricated with 2000 or more fibers in one cable. In all cases of fiber manufacture, the core of the fiber is coated with a silica cladding to offer protection against abrasion. A primary coating of plastic is placed over the cladding to provide extra structural rigidity and protection. In addition, a buffer jacket is used to protect the fiber against bending losses. There are three styles of buffer jacket: the tight buffer, loose buffer, and filled loose buffer jacket, as in Fig. 8.9. The tight buffer jacket is in contact with the primary coating, and because it is 0.25 to 1 mm in diameter, it stiffens the fiber so that the fiber cannot be easily bent to the point of breaking. In practice it is possible to bend a tight-buffered fiber to the breaking point, but this should not occur during normal handling procedures.

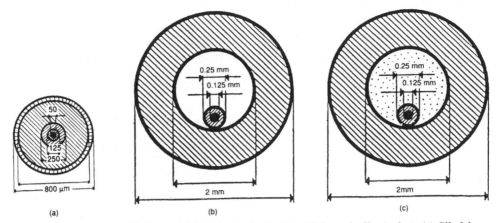

Figure 8.9 Optical fiber buffering. (*a*) Tight buffer jacket; (*b*) loose buffer jacket; (*c*) filled loose buffer jacket.

Loose buffering allows the fiber to move within an oversized, extruded hard plastic tube. Although it has an outside diameter larger than the tight buffering, it does offer better protection against excessive bending. Finally, at added expense, the loose buffer tube can be filled with a silicone-based gel, which provides moisture protection. The gel-filled loose buffer fiber is the preferred choice for long-distance links, whereas the tight buffer fiber is mainly used for LANs or short-to-medium-distance links (links in the 1- to 10-km range).

Optical fibers are normally used in pairs (i.e., one for GO and the other for RETURN). Wave division multiplexing (WDM) can reduce this to one fiber for both GO and RETURN for some applications (see Sec. 9.5.1). The economics of cable construction and installation are such that it is usually worthwhile to buy cable with more fibers than the initial design requires. A four-fiber cable is considerably less than twice the cost of a two-fiber cable, etc. Considering future capacity needs, plus spare fibers in case of failure, it is always a good policy to design in some extra fiber capacity at the outset. A six- or eight-fiber cable is usually the minimum considered for any interexchange or long-distance link. There are many possible configurations for organizing optical fibers in a cable. All configurations use a strength member made of steel or a very strong material such as Kevlar (the bulletproof jacket material) in order to prevent longitudinal stress. This is particularly important for cables which are to be used in overhead installations. Figure 8.10 shows typical styles of packaging for 6, 12, 50, and 2000 fibers. The cables can be made with no metal whatsoever, if the application warrants such a construction. Other designs incorporate several pairs of copper wires to enable a current to feed remote optical amplifiers or regenerators. There are other equally acceptable methods of cable construction. One is illustrated by the ribbon matrix style in Fig. 8.11. Rows of fibers are stacked to form a ribbon matrix. The ribbon style has the potential benefit of reducing the splicing time when a large number of fibers are to be spliced.

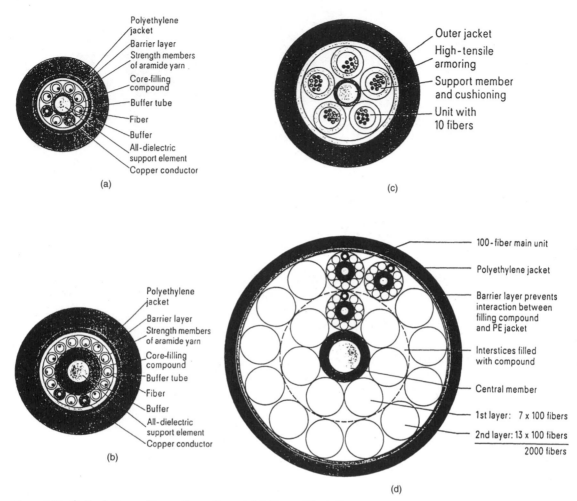

Figure 8.10 Optical fiber cable configurations. (*a*) 6 fibers; (*b*) 12 fibers; (*c*) 50 fibers; (*d*) 2000 fibers.

Finally, submarine optical fiber cables require some special characteristics. For long-term transmission stability, the fiber should be protected against water penetration and excessive elongation. The very high pressure exerted at depths of up to 8000 m means that the cable construction needs to be more rugged than that of land-based cables. Again, a variety of designs can be used, with a general preference for placing the fibers at the center with the strength member surrounding the group of fibers. An example of a submarine cable is shown in Fig. 8.12.

8.4.2 Splicing, or jointing

One of the most essential requirements of any cable is the ability to join (splice) two pieces together to provide a low-loss connection that does not appreciably

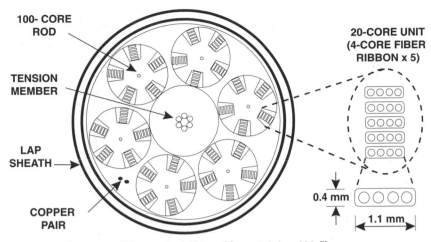

Figure 8.11 Four-core ribbon optical fiber cable containing 600 fibers.

deteriorate with time. The optical fiber clearly cannot be treated in the same manner as ordinary metallic cable. The technique used for splicing optical fibers is fusion splicing. Mechanical splicing is sometimes used indoors and for terminating (connectorizing) fibers. Both methods involve fiber-end preparation, alignment of the fibers, and retention of the spliced fiber in the aligned position.

The fusion splice is done by placing two fibers end to end and applying an electrical arc at the point of contact for a short period of time. The glass momentarily melts at that point, causing the two fibers to fuse together and resulting in a low-loss joint that should last for the lifetime of the fiber. This is considered to be the better method for fiber splicing, but it has the disadvantage of taking a relatively long time for each splice, and it requires very sophisticated and expensive equipment. The other method of splicing is to mechanically hold

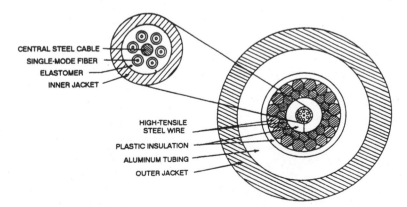

Figure 8.12 Submarine optical fiber cables.

the two fibers end to end and to use a gel at the interface. In the 1980s this method was used for temporary restoration of damaged links, and some companies used it for permanent splices; long-term performance in hostile outdoor environments has proved unsatisfactory in some cases, and mechanical splices are now mainly used only for indoor installations.

For either style of splicing, there are several aspects that are common to both. First, it is imperative that the two fibers to be joined have surfaces that are smooth, clean, and at 90° to the length of the fiber. Figure 8.13 emphasizes these requirements by indicating two unacceptable presplicing conditions. If a splice were made in either of these two situations, a high splice loss would be the result. Severe splice loss is always encountered unless extreme care and rigid splicing practices are observed. The ends of each fiber are always "cleaved" to provide a new, smooth, clean, 90° surface, whether the fiber is to be fusion spliced or mechanically spliced.

Fusion splicing. In the early days of optical fiber installation, the use of fusion splicing instruments required considerable skill and dexterity to ensure good low-loss splices. The fusion splicers now available are much easier to operate, and reliable, repeatable, very-low-loss splices are the norm. This is mainly attributable to the automated aspect of these machines. But, because they are built to operate within extremely fine tolerances, they are expensive. The ends of the two fibers to be spliced are first cleaved, then cleaned, preferably in an ultrasonic cleaner. Next they are clamped in the alignment mechanism and brought into position for the prefusion. The fiber ends are brought close to each other, but not touching. The electric arc is then activated, to melt one of the fiber ends and make it smooth and rounded. A microscope is used to give a clear view of the procedure, as illustrated in Fig. 8.14. The other fiber is treated in

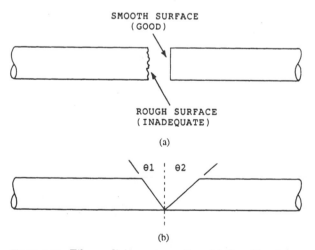

Figure 8.13   Fiber splicing preparation. (a) One fiber has an unacceptably rough surface. (b) Both fibers have inadequate end surface perpendicularity.

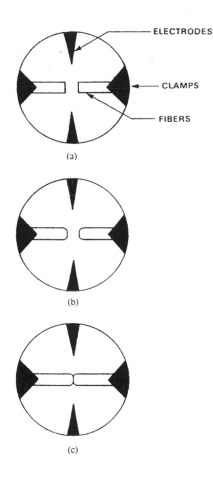

(a)

(b)

(c)

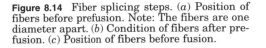

Figure 8.14 Fiber splicing steps. (*a*) Position of fibers before prefusion. Note: The fibers are one diameter apart. (*b*) Condition of fibers after pre-fusion. (*c*) Position of fibers before fusion.

the same way. The two fibers are then butted up against each other, and the arc is established for a few seconds to melt the two ends, causing fusion. High-quality fusion splicing machines automatically adjust the arc current to take into account the atmospheric pressure. The fiber ends are heated precisely to melting point (1600 to 2000°C). At high altitude the pressure is low, so at a sea-level setting the arc would not produce enough heat for a satisfactory splice. The early, cheaper fusion splicing machines needed manual alignment and arc control, resulting in inconsistent splice losses. Modern machines couple light into one fiber and use an optical feedback procedure to align the two fibers automatically, to maximize the light transmission and therefore minimize splice loss.

The alignment accuracy required for single-mode fibers is considerably greater than for multimode fibers because the core diameter of a single-mode fiber is much smaller (i.e., about 8 to 10 μm compared to about 50 μm for multi-mode). The two fibers to be spliced are brought to within a few micrometers of each other in an approximately aligned position. The two fibers are bent as shown in Fig. 8.15. Light is coupled into one fiber at the bend and an optical

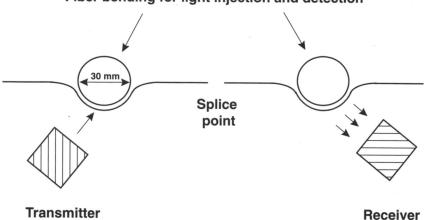

Figure 8.15 Fiber splice loss control.

signal injected into the core. This signal travels through the small air gap and into the other fiber core. A photodetector is used to measure the power level of light emitted at the bend in the second fiber. Using an automated step-by-step procedure of moving the fibers relative to each other, the maximum light transmission position is the position of best alignment. The main disadvantage of this method is the fact that the fiber has to be bent to a very small radius (about 1.5 cm) to achieve adequate light input coupling and output decoupling. The light injection method of attenuation measurement is only a relative and not an absolute measurement. The optical time domain reflectometer (OTDR) gives a readout of attenuation against distance along a fiber (see Fig. 8.16) and is indispensable for accurate measurements. This instrument sends pulses down the fiber, and light is backscattered from any discontinuity it encounters in its path. The OTDR displays on its screen a plot of fiber attenuation against distance as in Fig. 8.16. A small loss is observable at the connector interfacing the OTDR (or transmitter) to the fiber cable and at each splice point along the link. The accuracy of these instruments is so high that a 0.01-dB splice can be detected at a distance of more than 40 km from the exchange. Because of the extra work involved with such fiber end measurements, the OTDR is often used after splicing a complete route for the overall end-to-end attenuation measurement. Of course, by that time it is costly to go back and resplice any fibers that have excessive splice loss.

Some fusion splice machines have a video monitor attached to the alignment mechanism so that the operator can see the alignment as viewed in two directions (x-axis and y-axis). As fiber processing technology improves, imperfections such as lack of core concentricity within the fiber are becoming negligible, so observing good alignment of fiber exteriors is usually sufficient to indicate that the splice loss will be low. Ribbon fibers are becoming increasingly popular for use in large fiber cables. Multifiber splicing machines are available for speedy installation of these cables.

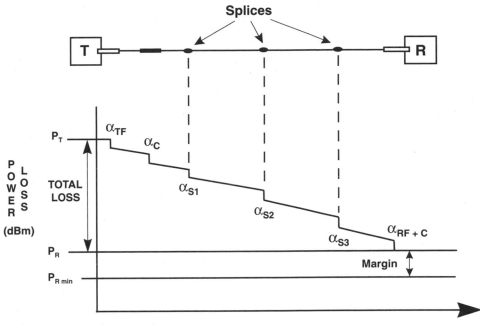

Figure 8.16 Attenuation versus distance using an OTDR. α_{TF} = transmit fiber loss, α_C = connector loss, α_S = splice loss, α_{RF+C} = receive fiber plus connector loss. Margin = received power – receiver sensitivity = $P_R - P_{R\,min}$

On completion of the splice, the fused region is then protected by a reinforcement member and fixed in place by either epoxy and/or heat-shrinkable tubing. The spliced portion of the cable is finally enclosed in a container with an air-tight seal to protect against moisture degradation. Fusion splice loss values from 0.01 to 0.03 dB are expected, and over 0.1 dB is unacceptably high.

The time taken to fusion splice one pair of newly installed fibers obviously varies considerably depending on the location, number of fibers in the cable, and experience of the personnel, but 15 to 30 min could be anticipated for a six-fiber cable. This is more than twice as long as the mechanical splice time. When the ribbon-constructed fiber cable is used, multiple mechanical fiber splicing is possible, which reduces the time per splice even further.

When the splices are completed they have to be placed in a *splice organizer*. As shown in Fig. 8.17, the excess length of each spliced fiber is coiled and carefully placed in position. This form of semipermanent containment provides protection of the splices against the elements and can be accessed at a later date for further work if necessary.

Mechanical splicing. This method is much quicker than fusion splicing, but the insertion loss is considerably higher. It is very convenient to use mechanical

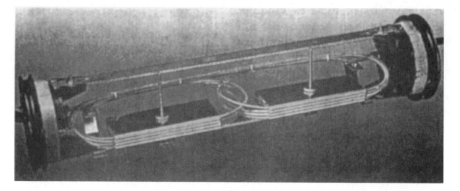

(a)

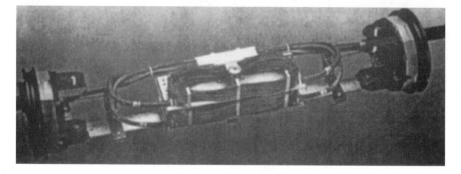

(b)

Figure 8.17 Typical splice organizers with (*a*) eight splicing modules or 80 fiber splices and (*b*) two splice mounts and, all together, 48 single splices. [*Reproduced with permission from Siemens Telcom Report 6 (1983), Special Issue, "Optical Communications," p. 56, Figs. 3 and 4.*]

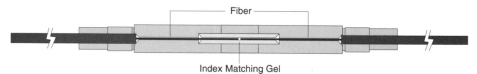

Figure 8.18 The Siecor CamSplice mechanical splice. (*Reproduced with permission from Siecor.*)

splicing for multimode or single mode applications such as indoor splices or emergency outdoor restoration. The Siecor CamSplice is, illustrated in Fig. 8.18. With a simple field fiber strip, cleave, clean, and alignment procedure, installation time is about one to three minutes. An index matching gel at the point where the two fibers butt up against each other provides a stable splice, and the fibers are held permanently in position. Average splice losses are typically 0.15dB after tuning (up to 0.5 dB before tuning).

Splicing problems. To appreciate the very tight tolerances that must be enforced to ensure good splices, various potential problems will now be addressed. First, high splice loss can be a result of splicing two fibers which have different fabrication characteristics. Figure 8.19a graphically illustrates the extra loss that can be expected if there is a mismatch in the *numerical apertures* of the two fibers. Notice how a difference of only about 5 percent can increase the loss by more than 0.5 dB. Similarly, if the *diameters* of the cores of the two fibers differ by only 5 percent, an additional loss of more than 0.5 dB can be anticipated (Fig. 8.19b). Second, extra loss can be incurred during the splicing procedure if there is an *offset* between the two fibers such that the core of each fiber is not precisely aligned with respect to the other. Figure 8.20a indicates how the misalignment of fibers increases the splice loss. A small misalignment causes a significant loss. For example, a misalignment of only 0.1 of a core diameter results in about a 0.5-dB loss. This is only about 0.1 μm for a single-mode fiber. The automatic alignment fusion splicers almost eliminate this problem. They can align the cores of the two fibers even if the core of one or both is eccentric in relation to the cladding. If one fiber is tilted with respect to the other, as in Fig. 8.20b, additional loss can be expected. As

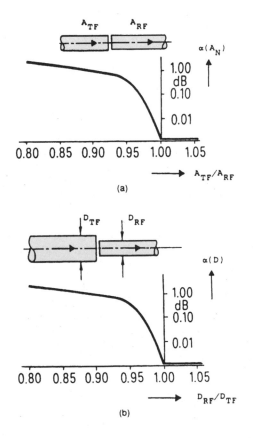

(a)

(b)

Figure 8.19 (a) Extra splice loss due to numerical aperture differential; A_{RF} = numerical aperture of receiver fiber, A_{TF} = numerical aperture of transmit fiber. (b) Extra splice loss due to diameter differential; D_{RF} = core diameter of receive fiber, D_{TF} = core diameter of transmit fiber.

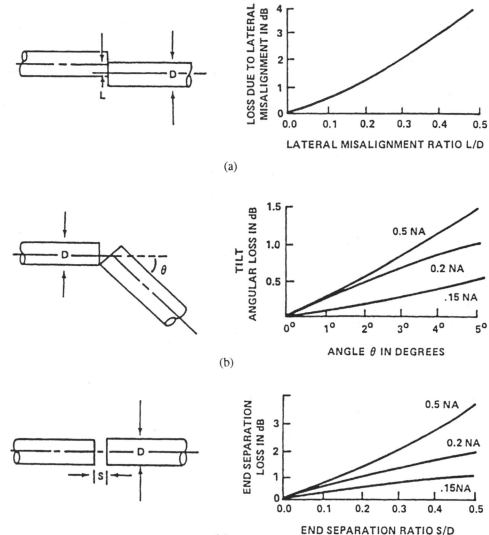

Figure 8.20   (*a*) Extra splice loss due to lateral misalignment; (*b*) extra splice loss due to angular tilt; (*c*) extra splice loss due to end separation. D = core diameter.

the graph shows, this loss can be severe. For fibers having a numerical aperture of 0.2 (multimode), an angular tilt of only 2° causes a loss of about 0.5 dB. The loss is less severe for a single-mode fiber (i.e., the numerical aperture is about 0.1). Finally, if there is an air gap between the two fibers, a relatively small gap produces a large loss. For example, the graph in Fig. 8.20*c* indicates, for single-mode fibers, an additional loss of about 0.5 dB for a gap of 0.5 of a core diameter. This is only about 5 μm for a single-mode fiber. An air gap is unlikely to occur in a fusion splice but could occur in a hasty mechanical splice. In normal circumstances, the index-matching gel should eliminate this problem.

The splice loss should ideally be kept to less than 0.1 dB. Because of the problems stated above, this is not always possible. The environmental conditions at the site of the splicing are perhaps the greatest cause of high splice loss. Ideally, splicing should be done in an air-conditioned truck especially set up for the procedures. In some parts of the world, this luxury is not always available, and splicing often has to be done in manholes or at roadside locations. In such circumstances, roadside dust is the worst enemy of the fiber splicing personnel. The presence of dust during a fusion or mechanical splicing procedure can easily produce splice losses in the 0.5- to 1-dB region. Such high losses are usually unacceptable. In very dusty regions it might be necessary to do one splice many times over until a satisfactorily low splice loss is obtained.

8.4.3 Installation problems

The intention of this section is not to relate the procedural details of placing the fiber cable but to discuss the advantages and disadvantages of the different types of installation. The installation methods can be simply split into two categories: overhead and underground. The choice is not always an easy one.

Invariably, the overhead installation is cheaper. However, the maintenance costs following installation can easily offset this initial advantage. Placing fiber on existing telephone cable poles makes it very vulnerable to extreme weather conditions such as hurricanes, tornadoes, or flooding. Fortunately these types of disasters rarely occur in most parts of the world, but in some countries they are an annual event. Even in areas that do not experience such severe conditions, heavy storms can knock poles down or accidents can occur, such as a truck hitting a pole. In places of political instability, deliberate disruption of the communication system is very easy with overhead cables. Even internal labor unrest could lead to sabotage, in which case overhead cables are easy targets. A more secure overhead type of installation is to use existing electrical power transmission lines for suspending the cable. This location presents a high-voltage hazard to deter saboteurs, but it is also a severe hazard for the personnel who have to install and maintain the cables.

Underground cables are not without their problems. First, some regions are so rocky that very expensive blasting would be necessary before laying underground cables. In many cities the underground ducts are full to capacity, and in such cases it could be very expensive to lay more ducting. Then there is the question, Is ducting required for all underground fiber cables? If ducting is not used, the cable could be damaged by rodents if not placed deeper than about 1.5 m. Also, the cable is more susceptible to damage from construction workers or agricultural machinery if ducting is not used. In summary, if local conditions are favorable, and installation cost is secondary to security or preferable to later high maintenance costs, fibers should be placed underground in ducts set in concrete. Optical fiber installation is certainly a high-precision activity. Even the cable pulling or pushing must be taken very seriously. When unreeling fiber cable from drums, the pulling tension must be carefully controlled, otherwise damaged or weakened fibers will cause problems in the future. Mechanized cable pulling is essential, and the amount of cable bending during pulling must be carefully monitored so that the maximum allowable tension is not exceeded. Since the mid-1990s, cable pushing, using compressed air to force fiber cable down ducts, has become an accepted and convenient way of keeping the cable tension below predetermined limits.

8.5   Fiber Optic Equipment Components

8.5.1   Light sources

As stated earlier, the two light sources available are the semiconductor LD and the LED. Both devices have small physical dimensions, which make them suitable for optical fiber transmission. As the term *diode* suggests, the LDs and LEDs are *pn* junctions. Instead of being made from doped single crystals, they now have exotic combinations of two or more single-crystal semiconductor materials. These heterojunctions are consequently called heterostructures.

The fundamental difference between an LD and an LED is the fact that the light from an LED is produced by spontaneous emission, whereas light from an LD is made by stimulated emission. This results in the laser having an output that is coherent and therefore has a very narrow spectrum, whereas an LED has an incoherent output and a wide spectrum. The selection of an LD or LED for an optical transmission system depends upon the following factors:

- Required output power
- Coupling efficiency
- Spectral width
- Type of modulation
- Linearity requirements
- Bandwidth
- Cost

LED. The semiconductor LED can be used in the surface-emitting or edge-emitting mode depending upon the type of fabrication, as indicated in Fig. 8.21. The surface-emitting style has good temperature stability and low cost. However, the coupling efficiency into the fiber is limited by its wide active area. The light power coupled from a commercially available LED into a single-mode fiber is about 100 μW. The light power output is incoherent (i.e., the output is

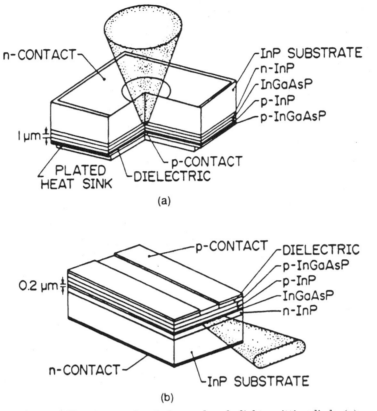

Figure 8.21 Structure and emission modes of a light-emitting diode. (*a*) Surface-emitting type; (*b*) edge-emitting type. (*Reproduced with permission from IEEE © 1988, Nakagami, T., et al., IEEE Communications Magazine, vol. 26, Fig. 3.*)

spread over a wide spectrum of about 40 nm). The operational bit rate is limited by the parasitic capacitance within the LED. The bit rate limit for LEDs is several hundred megabits per second.

The edge-emitting LED has improved performance compared to the surface-emitting type. The structure can achieve a higher coupling efficiency into a single-mode fiber, and the narrower active layer compared to the surface-emitting style has a smaller capacitance, which allows higher bit rate operation. The low cost and improved temperature characteristics of the edge-emitting LED compared to the LD have stimulated a lot of research to improve the devices so that they can be used for fiber-in-the-loop (FITL).

For an LED to achieve a reasonable transmission distance (more than 10 km between regenerators) at high bit rates (more than 622 Mb/s), single-mode fiber, operating at the zero dispersion wavelength, must be used. Multimode fiber operation significantly reduces the bit rate-regenerator distance product.

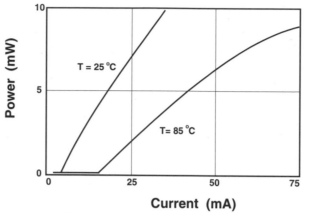

Figure 8.22 Characteristics of the laser diode.

Laser diode. The LD has evolved extremely quickly over the past decade. The development of the LD is central to the present-day long-distance capability of optical telecommunications. The electrical characteristics of an LD are illustrated in Fig. 8.22. When the current density within the active region of the diode reaches a certain level, the optical gain exceeds the channel losses and the light emission changes from spontaneous to stimulated (i.e., lasing). The threshold current at which this occurs is quite low in the double heterostructure semiconductor lasers and is typically 5 to 15 mA at 25°C. Figure 8.22 also shows how the threshold point shifts with temperature. This is a very undesirable characteristic because it means that the drive current must be increased as the temperature increases in order to maintain a constant output power. The internal power dissipation within the diode itself contributes to an increase in temperature, so a runaway situation can occur if some form of temperature control is not used. Also, aging deteriorates the laser performance. Furthermore, the wavelength of the optical output is also temperature dependent. To counteract these problems, it has been normal practice to mount the LD on a Peltier-effect thermoelectric cooler with a feedback circuit to stabilize the temperature, and another circuit is included to maintain a constant drive current.

Recent advances in laser technology using indium phosphide (InP) have resulted in LDs that do not require cooling. This is significant, because coolers are not only expensive but also require considerable power. Cooler-free lasers are very attractive for undersea link applications.

There are three major types of LD:

1. Fabry-Perot

2. Distributed feedback

3. Distributed Bragg reflector

The fabrication of the LD is similar to that of the LED. The main difference is clarified in Fig. 8.23, where the active layer is embedded in an optical resonator. Figure 8.23*a* shows a Fabry-Perot semiconductor LD, which is characterized by a resonator formed from two opposite end-surface mirrors at the natural cleavage planes of the laser crystal. This structure can support multiple optical standing waves. The length of the resonator is an integer multiple of the optical half-wavelength. The standing waves add in a constructive interference manner, providing an output spectrum that has multiple frequencies defined by the standing waves as shown in Fig. 8.23*a*. This Fabry-Perot LD output spectrum has a number of spectral lines spaced between 0.1 and 1.0 nm in optical wavelength and spreading across a spectrum of about 1 to 5 nm, which corresponds to a frequency range of 176 to 884 GHz (at 1.3 µm). In the literature the term *linewidth*, ω, is often used; it is the width of the output spectrum at the −3-dB power points. For the Fabry-Perot LD ω is about 2 nm, which, in terms of frequency, is equivalent to more than 350 GHz at a center wavelength of 1.3 µm. Because the Fabry-Perot lasers produce multimode outputs, they are most often used at 1.3 µm, where the fiber dispersion is very low.

The width of the Fabry-Perot LD output spectrum is considerably narrower than the LED linewidth of about 40 nm. Both the LD and LED devices are suitable for direct intensity modulation (i.e., transmitting 1s and 0s by simply turning the diode on and off). Optical systems having high transmission bit rate performance (gigabit-per-second rates) require lasers to have a very narrow linewidth.

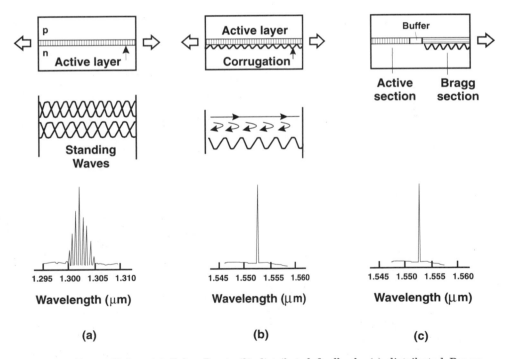

Figure 8.23 Laser diodes. (*a*) Fabry-Perot; (*b*) distributed feedback; (*c*) distributed Bragg reflector.

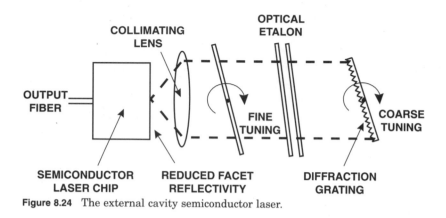

Figure 8.24 The external cavity semiconductor laser.

The distributed feedback (DFB) LD of Fig. 8.23*b* provides that improvement. It is achieved by fabricating a corrugated layer (grating) close to the active layer, and the light-emitting facet is processed with an antireflecting coating to suppress the Fabry-Perot mode of oscillation. The resulting output power spectrum has a single narrow line of less than 1 MHz (about 5×10^{-6} nm) in width. Such narrow linewidth devices allow the transmission of very high bit rates (more than 10 Gb/s). An alternative method of reducing the linewidth is to couple the light output from a Fabry-Perot LD into an external reflecting cavity as shown in Fig. 8.24. The enlarged cavity formed by directing the output beam to a diffraction grating about 20 cm from the laser and reflecting it back to the laser is a resonant structure. This technique can produce a linewidth of a few kilohertz, although its cumbersome size is a disadvantage.

The basic DBR laser is composed of two sections: (1) an active amplifier section, and (2) a passive Bragg grating section. As illustrated in Fig. 8.23*c*, a buried ridge structure is created that optically and electrically confines active and passive regions. A narrow (a few micrometers) high-impedance channel is formed between the two sections for electrical isolation. The passive region has a Bragg grating etched in the buried ridge waveguide. The dimensions of the grating determine the operating frequency of the DBR laser.

This type of laser has some attractive characteristics. First, it has a narrow linewidth of a few hundred kilohertz. Its switching time is less than 1 ns, which is comparable to the DFB laser, so it can be tuned at high rates. The frequency can be tuned over about 10 nm by applying a voltage to the *passive* Bragg section. Carrier injection into the DBR zone conveniently changes the lasing wavelength.

Ongoing laboratory research efforts are being directed toward reducing the linewidth and increasing the output power of semiconductor laser diodes. InP semiconductor DFB lasers fabricated with a buried ridge structure can be designed for several milliwatts of laser emission at wavelengths in the range 1.1 to 1.6 μm, covering the main windows of interest. Such lasers can be used without thermoelectric coolers and be directly modulated at 2.5 Gb/s. They can also be designed for high output powers above 150 mW at 1.48 μm for use as EDFA pump sources. A closer look at the temperature sensitivity of LDs is instructive. Figure 8.25 shows the spectra of a 1.55-µm DFB laser, at 6-dBm facet power for three different temperatures within a typical anticipated outdoor operating environment range. A 14-nm wavelength shift is observed over the full temperature range of −40 to +95°C.

The temperature dependence of the laser wavelength can be exploited, especially for DBR lasers, by using a thin deposited metal film over the DBR region of the laser passivation layer as a heater. 10 nm of tuning has been verified by this technique.

From a manufacturing viewpoint consistent technical parameters, high laser chip yield, and high reliability are essential. Laser diodes are therefore subjected to burn-in at, for example, 200 mA, 100° C conditions for 48 hrs. A normal use at 50 mA producing 5 to 10 mW of output power can therefore easily satisfy a typical 15-year lifetime.

Summary of the LD and LED sources. The LED has the following advantages when compared to the LD:

- Higher reliability
- Simpler drive circuit
- Lower temperature sensitivity
- Immunity to reflected light
- Low cost

These characteristics make LEDs suitable for short-distance applications. They are particularly attractive for LANs and subscriber loops where economy is a very important factor.

Compared to the LED, the LD has several important advantages:

- High output power
- High coupling efficiency

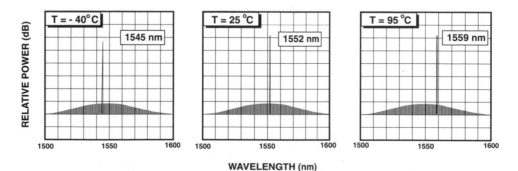

Figure 8.25 Effect of temperature on a 1.55-µm DFB laser wavelength.

- Wide bandwidth

- Narrow spectrum

In summary, it is clear that the narrower the linewidth, the higher the bit rate that can be transmitted and the greater the transmission distance before dispersion causes serious ISI. LDs are more complex, and therefore more expensive, than LEDs.

8.5.2   Light detectors

The light emerging from the end of an optical fiber link must be detected and converted into electronic pulses for further processing so that the transmitted information can be received. There are two types of detector: the APD and the PIN diode. The progress of the performance of these two types of optical detector followed a "cat and mouse" chase over the past decade, with both devices now having multigigabits-per-second capability.

APD. The early APDs were made from silicon and were used extensively in the 850-nm optical systems. Because silicon is effectively transparent at wavelengths greater than 1100 nm, another material was necessary for 1300- and 1550-nm systems. Germanium fulfilled the next generation of APD device requirements, operating up to 155 Mb/s. It was soon realized that gigabit-per-second data rates would rapidly become a reality in the mid-1990s, and a new class of APDs was investigated using III-V semiconductor compounds such as GaAs, InP, GaInAs, InGaAsP, etc. Although such APDs operate at a few gigabits per second, the hunger for ever-improving performance has led to other combinations of material such as $Al_xGa_{1-x}Sb$ for use in the 10-Gb/s region and above.

The operating principle of the APD can be described with the assistance of the simplified diagram of Fig. 8.26, which shows the APD to be a semiconductor diode structure having a p^+-doped region, followed by an n-doped region, followed by an n^+-doped region. There are many variations on this basic structure incorporating other doped layers of III-V compound semiconductors, but they all operate essentially as follows: The diode is negatively biased with a voltage of up to 100 V or more. When light from a fiber is incident on this diode, electron-hole pairs are generated. If the applied electric field is strong enough, accelerated free electrons generate new electron-hole pairs, and the process of multiplication continues, producing an avalanche effect. For each incident photon, many electron-hole pairs can be generated. As the multiplication factor is increased, the S/N decreases, so the chosen multiplication factor should not be too high. Typical values of multiplication factor for low-noise operation are 10 to 20 (10 to 13 dB), using a reverse bias of a few tens of volts to over 100 V. Furthermore, the multiplication factor is temperature-dependent, which means some form of compensation is required to stabilize the device. In addition, the high reverse bias causes a small current to flow even in the absence of incident light. This so-called *dark current* is undesirable and must be minimized, as it limits the minimum detectable received power.

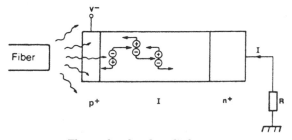

Figure 8.26   The avalanche photodiode.

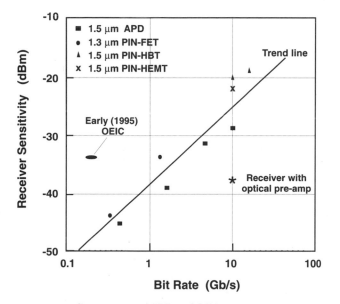

Figure 8.27   Comparison of APD and PIN receiver sensitivities.

The most important performance characteristic of an optical detector forming the receiver of an optical transmission system is its *receiver sensitivity*. This is generally defined as the minimum received power that will produce a BER of 10^{-9} at a particular bit rate. Figure 8.27 shows calculated and measured receiver sensitivities for direct detection using APDs, PIN-FETs, PIN-HBTs, and PIN-HEMTs, etc. As-expected, at the higher bit rates, the minimum detectable power increases (gets worse).

PIN diode. Again, silicon and germanium were used in the early PIN device designs but are being superseded by III-V semiconductors. As illustrated in Fig. 8.28, the device is basically a *pn* junction with an intervening "intrinsic" region usually known as a *charge depletion layer*. When light from an optical fiber is incident on the *p* region of the reverse-biased diode, electron-hole pairs are generated in the depletion layer. The electric field causes the electrons and holes to travel in opposite directions, and so produce a small current. The

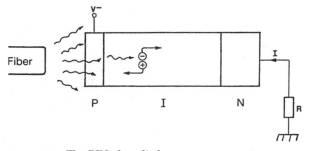

Figure 8.28 The PIN photodiode.

dimensions of the depletion region can be chosen so that the device has good sensitivity and short rise time, meaning high-frequency capability. If the thickness of the depletion region is increased, the sensitivity increases because of the higher probability of photon absorption, but this causes the travel time of the charge carriers to increase, reducing the upper frequency of operation.

The PIN diode operates with a *small* reverse bias and therefore does not contain any inherent gain, as in the case of the avalanche device. Its maximum possible gain is 1. The PIN diode does, however, have a wide bandwidth capability well in excess of 10 GHz. This means that in order to take full advantage of this wideband characteristic, a low-noise amplifier has to be placed after the PIN diode. The amplifier device originally used was the GaAs FET, but this has recently been superseded by the HEMT and the heterojunction bipolar transistor (HBT). These relatively new semiconductor devices have lower noise and higher gain characteristics than the FET in the microwave region of operation (i.e., 1 to 100 GHz).

The minimum detectable power (receiver sensitivity) of the PIN-FET diode is shown in Fig. 8.27, as a comparison with the APD. The performance comparison of a PIN diode and an APD is a little obscure because the APD has inherent gain whereas the PIN diode does not. Gain-bandwidth product (GB) is a necessary term to use in connection with APDs, because the receiver sensitivity deteriorates rapidly above a few gigabits per second for low GB values. APDs have now been fabricated with a GB in excess of 100 GHz. In practical systems, the minimum receiver input power should be at least 4 dB above the receiver sensitivity level to ensure error-free operation.

In conclusion, it is fair to say that the rate of improvement in the performance of optical photodetectors is keeping up with and is perhaps ahead of the bit rate and bandwidth requirements for telecommunication systems.

8.5.3 Polarization controllers

A perfectly symmetric, circular, monomode fiber allows two orthogonally polarized fundamental modes to propagate simultaneously. In reality, the fiber is not perfectly circular or symmetric, because manufacturing irregularities cause the geometry and internal strain to be slightly imperfect. These imperfections make the power in each mode unequal. The polarization stability is such that the power in each of the two modes does vary over a relatively long period of time (minutes or hours and not just a few seconds).

For a satisfactory communications performance, it is necessary for only one mode to be present during optical transmission. Although polarization-maintaining fiber might be cheaply fabricated in the future, other methods are presently used. There are several techniques that can be applied to achieve this goal. The fibers could be designed to be specifically slightly geometrically asymmetric or with anisotropic strain. Not only would this be expensive, but the attenuation would be at least double that of the best conventional fibers. A dual-polarization receiver has *polarization diversity*, which could be used so that the two polarization signals are received, separately detected, and then added, but this would cause the receiver sensitivity to degrade by 3 dB. Because the rate of change of power in each polarization is relatively slow, it is possible to use the conventional fiber in conjunction with a polarization compensator. Compensation can be achieved by physically squeezing the fiber, or by winding it on a piezo-electric drum.

However, the more attractive method is to use an optoelectric technique. Figure 8.29 shows a lithium niobate ($LiNbO_3$) polarization controller which is used after the laser diode to maintain one mode of polarization. First, the phase shifter adjusts the phase shift between the incoming TE and TM modes so that they are 90° apart. The mode converter then increases the ratio of power in the TE compared to the TM mode, after which the second filter restores the phase relationship between the two modes. The polarization converter of Fig. 8.29 is fabricated on "Z-cut" $LiNbO_3$, and a voltage of about 90 V is required to produce greater than 99 percent mode conversion. On X-cut $LiNbO_3$, the operating volt-

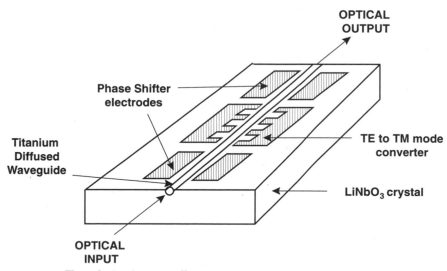

Figure 8.29 The polarization controller

age of the TE-to-TM mode converter can be reduced to about 10 V because of the increased overlap between the optical and electrical fields. Furthermore, interdigitated electrodes can improve the mode conversion efficiency compared to the finger electrode structures of Fig. 8.29. The $LiNbO_3$ crystal style of mode converters is attractive because these converters can be easily incorporated with other devices onto a single crystal to form optoelectronic integrated circuits. While $LiNbO_3$ is a good electro-optic material, semiconductor materials such as InGaAsP-InP also exhibit electro-optic properties. They are gaining popularity as the technology for the monolithic integration of active and passive optical devices matures.

8.5.4 Amplifiers

There are two main categories of optical amplifiers: semiconductor amplifiers and doped fiber amplifiers. In either case, the objective is to increase the regenerator spacing for a system that would otherwise be limited by fiber attenuation. If the distance is limited by dispersion, that is a different problem, which will be addressed later. Both types of amplifier can be used as in-line optical amplifiers to boost the optical signal without resorting to the regenerative style of repeater. Furthermore, the semiconductor amplifier can also be used as a preamplifier to boost the signal just prior to detection in the receiver. This technique increases the receiver sensitivity by about 10 dB, and therefore increases the regenerator spacing. The optical amplifier is a repeater, but it is a much simpler, smaller, bit-rate transparent, and cheaper repeater than the regenerative repeater. The optical amplifier does not yet eliminate the need for regenerative repeaters, except in special situations discussed later; it just increases the distance between them. Optical amplifiers introduce noise into the system, as all amplifiers do. Fortunately, the number of amplifiers that could be used before the noise causes an excessive BER problem translates, in some circumstances, to an unregenerated distance of thousands of kilometers. The system would be dispersion limited before amplifier noise caused a distance limitation. Even the dispersion limit is now being challenged by the *soliton* transmission described in Sec. 9.6.3.

Semiconductor amplifiers. An LD used as a light source is simply an amplifier with enough positive optical feedback to cause oscillation. This feedback is provided by making reflective facets at each end of the semiconductor chip. So, by removing the feedback using facets with antireflecting coatings, the laser oscillator can be converted to an optical amplifier. Two types of semiconductor amplifier can be made, depending upon the reflectivity of the coatings on each facet. First, when the facet reflectivities are lower than for a laser oscillator, but still allow some light to be reflected back into the active region, the amplifier is called a *resonant*, or *Fabry-Perot*, *amplifier*. Second, if the facet reflectivities are very low, light entering the device is amplified by a single pass, and the amplifier is called a *traveling wave* *amplifier* (Fig. 8.30). In practice, even with the best antireflective coatings there is a small amount of facet reflectivity, which means that most semiconductor amplifiers operate somewhere between a Fabry-Perot and traveling-wave amplifier. Gain values in the 30-dB range are readily achievable, with an output power in excess of 5 dBm for the traveling wave amplifier at 1.3 μm. The single-pass gain is different for each of the TE and TM polarization modes, implying that polarization-controlling measures are necessary for this type of amplifier. The gain difference between polarization states is worse for the Fabry-Perot amplifiers than for the traveling wave amplifiers. Higher single-pass output power amplifiers using elaborate semiconductor structures provide useful optical integrated circuit boosters for low-power laser diode transmitters, or simply compensation for the high-loss components formed by monolithic integration. Such amplifiers have a saturated output power in the 20-dBm region, and are therefore operable up to about 17 dBm with a similar fiber-to-fiber gain value (17dB), at 1.3 μm. These devices clearly have a bright future.

Doped fiber amplifiers. This type of amplifier is fabricated using a length of conventional single-mode silica fiber, doped with a rare earth metal such as erbium (Er). A concentration in the range of only 10 to 1000 ppm is sufficient to provide high gain. If an LD is coupled into the fiber, it acts as a "pump" to increase the energy, and therefore power level, of an incoming signal as it travels along the fiber. The pump light excites the erbium atoms to energy levels higher than their normal state, so when the signal light encounters the erbium atoms, stimulated light emission causes the signal power to increase gradually as the signal travels along the fiber. Following the stimulated emission process, the active medium relaxes to the ground state by amplified spontaneous emission (ASE) in the fluorescence band, which appears as unwanted noise at the signal wavelength. Figure 8.31 shows the typical construction of an Er-doped amplifier. The signal to be amplified is at 1.55 μm and passes through the direct path of a WDM coupler to the 15 m of Er-doped fiber. Experiments have been made with various

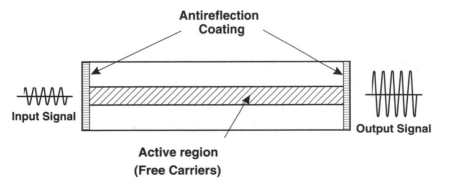

Figure 8.30 The semiconductor optical traveling wave amplifier.

lengths of doped fiber up to more than 100 m. The 0.98- or 1.48-µm DFB laser diode is the pump source which is launched into the coupler at a power level of about 50 to 150 mW. In practical systems, isolators are usually used after each amplifier to stop any reflections back into the amplifier from causing low-level lasing action. This would cause noise to be superimposed on the signal, thereby degrading the error performance. Also, there is loss in the coupler and output filter. These losses can be up to about 4 dB, which gives a fiber-to-fiber small signal gain of 30 to 40 dB. Depending upon the erbium concentration, doped fiber length, pump power, etc., the Er-doped fiber amplifier (EDFA) designs can have an output power of about +20 dBm over the bandwidth from 1.52 to 1.57 µm. The saturated output power from an EDFA is proportional to the pump power level.

Nonsilica fiber amplifiers. Combinations of silica, fluorozirconate, and other glass combinations have shown improvements in performance over EDFAs. Although the silica-based EDFAs are more mature, the less developed Er-doped fluoride fiber amplifiers (EDFFAs) will no doubt become prominent. The small signal gain of about 40 dB and an output power of more than 17 dBm with a noise figure of around 6 dB is good, but the major advantage lies in the gain flatness of less than 1.5 dB across a 30-nm band, and a similar S/N flatness. This performance is a significant improvement over EDFAs, which becomes particularly important for WDM applications.

So far, only third-window doped amplifiers have been mentioned. There has been a large amount of second-window (1.3 µm) fiber installed over the past decade, and telecom operators would prefer to choose the cheaper upgrade of these systems instead of replacing the fiber with 1.5-µm cable; 1.3-µm doped fiber amplifiers started to have large-scale commercial availability only in 1997. Fluorozirconate fibers doped with rare-earth metals such as praseodymium, neodymium, or dysprosium are producing good performances. The power level of the pump diodes needs to be rather high, and figures in the region of 300 mW or more are quoted. These nonsilica fibers cannot be arc-fusion spliced unless they have an intermediate silica fiber attached to the active fiber during amplifier manufacture. Because fiber-doped and semiconductor amplifiers are in their infancy, it can be anticipated that the bandwidth, gain, and saturation power level will all be gradually improved as the technology matures.

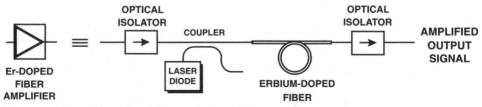

Figure 8.31 The erbium-doped fiber optical amplifier.

Raman fiber amplifiers. The Raman effect is a nonlinear interaction that occurs at relatively high power levels in suitable optical materials, including fibers. Small signal gains of 30 dB and output powers in the region of 27 dBm in the third window have stimulated further efforts. Raman amplifiers are also being considered for the 1.3-μm band.

8.5.5   Modulators

Modulation of a light source can be done by direct modulation of the dc current supplying the source, or by using an external modulator following the source. The direct modulation method has the advantage of having fewer components than the external modulator, but chirp problems often cause a designer to resort to the external modulator. Direct modulation above 15 Gb/s is achievable, but would usually be restricted to noncoherent detection systems, whereas external modulators are readily suitable for coherent detection systems at very high bit rates. Also, for WDM, the direct modulation method requires a larger channel spacing than the external modulation method.

Direct modulation. The simplest form of direct modulation is to change the laser biasing current above and below the threshold value to turn the laser on and off to produce optical 1s and 0s. Unfortunately, this ON-OFF intensity modulation causes wavelength chirping (wavelength change with laser bias current) which results in a broadening of the optical spectrum relative to the information bandwidth. Unless the operating wavelength is very close to the zero-dispersion wavelength, severe system degradations occur. This is clearly not acceptable for dense WDM.

Advantage can be taken of the wavelength dependence of the laser bias current to frequency modulate the laser. If the injection current is varied while the laser is operating above threshold, the wavelength can be changed by up to about 2 nm. This laser tunability can be used for FSK modulation at well into the gigabits-per-second rate.

Even multigigabits-per-second bit rate PSK can be accomplished by direct modulation, with a spectral width as narrow as the information bandwidth. Differential PSK has been used successfully without degradations due to thermal FM, by modulating the laser injection current with a bipolar signal. If an NRZ or RZ unipolar signal is used in the regular PSK mode, long strings of 1s or 0s cause large drifts in the optical frequency or phase due to temperature variations, resulting in severe distortion of the modulated signal. This problem is overcome by using a bipolar signal format. The NRZ signal is converted into a bipolar format because the bipolar pulses are better in terms of the thermal response of the laser. NRZ pulses cause a thermal FM within the laser, whereas the bipolar pulses are very short compared to the 1-μs thermal response time of the laser. Also, the polarity changes of the bipolar pulses allow a thermal improvement.

External modulation. External optical modulators do not suffer the excessive chirp of direct intensity modulators, although a small amount of chirp still exists. For this reason external modulators have attracted considerable interest for application in high-bit-rate and long-haul optical telecommunication systems.

There are two types of external modulator presently in use which exploit the electro-optic property of certain materials. Lithium niobate (LiNbO$_3$) has been established as a key material whose refractive index, and therefore phase, can be controlled by an applied electric field. As light propagates through the applied field region, it undergoes a phase change that is cumulative as it travels through the region. Figure 8.32 shows a simple lithium niobate modulator. The electrodes are about 2 cm long, and the applied voltage is less than 10 V to produce a 180° phase change. By simply switching the voltage ON and OFF by the 1s and 0s of the information-carrying bit stream, an intensity-modulated optical output is achieved.

An intensity modulator can be formed by an interferometric technique. Figure 8.33 illustrates the modulator based on the Mach-Zender interferometer. This device has a 3-dB Y-junction that splits the optical signal, and a second Y-junction at a specific distance from the first that combines the two signals "in phase." Metal electrodes are arranged as indicated along the two arms of the interferometer. When an electric field is applied to the electrodes, because of the opposite field directions, a differential phase shift is obtained between the signals in each arm. The two signals are subsequently recombined with a 180° phase shift (i.e., in antiphase) and dissipated in the substrate material, giving the OFF state of the intensity modulator. A voltage of about 4 V is used to define the OFF state.

The lithium niobate technology has advanced to the stage where external modulators can operate with bandwidths in excess of 40 GHz. This should satisfy the needs of most telecommunication systems designers for some years to come. Although the performance is indisputably good, the material, lithium niobate, is not optimal because it does not allow complete circuit integration with semicon-

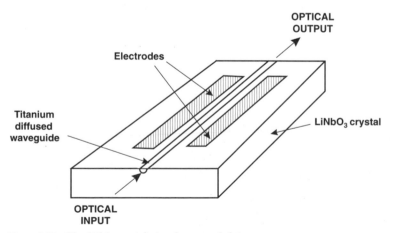

Figure 8.32 The lithium niobate phase modulator.

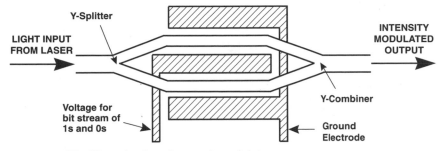

Figure 8.33 The Y-junction interferometric modulator.

ductor devices such as lasers, detectors, etc. Electro-optic external modulators made from III-V semiconductors are rapidly advancing, and bandwidth performances in excess of 20 GHz at a wavelength of 1.3 μm have been demonstrated. Other III-V semiconductor modulators have been constructed that make use of the electro-absorption effect in certain semiconductor materials. These also show promise for high-bit-rate transmission (more than 20 GHz at λ= 1.3 μm). Eventually, future high-bit-rate systems will probably use semiconductor external modulators, monolithically integrated with laser diodes.

8.5.6 Couplers

Couplers are devices that tap off some of the signal power from the main transmission path, usually for power monitoring or feedback purposes. This definition implies that there are some special cases. For example, when 50 percent of the main path power is tapped off, the device could be called either a 3-dB coupler or a power splitter and combiner. This special-case coupler can be fabricated simply by forming a Y-junction (Fig. 8.34a), which produces a device that is inherently wideband and can be used across the present operation range of 0.8 to 1.6 μm. Such devices are often described as *achromatic*. The splitter and combiner of Fig. 8.34a is a very important component because it can be fabricated using the planar technology. In other words, standard photolithographic techniques are used to define the waveguide pattern which forms the Y-junction within a larger integrated circuit. This style of signal splitter and combiner can also be used for multiple splitting and combining by incorporating several Y-junctions as shown in Fig. 8.35.

This substrate fabricated device (Fig. 8.34b) takes the same shape as the microwave equivalent, with the exception that the coupling region for the optical coupler is not a quarter wavelength as in the case of the microwave design. The proximity coupler is inherently wavelength dependent. Although this is a problem for some applications requiring wide bandwidth, it is a very useful, if not essential, characteristic for WDM applications (see Sec. 9.5). The coupling region can be symmetrical (in which case the wavelength transmission curve

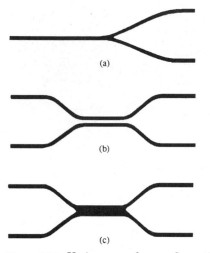

Figure 8.34 Various coupler configurations. (*a*) Y-junction coupler; (*b*) proximity coupler; (*c*) zero-gap coupler.

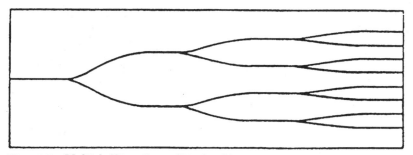

Figure 8.35 Multiple Y-junction splitter/combiner.

is periodic) or it can be asymmetrical (exhibiting a passband behavior). This wavelength (or frequency) dependence illustrates the ability of the directional coupler also to act as a filter.

If the coupling region gap between the two waveguides of the proximity coupler is reduced, the coupling strength between the two waveguides increases. In the special limiting case of zero gap, the two individual waveguides become one two-mode waveguide (Fig. 8.34*c*). This device is called either the *zero-gap coupler*, or the *multimode interference* (MMI) coupler, and has a periodic characteristic that is similar to that of the symmetrical coupler. The main difference is that the periodic passband peaks, which would correspond to the channel spacing when used for WDM, are less for the MMI coupler. In addition, the MMI coupler has the obvious advantage that there is no small, 2- to 4-μm gap to carefully control as is necessary for the proximity coupler.

When fabricated on an electro-optic material such as lithium niobate, with electrodes either side of the zero-gap region, the passbands become voltage-tunable.

This device has several applications such as modulators, switches, or WDMs. Tunability of WDM devices enables compensation for the fabrication tolerances and temperature sensitivity compensation. This is an important characteristic, because a transmission wavelength shift can cause an increase in the channel-to-channel crosstalk.

Many high-speed LANs include a passive star coupler (see Sec. 9.6.4). Ideally, the input power to any one of the N inputs of a transmissive $N \times N$ star coupler is divided equally between its N outputs. Absorption and scattering detract slightly from this ideal. Figure 8.36a indicates how the 3-dB symmetric proximity coupler operates as a transmissive 2×2 star coupler and how this is the basic building block for higher-order star couplers. Figure 8.36b is a 4×4 transmissive star constructed from four 3-dB couplers.

8.5.7 Isolators

Isolators are essential components for optical communication systems. They are essential interface components for reducing reflections and therefore noise, nonlinearities, and dispersion in reflection sensitive applications such as coherent

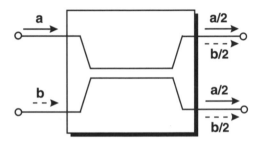

(a)

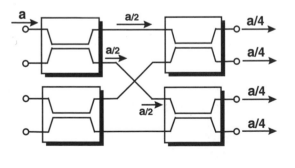

(b)

Figure 8.36 (a) A passive transmissive 2×2 star using a 3-dB coupler; (b) a transmissive 4×4 star using four 2×2 stars.

transmission systems and CATV. The most reflection sensitive optical components are lasers and amplifiers. Isolators are used at the output of LD sources to stop reflections from increasing the laser linewidth; 50 to 60 dB of isolation are required to ensure that reflections are adequately reduced. Such isolators have been fabricated using a bismuth-substituted iron garnet material that produces Faraday rotation. The classic operating principle of this isolator is to produce a unidirectional phase shift due to Faraday rotation, so that any reflected signal is highly attenuated by antiphase cancellation. Although these isolators produce the required isolation characteristics, they are relatively bulky devices that are not presently amenable to optical integrated circuits.

Hybrid integrated waveguide isolators that have 30 dB of isolation are emerging from the laboratory. The insertion loss of these devices is about 2 to 3 dB in the 1.5-μm wavelength band. Their operation is based on the 45° nonreciprocal waveguide rotator integrated on a silica-based planar lightwave circuit. Arrays of rotators are fabricated using a garnet waveguide, and a permanent magnet is placed over the array to form the isolator. These small devices (8 mm × 3 mm) are an important component in the process of miniaturizing photonic circuits.

8.5.8   Filters

Filters are essential components of all telecommunication systems, and optical systems are no exception. They are used in many places, in particular in WDM and coherent receivers (see Sec. 9.3). There are two categories of filter: fixed-frequency and frequency-tunable. Fixed frequency filters are discussed in more detail in Sec. 9.5.

Frequency-tunable filters. The important parameters of a tunable filter are channel bandwidth, channel spacing, and tuning range as described in the diagram of Fig. 8.37. The channel spacing should be as small as possible to allow the maximum number of channels to be tunable within the tuning range. The channel spacing limitation is based on the necessity to minimize the crosstalk between channels, which entails keeping the crosstalk penalty to less than 0.5 dB. The channel spacing is determined by the transmitter source linewidth, the modulation scheme, and the filter lineshape, which can vary from about 3 to 10 times the channel bandwidth.

Tuning speed, the time taken to tune between any two frequencies (channels), is another parameter of concern for tunable filters. In a video broadcast system, the random selection of video channels in a subscriber's television set can be made in a tuning time of several milliseconds. Although the tuning of multiaccess data networks and cross-connect systems is done rather infrequently, the tuning speed required is in the microsecond time frame. For packet switching of data having packet lengths of a few microseconds, tuning speeds of less than 100 ns are required.

There are several types of frequency tuneable filters. *Fabry-Perot* interferometric filters have been in existence for a long time. They work on the princi-

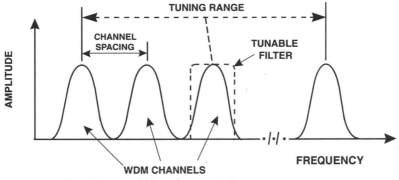

Figure 8.37   Tunable optical filter characteristics.

ple of partial interference of the incident beam with itself to produce periodic passbands and stopbands. Tuning requires mechanical movement of mirrors, so wavelength selection is relatively slow (i.e., more than milliseconds). Also, the Fabry-Perot filter tuning range is limited to about 10 percent of the center frequency. Typical single-stage filter insertion loss is about 2 dB.

Mode coupling tunable filters using acousto-optic, electro-optic, or magneto-optic effects have produced some useful tuning characteristics. As Fig. 8.38 indicates, the light enters the filter in the TE mode and is converted to the TM mode using either acoustic, electric, or magnetic fields. The mode coupling conditions are satisfied only by optical signals that have a very narrow wavelength. The two propagating modes are separated by a polarizer. The filter output has a bandpass characteristic without the inconvenience of multiple passband outputs generated by the Fabry-Perot filter. The filter output bandwidth is relatively wide (i.e., about 1 nm for both the acousto-optic and electro-optic filters). The acousto-optic filter can be tuned over the full 1.3- to 1.6-μm range, whereas the electro-optic filter has less than a 20-nm tuning range. The acousto-optic filter therefore has a much higher channel capacity than the electro-optic filter (i.e., several hundred versus about 10, respectively). Both types can be tuned relatively quickly. Furthermore, the acousto-optic filter has a unique property. If multiple acoustic waves are applied to the interaction region, multiple passbands can be simultaneously and independently created. This allows multiple channel selection, which might have some interesting future applications. The electro-optic filter, however, is faster than the acousto-optic filter, with tuning speeds in the nanosecond and microsecond ranges, respectively.

Semiconductor laser structures form the third main category of tunable filters. Biasing a resonant semiconductor laser structure below its lasing threshold produces a resonant optical amplifier, with reasonable selectivity. While this would be a problem for wideband amplifier designers, it is very convenient for tunable filters. The Fabry-Perot laser structures have multiple passbands whose center frequencies are periodically spaced, depending upon the physical dimensions of the cavity. The passband spacing, known as the free spectral range (FSR), defines the maximum range over which incoming signals can be tuned. Another useful filter parameter is the *finesse*, defined as FSR/Δf, where Δf is the filter

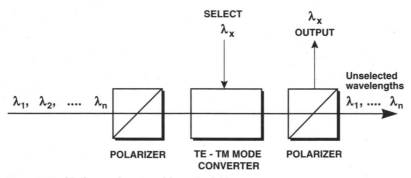

Figure 8.38 Mode coupling tunable optical filter.

bandwidth. A typical 300-μm-long Fabry-Perot laser would have an FSR of about 150 GHz (about 1 nm of tuning range), and a finesse of about 150. Because 1 nm is a relatively limited tuning range, DFB laser structures are preferred for tunable filters. In single-electrode DFB amplifiers, the passband narrows and the gain increases as the bias current is increased toward the threshold value. In a multielectrode DFB amplifier, the resonant frequency and gain can be controlled independently.

The frequency tuning in both Fabry-Perot and DFB semiconductor filters is produced by the change of injection current as the bias is varied. The injection current determines the carrier density, which subsequently defines the refractive index of the region. The refractive index decreases with increasing carrier density. The resonant wavelength, λ, is proportional to the refractive index, so that $m\lambda = 2nL$, where m is an integer, n is the refractive index, and L is the grating pitch for the DFB structure or cavity length for the Fabry-Perot structure.

The DFB filter tuning range of about 4 nm is still relatively small compared to other filter techniques. The width of the filter passband is less than 0.05 nm. This allows a capacity of 20 or more channels. The tuning speed is very good (i.e., in the nanosecond region). Also, the filter has more than 10-dB gain instead of a 2- to 5-dB loss for passive types.

Polarization control or polarization-maintaining fiber is needed because the gain for the TE orientation is about 10 dB more than for the TM orientation. Also, the input signal power level must be constant; otherwise the resonant frequency (filter passband) and gain are affected. There are additional problems, but one must remember that semiconductor tunable filters are in their infancy, and it can be anticipated that they will excel in the future as more research is devoted to improving their characteristics.

8.5.9 Photonic switches

Telecommunications exchanges throughout the world have undergone an upgrade from the slow electromechanical switches to the faster and more reliable digital electronic, stored-program-controlled (SPC) switches. When these switches are used in conjunction with optical fiber transmission systems, there is usually a conversion of the signal from electronic to optical and vice versa in order to perform the switching in the exchanges. As wideband connection to the subscriber becomes increasingly attractive and eventually becomes expected, the practice of switching at the optical level instead of at the electrical level will be essential. Optical add/drop multiplexers and optical cross-connects are also devices that can benefit from optical switches. Optical packet switching within time division multiple access (TDMA) systems will also need optical switches. Subpicosecond switching times are possible with switches that are activated by an optical signal.

In a different but closely related field of technology, computers promise to expand in power and speed capability by using optical switching. Optical computing substitutes photons for electrons, which travel through free space 1000 times faster than electrons transfer information through electronic media. Light beams replace wires, and information is transmitted in three dimensions. Because light beams do not interact with one another, multiple signals can propagate along the same path, increasing the information transfer by at least 10^6.

At present, a very useful application of photonic switching is for LANs using the star coupler arrangement of Sec. 8.5.6. WDM optical cross-connects including wavelength interchanging, broadcast and select, and optical protection switching in a $1 \times N$ protected optical transmitter system are important applications that require optical space switches integrated with other components. There are several methods of realizing a space switch, and the best choice depends on how many of the following desirable characteristics can be achieved:

- Low crosstalk
- Bit rate transparency
- Fast switching
- Low insertion loss (or preferably gain)
- Wavelength independence (operation over the EDFA range)
- Multiwavelength operation
- Scaleability
- Monolithic integration

The *electro-optic directional coupler* (Sec. 8.5.6) provides a space switch with some of the above attributes. Unfortunately, the crosstalk between switch outputs is only about –15 dB, which can be improved to –30 dB or more by cascading two couplers. However, the size of a resulting switch matrix is quite large and complex. *The Mach-Zehnder interferometer* switch is another possible design, which uses the destructive interference technique similar to that used in the modulator. Here, 3-dB multimode interference couplers are used instead of the Y-junctions to provide two outputs instead of just one. Again, crosstalk in the –15- to –20-dB region is a limitation to widespread use. Perhaps the most promising configuration for optical switches is *the semiconductor optical amplifier* (SOA) switch.

Figure 8.39 shows a SOA-based $N \times N$ space switch. Varying the current injection into the semiconductor *pn*-junctions of an SOA generates free carriers, which varies the loss/gain characteristics. ON/OFF loss/gain variations of about

45 dB are typical for each SOA, with net fiber-to-fiber gain for all paths of a 2 × 2 matrix in the ON condition (the amount of gain depending on the type and construction of the SOA). Any paths can be interconnected as required by switching on or off the appropriate SOAs. As the array gets larger, say to 4 × 4, the path lengths, and therefore insertion loss, increase and additional booster amplifiers might be necessary to establish path gain, at the expense of noise accumulation. Switching times of < 1 ns are typical. Monolithic integration should allow large switch matrices to be constructed.

8.6   Fiber Optic Equipment Subsystems

Several of the components in the previous section are starting to be incorporated into circuits that could be called subsystems. For example, the $N \times N$ switch array uses multiple couplers or amplifiers. The integration of these components into miniaturized circuits is essential for the usual reasons of cost reduction and reliability. This miniaturization has many driving forces, such as semiconductor materials development, telecommunications optical fiber deployment for networking, and increased bandwidth requirements, etc. Some of the directions in which this industry is moving are now very briefly addressed.

8.6.1   Solid-state circuit integration

The concept of integrating several optical devices and electronic circuits was first proposed in the early 1970s. It was not until the early 1980s that the technology on several fronts had matured enough to produce circuits with eye-catching performances. Monolithic integration of a variety of optical devices is often called *photonic integrated circuits* (PICs), and only in the 1990s did they accelerate their technological advancement. The development of methods to

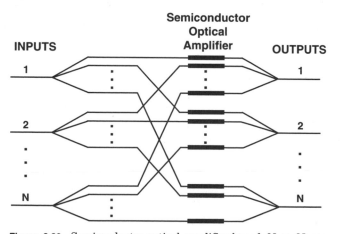

Figure 8.39 Semiconductor-optical-amplifier-based $N \times N$ space switch. ((*Reproduced with permission from IEEE © 1996, Renaud, M., Journal of Selected Topics in Quantum Electronics, vol. 2, no. 2, June 1996, Fig. 13.*)

produce very precise dimensions of semiconductor layers has been a key to this rapid advancement. The evolution of semiconductor growing techniques such as *metal organic vapor phase epitaxy* (MOVPE) and *chemical beam epitaxy* (CBE) has provided indispensable tools for PIC fabrication. In addition to optical device integration, for a complete system such as a coherent optical receiver, it is also necessary to integrate high-performance electronic devices such as laser drive circuits and HEMT or HBT amplifiers on the same substrate (i.e., monolithically). The resulting circuits are called *opto-electronic integrated circuits*, or OEICs.

Monolithic integration has already had phenomenal success for electronic circuits up to microwave frequencies. The benefits are well known. The main obvious manufacturing improvements derived are lower cost, higher reliability, higher productivity, compactness, and simple assembly. Another very desirable quality is the small deviation in design characteristics from one circuit to another. Also, circuit performance is enhanced by minimizing parasitic capacitances and inductances.

Lightwaves and microwaves are both electromagnetic waves, the major difference being the frequency. So, it is not surprising that some of the techniques used in designing microwave integrated circuits can be applied to optical integrated circuits. Optical components such as couplers, modulators, phase shifters, switches, etc., which are the necessary building blocks for an optical telecommunication system, have been designed and constructed using the electro-optic material $LiNbO_3$, as described earlier. When each component is fabricated individually and then connected into a system, there are many losses incurred by the waveguide-to-fiber transitions and connectors at the input and output of each component. If many of the components are fabricated on a single $LiNbO_3$ substrate, losses are reduced substantially. The $LiNbO_3$ material technology might be only an interim solution for optical integrated circuits.

The more complex semiconductor materials such as InP/GaInAsP and InAlAs/InGaAs are used as the device materials, and InP is becoming the material of choice for monolithic circuit integration. Much research work is currently being devoted to III-V semiconductor, and *multiple-quantum-well* (MQW) monolithic optical integrated circuits. An MQW is the solid-state physicist's terminology for a device constructed by sandwiching a low-band-gap material such as GaAs between two layers of high-band-gap material such as GaAlAs to produce what is called a *thin heterostructure*.

In summary, optoelectronics is progressing along the same road as the lower-frequency electronics industries by moving from discrete components to integrated circuits. At present, optical technology is passing through the hybrid integrated circuit phase en route to monolithic integrated circuit maturity. Although the technology is still relatively new, rapid development is taking place, and there are still many problems to be solved. One of the potentially more serious problems is light reflections at interfaces between components within a PIC.

Lithium niobate integrated circuits. The $LiNbO_3$-technology-based opto-electronic integrated circuits had several years head start on the semiconductor-based PICs. This is because materials technology, which was blossoming in the 1980s, was still not mature enough by the late 1980s to produce the excellent PIC performance required for long-distance communication systems. The more well-understood $LiNbO_3$ material technology received considerable research attention, which produced quick results, enabling it to provide the first commercially available optical integrated circuits in the late 1980s.

Optical waveguides can be formed at the surface of a $LiNbO_3$ substrate by evaporating or sputtering titanium strips onto the surface of the substrate. The evaporated strips are typically 600 Å thick and 5 μm wide. The waveguide dimensions are then defined by standard photolithographic techniques. The titanium is then diffused into the $LiNbO_3$ substrate by heating it to about 1050°C for 8 h. The waveguide is formed by the diffused titanium changing the refractive index by a small amount in the diffused region compared to the bulk material. The resulting waveguides are relatively low loss (≤0.3 dB/cm at 1.3 μm). These optical waveguides on $LiNbO_3$ allow the fabrication of the passive components used to form the main circuitry of a coherent optical receiver, such as polarization controllers, frequency translators, phase controllers, and couplers. Such $LiNbO_3$ integrated circuits do not contain the active devices such as the local oscillator or photodetectors as part of the planar circuit fabrication process. They are discrete semiconductor devices that are mounted as extra chips, or fed into the $LiNbO_3$ circuit using an optical fiber. By the low-frequency definition, this is a hybrid integrated circuit and not a monolithic IC.

Over the past few years there has been growing interest in forming Er-doped devices on $LiNbO_3$ to produce monolithic integrated circuits. Er-doping of the surface of a $LiNbO_3$ substrate can be done by implantation and annealing, or by diffusion of vacuum-deposited Er-layers. Er-doped $LiNbO_3$ waveguides have losses of about 1 dB/cm, but this can be improved by Er-doping titanium diffused waveguides. If optically pumped, a 5-cm Er-doped $Ti:LiNbO_3$ channel waveguide would provide a useful 10 dB or so of single-pass gain.

A variety of lasers can be fabricated by forming the necessary cavity or grating to cause oscillation. For example, simple free-running Fabry-Perot lasers with more than 50 mW have been fabricated. Acousto-optically tunable lasers have been made by integrating acousto-optically amplifying filters having a tuning range of more than 30 nm. DBR lasers using a surface grating for one of the laser resonator mirrors have produced a 1-mW power output having a remarkable linewidth of < 8 kHz.

Figure 8.40 illustrates an example of a DBR laser monolithically integrated with a Mach-Zehnder modulator. The Z-cut $LiNbO_3$ substrate is Er-doped over its entire area, and Ti waveguides are formed for the circuit components. The DBR laser is created by an 8-mm-long grating dry-etched 300 nm deep into the waveguide surface. The 46-mm waveguide becomes an Er-doped amplifier when a 1.48-μm pump input is introduced at the dichroic mirror. The mirror is 60 per-

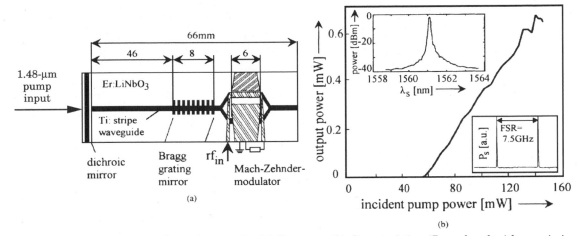

Figure 8.40 LiNbO$_3$ integrated circuit example. (*a*) Structure; (*b*) characteristics. (*Reproduced with permission from IEEE © 1996, Baumann, I., et al., Er-Doped Integrated Optic Devices in LiNbO$_3$, Journal of Selected Topics in Quantum Electronics, vol. 2, no. 2, Fig. 14, June 1996.*)

cent transmissive at the pump wavelength, but highly reflective at the laser output wavelength. Even the modulator has some gain, because its waveguides are also Er-doped. The main disadvantage of this circuit is the high pump power (145 mW) required to give an output of 0.63 mW. The bandwidth of this circuit is anticipated to be extended significantly by using X-cut substrates.

Semiconductor integrated circuits. It was not until about 1990 that semiconductor PICs and OEICs started to reach a satisfactory level of performance for serious commercial viability. It might be beyond 2000 before a substantial quantity of these circuits is in widespread use in optical telecommunication systems.

GaAs is a well-developed material technology, but its use as a PIC substrate material is limited to the short optical wavelengths of operation (i.e., in the 0.8-μm region). The less-well-established InP material is the more suitable substrate material for PICs operating at 1.3 and 1.5 μm.

Figure 8.41*a* is a diagram of an OEIC photoreceiver which was translated into a laboratory-fabricated, monolithically integrated circuit. It contains a p-i-n photodiode and a three-stage HBT amplifier with a feedback resistor and output buffer stage photodetector fabricated on an InP substrate. The molecular beam epitaxy (MBE) fabrication process was used, and as Fig. 8.41*b* illustrates, the multilayer structure has some common layers for the photodiode and the HBT.

To give an indication of the performance of this OEIC, Fig. 8.41*c* shows the graph of measured and simulated optical response against frequency. With an additional inductor tuning element, this circuit allowed a 19.5-GHz bandwidth demonstration. The HBT had an f_T and f_{max} of 67 and 120 GHz, respectively.

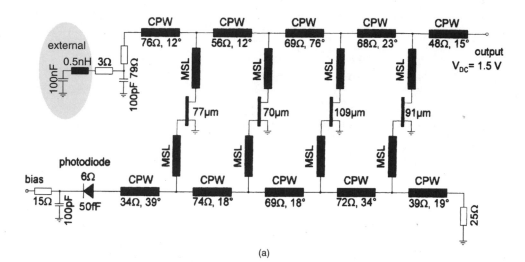

(a)

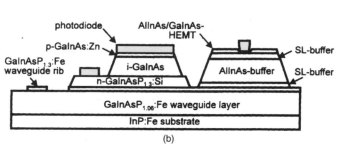

(b)

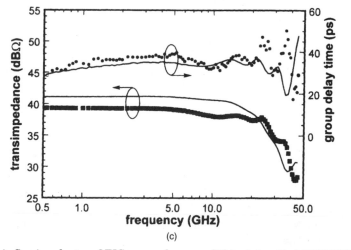

(c)

Figure 8.41 (*a*) Semiconductor OEIC example: monolithic integrated PIN-HEMT receiver. (*b*) Simplified equivalent circuit. (*c*) Receiver performance. (*Reproduced with permission from IEEE ©1996. Bach, H.-G., Journal of Selected Topics in Quantum Electronics, vol. 2, no. 2, June 1996, Figs. 1, 3, 7.*)

It could be anticipated that by employing HEMT devices with f_T/f_{max} values of about 90/200 GHz, OEICs can be fabricated so that 40-Gb/s detection is possible.

8.6.2 Wavelength converters

Wavelength (or frequency) conversion is important for achieving full scalability in WDM networks. Mixing an information-carrying signal with a fixed frequency local oscillator signal has been the hallmark of radio technology since the first days of radio transmission. The use of mixers in optical technology is consequently a natural evolution as telecommunications moves to higher frequencies of operation. Mixers are most well known for upconversion and downconversion, otherwise called frequency translation. In each case, a nonlinear device creates sum and difference output frequencies from two input frequencies. The best method of creating an optical device to do this job is still under debate. The main contenders for the wavelength conversion prize are:

1. Demodulation/remodulation (Fig. 8.42a)

2. Optical grating
   a. Cross-gain-modulation in semiconductor optical amplifiers
   b. Cross-phase-modulation in semiconductor optical amplifiers

3. Wave mixing (Fig. 8.42b)
   a. In semiconductor waveguides
   b. In semiconductor optical amplifiers

Wave mixing is a front runner because of its properties of transparency, chirp reversal, and high bit rate (over 10 Gb/s) capability.

The first candidate—demodulation followed by remodulation—can be viewed as a regenerator without retiming. The incoming optical signal at f_1 is

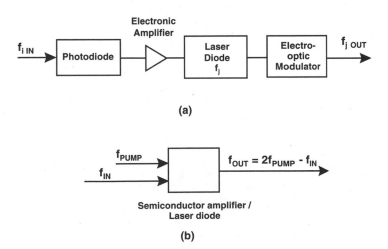

(a)

(b)

Figure 8.42 Optical wavelength conversion.

received by a photodiode detector which is used to remodulate a laser at a different frequency f_2. This scheme is known to work quite well, but is not an all-optical solution. The optical-to-electrical-to-optical conversions are undesirable because of bandwidth restrictions and only partial transparency.

There are various ways of using nonlinear effects in materials to create intermodulation products for frequency changing. The two types of wave mixing will be described briefly. Difference frequency generation is the optical equivalent of the electronic radio mixer. The wide bandwidth of these devices is very suitable for multiple input wavelengths as in WDM systems.

Lithium niobate waveguides have been used as the material for difference frequency generation. AlGaAs semiconductor waveguides have also been used to demonstrate wavelength conversion. The main problem with these devices is phase matching the two interacting input waves. Theoretical conversion efficiency of −4 dB is acceptable, but practical devices are still far from that value. Wavelength conversion using optical amplifiers has the benefit of conversion gain.

Four-wave mixing is an interesting prospect. Here, three waves are mixed to produce a fourth wave for the output. If any of the two or more waves that interact in a nonlinear medium contain information (instead of being just a single frequency), that information will be contained in the output waveform. Figure 8.43 shows the products of two or three interacting waves. If there are two equal amplitude inputs at f_1 and f_2, the resulting spectrum has four output waves $f_1, f_2, 2f_1 - f_2$, and $2f_2 - f_1$. The number of outputs increases dra-

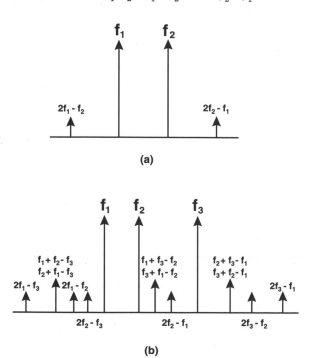

Figure 8.43 Four-wave mixing.

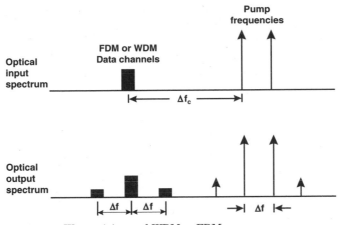

Figure 8.44 Wave mixing and WDM or FDM.

matically with the number of inputs. For three inputs there are fifteen outputs of which three are degenerate. Figure 8.44 shows a situation where a WDM data channel, say ten channels at 2.5 Gb/s, is mixed with two pump signals to change the WDM channel frequency. Clearly, care must be taken in choosing the pump frequencies so that there is enough separation between all the input and output frequencies.

Some references in the literature describe four-wave mixing as a process in which three input waves mix to produce a fourth desired signal (plus several unwanted signals). Elsewhere it is defined as two input signals mixing to produce four outputs, of which one is selected for the output.

Incidentally, four-wave mixing, which is useful for frequency converters, can be a major problem for nonconverting WDM systems where the frequencies of several channels interact to produce unwanted signals at other frequencies. Clearly, fiber chromatic dispersion and other nonlinear effects must be eliminated, or at least minimized, for WDM systems.

8.6.3 Wavelength routers

Another important device for all-optical networks is the wavelength router. It is the key component necessary for frequency reuse. A wavelength router has N input and N output ports. Each of the inputs can contain N channels. The outputs can each contain N different frequencies. This allows a broadcast effect, as illustrated in Fig. 8.45, where an input signal containing information at one wavelength is distributed to all outputs. If each input has a different frequency and a time sequencing allows a fixed interval of selection from each input, then all outputs will contain multiwavelength information.

A wavelength router can be realized using a symmetrical phased array (de)multiplexer device (see Sec. 9.5.1) with N input and N output ports. In this case, the geometry of the phased array is designed so that the free spectral range (FSR) equals N times the channel spacing. The outer frequencies away

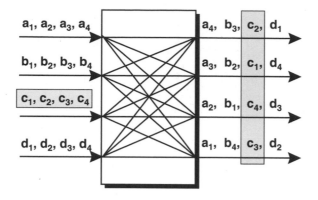

Figure 8.45 Wavelength router.

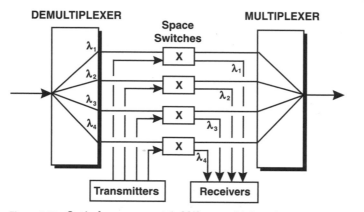

Figure 8.46 Optical crossconnect (add/drop multiplexer).

from the design center frequency suffer an additional 3-dB power loss as the spacing between the outer channels approaches the FSR.

8.6.4 Optical crossconnects (or optical add/drop multiplexers)

The terms *add/drop multiplexer* and *optical crossconnect* seem to be overlapping in optical communications terminology. Originally, an add/drop multiplexer was used to describe the electrically multiplexed signal adding and dropping channels within the SDH structure without demultiplexing the bit stream. In the optical domain, an add/drop multiplexer adds and drops one or more optical signals from a high-capacity route. Of course, each optical signal carries numerous electrically SDH- or PDH-multiplexed bit streams.

An optical crossconnect routes multigigabit streams between different input and output streams, without breaking the optical signals into smaller bit streams. Because some of the optical rerouted signals go to a branching route and the rest might continue along a main route, it can also effectively be an *optical* add/drop multiplexer.

Figure 8.46 is one possible configuration for an optical crossconnect or add/drop multiplexer using a phased-array multiplexer and demultiplexer, with space switches arranged to allow the reception of dropped channels and the transmission of added channels. Optical crossconnect is perhaps a more appropriate term to use here, because both the through-connected and add/drop signals are all-optical. The signals are added at different wavelengths from the dropped and through-connected signals to avoid wavelength contentions. The multiplexing in this case is therefore optical WDM, and not electrical multiplexing at the bit level. This is a key component that is necessary for all-optical networking. Section 10.10 expands on this subject.

Optical Fiber Transmission Systems

9.1 Overview

Optical fiber is gradually finding application in all aspects of telecommunication systems. The three broad categories, which relate to short, medium, and long distance, are referred to as:

1. Local loop (for subscribers) and LANs

2. Interoffice (or interexchange) traffic

3. Long haul (intercity traffic)

The distance, capacity (bit rate), and topology are the primary factors that influence these systems' designs and the associated economic viability of constructing and operating them. Also, the extent to which optical fiber is deployed depends on the cost relative to traditional copper-based cable. Optical fiber has already displaced copper for long-haul and interoffice traffic, and it is only a matter of time before the local loop becomes fiber. This chapter addresses the various technical factors that need to be taken into account to realize each of the above three system categories.

First, the optical fiber transmission systems technology has been evolving along two different paths:

1. Intensity modulated systems

2. Coherent systems

Figure 9.1 shows the overview of optical fiber transmission systems in terms of how they differ as the bit rate and transmission length increases, based on present and future anticipated technology. There have been several generations of

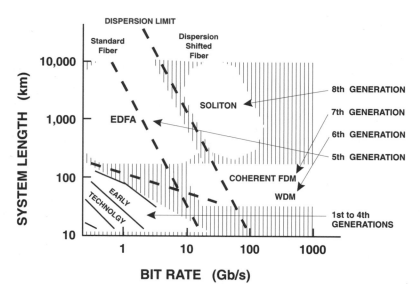

Figure 9.1   Eight generations of optical fiber communication systems.

systems so far. The early systems, before about 1990, did not have the luxury of optical amplifiers, and unregenerated distance was a major cost concern because regenerators were expensive. Optical amplifiers radically changed the landscape of fiber systems, and have now shifted the focus of attention to capacity (that is, bit rate).

First-generation systems were primarily limited in regenerator spacing and bit rate by high fiber loss and excessive chromatic dispersion in the fiber caused by the use of LEDs. The development of 1.3-μm-wavelength fiber systems produced the improved regenerator spacing and bit rates of the second-generation systems. These systems were still operating with multimode fibers, which limited the performance by interference between the propagating modes of the fiber (i.e., modal dispersion). The move to single-mode fibers operating at 1.3 μm gave the third generation a very impressive regenerator spacing and bit rate performance. The fourth generation benefited from shifting the operating wavelength to 1.55 μm, which offers the minimum achievable attenuation for silica fibers.

The fifth generation (which appeared in the early 1990s) included optical amplifiers, and today these systems vastly increase the distance between regenerators. Notice the optical amplifier is an optical repeater, and not a regenerator that involves conversion to the electronic domain for pulse reshaping and retiming. The sixth generation is WDM, and these systems became practicable in the mid-1990s. Placing multiple OC-192 bit streams at different wavelengths onto each fiber allows terabit-per-second throughput.

So far, all of the systems described use intensity modulation of the optical transmitter. The seventh generation will use phase modulation of the optical transmitter, and a coherent detection scheme, which improves the receiver sensitivity. The very narrow spectrum distributed-feedback diodes used in these systems will allow very high operating bit rates. In addition, the coherent optical communication systems will allow FDM, which will give an astounding increase in the channel capacity because of the more efficient use of the available optical bandwidth. The case in favor of using the coherent optical system in future designs appears to be clear-cut, but cheaper WDM designs will provide enough capacity for the near term. Soliton transmission will be the eighth generation, and will no doubt provide enough capacity for very broadband ISDN systems.

Perhaps one important surprise over the past five years has been the pace at which electronics has progressed to enable higher bit rate generation and processing. Only a few years ago, a bit stream of 565 Mb/s was considered to be state of the art, but 10-Gb/s bit streams have appeared remarkably quickly. This design advance has taken considerable pressure off the more exotic and expensive technical solutions for high-capacity systems. In other words, the development and deployment of FDM and soliton-based systems is not as urgent as it was once thought.

In the design of practical systems, there are many variables to take into consideration. First, what is the nature of the interconnection? As highlighted in Fig. 9.2, there are three categories of interconnection:

1. Point-to-point (link)
2. Point-to-multipoint (broadcast)
3. Multipoint-to-multipoint (network)

So far, the focus of attention has been on the point-to-point link, and it has been assumed that the ever-increasing hunger for more bandwidth inevitably leads to the need for coherent detection technology. Although coherent systems are technically more elegant than the structurally simpler direct-detection systems, cost does not yet favor the coherent technology. In fact, the case for employing direct-detection systems is still very strong. When monolithic photonic integrated circuits reach maturity, the cost of coherent detection systems should be low enough to make it the dominant technology for all of the above applications. Until that day, the presently cheaper direct-detection systems will no doubt continue to have widespread use.

9.2 Intensity Modulated Systems

Most optical fiber communications systems presently use the *intensity modulation* technique. As already stated, this is simply an ON-OFF transmission, whereby the light from the optical source produces 1s or 0s by the light being ON or OFF, respectively. This can be done by directly switching the source ON and OFF, or externally blocking the source to form 1s and 0s. If a laser diode has its current changed from zero to its operating value, the frequency of the output from the laser changes by a small amount. This *chirping* becomes a very significant problem for coherent detection systems using PM or FM.

OPTICAL FIBER

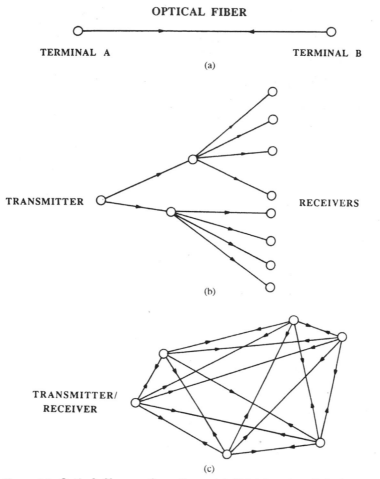

Figure 9.2 Optical fiber configurations. (*a*) Point-to-point: link (one transmitter and one receiver); (*b*) point-to-multipoint: broadcast (one transmitter and many receivers); (*c*) multipoint-to-multipoint: network (many transmitters and many receivers).

The bandwidth and therefore the number of channels transmitted by intensity modulation are determined by how fast the laser can be turned on and off, and by the dispersion associated with the chirp within the laser. The appearance of the DFB laser revived the life of the directly (intensity) modulated approach. At 1.3 and 1.55 μm, systems have been demonstrated to operate well above 10 Gb/s using intensity modulation.

The receiver sensitivity of any transmission system should be maximized as part of the overall system gain optimization. An optical intensity modulated system, also known as *amplitude shift keying* (ASK) modulation, uses direct detection, and the receiver sensitivity depends upon the minimum number of generated electron-hole pairs that can be translated into a detectable current. An ideal receiver would be able to observe the generation of a single electron-hole pair. For this ideal receiver one can calculate the minimum required pulse energy needed to achieve a specific error probability. This is called the *quantum limit*. Calculations show that the minimum number of photons that can be detected in an ideal direct-detection receiver for a BER of 10^{-9} is 10. Unfortunately, the ideal receiver does not exist, and early direct-detection receivers had electronic amplification following the photodetection, which added both thermal and transistor shot noise. These additional noise sources increased the required received power level (average number of photons per second) to 13 to 20 dB above the quantum limit (i.e., 200 to 1000 photons per bit). However, the introduction of optical preamplification, with large gain values, prior to photodetection has improved the sensitivity by about 10 dB. In this configuration, the signal and amplifier spontaneous emission noise, which is approximately proportional to the gain, is dominant over other noise sources (including the receiver thermal noise). FSK requires a higher number of photons per bit than ASK.

9.3 Coherent Optical Transmission Systems

The large bandwidth potential of coherent optical transmission systems has been known since the first days of optical fiber systems. However, in 1990 it was established that optical signal phase noise caused by the interaction of optical amplifier spontaneous emission and the nonlinear Kerr effect of fibers severely degrades the receiver sensitivity of coherent systems. A Kerr nonlinearity is the change in a medium's refractive index caused by the presence of optical radiation. Intensity modulated (noncoherent) systems do not suffer from this problem. Furthermore, at about the same time, optical preamplifiers narrowed the gap between the receiver sensitivities of coherent and intensity modulation to almost zero. So, while coherent systems are temporarily being eclipsed by the apparently adequate bandwidth of intensity modulated WDM systems, a day will surely come when the next level of bandwidth requirement will call on the services of coherent optical technology.

In optical communications literature, the term *coherent* refers to any detection process that uses a local oscillator. Note that this differs from the radio literature definition of coherent detection, which specifically refers to the phase of the signal involved in the detection of the IF signal. Coherent optical transceiver systems have many similarities to their microwave transmission counterparts. For example, a present-day coherent optical system design might have a PM in the transmitter, and a heterodyning technique in the receiver. The circuit block diagram of Fig. 9.3 highlights the main features of a coherent optical communications system.

In the coherent detection mechanism, a weak optical signal is mixed with a relatively strong local optical oscillator. In these circumstances, the calculation of receiver sensitivity is somewhat different from the radio system. Here, the electrical signal power is proportional to the product of the optical signal

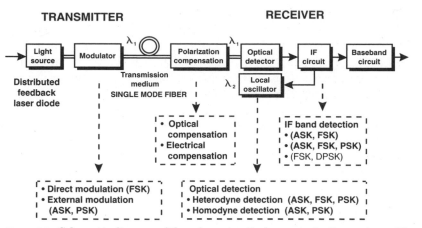

Figure 9.3   Schematic diagram of the coherent optical communication system with various configuration combinations.

and local oscillator power, and the receiver shot noise is proportional to the sum of the optical signal and the local oscillator power. This means that with sufficiently high local oscillator power, the shot noise in the coherent detection system receiver will be large compared to the thermal and active device noise power of the receiver electronics. In comparison, for the direct-detection system, the detector output power is proportional to the square of the incident optical signal power, and the receiver noise is mainly determined by the electronics following the detector. The coherent system should be able to approach the quantum-limit performance.

Analyses have been made for several optical systems to establish which type of modulation produces the best receiver sensitivity performance. The results of these calculations are summarized in Table 9.1 in terms of the number of detector photons required to produce a BER of 10^{-9}. Figure 9.4 graphically shows how the receiver sensitivities degrade with increasing bit rate. For coherent detection, these receiver sensitivities are the theoretical best possible (quantum-noise-limited) values. As a comparison, some practically measured receiver sensitivities are also plotted for direct detection. For these systems, there was no optical preamplifier incorporated prior to the photodetector.

TABLE 9.1   Receiver Sensitivities for Various Detection Schemes

Modulation/detection type	Number of photons per bit for BER = 10^{-9}
Direct detection	
Quantum limit	21
Practical receiver	38
ASK heterodyne	72
ASK homodyne	36
FSK heterodyne	36
PSK heterodyne	18
PSK homodyne	9

As indicated in Fig. 9.4, PSK homodyne detection in the receiver is theoretically the best technique, and this is borne out by experiment. When used with PSK (specifically differential PSK) in the transmitter, the best system performance is obtained. Good performance is achieved by using either the heterodyne or homodyne coherent detection technique in the receiver. The difference between the heterodyne and homodyne receiver is simply that the *heterodyne* receiver has a local oscillator whose frequency is *different from* that of the transmit frequency, whereas the *homodyne* technique has a local oscillator whose frequency is *equal to* that of the transmit frequency. The homodyne technique theoretically produces a receiver sensitivity 3 dB better than heterodyne, but is more complex to build, because the local oscillator has to be phase locked to the incoming optical signal. As indicated in Table 9.1, it has been calculated that only nine photons are required at the receiver for perfect homodyne reception of 1 bit of transmitted information at a BER less than 10^{-9}.

When compared to intensity modulation with optical preamplification, the coherent system therefore has an improvement, but only about 6 dB at best. That alone is not sufficient to warrant the use of this more expensive technology. The main advantage of coherent systems resides in the numerous channels that can be multiplexed in an FDM style over the 25,000-GHz or so bandwidth available in the 1.45- to 1.65-μm band (Fig. 9.5). Within each channel, thousands of telephone channels can be established, which means the capacity of this type of system is potentially enormous. The components required for coherent optical systems need considerably improved characteristics compared to intensity modulation systems.

The various possible combinations of system configuration for coherent optical fiber transmission are detailed in the block diagram of Fig. 9.3. Considering the *light source*, the laser diode FM-to-PM noise must be minimized by making the linewidth as narrow as possible. Ideally it should have zero linewidth, which of course is impossible. The minimum acceptable linewidth requirements depend on the bit rate and the *modulation scheme*, as summarized in Fig. 9.6. This figure highlights the fact that the more sophisticated, higher-quality performance modu-

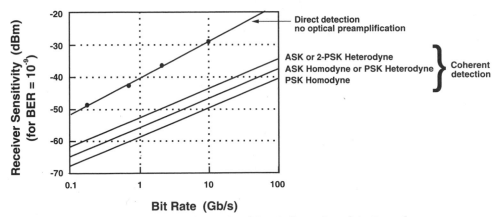

Figure 9.4 Graph of receiver sensitivity against bit rate for various detection schemes.

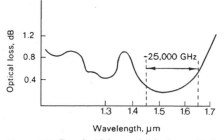

Figure 9.5 Bandwidth available for optical multiplexing.

lation schemes require as small a light source linewidth to transmission bit rate ratio as possible. It is also desirable for the IF bandwidth to be equal to the bit rate, as is the case for homodyne PSK. Note that the intensity modulation and direct-detection scheme is very good in this respect. The other modulation schemes such as heterodyne ASK, FSK, PSK, and homodyne ASK and PSK have also been included in this figure. The first systems to be investigated incorporated ASK, but these were soon superseded by the more sophisticated FSK and PSK systems.

The heterodyne ASK and FSK can tolerate an LD spectrum linewidth to bit rate ratio of 10^{-1}. In a 622-Mb/s system, for example, the required source linewidth of about 60 MHz is easily attainable with a basic DFB LD. The PSK homodyne requirement of 10^{-4} is a much more difficult specification to fulfill, especially at low bit rates. This requirement favors high-bit-rate systems, because the allowable laser linewidth increases linearly with signal bit rate. For the 622-Mb/s PSK homodyne system, the source linewidth must now be about 60 kHz. This is a very narrow linewidth, but it can easily be achieved by external-cavity DFB-LDs. At 10 Gb/s the linewidth can be relaxed to the more easily achievable value of about 1 MHz, although this high bit rate makes the receiver circuitry more difficult and complex.

The *transmission medium* decision is whether to use conventional fiber or polarization-holding fiber. If the polarization fiber can be purchased at a price and attenuation value not significantly higher than the conventional fiber, it is preferable. If conventional fiber is chosen, some form of *polarization compensation* is necessary, either at the optical or electrical level. Coherent systems at 1.55 μm should have dispersion-shifted fiber. Non-dispersion-shifted fiber would significantly affect the maximum link length.

The *demodulation* process is obviously the inverse of the modulation, with homodyne PSK being the best choice for high performance. An *IF band detection* is then necessary to convert the microwave IF signal down to the baseband bit stream. This can be done either by envelope detection or the better-quality heterodyne detection.

Observing the modulation scheme in more detail, the most attractive technique, as indicated in Table 9.1, is the homodyne PSK detection system. It is theoretically superior to all other detection schemes and has considerable potential for future systems. An optical phase-locked loop (OPLL) is universally

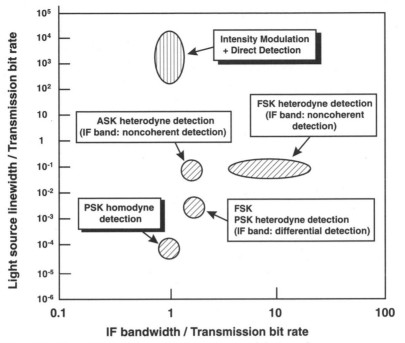

Figure 9.6 Linewidth requirements for various modulation schemes.

accepted as being essential for PSK homodyne detection systems. There are several problems associated with the practical implementation of PSK homodyne detection. First, the carrier component of the modulated transmit signal is suppressed, so what does the PLL lock onto? This suppression is evident by observing that in Fig. 9.7a the digital 1s and 0s give a net average value of zero. Second, for optimum homodyne detection, the data and the local oscillator signals must be locked in-phase (Fig. 9.7b). Unfortunately, a conventional PLL locks the local oscillator at 90° to the data (reference) signal, as in Fig. 9.7c. Third, the data signal and local oscillator signals are mixed at the photoconductor to produce the required beat frequency. The detector output current contains an additional, unwanted dc signal which must be eliminated; otherwise it interferes with the PLL operation. These three problems can be solved by using either a nonlinear loop or a balanced PLL.

9.3.1 Nonlinear PLL

There are two main types of nonlinear PLL: Costas loop and decision-driven loop. In the Costas loop, the phaselock current is multiplied by the signal current from the data photodetector, whereas in the decision-driven loop, the phase-lock current is multiplied by the output signal of the data receiver. The decision-driven loop, homodyne optical receiver is theoretically superior in performance to the Costas loop receiver.

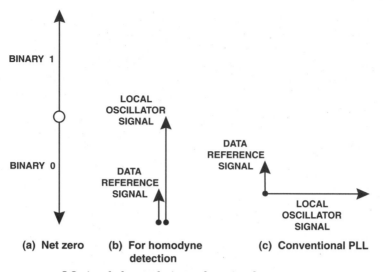

Figure 9.7   LO signal phase relative to data signals.

Figure 9.8 illustrates a homodyne optical receiver based on the decision-driven loop. The received signal power is split by a 90° optical hybrid. One part of this signal is sent to the data detector and the other part is sent to the PLL. This implies an extra power penalty due to the loss in the 3-dB coupler. Because the average value of the phase-lock current from the loop photodetector is zero (as $\theta = 90°$), it must be processed nonlinearly prior to use in phase locking. This is achieved by multiplying the output from the 1-bit delay circuit (phase-lock current) by the output from the data detector (data output signal). Notice the RC circuits, which are included for ac coupling to eliminate the dc component. For this decision loop optical homodyne receiver circuit, the required laser linewidth-to-bit-rate ratio is calculated to be about 3×10^{-4}.

9.3.2   Balanced PLL

The block diagram for the balanced PLL receiver is shown in Fig. 9.9. The optical power is split by the 3-dB coupler. This is a 180° coupler, as opposed to the usual 90° coupler. In the other arm of the coupler is fed the optical output from the voltage controllable laser local oscillator. The signals are detected by the photodetector and passed to an amplifier whose output is fed back via the loop filter to the VCO, forming the PLL. The dc component is eliminated by the balanced receiver. The linewidth-to-bit-rate ratio for this circuit is about 10^{-5} to 10^{-4} (for a 1-dB power penalty at a BER of 10^{-10}), and it incorporates extra post-detection processing to reduce the data-to-phase-lock crosstalk (i.e., interference between the data detection and phase-lock branches of the receiver).

Comparing the performance of the decision loop receiver with the balanced loop receiver reveals that the balanced loop receivers suffer data-to-phase-lock crosstalk noise, which must be canceled. This is in addition to the usual laser phase and shot noise. From the linewidth point of view, the decision loop receivers are better than balanced loop receivers without data-to-phase-lock crosstalk cancellation. If data-to-phase-lock crosstalk is canceled, the linewidth required by the balanced-loop systems approaches that of the decision loop systems. Photonic and electronic integrated circuits should eventually neutralize any advantage one system might have over the other.

In summary, the laser source for homodyne PSK requires a minimum spectrum-width-to-bit-rate ratio that depends upon the receiver configuration. At present, the binary PSK homodyne detection, balanced OPLL receiver is a good option, requiring the minimum linewidth-to-bit-rate ratio

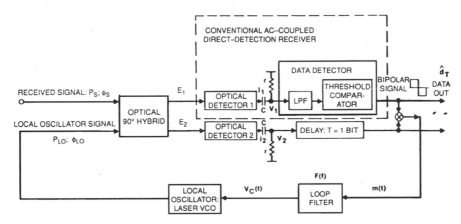

Figure 9.8   Decision-directed PLL optical homodyne receiver.

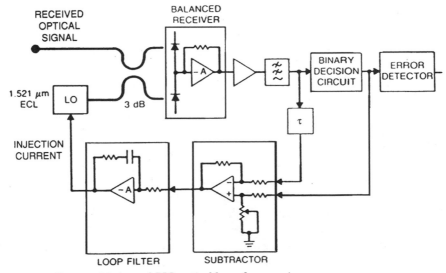

Figure 9.9   Improved balanced PLL optical homodyne receiver.

of the order 10^{-4}. Homodyne PSK systems have experimentally verified the theory by exhibiting receiver sensitivities quite close to the quantum limit of nine photons per bit.

The only question that now remains is, When will these systems be implemented into the networks by becoming cost-effective compared to existing direct-detection systems?

9.4   Regenerators and Optical Repeaters

Optical repeaters use one of two methods: (1) regeneration of the bit stream and (2) optical amplification. It is important not to confuse the two. Because of the distances involved and the point-to-point nature of the inter-CO (interexchange) routes, neither optical repeaters nor regenerators are required. However, for the distances involved in long-haul links, optical repeaters are necessary and, if high-dispersion fiber is used, regenerators are necessary. Many LANs also require optical repeaters because of multiple splitting of the light source power. Since optical amplifiers have become commercially available, they are invaluable in both of the above applications. Optical amplifiers have noise accumulation, which means that pulse regeneration is needed after a certain distance. Fortunately, this distance is longer than 10,000 km, which makes it suitable for transoceanic distances.

9.4.1   Regenerative opto-electronic repeaters

The block diagram of Fig. 9.10 shows a typical sequence of events in a conventional optical fiber regenerator. The optical signal is converted to an electrical signal by the optical detector (APD or PIN diode). The resulting electrical pulses are amplified, equalized, and then completely reconstructed by a standard pulse regenerator and timing circuit. A measure of the BER is established at this stage of the regenerator. The clean pulses are then used to drive a LD light source that is ready for onward transmission down the optical fiber. Until recently, this method has been used exclusively in almost all long-distance systems. The availability of the EDFA has radically reduced the number of regenerators required. The regenerators are placed in manholes and are relatively small for a small fiber cable having only eight or ten fibers. For large cables having several hundred fibers, the regenerators would occupy several cubic meters, necessitating large manholes. Furthermore, before optical amplifiers existed, the power required for the regenerators was a considerable engineering problem, particularly for submarine links. Because power is not usually available at manholes, copper wires must be included in the construction of the cable to power the regenerators remotely from the nearest line terminal equipment. Note that in non-WDM systems, each optical channel requires two regenerators, one for GO and the other for RETURN. By using regenerative repeaters, the effects of fiber dispersion (intersymbol interference) are cleaned up at each regenerator.

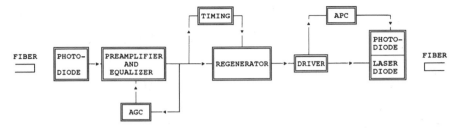

Figure 9.10 Block diagram of an optical regenerator (without optical amplification).

9.4.2 All-optical repeaters

The commercial availability of optical amplifiers in the early 1990s revolutionized long-distance communications designs, and also made optical LANs economically viable. Optical amplifiers, which do not require the signal to be converted to an electrical signal and then back to optical, offer a smaller and cheaper approach to counteract fiber attenuation. If the signal of a long-haul route is periodically boosted with optical amplifiers, the length of the link is limited by the cumulative fiber chromatic dispersion or amplifier noise. The dispersion can, to a large extent, be eliminated by using dispersion-shifted fiber and transform-limited pulses. A transform-limited pulse is the shortest laser pulse achievable for the available laser oscillating bandwidth. One of the very attractive features of the optical amplifier repeater is the fact that it is bit rate and modulation format independent. Multigigabit-per-second signals have been demonstrated over distances of several thousand kilometers, which gives an indication that eventually the optical amplifier repeater will completely eliminate the opto-electronic regenerative repeater.

9.5 Optical Multiplexing

WDM is exactly the optical equivalent of the electrical *frequency division multiple access* (FDMA) seen in radio technology. In some respects, it is unfortunate that the optical fraternity chose to use the term wavelength instead of frequency to characterize optical signals, because it can lead to some confusion. Admittedly, the wavelength numbers (in nm or μm) are more manageable than frequencies in terahertz (THz) or petahertz (PHz), but frequency would be more consistent with the electronics world.

In the optical communications literature, references are frequently made to WDM and FDM. They can easily be confused with each other, so we shall start with a definition of each. WDM is the term used for multichannel operation at *several optical frequencies*. WDM involves placing several digital bit streams on different optical carriers. At present, each bit stream intensity modulates a different light source to form a channel at the wavelength of the light source. The light sources have wavelengths that are sufficiently separated from each other so as not to cause adjacent channel interference. These optical signals are combined and transmitted down a fiber, and by an array of optical couplers and filters (only passive devices) at the receiver, channels are individually isolated so that after demodulation each bit stream can be electronically processed.

FDM is a term that was previously used in analog transmission. Optical FDM systems use the heterodyne or homodyne techniques to multiplex many electronic (bit stream) channels onto a single optical carrier whose frequency is in the 1.5- or 1.3-µm bands. Each channel to be frequency division multiplexed is already a digitally time-division-multiplexed signal containing thousands of voice channels and/or many video channels, depending on the hierarchical level (e.g., OC-48, 2.488 Gb/s). The total optical bandwidth used is determined by the modulating baseband signal (i.e., microwave or even infrared).

A system designed to take advantage of a combination of both WDM and FDM techniques would produce an exceptionally large capacity system.

9.5.1 Wave division multiplexing

WDM was first introduced 20 years ago, but failed to gain commercial viability because it was cheaper to increase capacity simply by increasing the bit rate. When the EDFA arrived on the scene, the transmission economics changed significantly. Because the EDFA was broadband, and could therefore amplify many optical signals simultaneously, WDM began to factor into the design equations. The first EDFA long-haul systems were installed in 1995 and used just one wavelength, which carried a 5-Gb/s bit stream. In a WDM system, one EDFA can be used in place of many opto-electronic regenerators.

The point-to-point long-distance transmission systems are rapidly becoming standardized, with each pair of optical fibers incorporating WDM and each wavelength carrying an intensity modulated signal containing a bit stream of 10 Gb/s or more. Depending on the fiber used, these systems still need regenerators if the links are so long that dispersion becomes large enough to need correcting. Upgrades to larger capacity can be done by increasing the bit rate or expanding from, say, 8 wavelengths to 16 or 32. Of course, the bit rate of the regenerators also needs upgrading. In summary, the point-to-point WDM systems are well on the way to solving major trunk route and metropolitan area transmission bottlenecks. (Section 9.6 details a typical transmission system.) Unfortunately, the same cannot be said for the transmission systems of telecommunications networking, as discussed in Chap. 10.

Figure 9.11 shows a typical WDM system, which in this case has just 17 channels. Each channel carries a 20-Gb/s stream, giving a total capacity of 340 Gb/s. Eight DFB lasers and nine external cavity lasers were used in this test. As usual, polarization controllers were incorporated and $LiNbO_3$ external modulators used. The channel separation in this case is 0.8 nm, and this figure can be further reduced as laser bandwidths improve. An important characteristic of WDM is that the information signals superimposed on each optical wave do not broaden the bandwidths of those optical signals. Each wave is simply switched on and off to form the 1s and 0s. The situation is different for FDM, where the optical signal bandwidth is increased to the baseband bandwidth. This places a limit on the minimum spacing of wave division multiplexed, FDM optical signals. The 1.5-μm band offers over 100 nm of available optical bandwidth, and the 1.3-μm band has a similar bandwidth. Theoretically, if 200 nm of bandwidth could be used with 0.5-nm spaced channels, the total WDM channels would be 400. At present, optical amplifier characteristics do not allow this full bandwidth to be exploited. For example, EDFAs for use in the 1.5-μm band have only about 35 nm of usable bandwidth. Even that bandwidth is reduced significantly unless gain equalization is used to improve the optical gain flatness. Another problem is dispersion. Single wavelength dispersion compensation can be very effective even for long-haul links (Sec. 9.6.3), but as the number of wavelengths of a WDM system increases,

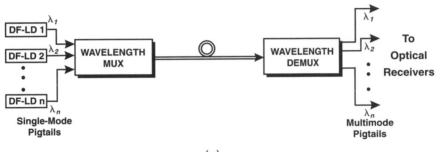

(a)

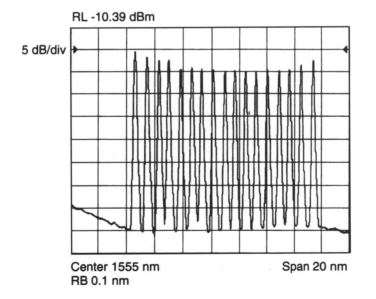

(b)

Figure 9.11 (a) Optical WDM transmission. (b) WDM channel spacing. (*Reproduced with permission from IEEE © 1995, Chraplyvy, A. R., et al., IEEE Photon. Technol. Lett., vol. 7, no. 1, January 1995.*)

so does the complexity and effectiveness of the dispersion management. No doubt, this 200 nm of bandwidth will not remain only partially used for very long. Even the demonstration of 17×20 Gb/s WDM shows the viability of a transmission of over four million voice channels, on one pair of dispersion-managed fibers, over a distance of 150 km.

An example of the optical (de)multiplexers that could be used in such a system is shown in Fig. 9.12, and is called the *phased-array* or *PHASAR (de)multiplexer*. When the beam in the transmitter waveguide reaches the free propagation region, it diverges from the object plane. At the input aperture, the light is coupled into the waveguide array. This single-wavelength light is therefore split up, and travels through each of the individual waveguides in the array. The physical lengths of the waveguides in the array are chosen so that the optical lengths of adjacent waveguides at the center frequency of the demultiplexer differ by an integer multiple of 2π. At the output aperture, the light from the waveguides converges in the free propagation region so that an image of the input wave at the object plane appears at the center of the image plane. At any other frequency above or below the center frequency, the different optical lengths of the waveguide array cause the energy at those other frequencies to be focused at different points on the image plane. The receiver waveguides can therefore be specifically located so that each waveguide collects light at a specific frequency (wavelength). This is a reciprocal device, so its multiplexing behavior is described as the reverse process of several different frequency input waveguides feeding one output waveguide, which carries a multiple-frequency signal.

For semiconductor-based devices, the loss associated with each channel passing through this multiplexer/demultiplexer is about 2 to 3 dB. Crosstalk between channels mainly arises through fabrication imperfections, and is typically −30 to −35 dB. These crosstalk values determine the channel bandwidth to be about 30 percent of the channel spacing. For a channel spacing of 100 GHz (0.8 nm at 1.5 µm), the channel bandwidth is therefore 30 GHz. Channel spacing can be quite small. For example, an eight-channel WDM receiver, each channel having a useable bandwidth of 10 GHz and using the phased-array (de)multiplexer, was demonstrated as early as 1996.

Another, simpler application of WDM is to enable full-duplex communication between two points using only one optical fiber instead of a pair of fibers. If, for example, one wavelength at 1.50 µm is used for GO and another at 1.60 µm for RETURN, only one fiber is necessary. This has interesting implications for future cost reduction of local loops using fiber to the home (FTTH).

9.5.2 Frequency division multiplexing

The enthusiasm for FDM has waned since the euphoria of the late 1980s. The rise of the EDFA and the huge capacity possible with the simpler and cheaper WDM, combined with high-bit-rate data streams, have reduced the need for the FDM solution for high capacity. No doubt, FDM will reemerge in the future. This might happen, for example, when 32 WDM signals each containing an OC-768 at 40 Gb/s (totaling 1.25 Tb/s throughput) do not provide enough capacity.

For point-to-point systems, FDM is considered to be technically superior to WDM because more channels can be multiplexed within the available optical bandwidth. FDM is often described as *dense optical multiplexing*. To appreciate its potential, some hypothetical FDM capacity figures must be assessed. The channel spacing for FDM can be typically 5 GHz (or 0.04 nm at 1.5 μm), which is considerably less than the 250 GHz (or 2 nm) for WDM. Also, the losses in the components are less for FDM. Figure 9.13 illustrates a point-to-point FDM system, where the transmitted optical signal is built up in a manner similar to the old analog microwave radio system. The total number of bit streams that can be multiplexed depends on the bit rate of each stream. Clearly, more bandwidth is going to be consumed by heterodyning a 10-Gb/s bit stream than a 100-Mb/s stream. At the receiving end the optical signal is detected, power is divided, and the individual bit streams are recovered by IF heterodyning.

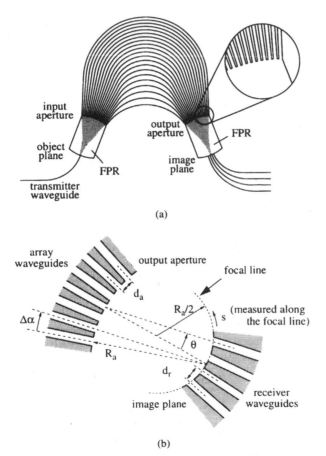

Figure 9.12 Optical multiplexer for WDM. (*a*) Layout of the PHASAR demultiplexer; (*b*) geometry of the receiver side. (*Reproduced with permission from IEEE © 1996, Smit, M., Journal of Selected Topics in Quantum Electronics, PHASAR-Based WDM-Devices: Principles, Design and Applications, June 1996, Fig.1.*)

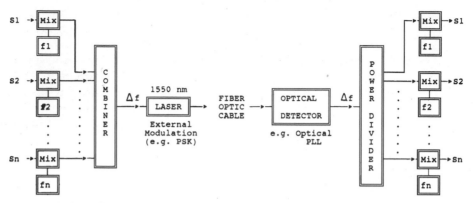

Figure 9.13 Optical FDM.

The full capability of optical FDM will be realized only if a baseband of up to 12,500 GHz or so can be used to modulate the LD. This corresponds to an optical $\Delta\lambda$ of about 100 nm, which could constitute the upper sideband of a PSK-modulated optical carrier covering 1500 to 1600 nm (or 187,500 to 200,000 GHz). To allow a large regeneratorless distance, it would be necessary for the zero dispersion to be almost flat across the full 100-nm optical bandwidth.

To fantasize, the 12,500-GHz baseband could be constructed by 2500 5-GHz-spaced signals, each containing 2.488-Gb/s (STM-16 or OC-48) bit streams, giving a total of $32,256 \times 2500 = 80.6$ million voice channels (or about 625,000 video channels) on one pair of fibers. However, 12,500 GHz far exceeds the present and foreseeable future capability of microwave and infrared technology, not to mention present voice and video channel capacity requirements.

The following is an example of a realistic system that could be designed using present-day technological capability.

Example. In Fig. 9.13, consider S1 = S2 = S3... = Sn = 155 Mb/s (i.e., *2016* voice channels per input). Also, consider 40 GHz to be presently an acceptable frequency for obtaining high-quality microwave components (e.g., for modulating the LD).

The number of FDM channels depends on the modulation technique, which defines the bandwidth efficiency; that is, if the bandwidth efficiency = 1 b/s/Hz, $f_1, f_2,..., f_n$ would be typically 2.0, 2.2, 2.4, 2.6,..., 40 GHz ($\Delta f = 38$ GHz); $\rightarrow$ 190 CH $\rightarrow$ $2016 \times 190 = 383,040$ voice channels on one 1550-nm optical carrier using only 38 GHz, or about 0.3 nm, of *optical* bandwidth.

The 512-QAM with the TCM modulation technique provides a bandwidth efficiency of 7.5 b/s/Hz. So, $f_1, f_2,..., f_n$ would be typically 2.03, 2.06, 2.09, 2.12,..., 40 GHz ($\Delta f = 38$ GHz); $\rightarrow$ 1266 CH $\rightarrow$ $2016 \times 1266 = 2,552,256$ voice channels on one 1550-nm optical carrier using only 38 GHz, or about 0.3 nm, of optical bandwidth.

Even present-day technology could use up to 20 of the 0.3-nm (2,552,256 voice channels) FDM signals in a WDM system having 2-nm channel spacing.

This combination of FDM and WDM would provide about 51 million voice channels on one pair of fibers. This large quantity of voice channels far exceeds existing or even future capacity requirements. However, the corresponding 380,000 video channels is an interesting prospect for applications such as feeders for broadcasting numerous TV channels. This number of video channels is calculated using 10 Mb/s per video channel. This is MPEG-2-quality video, which could be distributed to customers via the telecom network using an FDM-WDM system, provided optical fiber is installed in all sections of the network. That is presently a very large proviso.

9.6 Systems Designs

Network availability is becoming increasingly important for telecom operators, and optical fiber ring structures are fulfilling that requirement. Combining the SONET enhancements (such as optimized costs, improved flexibility, and improved transmission quality) with ring structures is allowing networks to evolve into far superior configurations than was possible in the early 1990s. A generic transmission network structure has been created (as shown in Fig. 9.14), and it is receiving worldwide acceptance. Unlike the star-shaped backbones of old, the new backbone topology is meshed and has high-capacity cross-connect nodes. The first generation of cross-connects are electronic synchronous cross-connects, and they interconnect digital bit streams; that is, bit streams in the electronic domain. The next generation of cross-connects will interconnect optical signals without the optical-to-electrical conversion. The regional and access network contains ring structures, and includes SONET/SDH add/drop multiplexer nodes. Some of the rings in the access network are closed or star connections made by SONET/SDH digital microwave radio systems, where terrain precludes the use of optical fiber.

Let us consider the three main categories of fiber transmission systems: (1) long haul, (2) interoffice (interexchange) routes, and (3) local (subscriber) loop. The technology is moving so fast that engineering designs for each of these categories are changing rapidly. As usual, cost considerations dictate the design and components used for a particular link. The interoffice routes are perhaps the most standardized designs which use well-proven regeneratorless technology. A massive research effort is under way to eliminate the regenerator from long-haul links and therefore reduce their cost. Many engineers once considered this to be the last major hurdle to overcome to enhance global optical fiber connectivity. Then came global networking, which changed that viewpoint (see Chap. 10).

The capacity requirements of interoffice and long-haul links seem to be increasing annually. Upgrades can be made wherever possible, simply by increasing the bit rate if optical amplifiers are in place instead of regenerators. Optical fiber in the local loop is the subject of intense debate in the developed world. The cost of delivering fiber to every customer is enormous, but the services that could be provided and the subsequent revenue generated is even greater. Some of the link designs presently in use and proposed for the future will be discussed in the following subsections.

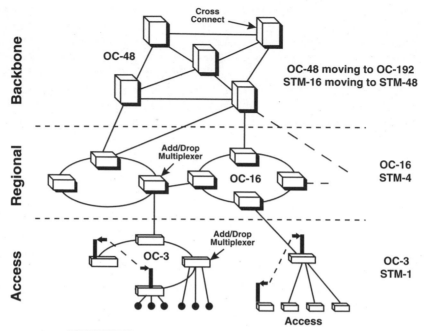

Figure 9.14 SONET/SDH transport network structure.

9.6.1 SONET/SDH rings

Ring structures are central to the significant improvement in availability experienced by new optical systems installations. The concept is simple. A ring structure offers the route diversity that a linear or star topology does not provide. There are three main types of optical fiber rings. As illustrated in Fig. 9.15, they are:

- Unidirectional
- two-fiber bidirectional
- four-fiber bidirectional

Figure 9.15a shows the *unidirectional path-protecting ring*. This is a two-fiber ring architecture in which traffic moves around one ring in a clockwise direction and the other ring counterclockwise. At each node, information entering the rings is simultaneously transmitted in both directions. Each node receives signals coming from both directions, and the better of the two signals is selected for reception. At each node, any through traffic (not dropped at the node) passes directly through from the node's receiver to its transmitter. If the optical fiber cable is cut or in some way damaged, a fast self-healing process is initiated. Each node still continues to transmit in both directions, but clearly traffic sent toward the break will not reach the receiver. At the nodes each side of the break, the receivers at those two nodes will experience a loss of signal on the break side of each node. Therefore, the receivers will each switch (if not already set to that position) to receive the signal incoming from the nonbreak side of the node. At the time the break occurs some information might be lost if one or both of the receivers at the nodes nearest to the break are set to receive information along that sector. Approximately 50 ms of data, at worst, will be lost until the node receivers are switched. If, by good luck, information on that sector is not chosen for reception at the time of the break, then no information will be lost. Complete node failure clearly disrupts traffic from dropping or inserting to and from destinations accessed by that node. Through traffic would also be blocked. Partial node failure would allow either through traffic or add/drop functions. In practice neither type of node failure should occur, because each node would be either 1+1 (ideally) or 1+N (more cost-effectively) protected. In other words, the optical line terminals incorporating optical transmitters and receivers, the cross-connects, and the add/drop multiplexers would have some level of redundancy.

Figure 9.15b shows the *four-fiber bidirectional line-protecting ring*. As the name suggests, the difference between this ring structure and the unidirectional rings is that within each fiber pair, the rotation of traffic on one fiber is always clockwise, and on the other fiber traffic always progresses counterclockwise. This means that traffic between each node is effectively a point-to-point connection with 1:1 protection. There are three main failure modes for this topology: ring failure, span failure, and node failure. If a complete break in the cable occurs in the path between nodes C and D (ring failure), a loopback switch activates at both nodes C and D to reroute the traffic to the required nodes by using the inner (protection) rings as well as the outer (working) rings. If only one of the fiber pairs, or one transmitter or receiver, fails on the traffic-carrying ring (span failure), a switchover to the protection pair will occur for that span. This is just the same style of protection offered by the older nonring type of backbone or point-to-point configurations. Span switching has priority over ring switching. This is because span switching simply hops the traffic from only one portion of the ring onto an adjacent host standby circuit, whereas ring switching involves traffic rerouting, affecting the rest of the ring. An unprotected node failure can be serious.

If node C fails, traffic clearly cannot reach node C destinations unless the other nodes provide connection through a mesh topology. As soon as a node failure occurs, an alarm indication signal (AIS) must be inserted into the frame of bit streams destined for the failed node. Otherwise, misconnections would occur, where traffic looped back at nodes B and D (adjacent to the failed node) would incorrectly appear at node A to be coming from the failed node C. Again, node protection is necessary to avoid this situation.

There is also the *two-fiber bidirectional line-protecting ring*. This ring structure can provide only ring protection, and not span protection. The bandwidth for any span is only one half of the four-fiber ring. Of course, the cost is less for the two-fiber ring installation, but not one half of the cost, making the four-fiber ring more cost-effective for large traffic volumes.

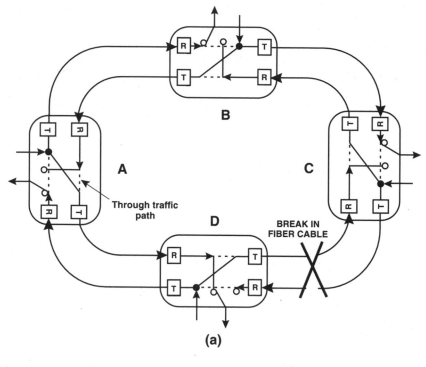

(a)

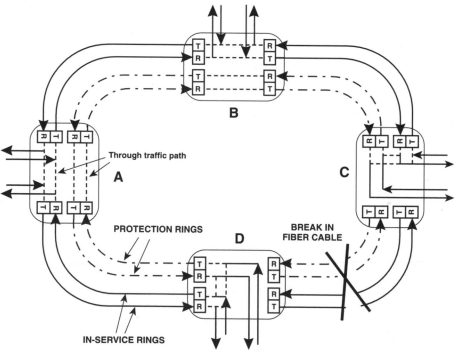

Figure 9.15 SONET/SDH ring topologies. (*a*) Unidirectional path-protected ring; (*b*) dual bidirectional line-protected rings.

Performance comparison. Bandwidth and availability are two important considerations when deciding on which ring structure to use. The four-fiber bidirectional ring has the best availability, but not necessarily the best bandwidth. In the extreme case of all traffic focusing on one node, the ring capacity is that of one span. This is the same capacity as a unidirectional ring having similar span capacity. At the other extreme, when traffic flows only between adjacent nodes, the available bandwidth can be reused between each span (each pair of nodes). In this case, the capacity is the product of the span capacity and the number of nodes.

In summary, four-fiber (and sometimes two-fiber) bidirectional rings offer high capacity and high availability. Multiple-span switching can occur if more than one span fails. This improves availability significantly. These rings are popular for interoffice traffic. Unidirectional rings have lower complexity and provide a cost-effective solution for access networks.

Multiple rings. Interconnection of multiple rings offers an interesting type of multiple route diversity which can maximize circuit availability. Figure 9.16 shows just two interconnecting rings, but there could be many. Two of the nodes are common to each of the two rings and are colocated in the same building. If any one of the nodes fails, including one of the common nodes, traffic can still reach its destination. This configuration is effectively a single common node having 1+1 protection. The versatility and benefits of ring structures are so great that they are now being used in all new installations or major network upgrades throughout the world.

9.6.2 Interoffice (interexchange) traffic

Today in the United States and other developed countries, 2.488-Gb/s OC-48 bit streams on 1.3-μm, *single-mode* optical fiber systems are widely used to

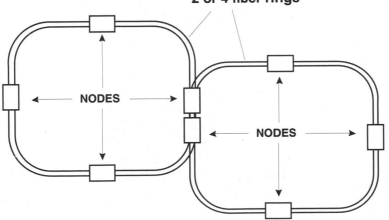

Figure 9.16 Multiple rings.

interconnect COs. A typical 2.488-Gb/s optical fiber line terminal equipment (LTE) is shown in Fig. 9.17. The 2.488-Gb/s system is chosen because this provides a sufficiently large number of channels (32,256). In the transmitter part of the LTE, the incoming signal from the digital multiplexer has the NRZ code.

The signal is then scrambled by the circuit shown in Fig. 9.18. In the relatively low bit rate PDH systems, transmission signals had sophisticated line coding to facilitate timing, clock recovery, and error detection, but SDH has bit rates well into the gigabit-per-second realm. Clock recovery is the main requirement of the line code for optical systems and this can be arranged by simple unipolar NRZ or RZ line codes, provided scrambling is included to randomize the bit level.

There are two possible types of scrambling: serial and parallel. The very high bit rate signals above, say, OC-12 can be parallel-scrambled at the STS-1 level prior to forming the STS-N frame. Fortunately, electronics is progressing fast enough for serial scrambling to be possible at gigabit-per-second rates. However, the more expensive GaAs circuit technology would be needed at such high rates.

After scrambling, the bit stream is fed to the transmitter for converting the NRZ bit stream into optical ones and zeros by intensity modulating a laser diode. After the optical signal has passed through the fiber it is received at the optical line terminal in the CO.

The reverse process occurs in the receiver. An optical preamplifier enhances the signal prior to photodetection of the light pulses. The output of the photodetector circuit is an NRZ signal, which is then descrambled and sent to the demultiplexer for further processing.

9.6.3 Long-haul links

There are several interpretations of the term *long haul*, but it is broadly described as any distance in excess of interoffice routes. Long-haul links are specifically trunk routes that can be intercity connections, a country backbone network, an international link, or a transoceanic link. Ideally it would be preferable to span any of these links without the use of optical amplifier repeaters or regenerators. If the attenuation of optical fiber cables ever gets

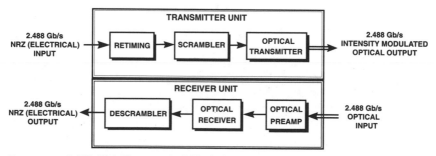

Figure 9.17   2.488-Gb/s line terminal block diagram.

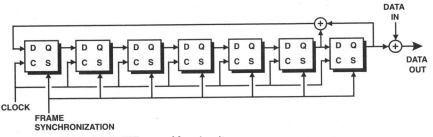

Figure 9.18 Typical SONET scrambler circuit.

down to 0.01 dB/km or less, as theoretically predicted, this would become a reality. In the near- to medium-term future, this is unlikely. From the link budget equation the present minimum attenuation of about 0.2 dB/km at 1.55 μm dictates the distance between optical amplifiers. This can be improved by increasing either the optical source power level or the receiver sensitivity. The receiver sensitivity limit is fixed by the laws of physics, and is already being approached by laboratory experimental systems. Optical power sources are continually being improved as time goes by. LD technology is steadily increasing the optical output power but, again, the laws of physics step in and supposedly limit the maximum power that can be used in very high bit rate systems (over 40 Gb/s) to a little more than 10 mW because of scattering phenomena. Recent developments dispute that initial proposal, and it is now believed that pulse compression can occur at transmitting power levels higher than about 10 mW. The benefits of this are far reaching and will be discussed in the next section, on *soliton* transmission.

Optical amplifiers are indispensable. New long-haul installations using 1.55-μm dispersion-shifted fiber and EDFAs give the best performance. A considerable amount of standard (non-dispersion-shifted) fiber had already been installed before EDFAs became available. In this case upgrades have been made by replacing many regenerators with EDFAs. After several hundred kilometers, dispersion had accumulated to the point where regeneration was necessary. Regenerative repeaters effectively reset the fiber dispersion effects at each regenerator.

Dispersion management or mapping can be employed to extend the distance between repeaters to more than 10,000 km. If a fiber is used which has zero dispersion at, for example, 1580 nm and an optical signal is transmitted at 1558 nm, dispersion is about −2ps/km·nm. If, after traveling a distance of about 900 km, the signal passes through 100 km of 1310-nm zero-dispersion fiber, the dispersion it experiences at 1555 nm is *positive* and cancels out the previous negative dispersion (see Fig. 9.19). This technique has achieved transmission over transoceanic distances without regenerators, and has been so successful that WDM can be incorporated.

In experiments for a trans-Pacific optical fiber link, AT&T laboratories verified that 20 WDM channels each containing a 5-Gb/s signal could be transmitted over 9100 km. A single-wavelength design enables the dispersion compensation to be managed to create a net zero over the desired distance. For multiple wavelengths (WDM), the dispersion is progressively worse for channels at wavelengths other than the perfect compensation wavelength. Dispersion is not the only concern in long-haul systems. The EDFAs' gain flatness is of no importance to a system operating on just one wavelength, but for WDM, the gain flatness is a problem when many EDFAs are cascaded. For example, if the gain varies by only 1 dB across a WDM operating range of 1550 to 1565, without any gain equalization, after a signal passes through a chain of 200 EDFAs the flatness could vary by 200 dB.

This is clearly too much variation, and passive equalizers must be introduced at several points along the route. They are simply filters that have an attenuation curve the inverse of the EDFA gain curve. In the trans-Pacific simulation, the EDFAs were pumped at 1.48 μm and operated at about 9-dB net gain, with about 5 dB of gain compression (into saturation). One important problem in a chain of EDFAs is known as EDFA *polarization hole burning*, which is due to a polarized saturating signal being launched into the erbium-doped fiber. Resulting amplified spontaneous emission accumulation must therefore be suppressed. This has been successfully achieved using a polarization scrambling process, which varies the two polarization states at a rate (about 1.3 kHz) faster than the EDFA can respond, and therefore reduces the impairment caused by polarization hole burning. Ideally EDFAs should operate in the linear region, and not into gain compression.

Another long-haul demonstration used a typical circulating loop style of optical WDM transmission test setup. Fiber spans 45 km long were used in the test to demonstrate 20 channels each containing 5-Gb/s streams transmitted over 9100 km. Channel spacing was about 0.6 nm and, by including a (255,239) RS, FEC encoder, all 20 channels were measured to have a BER better than 10^{-10} after 9100-km transmission. Passive gain equalization and dispersion management were essential for achieving such results. Incidentally, the end-of-life BER for a submarine optical fiber link is usually considered to be 10^{-10}.

In addition to the traditional ploughing in of fiber cable, some interesting innovations are being implemented for installing long-distance optical fiber routes. For example, suspending fiber cable from high-voltage power lines has been successful, and wrapping fiber cable in a helix around high-voltage cable is another option. Perhaps the most appealing method now being assessed is for new high-voltage transmission line cables to be fabricated with some fibers running down the center of the ground wire. This is particularly attractive for new power line installations in developing countries.

Soliton transmission. Possibly the most interesting innovation in optical communications to appear so far is optical soliton transmission. Although this method of transmission has been theoretically known for some years, it was only in about 1989, when optical components such as Er-doped fiber amplifiers became readily available, that efforts to achieve soliton transmission succeeded. What are solitons? They are dispersionless pulses of light. They exist only in a

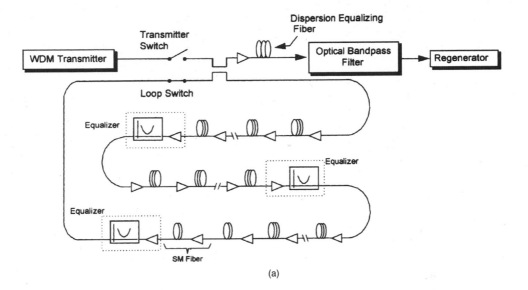

(a)

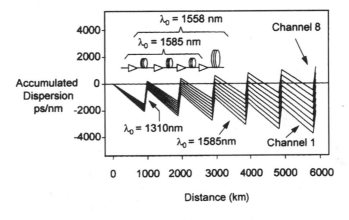

(b)

Figure 9.19 Dispersion management. (*a*) Block diagram of the equipment. (*b*) Accumulated chromatic dispersion versus transmission distance for eight channels of a WDM transmission experiment. The majority of the amplifier spans use negative dispersion fiber with $\lambda_0 \approx 1585$ nm and $D \approx -2$ ps/km•nm. The dispersion is compensated every 1000 km using conventional single-mode fiber (i.e., $\lambda_0 = 1310$ nm). (*Reproduced with permission from IEEE © 1996, Bergano, N. S., and C. R. Davidson Wavelength Division Multiplexing in Long-Haul Transmission Systems, Journal of Lightwave Technology, vol. 14, no. 6, June 1996, Fig. 1.*)

nonlinear medium, where the refractive index is changed by the varying intensity of the light pulse itself. There are at least two classes of solitons: temporal and spatial. The temporal soliton is the candidate for ultralong-distance optical communication systems.

A soliton is a transmission mode of optical fiber that is the only stable solution to the fundamental propagation equation. Solitons cancel out the chromatic dispersion of optical fibers. When operating at the zero-dispersion-shifted value of 1.55 μm, a pulse will be spectrally broadened because of silica's nonlinearity. The mere presence of a light pulse that has a varying light intensity over its envelope causes the refractive index of the material to be higher at the pulse's peak than at its lower-intensity tails. As illustrated in Fig. 9.20a, the peak is retarded compared to the tails, which lowers the frequencies in the front half of the pulse and raises the frequencies in the trailing half. For the soliton to propagate, a small amount of chromatic dispersion is necessary. If the system operates at a wavelength slightly longer than the zero-dispersion wavelength, there would be a tendency for pulse spreading to occur, with the higher frequencies of the pulse leading and the lower frequencies being retarded (Fig. 9.20b). As one can see, the effects of fiber nonlinearity are balanced by the fiber's chromatic dispersion, producing the soliton of Fig. 9.20c. A soliton is therefore truly nondispersive in both the frequency and time domains. Furthermore, it has already been demonstrated that solitons remain stable after passing through many amplifiers, provided the distance between amplifiers is short enough so that the pulse's shape is not disturbed. A practical demonstration of this theory was recently successfully completed over a 1,000,000-km optical fiber with a 10-Gb/s pseudorandom ASK soliton stream. This signal was transmitted 2000 times around a 500-km recirculating loop incorporating Er-doped amplifiers.

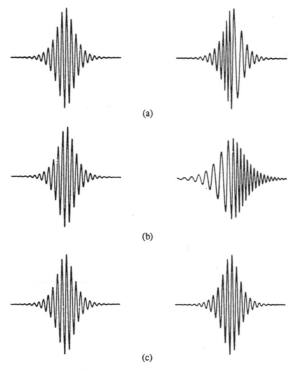

(a)

(b)

(c)

Figure 9.20 Soliton formation.

9.6.4   Local access networks (subscriber networks)

Optical fiber deployment in the telecommunications network is progressing rapidly. Although the customer local loop is the shortest distance in the network, it will be the last part of the optical system to be connected. This is clearly because of the enormous quantity of customers. Although the distance from each customer to the nearest CO is small (a few kilometers maximum), the huge number of customers means that very large quantities of fiber are necessary. But, even more problematic is the fact that because every customer is in a different location, the time and cost required to make all of the installations is formidable.

It is only when every subscriber is connected to the CO with fiber that a true B-ISDN will become a reality. At that time, there can be high-bit-rate (more than 2 Mb/s) interactive computer links from the home to a subscriber or business anywhere in the world. Video transmission with high-fidelity sound will be commonplace. In several countries in the developed world, there are already plans for installing optical fiber into the subscriber loop. At present, most of the fiber to the end user is in the form of experimental connections or connections to large businesses that use high-bit-rate computers.

First, as in Fig. 9.21, the CO (exchange) is connected to its remote nodes (or remote line concentrators, or remote line units). These remote nodes distribute cables out to the service access points, which then extend the service cables the last few meters to the customer premises. The objective is eventually to use optical fiber cable for all three of these segments, which is FTTH. Economic considerations indicate that the last segment from the service access point to the customer premises will be the last to be installed. The phrase *fiber to the curb* has been coined to describe the interim solution before the local loop becomes completely optical fiber. There are several interconnection options, described as topologies, when considering the local loop, and each segment might have a different topology.

Figure 9.22 shows the three categories of topology: the star, bus, and ring. The star configuration is generally considered to be the best topology for the service access point to the customer premises segment because it provides the highest possible bandwidth to the customer while maintaining a high degree of privacy and security. In the feeder and distribution segments, the ring or star topologies can be used for voice and data services, bus or star topologies can be used for distributive video, and the star topology is preferred for the point-to-point, high-bandwidth, switched video services. There are many pos-

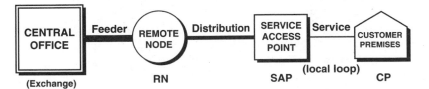

Figure 9.21 Subscriber loop (local access).

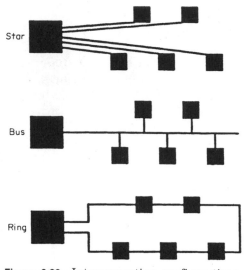

Figure 9.22 Interconnection configurations. (*Reproduced with permission from IEEE © 1989, Lin, Y.-K. M., D. R. Spears, and M. Yin, Fiber-Based Local Access Network Architectures, IEEE Communications Magazine, October 1989, pp. 64-73.*)

sible combinations of star, bus, and ring topologies for the local loop (local access network), and several examples are illustrated in Fig. 9.23.

The star-star configuration offers the highest flexibility for supplying voice, data, and video (broadcast or interactive) services. At increased expense and complexity, the reliability can be enhanced by supplying route diversity as in Fig. 9.23*b*. The bus-star and bus-bus topologies of Fig. 9.23*c* can be used for video program distribution as well as voice and data communications. Additional measures would be required to ensure security and privacy. Figure 9.23*d* is a ring-star configuration, where the ring in the feeder segment offers some protection against equipment failure or cable cuts by effectively providing route diversity. Figure 9.23*e* combines the bus-star for video distribution and star-star for voice and data communications.

Having decided upon the topology or topologies to use for the interconnections between the CO and the subscribers, the next technical problem to solve is the *style of transmission*. Should the system incorporate WDM, or FDM with heterodyne/homodyne (coherent) detection? Figure 9.24 shows an early optical fiber systems application which used WDM to provide each customer with a dedicated transmit and receive wavelength. This technique allows a single optical fiber feeder to be used between the CO and the remote node or service access point. No wavelength filtering is necessary at the customer premises because each customer receiver has only one allocated wavelength specifically assigned by the CO. The number of subscribers that can be connected by this system is relatively small. For a given available optical bandwidth (e.g., 1200 to 1300 nm) the number of usable channels is limited by the filter characteristics of the WDM

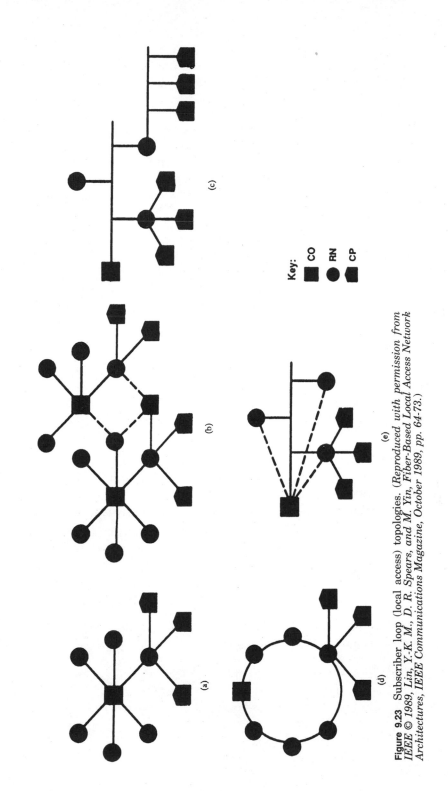

Figure 9.23 Subscriber loop (local access) topologies. (*Reproduced with permission from IEEE © 1989, Lin, Y.-K. M., D. R. Spears, and M. Yin, Fiber-Based Local Access Network Architectures, IEEE Communications Magazine, October 1989, pp. 64-73.*)

Key:
CO ■
RN ●
CP ⬟

devices. If the filter passband is typically 0.4 nm, the minimum value of channel separation is about 2 nm. Narrower channel spacing could cause intolerable adjacent channel crosstalk. Therefore the maximum number of usable channels in this case would be about 50.

FDM uses a heterodyne multiplexing technique at the transmitting end to form a composite signal to be transmitted. At the receiving end, a heterodyne or homodyne receiver would be used. Because this can enhance the channel spacing to about 5 GHz (0.04 nm), the number of usable channels is therefore increased to 2500 for an optical passband of 100 nm (e.g., 1.25 to 1.35 µm); however, as mentioned earlier, the microwave and infrared technologies are not yet sufficiently developed to exploit such a large number of channels. Also, FDM technology is more expensive than WDM.

Examples of coherent systems are shown in Fig. 9.25. In Fig. 9.25a, this multichannel network could be used for CO interconnections or LAN to metropolitan area network (MAN) computer interconnections. Each transmitter has a specific optical frequency and is connected to the passive star coupler, which combines the transmitted signal and distributes the composite signal to all receivers. Each receiver can be tuned to any of the transmitted frequencies, thereby allowing communication between any of the transmitters and receivers within the network. Figure 9.25b is a distribution network in which many optical signals are combined by a star coupler and then passively split to a large number of subscribers. Each subscriber can tune to one of the many transmitted signals using a coherent detection receiver. This network has applications for TV distribution or B-ISDN and is often called a *broadcast and select network*.

9.6.5 Local area networks

LANs, which are primarily associated with computer interconnectivity, are evolving rapidly. These are networks within a building or group of buildings such as business premises, factories, educational institutions, etc. The bit rate

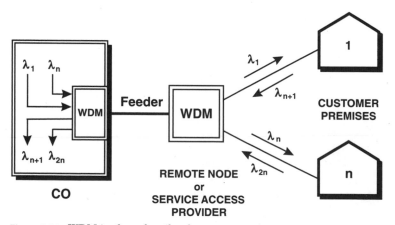

Figure 9.24 WDM in the subscriber loop.

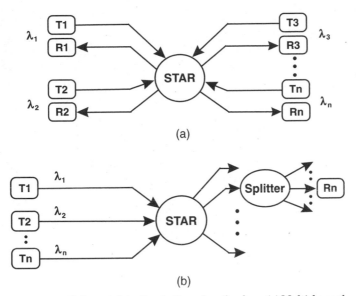

Figure 9.25 Coherent detection in the subscriber loop. (*a*) Multichannel network; (*b*) distribution network.

required for each application can differ significantly. For example, personal computer interconnection requires at least 1 Mb/s. Higher-performance workstation and file-server interconnection requires about 10 Mb/s. In the 100-Mb/s range are networks such as the *fiber distributed data interface* (FDDI). For supercomputer and complex graphics, bit rates in excess of 1 Gb/s are necessary. The LAN has some extra complexities not experienced in the long-distance point-to-point systems. Optical fiber cable is not simply a direct substitute for copper cable to upgrade the capacity of a wire-based LAN. While the fiber-based systems have abundant transmission bandwidth, they also have a severe signal power limitation. The primary objective of most LANs is for any one of the network users to be able to communicate with any, or all, of the other users simultaneously. This is a significant increase in system complexity compared to the point-to-point connections predominant in most telecommunication systems. The LAN has an interconnectivity which is both point-to-point and simultaneous-broadcast in nature. This is accomplished without the use of switches.

There are presently two favored topologies for optical fiber LANs: the star network and the unidirectional bus network (Fig. 9.26). In order to exploit fully the available bandwidth of an optical fiber system, a LAN must not be (1) power limited, or (2) suffer an electronic bottleneck. The power transmitted by one user is split many times by, for example, an $N \times N$ star coupler, in order to have the required broadcast capability. Depending upon the topology, number of users, distance between users, etc., it might be necessary to incorporate optical amplifiers to boost the power prior to splitting, or before each receiver. The electronic bottleneck problem is particularly severe in the ring topology, where each node must process all of the traffic for the network. The electronic processing capability of the entire network is therefore limited by the electronic processing capability of one node. The obvious way to surmount this problem is to use multiple wavelengths instead of just one. WDM or the more sophisticated FDM techniques are being successfully applied to large user LANs.

9.7   Optical Fiber Equipment Measurements

The measurements made on telecommunications equipment are usually clearly divided into external plant equipment and exchange equipment. In the case of optical fiber transmission, for some measurements, cooperation is normally required between both outside plant and transmission personnel. The optical fiber line terminal equipment is obviously the domain of transmission engineers or technicians, but restoration of a fiber fault also requires outside plant personnel.

9.7.1   Cable breaks

If an optical fiber cable is cut because of building construction, severe weather conditions, earthquakes, or even sabotage, the location of the break must be established. Sometimes the location is obvious, but often it is not. The *optical time domain reflectometer* (OTDR) is used for this purpose. The optical fiber is disconnected from the line terminal equipment in the exchange, and connected instead to the OTDR instrument. A break in the fiber gives a plot that suddenly drops to infinite attenuation at the break point. The distance from the exchange to the break can then be measured, with a precision of less than 1 m.

When repairing the break, it is useful to have a voice link between the OTDR operator and the technician at the repair site. In remote locations a pair of VHF radios might be necessary. Fiber talk-sets are more useful, as they are hands-free devices that operate over 80 km or more on one single-mode fiber. The set either connects to the ends of the fiber, or midspan access can be made using a clip-on coupler. When the fiber has been spliced, or even just prior to splicing or mechanically fixing it in position, the OTDR operator can inform the splicer when the

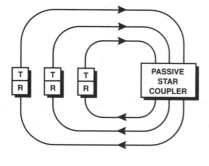

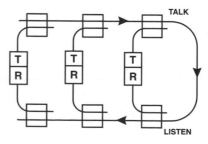

(a)   **(b)**

Figure 9.26   LAN topologies.

fiber alignment is adequate. On completion of the fusion or mechanical splice, the OTDR operator can indicate to the splicer whether the splice loss is adequately low, or whether the splice needs to be redone. Fiber restoration followed by measurement of the received power at the distant exchange could also be carried out to establish whether the splice is good or bad by comparing the power received after the splice with the value obtained just after cable installation.

9.7.2 Line terminal measurements

The optical fiber LTE is simple to operate and maintain compared to other transmission systems such as microwave radio. After the usual supply voltage checks, the main measurements to be made are:

1. Optical output power

2. Optical received power

3. Optical output pulse waveform (extinction ratio)

4. BER and/or BBER

5. Electrical output pulse waveform

6. Alarm indication and protection switching

The *optical output power* is simply a periodic check to ensure that the LD is healthy. The transmitter output is disconnected from the optical fiber line, and a hand-held power meter can be connected to the transmitter output. Obviously, if the output power is rated too high for the power meter, an attenuator must be inserted to protect the power meter. A typical output power value is 0 to10 dBm, but future LDs might be 26 dBm or higher. A similarly simple measurement of *optical received power* is made by disconnecting the fiber into the receiver to check that the anticipated received power level is correct and there has been no degradation over a period of time (weeks or months). Incidentally, on very short spans, the received power might be too high for the photodetector (typically more than −12 dBm) and an attenuator might need to be permanently installed to protect the photodetector.

As the bit rates move higher, an important parameter for optical systems is the *extinction ratio*, which is defined as $10 \log_{10} (A/B)$. A is the average optical power level for ones, and B for zeros. A typical value specified for the extinction ratio is 10 dB. BER is a relatively easy measurement to make, and converting to the more unfamiliar block-error-based *error-second ratio* (ESR) measurement will take some time. However, the latest SDH equipment makes in-service measurements of BER or ESR very easy. Jitter, which is increasingly important at the higher bit rates, is often not so easy to measure. The eye diagram, which has been used as a subjective test for many years, is now becoming more scientific. This is because ITU-T has recommended a mask for STM-n rates as indicated in Fig. 9.28*b*.

SDH equipment has extensive surveillance features, which are carried in the frame overhead bits. The resulting TMN gives easy in-service monitoring of

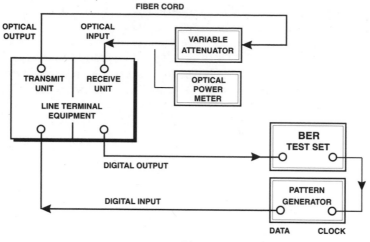

(a)

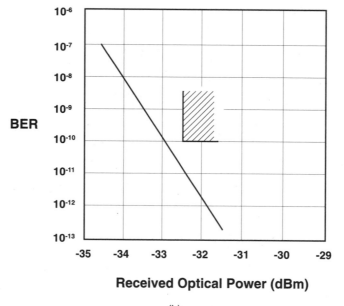

(b)

Figure 9.27 (*a*) BER measurement. (*b*) Typical graph of BER against optical received power for APD direct detection.

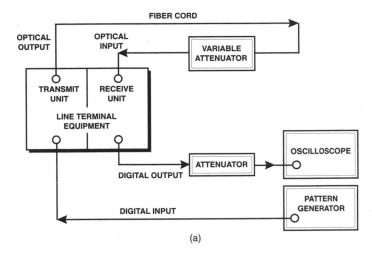

(a)

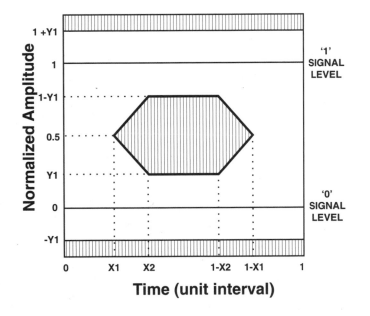

Rate (Mb/s)	*X1*	*X2*	*Y1*
155.52	0.15	0.35	0.20
622.08	0.25	0.40	0.20
2488.32	0.40	0.40	0.25

(b)

Figure 9.28 (a) Electrical output pulse waveform measurement. (b) Electrical output pulse waveform mask for several measurement SONET bit rates.

the main parameters which indicate how well the system is functioning. Loss of incoming or outgoing signals to a network element, and AIS and BER degradation early warnings are all reported on monitors in various locations around the network.

There is still a large amount of PDH equipment installed and, during periodic maintenance or restoration after fault location, many of the measurements which have to be made are out-of-service measurements. The out-of-service BER measurement of the LTE is performed as indicated in Fig. 9.27a. The optical output is looped back to the optical input and an attenuator is included in the loop to simulate the fiber link and also to prevent damage to the optical detector. The pattern generator, in the case of a 2.488-Gb/s system, provides a PRBS ($2^{23} - 1$), NRZ, 2.488-Gb/s signal into the LTE, and an error detector measures any errors. For the overall system BER performance to be adequate, it is necessary to ensure the lowest possible BER for the back-to-back LTE (i.e., Tx to Rx loopback) without the presence of the link fiber or regenerators. For a 2.488-Gb/s, intensity-modulated, 1.3-μm LD transmitter and an APD detector in the receiver, a good line terminal BER should typically be better than 10^{-12} for a received optical input signal of about −31 to −32 dBm (without preamplification). Figure 9.27b shows the desired graph of BER against optical received power level. Some systems have a built-in error monitor which can be conveniently checked at this juncture.

The 2.488-Gb/s *electrical output pulse waveform* can be checked using the test setup of Fig. 9.28a. With the LTE in the loopback mode, the attenuator is adjusted for −31-dBm optical input power. The pattern generator feeds an all-1s or all-0s signal into the LTE at 2.488 Gb/s, and the output waveform is observed after a complete path through the LTE. The 1s and 0s should conform to the masks of ITU-T Recommendation G.957 (see Fig. 9.28b).

Finally, the usual LTE alarm indications such as those in the following typical list are checked for correct operation:

1. Power supply failure

2. Loss of incoming 2.488-Gb/s signal

3. Loss of outgoing 2.488-Gb/s signal

4. Loss of incoming optical signal

5. Loss of outgoing optical signal

6. Loss of frame alignment

7. Degradation of BER above 10^{-10} or even 10^{-12}

8. CPU failure

9. Increase in LD drive current

10. Detection of AIS signal

11. Remote alarm detection

Chapter 10

Data Transmission and the Future Network

Technology is advancing at a frantic pace. Data communications is no longer restrained by the minute bandwidth of a telecommunications voice channel (less than 4 kHz). Eventually, when optical fiber is fed all the way to the customer premises, bandwidth problems will disappear. In the meantime, bandwidth is always an important consideration. Also, data transmission discussions go beyond the bounds of the transmission media and terminal equipment, and it is necessary to have a feel for the network as a whole. In this chapter more than any other, the reader will appreciate that the industry is moving toward a digital telecommunications engineer or technician job description, rather than separate switching and transmission disciplines.

As the convergence of telecommunications and data communications accelerates, the term *multimedia* is used to describe applications that incorporate a combination of voice, data, and video. For example, workstation teleconferencing, image retrieval or transfer, and voice electronic mail are multimedia applications that require a wide range of bit rates. The increased acceptance of the latest generation of packet switching instead of conventional circuit switching is central to the convergence of data communications with telecommunications. Some of the advantages of packet switching are (1) variable bit-rate service capability, (2) multipoint-to-multipoint operation, (3) service integration, and (4) resource sharing. These facilities result in a lower switching cost per customer.

The high-bit-rate local, metropolitan, and wide area networks (i.e., LANs, MANs, and WANs), which make up the Internet, are expanding their horizons by becoming broadband multimedia networks instead of just data networks. This has necessitated the development of fast packet-switching protocols. *A protocol is a set of rules that control a sequence of events which take place between equipment or layers on the same level.* The asynchronous transfer mode (ATM) is now being used in this respect to allow voice, data, and video multimedia operation. These protocols speed up packet-switching procedures by eliminating or limiting error control and flow control overheads. Eventually, such protocols will probably merge with the Internet protocol.

When one first encounters the subject of data communications, there appears to be a plethora of standards, and some believe there are too many. In the literature, it seems that references are made to ISO, ANSI, or IEEE Standards or ITU-T Recommendations in every sentence. As one becomes more familiar with data communications at the engineering or technician level, the need for very stringent standards quickly becomes apparent. With so many manufacturers in the business, it is absolutely essential that equipment is fabricated to ensure compatibility with all the other equipment with which it is required to interface within the network. In the early days of personal computers it was almost impossible to interface one computer with another. Each manufacturer was trying to establish its own de facto standard and therefore corner a major share of the market. With powerful software, this problem is now largely overcome. The necessity for standardization is very important to the consumer, and telecom administrations are adopting international standards to make life easier, even though many engineers might not appreciate the fact.

10.1   Standards

Standards aim to specify technical characteristics that allow compatibility and interoperability of equipment made by many different manufacturers. In addition, they recommend certain minimum levels of technical quality based on the state of the art of present-day technology.

The International Standards Organization (ISO), which was founded in 1947 and has its headquarters in Geneva, Switzerland, has become a dominant leader in the quest for global standardization of data communications. Together with the ITU-T section of the ITU, the ISO has developed a significant number of standards that are gradually receiving global acceptance.

The ISO (with ITU-T collaboration) has established a seven-layer network architecture as illustrated in Fig. 10.1. Each layer can be developed independently, provided the interface with its adjacent layers meets specific requirements.

Layer 1: The physical level.   This layer is responsible for the electrical characteristics, modulation schemes, and general bit transmission details. For example, ITU-T Recommendations V.24 and V.35 specify interfaces for analog modems, and X.21 specifies the digital interface between data terminal equipment and synchronous mode circuit-terminating equipment.

Layer 2: The link level.   This level of protocol supplements the Layer 1 raw data transfer service by including extra block formatting information to enable such features as error detection and correction, flow control, etc.

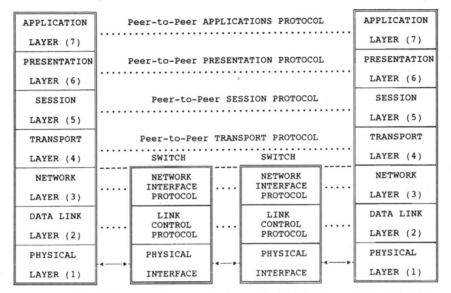

Figure 10.1 The OSI seven-layer protocol model. (*Previously published by Peter Peregrinus Limited, Data Communications and Networks 2, by Brewster, IEE Telecommunication Series 22, 1989.*)

Layer 3: The network control level. The function of this layer is to provide addressing information to guide the data through the network from the sender terminal to the receiver location, in other words, call routing. ITU-T Recommendations X.21 and X.25 are the protocols involved. X.21 refers to a dedicated circuit-switched network, and deals with the signaling and data transmission from terminals to switches. The X.25 protocol refers to network signaling, call routing, logical channel multiplexing, and flow control for terminals operating in the packet mode and connected to public data networks by dedicated circuits.

Layer 4: The transport level. The fourth to seventh layers are devoted to network architecture. The transport layer is concerned with end-to-end message transport across the network. It takes into account the need to interface terminals with different networks (circuit switched or packet switched). Further error protection might be included at this layer to give a specific quality of service. Details can be found in ITU-T Recommendations X.214 and X.224 and ISO 8072. Layers 1 to 4 relate to the complete communication service.

Layer 5: The session level. Layers 5, 6, and 7 relate to applications. Layer 5 protocols are concerned with establishing the commencement (log-on) and completion (log-off) of a "session" between applications. It can establish the type of link to be set up, such as a one-way, two-way alternate, or two-way simultaneous link. ITU-T Recommendations X.215 and X.225 and ISO 8326 provide details of these protocols.

Layer 6: The presentation level. This layer is necessary to ensure each subscriber views the incoming information in a set format, regardless of the manufacturer supplying the equipment (e.g., ASCII representation of characters). Also, screen presentation of size, color, number of lines, etc., must have uniformity from one supplier to another. ITU-T Recommendations X.216 and X.226 and ISO 8823 refer to these protocols.

Layer 7: The application level.   Finally, Layer 7 is concerned with the interface between the network and the application. For example, the application could be a printer, a terminal, file transfer, etc. In this respect, ITU-T Recommendation X.400 relates to message-handling services and ISO 8571 to file transfer and management.

These seven layers might seem somewhat academic on first reading, but they are of value. They are what is known as the *Open Systems Interconnection* (OSI). It must be emphasized that OSI is not a protocol and does not contain protocols. OSI defines a complete architecture that has seven layers. It defines a consistent language and boundaries establishing protocols. Systems conforming to these protocols should be "open" to each other, enabling communication. OSI is nothing new per se. It just standardizes the manner in which communication is viewed. In fact, the protocols at all seven levels of OSI can actually be applied to any type of communication regardless of whether it is data or speech. For example, one can make an analogy between the seven levels and the processing of a conventional telephone call, as follows:

1. *Physical layer*. Concerns the sounds spoken into the telephone mouthpiece and heard in the receiver earpiece.
2. *Link layer*. Involves speaking when required and listening when necessary. A repeat is requested if something is misunderstood. The person at the other end is told to slow down if talking too fast.
3. *Network layer*. Concerns dialing the number and listening for the connection to be made. If the busy signal is given, dial again. On completion of the call, disconnect by replacing the handset.
4. *Transport layer*. Involves deciding which is the most cost-effective way to make the call (which carrier to use).
5. *Session layer*. Concerns deciding if one call will suffice or whether several will be necessary. Will more than one person need to be included? If so, who will control a subsequent conference type of call? Who will set up the call again if cut off?
6. *Presentation layer*. Establishes whether or not the two parties are speaking the same language.

7. *Application layer*. Concerned with: Am I speaking to the right person? Who will pay for the call? Is it convenient to converse now or should the call be placed again later? Does the called party have the means to make notes on the conversation?

The Standards presented above are summarized by the ITU-T in its Recommendations that are published and frequently updated. The various different series of Recommendations follow:

V Series: "Data Communication over the Telephone Network"

V.1—V.7	General recommendations
V.10—V.34	Interfaces and voice band modems
V.36—V.38	Wideband modems
V.41—V.42	Error control
V.50—V.57	Transmission quality and maintenance
V.100—V.230	Interworking with other networks

X Series: "Data Communications Network Interfaces"

X.1—X.15	Services and facilities
X.20—X.36	Interfaces (X.21 is central)
X.50—X.82	Transmission signaling and switching
X.92—X.144	Network aspects
X.150	Maintenance
X.180—X.181	Administration arrangements
X.200—X.296	Open Systems Interconnection
X.300—X.340	Internetworking between networks
X.350—X.353	Satellite data transmission systems
X.400—X.485	Message handling systems
X.610—X.691	OSI networking and system aspects
X.700—X.790	OSI management
X.800—X.833	Security
X.851—X.882	OSI applications

Sometimes two or three Recommendations will address the same issue, but in a totally different manner, and they will be called, for example:

X.21
X.21 bis (meaning second recommendation)
X.21 ter (meaning third recommendation)

G Series: "Digital Networks Transmission Systems and Multiplexing"
G.700—General aspects of digital transmission systems, terminal equipments (as in Chap. 2)

- PCM

- Speech compression (encoding)

- Higher-order multiplexers or systems
- Synchronous digital hierarchy

G.800—Digital networks

- Network performance (quality and availability targets)
- Design objectives for PDH and SDH systems

G.900—Digital sections and digital line systems

- FDM
- Coaxial and optical fiber cable systems

I Series: "Integrated Services Digital Network (ISDN)"

I.100-I.150	General
I.210-I.259	Service capability, teleservices, bearer and supplementary services
I.310-I.376	Overall network aspects and functions
I.410-I.465	ISDN user-network interfaces
I.500-I.580	Internetwork interfaces
I.731-I.751	B-ISDN equipment aspects (ATM)

10.2 Data Transmission in an Analog Environment

The principles involved in data transmission are basically the same as for pulse transmission, already encountered in the earlier chapters. A brief review of the problems and how they are overcome for data transmission in an analog environment will follow.

10.2.1 Bandwidth problems

Ever since the start of data communications, perhaps the greatest impediment to progress has been the constant battle to solve the problems associated with operating with very limited bandwidth. The public telephone network was not designed to deal with isolated, individual pulses, so it is even less able to cope with high-speed bit streams. Transmitting pulses through the *analog* telephone bandwidth of 300 to 3400 Hz causes severe pulse distortion because of (1) loss of high-frequency components, (2) no dc component, and (3) amplitude and phase variation with frequency. Chapter 3 dealt with the effects of pulse transmission in a bandlimited channel.

Ultimately, the bandwidth problem will simply disappear when optical fiber is taken to every home. Because that situation is probably at least 10 to 15 years away, what can be done in the meantime? Packet-switching networks are springing up and special leased lines are available, but the fact remains that in a mixed analog/digital environment, some temporary solutions are needed. As many Internet users will confirm, retrieving data can be painfully slow, depending on the data throughput. The Internet itself will grind to a halt if increased bandwidth is not made available at a rate that can cope with the rapidly expanding Internet population.

To explore the fascinating world of data communications it is instructive to start by considering the translation of digital signals to a condition suitable for transmission over analog voiceband telephone circuits. This is performed by the famous device called the *modem*. Although these devices will become a dying breed by the time broadband ISDN is universal, they are worthy of a brief mention as they are still very much alive today. The asymmetric digital subscriber line (ADSL) equipment is also often referred to as a modem and will be discussed in detail later.

10.2.2 Modems

A variety of voiceband modems has evolved, each operating at a specific bit rate. The ITU-T V series of Recommendations relates to these modems, and Table 10.1 summarizes the important details. In addition, Recommendation V.24 specifies the various data and control interchange circuit functions operating across the modem-terminal interface. Also, V.25 specifies how automatic answering or calling modems should operate. The original stand-alone box modems are now mainly replaced by chip or card modems that are offered as an extra item within a PC. This temporary situation will exist only until the full deployment of broadband ISDN.

Modulation methods. For data transmission beyond 10 to 20 km (in the trunk circuits), line distortion necessitates translating the digital baseband signal into the usable portion of a voiceband telephone channel. This is done by using

TABLE 10.1 Summary of V-series ITU-T Recommendations for Modems

ITU-T Rec.	Transmission rate, b/s	Modulation type	Additional information
V.21 (FDM)	300	FSK	2-wire full duplex
V.22 (FDM)	1200	PSK	2-wire full duplex
V.22bis(FDM)	2400	QAM	2-wire full duplex
V.23	1200/75	FSK	2-wire full duplex (1200/75); half duplex (1200/1200)
V.26,V.26bis	2400	PSK	Half duplex
V.26ter			2-wire full duplex + echo cancellers
V.27,bis,ter	4800	PSK	Half duplex
V.29	9600	QAM	Half duplex
V.32			2-wire full duplex + echo cancellers
V.32 bis	14400	QAM (+ trellis)	2-wire full duplex
V.33	14400	QAM (+ trellis)	4-wire leased
V.34	28800	QAM (+ trellis)	2-wire full duplex; asymmetric symbol rates

the baseband signal to modulate a carrier frequency centered within the voiceband range. FSK is the simplest method used for low-bit-rate modulation in modems up to 1200 b/s (ITU-T Recommendations V.21 and V.23). For high bit rates, PSK is suitable. Four-state differential PSK is used for 2400-b/s modems (V.26), and this is extended to eight-state PSK for the 4800-b/s modem (V.27). The higher-bit-rate modems employ the more bandwidth-efficient QAM as designated by V.29/V.32 for 9600 b/s, V.32 bis for 14,400 b/s, and V.34 for 28.8 kb/s. Multilevel trellis coding is used on V.32 bis and V.34, with fall-back rates down to 7.2 and 2.4 kb/s respectively, on poor connections. Other performance-enhancing techniques discussed in Chap. 3, such as Viterbi detection, have been incorporated in recent modem designs.

Higher-speed modems are coming to the market (e.g., 56 kb/s), which are getting close to the 64-kb/s ISDN B-channel. However, standardization (and therefore compatibility) can be a problem when a new modem first comes to market, because some manufacturers are ahead of the standardization process. No doubt, all manufacturers would like their designs to become the de facto industry standards.

Two-wire modems.   If two-wire transmission is used instead of four-wire, for full-duplex operation there must be a mechanism for organizing the two directions of transmission. FDM is an obvious choice, whereby the voiceband is split up into two halves, one for GO and the other for RETURN. This is done for 300 b/s (V.21), 1200 b/s (V.22), and 2400 b/s (V.22 bis using 16-QAM), but bandwidth constraints require other techniques for the higher bit rates. Echo cancellation is preferable to FDM for high bit rates, in which case the full voice bandwidth is used for both the transmit and receive directions. In the 9600-b/s modems (V.32), an adaptive canceller is used in each receiver to eliminate interference caused by its own transmitter.

The future.   An extension of modem technology is the asymmetrical digital subscriber line (ADSL). As described in Sec. 10.7, this is a technical stunt that allows several megabits per second of data to travel through the POTS copper wire infrastructure. This is clearly a major leap from the 9.6- to 56-kb/s progress seen during the past decade.

10.3   Packet Switching

Since its introduction in the early 1970s, packet switching has received widespread acceptance. Public networks have been constructed in most developed countries and many developing countries. The internetwork ITU-T X.75 protocol provides for interlinking of national networks at an international level. The ITU-T X.25 Recommendation is the original standard for packet-switching architecture.

Packet switching has several advantages over conventional circuit-switched networks. Figure 10.2 highlights the difference between the two. The circuit-switched network maintains a fixed bandwidth between the transmitter and receiver for the duration of a call. Also, the circuit-switched network is bit stream transparent, meaning it is not concerned with the data content or error-checking process. This is not the case for packet switching, where bandwidth is allocated dynamically on an "as required" basis. Data is transmitted in packets, each containing a header that contains the destination of the packet and a tail, or footer, for error-checking information. Packets from different sources can coexist on the same customer-to-network physical link without interference. The simultaneous call and variable bandwidth facilities improve the efficiency of the overall network. The buffering in the system which allows terminals operating at different bit rates to interwork with each other is a significant advantage of packet switching. The obvious disadvantage is the extra dimension of complexity with respect to the switches and network-to-customer protocol.

Furthermore, in certain circumstances, packet switching has several advantages over other methods of data communication:

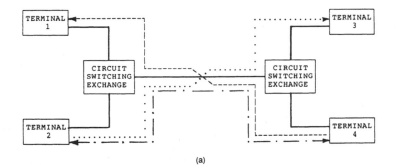

(a)

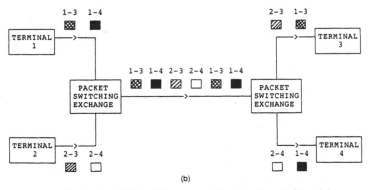

(b)

Figure 10.2 Packet switching (b) compared to circuit switching (a).

1. Packet switching might be more economical than using private lines if the amount of traffic between terminals does not warrant a dedicated circuit.

2. Packet switching might be more economical than dialed data when the data communication sessions are shorter than a telephone call minimum chargeable time unit.

3. Destination information is contained in each packet, so numerous messages can be sent very quickly to many different destinations. The rate depends on how fast the data terminal equipment (DTE) can transmit the packets.

4. Computers at each node allow dynamic data routing. This inherent intelligence in the network picks the best possible route for a packet to take through the network at any particular time. Throughput and efficiency are therefore maximized.

5. The packet network inherent intelligence also allows graceful degradation of the network in the event of a node or path (link) failure. Automatic rerouting of the packets around the failed area causes more congestion in those areas, but the overall system is still operable.

6. Other features of this intelligence are error detection and correction, fault diagnosis, verification of message delivery, message sequence checking, reverse billing (charging), etc.

10.3.1 Packet networks

The intelligent switching nodes within packet-switching networks, in general, have the following characteristics:

1. All data messages are divided into short blocks of data, each having a specific maximum length. Within each block there is a header for addressing and sequencing. Each packet usually contains error control information.

2. The packets move between nodes very quickly and arrive at their destinations within a fraction of a second.

3. The node computers do not store data. As soon as a receiving node acknowledges to the transmitting node that it has correctly received the transmitted data, by doing an error check, the transmitting node deletes the data.

For packet switching, a data switching exchange (DSE) is required, which is a network node interlinking three or more paths. Data packets move from one DSE to another in a manner that allows packets from many sources and to many destinations to pass through the same internode path in consecutive time sequence. This, of course, is in contrast to the circuit-switched network that seizes a specific path (link) for the duration of the message transfer. Figure 10.3 shows how the DTE, data circuit-terminating equipment (DCE), and DSE are related within a packet network. The boundary of the network is usually defined as the point where the serial interface cable is connected to the DCE. This point is also often referred to as the *network gateway*.

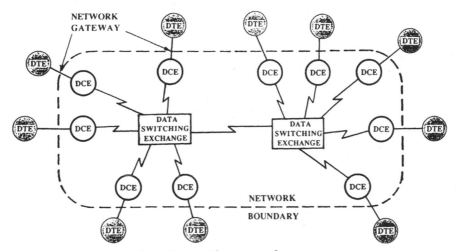

Figure 10.3 An example of a packet-switching network.

10.3.2 The X.25 protocol

ITU-T Recommendation X.25 is recognized by the data communications fraternity as one of the most significant milestones in the evolution of networking architecture. Within the X series of recommendations, X.25 specifies the physical, link, and network protocols for the interface between the packet-switching network and the DTE/DCE at the gateway. Figure 10.4 clarifies the packet-switching network architecture by comparing it to the analogous situation for a regular telephone local loop.

The DTE is connected to the DCE, which is equivalent to the subscriber telephone connection to the central office. X.25 does not describe details of the DTE or the packet data network equipment construction, but it does specify, to a large extent, the functions the DTE and packet network must be able to perform. The X.25 Recommendation specifies two types of service that should be offered by a carrier: (1) virtual call (VC) service and (2) permanent virtual circuit (PVC) service, as described in the following.

Virtual call service. Prior to the transmission of any packets of data in a VC service, a virtual connection must first be set up between a calling DTE logical channel and the destination DTE logical channel. This is analogous to placing a telephone call before beginning the conversation. Special packets having specific bit sequences are used to establish and disconnect the virtual connection. Having made the virtual connection, the two DTEs can proceed with a two-way data transmission until a disconnect packet is transmitted. The term virtual is used because no fixed physical path exists in the network. The end points are identified as logical channels in the DTEs at each end of the connection, but the route for each packet varies depending on other packet traffic activity within the network.

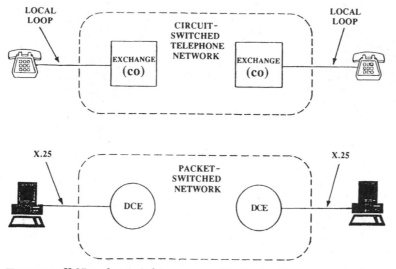

Figure 10.4 X.25 packet switching compared to the local loop.

Any DTE can have many active logical channel numbers simultaneously active in the same X.25 interface. Having established a virtual connection, for example between DTE logical channel X and DTE logical channel Y, the header in each data packet will ensure the data from X reaches its destination at Z via the intelligent nodes N in the network. An example of packet routing in virtual circuits is illustrated in Fig. 10.5, using logical channel numbers. The network N stores the following information for the duration of the call setup and then deletes it when the call is cleared:

- 16 on link XN, connected by node N to 24 on link NZ
- 33 on link XN, connected by node N to 8 on link NY
- 2 on link ZN, connected by node N to 17 on link NY

In this manner, packets originating at any terminal are routed to the desired destination.

Permanent virtual circuit service. A PVC service is similar to the VC in that there is no fixed, physical end-to-end connection, and the intelligent network provides the data transfer between the logical channels of any two DTEs. A PVC is analogous to a private line service. Other PVC and VC similarities are (1) the network has to deliver packets in the precise order in which they were submitted for transmission, regardless of path changes due to network loading or link failure, and (2) there is equivalence of network facilities, such as error control and failure analysis.

The difference between the PVC and the VC lies mainly in the method of interconnection. A PVC is set up by a written request to the carrier providing the packet-switching network service. Once a PVC has been set up, the terminals need only transmit data packets on logical channels assigned by the carrier. Each DTE can have many logical channels active simultaneously within the X.25 interface. This means that some channels might be used as PVCs and others used as VCs at the same time. Another written request is required to break the connection. Note that there are no unique packets transmitted by the DTEs to set up or break a connection. Finally, the cost of a PVC compared to a VC is analogous to private line service versus a long-distance telephone service.

Packet formation—frame format.   Figure 10.6 illustrates the X.25 DTE and DCE data packet format. At least the first three octets, as in Fig. 10.6a, are for the header. The first octet contains the logical channel group number and the general format identifier. This is followed by the second octet, which gives the logical channel number information. Each X.25 gateway interface can support a maximum of 16 channel *groups* and each group can contain 256 logical channels, totaling 4096 simultaneous logical channels for each gateway. These can be distributed between VC and PVC services as set up by agreement between the customer and equipment provider. The third octet in the header contains two 3-bit data packet counters, P(S) and P(R), an M-bit, and a zero for the first bit. When M is a 1, this is an announcement that further data packets will follow and these packets are considered to be a unit or entity. When M is 0, the unit is finished.

The data packet counters, P(S) and P(R), start at 000 and count up to 111, and then start back at 000 again. A P(S) counter is included only in data packets,

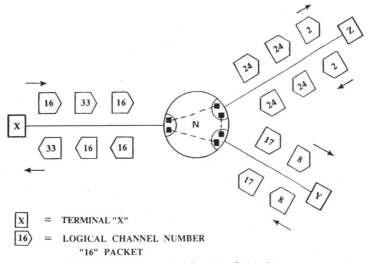

$\boxed{X}$ =   TERMINAL "X"

$\boxed{16\rangle}$ =   LOGICAL CHANNEL NUMBER
              "16" PACKET

Figure 10.5 Packet-switching routes for virtual circuits.

and the P(S) is sent across the interface from DTE to DCE and from DCE to DTE at the network gateway. P(R) is defined as the number of the next expected value of P(S), in the next packet coming from the other direction, on that particular logical channel. To make sense of that, observe Fig. 10.7. This is a time sequence for a full-duplex transmission of packets from DTE to DCE and vice versa. Each block is a packet, and the values of P(S) and P(R) are shown for each packet header. P(S) is easy, because it simply increases in numerical order. P(R), on the other hand, is more complicated. It has values that are updated only after the previous P(S) has been received, free of errors.

The user data then follows the header and there are 128 octets of user data in a standard X.25 data packet (modulo 8). Figure 10.6*b* shows, by contrast, the header for a modulo 128 scheme. In this case, the P(S) and P(R) counters are each 7 bits long, and the header has at least four octets instead of three; and data packet sequences are numbered 0 to 127 instead of 0 to 7.

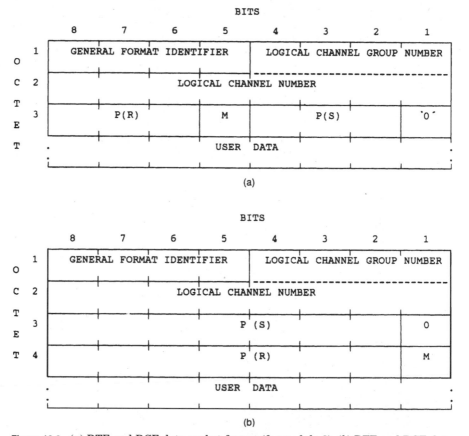

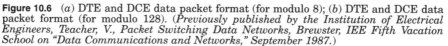

Figure 10.6 (*a*) DTE and DCE data packet format (for modulo 8); (*b*) DTE and DCE data packet format (for modulo 128). (*Previously published by the Institution of Electrical Engineers, Teacher, V., Packet Switching Data Networks, Brewster, IEE Fifth Vacation School on "Data Communications and Networks," September 1987.*)

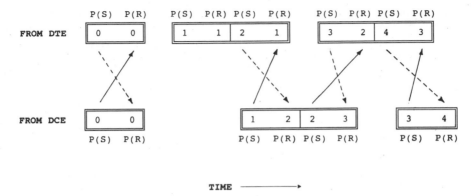

Figure 10.7 P(S) and P(R) values for a data packet sequence.

Although most packets contain data, a few packets are reserved for overhead functions such as control, status indication, diagnostics, etc. A data packet is identified by placing a zero for the first bit of the third octet.

Call request. The procedure for setting up and disconnecting a VC two-way simultaneous data transfer is shown in Fig. 10.8. This sequence of events resembles the various steps involved in placing and terminating a telephone call on a circuit-switched network. The contents of the packet header required to fulfill this procedure are clearly more complex than those of a typical data packet. Figure 10.9a shows the packet format for the call request/incoming call. Notice that the third octet in the header has a packet-type identifier. Figure 10.9b details the values of the third octet for the various identifiers for call request/incoming call, call accepted/call connected, clear request/clear indication, and DTE-initiated clear confirmation/network-initiated clear confirmation.

The header also contains destination addressing and VC/PVC selection information. The numerical values assigned to the logical channels at each end are never the same. The value assigned to the logical channels is chosen from a pool of numbers available when the call is set up.

X.25 summary. Figure 10.10 summarizes the three levels of the X.25 standard. This diagram shows the various stages in transmitting and receiving data blocks. First, packets are formed that include headers, and then frames are constructed at the link and then at the physical level, ready for sending to the line. On receipt of frames, the link level checks for errors, after which the data blocks are extracted from the packets. There is an enormous amount of information concerning the X.25 Recommendation and its implementation. Only a brief outline has been presented here.

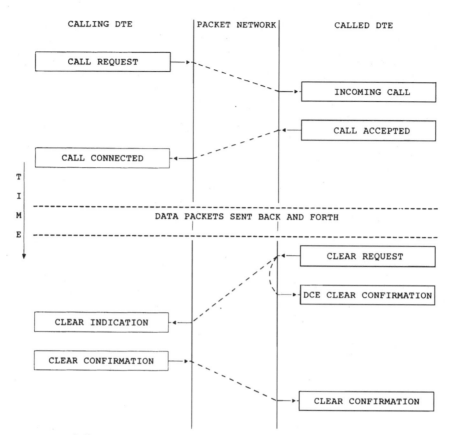

Figure 10.8 Call request/incoming call sequence for a virtual call.

10.3.3 Network interface protocols

Figure 10.11 summarizes how the various ITU-T recommendations apply to a packet-switching network and its associated equipment. X.25 specifies how the packet mode DTE is related to the packet-switching network. X.25 also specifies the relationship between the packet assembly and disassembly [PAD (X.3)] equipment and the packet-switching network. X.28 relates the X.3 PAD to the character mode (nonpacket mode) DTE. X.29 specifies the relationship between packet mode DTEs when communicating with character mode DTEs. This protocol also specifies the interconnection of a PAD to a packet mode DTE or a PAD to another PAD.

10.3.4 Optical packet switching

The throughput of information in the above packet-switching networks is quite adequate for text messages. However, with the ever-increasing computing power available from workstations and PCs, it is desirable to be able to send graphics through networks. To achieve graphics transfer in a reasonable amount of time, the networks need to be able to manipulate data at multigigabits-per-second bit rates. Optical packet switching promises to offer the solution to this problem.

A lot of research work is being directed toward improving multiwavelength N × N passive optical star components as a step toward the all-optical network. Already, the potential for terabits-per-second optical switches is on the foreseeable horizon. Optical packet-switching networks are still in their infancy, but it might not be long now before these networks emerge on the market with a very-high-performance capability.

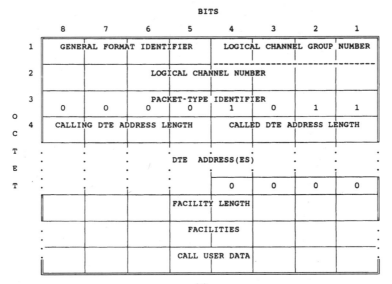

(a)

DTE TO DCE	DCE TO DTE	OCTET #3
CALL REQUEST	INCOMING CALL	0 0 0 0 1 0 1 1
CALL ACCEPTED	CALL CONNECTED	0 0 0 0 1 1 1 1
CLEAR REQUEST	CLEAR INDICATION	0 0 0 1 0 0 1 1
DTE-INITIATED CLEAR CONFIRMATION	NETWORK-INITIATED CLEAR CONFIRMATION	0 0 0 1 0 1 1 1

(b)

Figure 10.9 (*a*) Packet format for a call request/incoming call. (*b*) Identifier for call setup and clearance. (*Adapted from the Institution of Electrical Engineers, Teacher, V., Packet Switching Data Networks, Brewster, IEE Fifth Vacation School on "Data Communications and Networks," September 1987, p. 2/10.*)

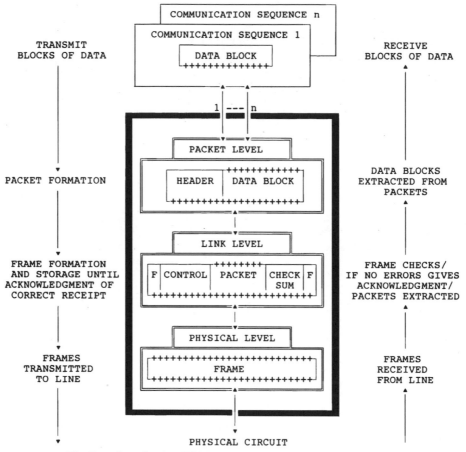

Figure 10.10 The first three levels of X.25.

10.4 Local Area Networks

The LAN was originally designed for interconnecting computers within an office. Since its inception, it has combined its original role with operation within the telecommunications network. The term *local* is then expanded to *wide* and *metropolitan* area networks. To understand the problems associated with linking LANs up to a telecommunications network, a description of the LAN is necessary.

The first generation of LANs had data rates in the region of 100 kb/s to 10 Mb/s and operated with wire cables over distances of a few kilometers. The most successful category of first-generation LANs is the carrier sense multiple access with collision detection (CSMA/CD) scheme. *Ethernet* is the term widely used to describe any CSMA/CD LAN. However, Ethernet is actually a proprietary standard founded by DEC, Intel, and Xerox which has kept its name because it was initially the major player in computer networking. Ethernet is an example of a de facto standard, although it is only slightly different from the ISO Standard 8802-3, and it allows distributed computing that encompasses file and printer sharing. The 10-Mb/s Ethernet capacity, which was initially over coaxial cable, provides good simultaneous user access to files in a file server; clearly, however, the more users, the slower the system. It might be necessary for each different department of an organization to have its own LAN, but with an interconnection to the other LANs via bridges. Figure 10.12 shows three different styles of LAN, each interconnected to the other two by bridges. For optimum operation, the amount of traffic flowing between LANs across the bridges should be small compared to the traffic within each individual LAN.

The second generation of LANs operate at about 50 to 150 Mb/s over distances up to 100 km, using optical fiber cables. This is known as the MAN and offers a wider range of services than the lower data rate LAN. Multimedia services such as voice, data, and video require even greater bit rates. The third-generation LANs are now evolving with bit rates up to and beyond 1 Gb/s. The well-established term *LAN* is now modified to *computer networking* in many texts.

10.4.1 CSMA/CD (Ethernet)

The CSMA/CD network is characterized by a single, passive, serial bus to which all nodes are connected. Each node detects the presence or absence of data on the bus. Consider the bus of Fig. 10.13 with n nodes connected. If 1 intends to transmit to 2, the bus must first be idle, at which point in time the

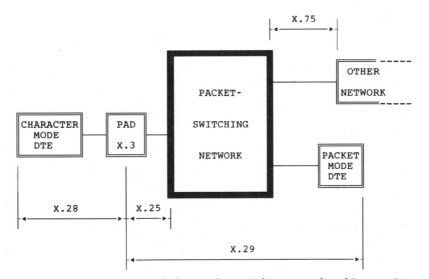

Figure 10.11 Interface protocols for a packet-switching network and its associated equipment.

CSMA/CD

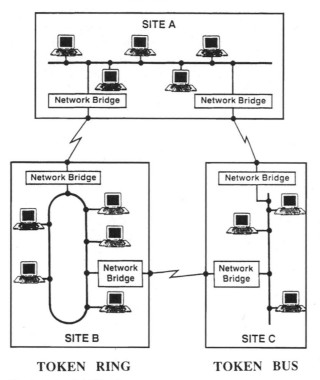

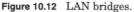

TOKEN RING   TOKEN BUS

Figure 10.12 LAN bridges.

data is transmitted in both directions from 1. On completion of its journey, the data is absorbed by the terminations at each end. En route, any node can copy the data into its storage buffers, provided its destination indicated in the packet's header matches the node's address. Suppose now as the data approaches 3, 3 decides to transmit because the bus is still idle from its perspective. There will be a resulting collision between data sent from nodes 1 and 3. This will be evident to 3 as soon at it starts receiving 1's packet. Similarly, it will only be evident to 1 as soon at it starts receiving 3's packet. The maximum detection time of a collision is when a collision occurs between the two nodes at the outer extremities of the bus and could be equal to twice the propagation time of the signal across the length of the bus. This is called the *slot time* (S), which is the time taken for any node to be certain of the transmission of the smallest packet size without collision. Any smaller packets that are received are discarded as collision fragments.

The operating bit rate B, distance D, and minimum packet size P are related by the simple equation: $P = BD$. This means that if the packet size is maintained at, say, 50 octets, an increase in the bit rate necessitates a reduction in the maximum length of the LAN. Calculations and experience determine these

LANs to be able to transmit at 10 Mb/s over a length of about 2.5 km, constructed by several segments interconnected by repeaters. Each segment is approximately 180 to 500 m. This length is determined by the need for collision detection. A typical large-scale Ethernet configuration with five segments and several repeaters is shown in Fig. 10.14.

For segments that are far apart, repeaters can be included to maintain a good signal. The maximum repeater distance is 500 m and only two repeaters are allowed between nodes, given a maximum point-to-point length of 1500 m. Conversely, stations (nodes) should not be spaced closer than 2.5 m and each segment must not exceed 100 nodes.

The collision problem associated with bus LANs is eliminated by moving to the cyclic control mechanism such as the token ring structure. Also, the installation of optical fiber cable is better suited to ring rather than to bus topologies.

10.4.2   Token rings

The *token ring* is a ring structure that interconnects many nodes (computer terminals). The token is a sequence of bits that circulates around the ring (Fig. 10.15). In its quiescent state, the token rotates around the ring continuously. When a user at one of the nodes wishes to transmit data, that node *captures* the token and then transmits a packet of data. On capture, the token might be seized or have 1 or more bits altered to form what is often referred to as a *connector*. The size of the packet transmitted depends on the time allocated, as defined by the type of token ring. The user receiving the packet can modify the token to establish contact, after which the token and packet are usually returned to the sender. The token is then released and captured by another user who wants to transmit. If no other user wants to transmit, the token can be captured again by the previous user for the next packet of data transmission. There is usually a hierarchy of priority for users that want to capture the token. Just after a node has finished transmitting a packet, it places its priority level in the token, for circulation. Nodes with higher priority levels can capture the token and the original user has to wait.

The token ring suffers less congestion than other LAN structures. Unlike Ethernet, there is no bandwidth wasted on collisions. The main disadvantage

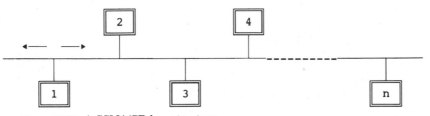

Figure 10.13   A CSMA/CD bus structure.

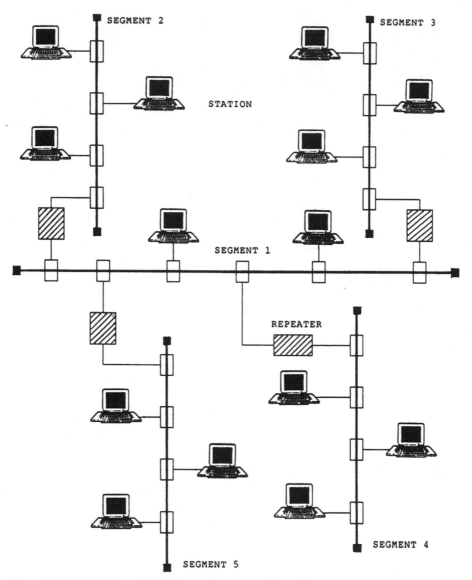

Figure 10.14 A typical multiple Ethernet system configuration.

of the token ring is the delay in accessing the ring, although the priority system ensures that the more important nodes get express service. In token ring systems where the token is released only on return of the outgoing packet, there can be a bandwidth problem for long, high-bit-rate rings. When a 10-Mb/s system operates on a 1-km ring, the ring delay is about 5 μs, which is equivalent to 50 bits. If the ring size is increased to 100 km and the bit rate to 100 Mb/s, as is typical for a MAN, the delay is about 50,000 bits. In such systems, early token release (i.e., immediately after packet transmission) is essential. Of course, a priority scheme cannot work in these circumstances. In this case, priority is based on the token rotation time (TRT). In an idle ring the token rotates very rapidly, and the rotation time increases as the ring becomes loaded with data traffic. Each node has a window of rotation times during which it can transmit, and if the rotation time is too slow for its priority, it must wait for the ring to be less congested before transmitting. A node decides whether or not transmission can take place by comparing the actual token rotation time with its target token rotation time (TTRT).

An unavoidable time wastage in the token ring LAN is the time between one node releasing the token and the next node capturing it. Again, this time period increases with the length of the ring.

Token ring technology started to mature in the mid-1980s, at which time IBM introduced its own version of a 4-Mb/s token ring based on IEEE Standard 802.5. During those early years, almost all token ring networks operated over shielded twisted-pair cable.

10.4.3 10Base-T (Ethernet)

A competition has emerged between Ethernet and token ring systems. In 1990, the IEEE plenary confirmed acceptance of Standard 802.3, which allows Ethernet 10-Mb/s data operation over *unshielded* twisted-dual-pair cable instead of coaxial cable. Its major attraction is the fact that it can be used to transmit data over *existing telephone cabling*. This standard, called 10Base-T, rapidly gained momentum and user acceptance. Using telephone lines for LAN interconnection opens up a new dimension of versatility, not to mention reduced installation time and overall cost.

The 10Base-T standard allows a maximum link distance of 100 m, using 24 AWG gauge wire, at an impedance within the range 85 to 110 Ω. The standard

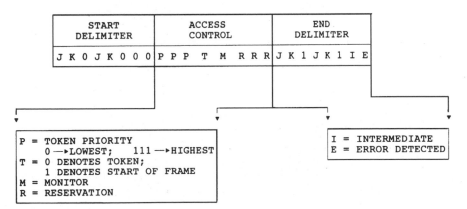

Figure 10.15 Structure of the token.

specifies functions such as link test, minimum receive threshold, transmit voltage levels, and high-voltage isolation requirements.

A very important aspect of the 10Base-T standard is that it can support star topologies. The star is desirable because each node is connected into the network by a separate cable, that is, between the desktop and the hub in the wiring closet. More cable is used than for other topologies, but that is more than compensated for by the improvement and simplification of centralized management and control of the network. For example, a fault can be immediately traced to a node and isolated without affecting other nodes. This degree of control cannot be achieved by bus or ring topologies. This U.S.-based standard will no doubt also penetrate the communications administrations in other parts of the world.

The competition between Ethernet and token ring systems led to the 4-Mb/s token ring manufacturers introducing 16-Mb/s token ring systems that operate on unshielded twisted-pair cables. Line noise, otherwise known as jitter (which is inherent in cyclic ring structures), severely affected performance of the first systems. VLSI circuits have been designed and manufactured to reduce jitter. Meanwhile, the competition between Ethernet and token ring designs continues.

10.4.4   Fiber distributed data interface

The transmission rate of LANs has been increasing steadily over the past decade. The limited bit rates of twisted-pair cable (shielded or unshielded) have already been exceeded with today's data throughput requirements. The next obvious step was to move to optical fiber cable, which has enormous bandwidth capability. FDDI was born in 1982, when ANSI formulated a standard for interconnecting computer mainframes using a 100-Mb/s token ring network.

Since then FDDI has received wide acceptance and now enjoys the distinction of a rapidly expanding market. Today's FDDI networks are usually dual counterrotating 100-Mb/s token rings, with early token release (Fig. 10.16). The two rings are called primary and secondary. This topology has evolved because it has the important advantage of isolating and bypassing faults. FDDI has now been adopted by the ISO. The standards governing FDDI operation are as follows:

1. Physical media dependent (PMD)

2. Physical protocol (PHY)

3. Medium access control (MAC)

4. Station management (SMT)

The PMD and PHY correspond to the OSI model Layer 1 (the physical link layer) which was developed by the ISO. MAC is a sublayer corresponding to

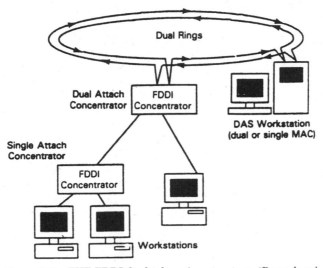

Figure 10.16   THE FDDI dual token ring structure. (*Reproduced with permission from IEEE, Strohl, M. J., High Performance Distributed Computing in FDDI CPE Directions, © 1991 IEEE LTS, May 1991, Fig. 1.*)

the lower half of Layer 2, which is the data link layer of the OSI model. SMT overlays the PMD and MAC layers.

The PMD gives details of the FDDI hardware. Each node within a ring is interconnected by a pair of multimode fibers, whose specifications are 62.5-μm core diameter and 125-μm cladding diameter, operating at a 1300-nm optical wavelength and a data frequency of 125 MHz. These characteristics allow the use of cheaper LED light sources instead of the more expensive laser. PIN diodes are used for detectors. The standard indicates a link length limit of 2 km between nodes imposed by the chromatic dispersion of the LED. The maximum circumference of the rings is 100 km, and a maximum of 500 nodes can be connected to each ring.

Connection to the ring is made by either a single or dual node. Single nodes are attached to the ring via a concentrator, whereas a dual node can be attached directly to the ring. The concentrators and dual nodes require a bypassing facility so that the rings are unaffected by their presence when they are not transmitting or receiving. Figure 10.17 shows an example of an optical bypass at a dual node. The diagram indicates the situation for the bypass condition, where the equipment at the node is disconnected from the dual ring. Notice that PHY A and PHY B are connected to each other, allowing a self-test mechanism. When required, the switches are activated to allow PHY A and PHY B to be connected to the rings.

PMD also outlines the peak power, optical pulse characteristics, and data-dependent jitter. For an FDDI network, these parameters allow a guarantee

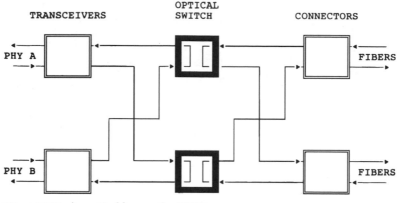

Figure 10.17 An optical bypass for FDDI.

of the worst-case BER to be 2.5×10^{-10} and a normal operational BER of 1×10^{-12}.

The PHY layer provides the rest of the physical layer facilities, that is, the upper half of Layer 1 of the OSI model. Data is transmitted on the ring at 125 MHz. The actual information transmission rate is limited to 100 Mb/s. A 4B/5B nonreturn-to-zero invert (NRZI) code is used to perform the conversion. This code simply encodes 4 bits of information into a 5-bit pattern. NRZI is the line code in which a binary 1 is represented by a transition at the beginning of the bit interval and a binary 0 is represented by an absence of transition at the beginning of the interval. At each node, elastic buffers of at least 10 bits are used to regenerate and synchronize the clock. There is no centralized clock.

The MAC sublayer is where the token-passing protocol for FDDI networks is defined, together with packet information, addressing, and recovery mechanisms. MAC controls the data flow on the ring. The structures of the token and packet are presented in Fig. 10.18. Each frame starts with a preamble that contains idle symbols for clock synchronization, a starting delimiter, and frame control bits. The frame control is an 8-bit word for denoting whether the sequence is a token, MAC frame, management frame, type of management frame, etc. The token is completed by an end delimiter, which is one or more digits to signify the termination of the sequence. In the packet containing the data, there are some additional bits included. Following the frame control bits, there is the destination address and then the source address. Following the data is the frame check sequence, end delimiter, and, finally, the frame status. The frame status indicates any detected errors, address recognition, and frame copied. The maximum packet size is 4500 bytes.

The flow of data on the ring is controlled by the MAC. When a start delimiter is placed on the ring, the MAC monitors each packet to establish its destination. If the packet is addressed to another station, the MAC repeats the packet on the ring and observes any errors that might have occurred. A packet addressed to the MAC's station is copied into its storage buffers and is simultaneously repeated on the ring. If the packet was transmitted by the MAC's station itself, it is absorbed instead of being retransmitted, and only a small portion of a packet (e.g., 6 bytes) is repeated.

When the MAC receives a token and it has data to transmit, it removes the token from the ring. The MAC then goes into the transmit mode, packetizes its data, and places it on the ring. This continues until it has transmitted all of its data or the station token holding time expires. At this stage, the MAC places a token on the ring so that other stations can simultaneously access the ring.

As indicated in the token ring section, TTRT is an important parameter in determining the order of access to the ring. Each station sends claim frames with their respective TTRT bid values, and each station stores the value of the TTRT in the most recently received claim frames. Any claim frames with higher TTRT values than its own are absorbed. A station can initialize the ring and transmit data only when it has received its own claim frame after having gone around the ring and become the lowest TTRT bid value.

There are two modes of FDDI operation: asynchronous and synchronous. The normal token ring mode of operation is the asynchronous mode. Synchronous bandwidth requirements are fulfilled first, and then any bandwidth remaining is dynamically allocated to asynchronous operation. Synchronous operation is used for predictable applications. When the ring is initialized, a TTRT is agreed to by all stations. If the token returns earlier than the TTRT, that station can transmit either synchronous or asynchronous data. If it returns later than the TTRT, it can send only synchronous data. Synchronous operation is managed by the SMT, which is the decision maker in bids for bandwidth. The sum of the bandwidth allocations, the maximum packet length time, and the token circulation time must not be greater than the TTRT. There are some circumstances in which the TRT will exceed the TTRT, but the TRT will always be less than twice the TTRT.

The asynchronous mode can have either *restricted* or *nonrestricted* tokens. Nonrestricted tokens are used in normal circumstances, and bandwidth is shared equally between stations. A nonrestricted token is always used in the FDDI network initialization. The token can be restricted in cases when two or several stations want to use all of the available asynchronous bandwidth. Network management establishes which stations are making the request, and

TOKEN

Preamble	Starting Delimiter	Frame	End

FDDI FRAME

Preamble	Starting Delimiter	Frame Control	Destination Address	Source Address	DATA	Frame Check Sequence	End Delimiter	Frame Status

Figure 10.18 Token and packet structures for FDDI.

one of the stations creates a restricted token. After transmission, the station that issued the restricted token returns the ring to its normal mode of operation by issuing a nonrestricted token.

The SMT defines the beacon token process, which is used to locate ring failures. When a station realizes there is a ring failure, it enters the beacon process and starts to transmit beacon frames continuously. Beacon frames are special bit sequences, unique to each station, which are recognizable by all stations as indicating the presence of an upstream fault. If that station receives a beacon from an upstream station, it will stop transmitting its own beacon and start repeating the upstream station beacon. This process continues until eventually the only station transmitting a beacon will be the one that is immediately downstream from the failure. When the failure has been repaired, a beaconing station will receive its own beacon. This is a signal to that station to stop beaconing and return to normal by issuing a claim.

SMT specifies how the other FDDI layers are coordinated, which means that it overlaps the PMD, PHY, and MAC layers. SMT also allows for some error detection and fault isolation. The SMT standard also includes three aspects of management: connection management (CMT), ring management (RMT), and frame-base services and functions. CMT takes care of the PHY layer components and their interconnections at each station so that logical connections to the ring can be accomplished. Also, within a station, the CMT manages the MAC and PHY configurations. Finally, CMT monitors link quality and deals with fault detection and isolation at the PHY level. RMT manages a station's MAC layer components and also the rings to which they are attached. RMT also detects faults at the MAC layer (e.g., duplicate address identification or beacon identification problems). The frame-based services and functions perform higher-level management and control of the FDDI network such as collection of network statistics, network fault detection, isolation, and correction, and overseeing the FDDI operational parameters.

As mentioned already, the FDDI standard specifies operation over a distance up to 2 km. A survey has shown that 95 percent of FDDI network connections are within 100 m of the wiring closet. The question arises, If only 100 m of cabling are required, why not use inexpensive twisted-pair cable instead of the more expensive fiber? This question is particularly relevant because the FDDI token ring LAN is experiencing strong competition from the Ethernet LAN.

10.4.5   100Base-T

In competition with FDDI, 10Base-T technology has been upgraded for 100-Mb/s operation and is known as fast Ethernet or 100Base-T/100Base-X. The new standard is downward-compatible with the 10-Mb/s (10Base-T) operation, and CSMA/CD is retained. Both 10-Mb/s and 100-Mb/s signals can exist on the same line, and the upgraded 100-Mb/s nodes can also communicate with the slower 10-Mb/s nodes. The fast Ethernet can only operate over a maximum distance of 100 m on twisted-pair cables, and 400 m over optical fiber.

Standards 100Base-TX and 10Base-T4 are twisted-pair-cable-based. 100Base-T4 requires Category 3 cabling, that is, voice-grade unshielded twisted-pair (UTP) cable. All four wire pairs are used, and consequently a two-pair installation is insufficient. Data transmission uses the 8B/6T block code, where each byte is converted into a six dc-balanced ternary symbol block. Consecutive blocks are placed on three of the four pairs so there is a 25-MHz rate ternary stream on each active pair. 100Base-FX operates on 62.5/125-μm multimode optical fiber similar to FDDI. The 100Base-X category has many FDDI characteristics, such as 4B/5B block coding, full duplex signaling, scrambling, and then MLT3 (three-level Manchester) line coding for copper cabling.

10.5   WANs and MANs

LANs are classified as *intra*premises communications, whereas WANs are *inter*premises communications. By the very nature of the WAN, it is implemented in cooperation with the public network operators, for example, using leased lines, dial-up PSTN connection, or X.25 packet-switching. Data rates of 2 Mb/s are popular, particularly with business users, and higher-speed switching is becoming more readily available from many service providers. The interconnection of high-speed LANs within a WAN is still impractical in today's public network environment because of bandwidth and cost constraints, although interconnection via modems and low-bit-rate leased lines is the basis of the global Internet. The term *MAN* has been devised for wide area interconnection of LANs within the public network. A standard has been developed by the IEEE, known as IEEE 802.6, which does not require any change to existing computer hardware or software and allows broadband connectionless data and integrated isochronous (voice) communication services.

The MAN was originally intended for data, but voice and video traffic can be included nowadays. MANs provide two-way communication over a shared medium, such as optical fiber cable, and satisfy the requirements of commercial users, educational institutions, and the home. The MAN is intended to operate up to about 150 Mb/s over links up to 100 km, preferably using optical fiber cable. Although a MAN is a shared medium, privacy is ensured by authorization and address screening at the source and destination. In this manner, either accidental or intentional third-party surveillance or intervention is eliminated.

The MAN architecture is based on the dual bus topology in a form known as the distributed queue dual bus (DQDB) shown in Fig. 10.19. This simple MAP is the basis of the IEEE 802.6 Standard. The DQDB structure can support both synchronous and asynchronous data, voice, or video traffic simultaneously. The DQDB MAN has two contradirectional buses that physically look like the dual-ring structure, broken at the node providing the frame generators. Each node is attached to both of the buses. The logical DQDB can be redrawn as the physical ring of Fig. 10.20. The end points of the two buses are colocated in this ring configuration.

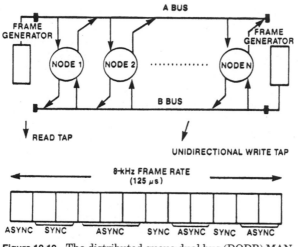

Figure 10.19 The distributed queue dual bus (DQDB) MAN. (*Reproduced with permission from Telecommunications. Zitsen, W., Metropolitan Area Networks: Taking LANs into the Public Network, Telecommunications, June 1990, pp. 53-60.*)

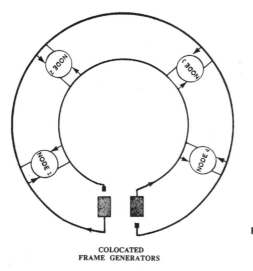

Figure 10.20 DQDB frame generation.

Access to the DQDB network by a user at any node is achieved in a manner that is very equitable, where requests by users are honored in their arrival order. This is often referred to as a perfect queuing structure, and Fig. 10.21 illustrates how this is achieved. Frame generators at each end of the two buses continuously transmit empty frames of data, in the form of packets, in opposite directions along the two buses. The frame generators are synchronized to the telephone network 8-kHz reference frequency so that if there is a break in either bus, the other frame generator can immediately take over. All packets are completely absorbed at the end of their trip along a bus. All nodes can read the data passing by it or write over it as necessary, but they cannot remove data. Each packet contains a bit indicating whether the information-carrying portion of the packet is full or empty. Each packet also carries a request bit time slot. When a user at any node wants to transmit, the node establishes which bus must be accessed in order to carry its information downstream to the desired destination. The user then places a request bit in the next available request bit time slot.

Several packets might pass the user's node before a request time slot becomes available. In this case, each time a full request time slot passes, the node's request queue is incremented. Eventually, when a request time slot is available, the node inserts a bit, indicating it has joined the queue. As more packets pass by, further user requests might be observed that were initiated prior to its own request, so the user has to wait for their completion. At any time, a node might have only one request uncompleted. A node having placed a request will eventually see empty information frames pass by, which will be filled by other nodes downstream that joined the queue at an earlier point in time. Each empty packet that passes decrements the node's queue number by 1, and when the number is 0, the node can transmit its data. If the system is idle (i.e., a condition in which no other users are transmitting or in the queue), clearly a user can transmit immediately. In these circumstances data from one node can be transmitted in packets one after the other. Comparing the DQDB to the token ring structure, the DQDB does not suffer from a time loss known as the *token slack time*. Furthermore, because the DQDB structure is not a closed ring, there are no "response bits" required. DQDB has a capacity utilization of almost 100 percent. Because the DQDB structure provides user interconnection without switches, it is referred to as a *connectionless*, or *seamless*, system.

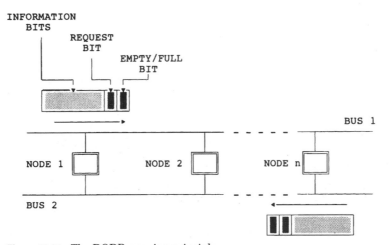

Figure 10.21 The DQDB queuing principle.

In order to connect MANs over large distances, high-data-rate leased lines are presently used. These existing MANs are building blocks for what will be the future B-ISDN, as described in the next section. The trend is an evolution toward a national and eventually an international *broadband* telecommunications network.

10.6 Narrowband ISDN

ISDN has been revolutionary in that it is a departure from the traditional telecommunications, often known as POTS (plain old telephone sets, or services). Exciting new services were envisioned. Unfortunately, after the initial optimism, it was soon realized that the 64-kb/s ISDN system was so low on bit rate (bandwidth) that only a limited number of extra services would be possible. The transmission of video signals is the obvious point that springs to mind. The 64-kb/s system supports only low-quality motion-video signals and is now referred to as *narrowband ISDN*.

As disappointing as narrowband ISDN might be to many people, the important point is that the wheels of change were set in motion. The experience gained and the problems solved during the narrowband, copper-wire-based ISDN era are paving the way for the next generation of ISDN, that is, broadband optical-fiber-based ISDN, which should certainly fulfill the promise of vastly improved services. To trace the evolution of what is indeed becoming a global telecommunications revolution, the details of narrowband ISDN will be discussed first. Many books have been written on this subject alone, but it is the intention here to highlight the major aspects in as short a space as possible.

What is ISDN? During the evolution of digital networks, PCM links were initially introduced for interoffice traffic. Then digital switches were installed to form what was called an integrated digital network (IDN), supporting telephony at 64 kb/s. It was the realization that an IDN could be organized to provide data-related services in addition to telephone services that led to the ISDN. ISDN can be defined as a *network in general evolving from a telephony IDN that provides end-to-end digital connectivity to support a wide range of services, including voice and nonvoice service, to which users have access by a limited set of standard multipurpose user-network interfaces*. The ISDN standardization process took more than a decade, and the ITU-T estimates that more than 1000 person years of international meeting time was spent on the process. End-to-end digital connectivity, which extends digital transmission to the customer, is the essence of ISDN. This digital connection from the main network to the customer is called the integrated digital access. The ITU-T has defined ISDN access in terms of channels, whose main types are:

- *B Channel.* 64 kb/s for carrying user information such as 64-kb/s voice encoded information or data information that is either circuit or packet switched.

- *D Channel*. Either 16 or 64 kb/s, primarily for carrying signaling information for circuit switching. It might also carry packet-switched information.

The ITU-T also recommends two types of access. The first is the *basic rate access*, which contains 2B + D (16 kb/s) = 144 kb/s. Each B channel can have a different directory number if required and both channels might carry voice or data up to 64 kb/s on copper wire cable. The second type is the *primary rate access*, which has either 30 B + 1 D channels (European networks) or 23 B + 1 D channels (North American networks). In some cases the D channel in one structure can carry the signaling for B channels in another primary rate structure without using a D channel. Depending on the situation, the D channel time slot might be used to provide an extra B channel (e.g., 24 B for a 1544-kb/s interface). The bit rates are $(30 \times 64) + 64 = 1.984$ Mb/s and $(23 \times 64) + 64 = 1.536$ Mb/s, respectively. Because the D channels are for signaling, the actual data rates are therefore 1.920 Mb/s, known as the H_{12}, and 1.536 Mb/s, known as the H_{11} channel (Table 10.2). If signaling is required for the H_{11} channel, it is carried in a D channel on another interface structure within the same user-network access arrangement. Notice these bit rates are not identical to, but are compatible with, the two PCM formats. This access is mainly for connecting digital PBXs to the ISDN. An important aspect of ISDN is the isolation of signaling from the subscriber data. Signaling for ISDN is specified by the ITU-T to be common channel signaling. ISDN is organized in the ISO layered structure for OSI. Layer 1, the physical layer, is specified by ITU-T Recommendation I.430 for the basic rate access and I.431 for the primary rate access. Layer 2, the link layer, and Layer 3, the network layer, are defined for both the basic and primary accesses by ITU-T Recommendations I.441 and Q.931, respectively.

10.6.1   ISDN call procedure

The procedure for establishing an ISDN call is indicated in Fig. 10.22. First, the customer requests a call, and the calling terminal creates a message for the call request, that is, SETUP. This is a Layer 3 process. This message is sent via the terminal Layer 2 to Layer 3 of the digital switch. The switch then makes the routing to the called customer by a Layer 3 call request (SETUP). If the call is across town, for example, this message maintains a route through

TABLE 10.2   Channel Bit Rates for N-ISDN

Narrowband	kb/s
D	16 or 64
B	64
H_0	384 (6×64)
H_{11}	1536 (24×64)
H_{12}	1920 (30×64)

the switch while the switch makes contact with the destination end switch, using the ITU-T no. 7 signaling system. Once the SETUP message is received at the called terminal, an ALERT message is returned to the caller as a SETUP acknowledgment. If the called party is available, the call acceptance message (CONNECT) is sent to the calling terminal. When the connection is completed through the network, dialogue in the form of voice or data transfer can take place. On completion of the communication phase, a CLEARDOWN can be initiated at any time by either party.

Figure 10.23 shows the basic rate user-network interface configuration as defined by the ITU-T. This configuration uses the concepts of reference points and functional groupings. In this figure, NT1, NT2, TE1, and TE2 are functional groupings, whereas R, S, and T are reference points. The network termination 1 (NT1) terminates the two-wire line from the local network on its network side and supports the S interface on the customer side. The network termination 2 (NT2) distributes the access to the customer network such as PBXs and LANs. Terminal equipment (TE1) has an I-series customer network interface, whereas terminal equipment (TE2) has a non-ISDN X- or V-series interface. The *terminal adapter* (TA) is an important item that adapts the non-ISDN interface of TE2 to the ISDN interface of NT2.

The reference points R, S, and T are not specifically physical interfaces but might be points at which physical interfaces occur. The S reference point is the

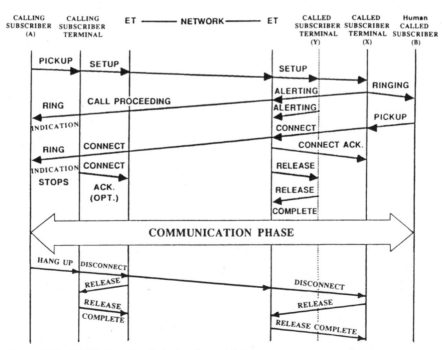

Figure 10.22   ISDN circuit-switched call procedure.

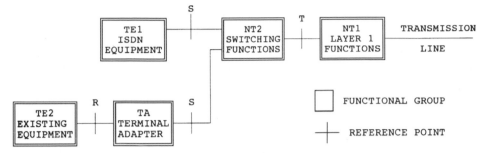

Figure 10.23 User-network interface reference configuration.

ITU-T, ISDN user-network interface. NT2 might not be present, which makes S and T colocated. In the United States, the NT1 functions are provided by the customer, whereas in Europe the service provider owns the network termination even though it is placed on the customer premises.

10.6.2 User-network interface

Transfer of information from Layer 2 of the terminal to the network is the responsibility of Layer 1. Conventional twisted-pair cable is used from the terminal to the NT1 and from the NT1 to the local CO. If errors are introduced by the twisted-pair wires, Layer 2 takes care of error detection and correction. The following characteristics are required for Layer 1 to support Layer 2.

The Layer 1 frame structures across the S interface differ depending on the direction of transmission, as indicated in Fig. 10.24. Each frame is 250 µs in duration and contains 48 bits. The transmission rate is 192 kb/s. From the terminal to the network, each frame has a group of bits that is dc-balanced by its last bit, L. The B- and D-channel groups are individually balanced because they might come from different terminals. There are bits allocated to perform the housekeeping functions of frame and multiframe alignment. From the network to the terminal, there is also a D-echo channel in each frame, in addition to the B and D channels. This D channel returns the D bits to the terminals. The complete frame is balanced using the last bit (L bit) of the frame. The M bit is used for multiframe alignment. The S bit is a spare bit.

Signaling. ITU-T Recommendation I.441 describes the Layer 2 procedures for a data link access, which is often known as the link access procedure for the D channel (LAP D). The Layer 2 objective is to perform an error-free exchange of I.441 frames between two physical medium connected endpoints. The call control information of Layer 3 is contained in the Layer 2 frames. I.441 frames are arranged in octet sequences as shown in Fig. 10.25. A separate layer address is used for each LAP in layer multiplexing. The address contains a terminal endpoint identifier (TEI) and a service access point identifier (SAPI). The SAPI defines the service intended for the signaling frame, and the TEI is

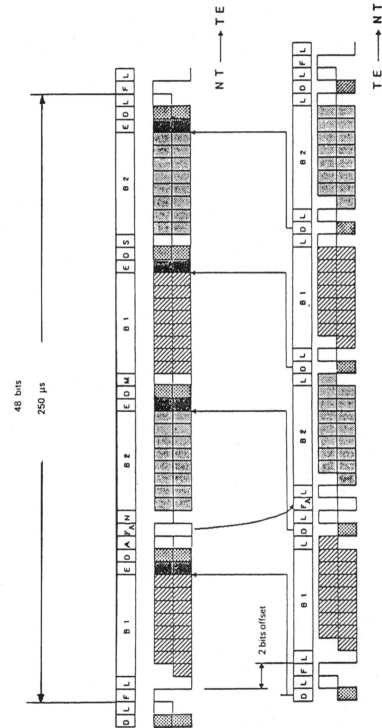

Figure 10.24 Layer 1 basic access frame structure. D = D-channel bit, F = framing bit, B1 = B-channel 1 (8 bits), A = bit used for activation, F_A = frame alignment bit, N = bit set to a binary value [N = $\overline{F_A}$ (NT to TE)], B2 = B-channel 2 (8 bits), M = multiframe alignment bit, S = spare bit for future use. (*Reprinted with permission from Design and Prospects for the ISDN by Dicenet, G. Artech House, Inc., Norwood, MA, 1987. http://www.artech-house.com.*)

set for each individual terminal on installation. These two addresses identify the LAP and form a Layer 2 address. A terminal always transmits these addresses in its frames and will not accept received frames unless they contain this address. The two link control octets contain frame identification and sequence number information. The next sequence of octets is for Layer 3 information. This is followed by two octets for frame checking and error detection.

To clarify Layer 2 operation, consider the following call attempt by a terminal. For a completely new call, the caller makes a request for service, whereupon Layer 3 requests service from Layer 2. Layer 2 can offer service only when Layer 1 is ready, so a request to Layer 1 is made by Layer 2. Layer 1 initiates the start-up procedure. Subsequently, Layer 2 initiates its start-up procedure, which is called setting up a LAP to ensure correct sequencing of information frames. When the LAP has been formed, Layer 2 can carry Layer 3 information. All information frames received by the called party must be acknowledged to the calling party. If a frame is lost or corrupted, this will be detected by the called terminal within its frame timing interval, and there will be no acknowledgment sent to the calling terminal. Retransmission then takes place, and if several retransmissions occur without acknowledgment, Layer 2 presumes that there is link failure, in which case it tries to reestablish the LAP. ITU-T Recommendation Q.931 specifies the call control procedures. The Layer 3 sequence of messages required to establish and disestablish a call is shown in Fig. 10.26. The calling terminal sends call establishing information such as calling number, type of service required, and terminal information to ensure compatibility. Once the call is ready for setup, the interexchange, ITU-T no. 7 signaling system is initiated. A call SETUP message is sent from the

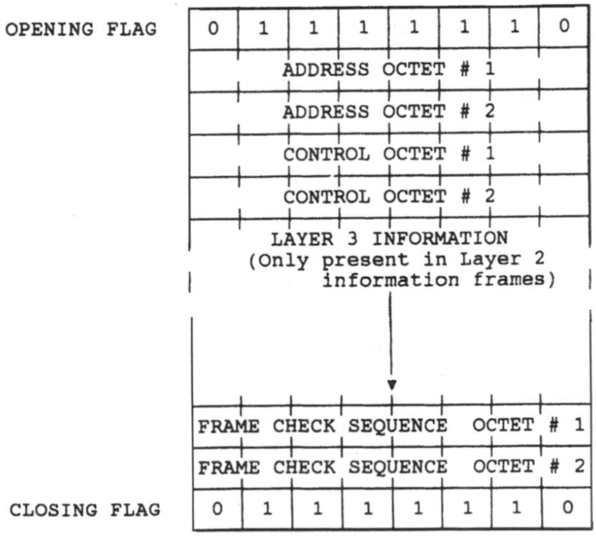

Figure 10.25   Layer 2 frame structure.

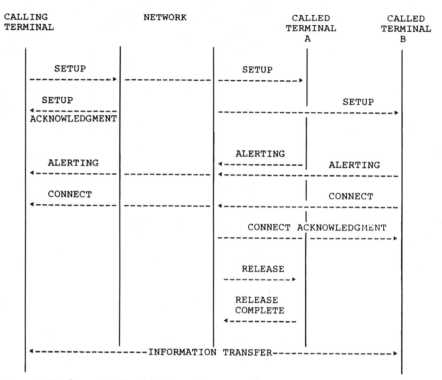

Figure 10.26 Layer 3 call establishment.

calling to the called user via the broadcast data link. This message can be scrutinized by all terminals at the called end to establish if they are compatible with the calling equipment. The ALERT message is returned to the network by all available compatible terminals. Simultaneously, an incoming call indication (ringing) is activated by the local terminal. On answering, a terminal sends a CONNECT message to the network and the call will be given to the first terminal to respond with a CONNECT message. The network connects that terminal to a B channel and returns a CONNECT ACKNOWLEDGMENT giving B channel information. A RELEASE message is given to all other terminals that responded to the call. When the called terminal has received the CONNECT message, the network informs the calling terminal by also sending it a CONNECT message. Call charging starts and the communication then proceeds. This whole procedure is accomplished using a sequence of octets, designed to fulfill the protocol requirements.

In conclusion, the principal features of ISDN can be summarized as a network that has:

- User-to-user digital interconnection
- Both voice and nonvoice services

- A limited number of connection types and multipurpose interfaces
- Circuit- and packet-switched connections
- Network intelligence

Figure 10.27 summarizes the ITU-T ISDN functional and corresponding protocol architecture. The ITU-T ISDN functional architecture of Fig. 10.27*a* includes functional components such as exchange terminations (ETs), packet handlers (PHs), integrated services PBXs (ISPBXs), NT1s, etc., that perform the switching, multiplexing, call processing operations, etc. The functional architecture can be divided into two parts—access and network. The access part of the ISDN is that portion seen and controlled by the user, often referred to as the private ISDN. The network part contains the portion of functional entities controlled by the service providers and is often referred to as the public ISDN. In each country implementing ISDN, depending on political and regulatory viewpoints, the line between user and service provider (access and network parts) can wander, even though the ITU-T functional architecture does not change.

Figure 10.27*b* summarizes the corresponding ITU-T ISDN protocol architecture for the access and network parts. This diagram indicates the various protocols used in a network containing both private and public ISDNs. As indicated by this diagram, ISDN protocols are a fairly complex business. This is not the full story, however. There are a number of additional supplementary services that have been standardized. Subscriber loops, which differ greatly from one country to another, have been difficult to standardize completely and modifications are still in progress. For example, the basic rate access interface has several reference points. In the United States, the *U reference* point has been chosen as the dividing line between the public and private ISDNs, whereas in Europe it is the S/T-reference point. The object of all this effort is to ensure satisfactory end-to-end interworking and global feature transport.

10.6.3 N-ISDN services

The time taken to establish a widespread ISDN is quite lengthy, and in many developed countries full implementation is still not complete. In Europe, there were three phases of implementation as indicated in Table 10.3. These services have been established with a view to open network provision. N-ISDN services can be categorized as bearer or supplementary services.

These services include such innovations as low-speed videoconferencing. At a bit rate of 128 kb/s, the resulting video is satisfactory for working-level meetings where the participants know one another but not, for example, for sales presentations where there is a lot of motion activity. ISDN services extend to the mobile unit or portable telephone, even to the extent of connecting a PC to allow data transfer. Facsimile is evolving to the group 4 level, which allows for photo quality images to be transferred.

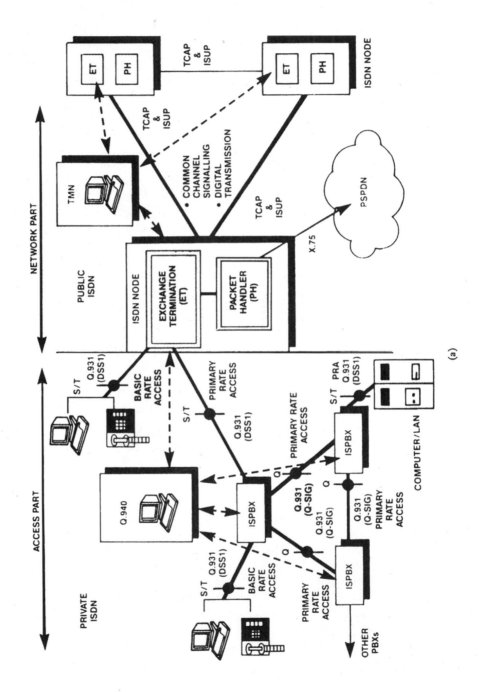

(a)

556

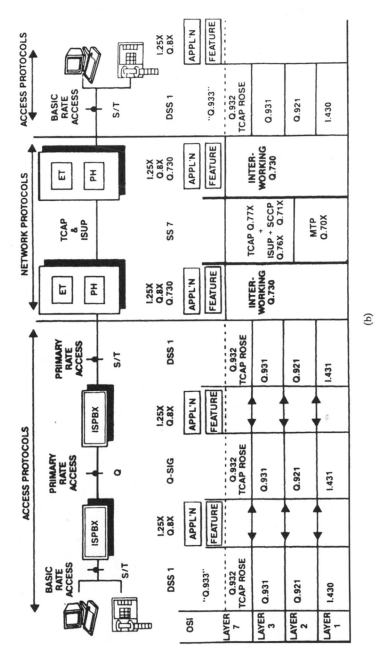

Figure 10.27 (a) ITU-T ISDN functional architecture. (b) ITU-T ISDN protocol architecture. (*Reproduced with permission from Telecommunications. Thomas, M. W., ISDN: Some Current Standards Difficulties, Telecommunications, March 1991, pp. 33-43.*)

(b)

TABLE 10.3   ISDN Services Implementation

	ISDN services
Phase 1	
Bearer services	Circuit mode 64-kb/s unrestricted bearer service
	Circuit mode 3.1-kHz audio bearer service
Supplementary services	Calling line identification (CLI)
	Calling line identification restriction
	Direct dialing-in
	Multiple subscriber number
	Terminal portability
Phase 2	
Bearer services	Circuit mode 64-kb/s unrestricted bearer service on reserved or
	permanent mode packet bearer service case A and case B
Supplementary services	Closed user group
	User-to-user signaling
	Reverse charging
Phase 3	
Bearer services	Circuit mode speech
	Circuit mode 2×64 kb/s unrestricted
Supplementary services	Advice of charge services
	Number identification service
	Call waiting
	Completion of calls to busy subscriber
	Conference service
	Diversion service
	Freephone
	Malicious call identification
	Subaddressing
	Three-party service

Facsimile (fax) grew rapidly in the United States, but the emergence of electronic mail (e-mail) is supplementing and perhaps to some extent overshadowing fax. The United States ISDN marketplace targets two groups: voice-, and data- or computer-oriented, and there is some overlap between the two. Eventually, the two should merge with full ISDN deployment.

In the United States, network terminators are considered part of the CPE, whereas in the rest of the world they are not. Regardless of the newly available bandwidth, it has been established that the CPE is critical to the success and acceptance of ISDN. Table 10.4 indicates the customer requirements for various types of CPE.

10.6.4   High-bit-rate digital subscriber line

High-bit-rate digital subscriber line (HDSL) is a revival of PCM technology, adapted to the subscriber loop. It allows 1.544 or 2.048 Mb/s data to be transmitted on a regular twisted copper pair of wires already installed to customer premises. PCM over twisted pairs required repeaters every 1 km or so. This

TABLE 10.4   CPE Customer Requirements

CPE	Customer requirements
PC with voice and data adaptor	Integrated telephone
	Keyboard dialing
	Modem replacement
	Multiple data channels
Stand-alone terminal adaptor to interface	Diverse terminals
	Host computers
	Data communication switches
	Group 2 and 3 Fax
	Analog telephones
	Special equipment
ISDN telephone	Interworkable with analog telephones
	High-quality voice
	Calling number display
	Voice conferencing
	Single-button feature activation
	Electronic key telephone service
ISDN telephone with adaptor	Modem replacement
	Data call on keypad
PC workstation	Same as PC with voice and data adaptor
	Integrated call management
Fax-Group 4	Transmit images on paper

distance can easily be doubled by using a more bandwidth-efficient line code, and 2B1Q is used instead of B3ZS or HDB3. In this line code, 2 bits are converted into one quaternary line signal, which halves the line symbol rate. Whereas PCM used two pairs of wires, one for each direction of transmission, HDSL uses just a single pair for bidirectional information flow. Echo cancellation is used at both ends to separate the transmit and receive signals. The 1.544-Mb/s signals require only 784 kb/s on the line, and the 2.048-Mb/s signals need 1.118 Mb/s.

The resulting factor-of-2-to-3 improvement in transmission distance means that in some areas about 80 percent of customers can be serviced by HDSL without using a repeater. This figure increases to more than 95 percent with the inclusion of one repeater in those 15 percent or so cases.

10.7   Asymmetric Digital Subscriber Line

Unshielded twisted-pair copper cable used for POTS has been given a stay of execution. The high cost of fiber to the home or even to the curb has given the copper pair infrastructure a life extension of several years. Innovations in the modem laboratories now allow the transmission of several megabits per second over regular copper pairs, which opens the door for video on demand and wideband Internet connections. Figure 10.28 shows the ADSL concept.

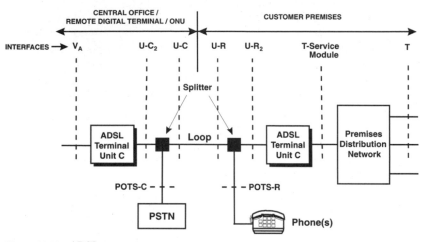

Figure 10.28   ADSL concept.

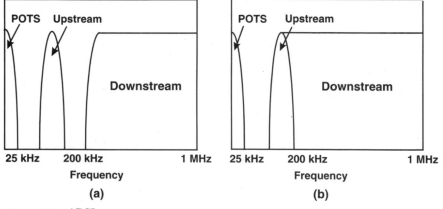

Figure 10.29   ADSL spectrum.

Asymmetric, as indicated in Fig. 10.29, describes the difference between broadband downstream transmission (1 to 9 Mb/s) and narrowband upstream transmission (64 to 800 kb/s). This technique eliminates the main bit-rate-limiting problem for copper pairs known as near end crosstalk (NEXT) where the signal transmitted along a cable on one pair interferes with incoming signals on adjacent or other pairs. In doing so, it creates a far end crosstalk (FEXT) problem where signals interfere with each other as they proceed along different pairs, so that by the time they reach the far end, the crosstalk is received and can cause severe distortion. The maximum bit rate for FEXT-limited transmission is still in the megabits-per-second region for 0.5-mm (24 gauge) wire. The wire attenuation, which increases with frequency as the skin effect resistance becomes dominant, is still a limiting factor for distance, and Fig. 10.30 shows the attenuation against frequency curve. In the United States,

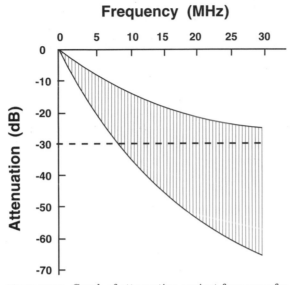

Figure 10.30 Graph of attenuation against frequency for most copper cable gauges.

about 50 percent of lines extend to 12,000 ft (3.66 km) and 80 percent to 18,000 ft (5.49 km), with splices at about 500 ft (152 m) on average. In Europe, 5.5 km would reach more than 90 percent of customers.

An important requirement is that ADSL shall coexist on the same line as existing telephony. Telephony uses tens of volts for ringing and other pulses, whereas ADSL signals at 25 kHz and above are less than 1 mV. Separation filters must have very high stopband attenuation to minimize the echo and sidetone interference of ADSL on telephony. Notice how the use of asymmetric echo cancellation (Fig. 10.29*b*) allows the upstream and downstream spectra to overlap and make better use of the lower loss region for some of the downstream transmission.

The major technology required for ADSL is highly bandwidth efficient modulation. Three contenders were identified as candidates: QAM, carrierless amplitude/phase modulation (CAP), and discrete multitone modulation (DMT). CAP is very similar to QAM in its spectral shape, the difference being that the two half-rate bit streams are each passed through a digital transversal filter that has the same amplitude response but an orthogonal (differing by $\pi/2$) phase response instead of each stream being mixed with a sine and cosine carrier. Bandwidth efficiencies in the region of 8 b/s/Hz are used for several megabits-per-second downstream performance. This is equivalent to 256-QAM or higher.

DMT is a type of multicarrier modulation that divides time into symbol periods, each carrying a specific number of bits. Serial-to-parallel converters are used to split up the bit stream for modulation on several separate carriers. Groups of bits are assigned to different carriers and the bits for each carrier set the carrier's amplitude and phase for the duration of the symbol period.

This is equivalent to several QAM systems operating in parallel, and the level of QAM on each carrier depends on the number of bits assigned to the carrier. This multicarrier modulation requires the carriers to be orthogonal, and fast Fourier transform (FFT) techniques are used to achieve DMT.

The number of bits that can be sent on each carrier varies as the noise or interference at that frequency changes. This allows adaptive bit rate variation, as shown in Fig. 10.31, to optimize performance. The symbol rate might vary from, say, 2 or 3 b/Hz to as much as 15 b/Hz depending on the frequency. DMT has received the vote of confidence from manufacturers and standards bodies alike. It has an inherent immunity to impulse noise, and larger bandwidth capability and higher transmission rates, particularly for the shorter loops. DMT has some disadvantages. The FFT processing causes transmission delays that might violate ISDN specifications (<1.25 ms). Echo cancellation is not trivial for DMT. In addition, FEC, which is used to enhance performance, is more complex for DMT than for CAP or QAM.

Considering that ADSL is a relatively new technology, its performance is already very impressive. Using 0.5-mm cable, 2 Mb/s over 5.75 km, 4 Mb/s over 4 km, and 6 Mb/s over 3.6 km for downstream transmission ranges are known to be successful. These figures are based on a pessimistic noise source including crosstalk and radio frequency interference (RFI) and include a 6-dB safety margin. The range performance values can be increased by additional signal processing, again at the expense of transmission delay. For example, a performance enhancement of 5 to 6 dB can be achieved by the combined use of (1) trellis coding, (2) echo cancellation, and (3) optimized transmit power distribution.

Main problems.   The existing cable plant in many areas might not be adequate for ADSL. The average line has been estimated to have 20 or more splices. Corrosion at poorly spliced joints or splices affected by water could seriously degrade the high-frequency performance. Bridge taps are another problem. They are unterminated spur lines that could be a few meters or thousands of meters. Each bridge tap places an attenuation notch at the frequency associated with the length of each branch. Finally, 20 percent of all U.S. residential lines have installed loading coils on long lines to lower the attenuation at the higher end of the voiceband. Unfortunately, that creates a low-pass filter that has a sharp cut-off just above the voiceband at 4 kHz.

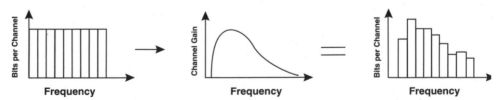

Figure 10.31 Discrete multitone modulation bit allocation.

Remember, in this application, asymmetric means broadband in only one direction, i.e., not full B-ISDN, so ADSL is not a simple modem upgrade.

10.7.1 Very-high-bit-rate digital subscriber line

The success in this field has even given some credibility to the prospect of sending an STS-1 (51.8 Mb/s) down a POTS line. The extension of ADSL to a very-high-bit-rate digital subscriber line (VDSL) is therefore a natural trend to get the ultimate performance possible out of copper wires.

The economics of FTTH are such that hybrid fiber-copper will prevail for some time. FTTC takes fiber to the service access point or last joint box in the cable from the CO. The last 50 to 300 m to the customer premises is a suitable application of VDSL. Data rates as high as 25 to 50 Mb/s are being considered. STS-1 would allow ATM compatibility. If fiber is taken out only to the end of the feeder (or cabinet), VDSL transmission over only 1 km or so would be needed. It is anticipated that 25 Mb/s would work in most cases over this distance, and possibly even higher data rates.

10.8 Broadband ISDN

10.8.1 B-ISDN services

The demand for high-quality video (HDTV) and high-speed data information exchange is the driving force behind the inevitable implementation of the B-ISDN. ITU-T Recommendation I.121 classifies B-ISDN services into *interactive* or *distribution* services as follows.

Interactive services. These consist of the following:

1. Conversational services
2. Messaging services
3. Retrieval services

Interactive implies a two-way information exchange, which can be between two customers or a customer and a service provider. Distribution services are primarily one-way information exchanges from a service provider to the customer.

Conversational services are the classic two-way communication with the exception that B-ISDN would include video, data, and text as well as voice. Video telephony is a major application that will probably be employed first of all in business offices in conjunction with PCs. Full-motion, live video will be very advantageous for users such as sales and technical personnel who need to discuss diagrams and charts, etc. As prices decrease, video telephony will become a popular service to the home. One drawback is that obscene or malicious phone calls would take on an extra dimension of reality that could create a serious social problem.

Video telephony can be expanded to provide a cheaper service to match existing video teleconferencing by using larger camera area coverage for multiple participants and larger screens at each venue. Or, multiple video telephones could be interlinked in a conference situation by subdividing each participant's screen into several windows.

Document transfer can also be placed in the conversational category for applications such as high-resolution fax or transfer of documents between workstation users.

Messaging services are user-to-user communications that use storage units for store-and-forward, message handling, and mailbox functions. These are not real-time conversation services, which means they make less demands on the network because both users do not have to be present simultaneously. This is analogous to the X.400.

An attractive new B-ISDN messaging service is video mail, which is analogous to the existing e-mail and voice mail. The main difference is that instead of allowing only text and graphics in document form, video mail would be full motion. This is the equivalent of mailing a video cassette. Electronically mailing a document with this type of messaging service capability could include a combination of text, graphs, voice, and video. That would be multimedia communications at the highest level.

Retrieval services allow a user to access information stored in a general public data bank. The information can be retrieved by user control, which is analogous to the N-ISDN videotex service. It is an interactive system that has a general-purpose database, for both private and business requirements, and can operate over the PSTN or an interactive metropolitan cable TV system. The N-ISDN information is in the form of pages of text (e.g., stock market information). The B-ISDN retrieval service could also include sound passages, high-resolution images, and video footage in addition to the existing text and graphics. This service could be very useful for applications such as educational remote learning. Video retrieval could be extended to full-length movie films from a video library. This would work by directly linking up the library to the user's video cassette or CD recorder. Again, multimedia retrieval of documents containing a mixture of video, sound, text, high-resolution graphics, and even software would be possible.

Distribution services.   These are also referred to as broadcast services and might exist as follows:

1. *Without* user individual presentation control (broadcast), for example, CATV, electronic newspaper

2. *With* user individual presentation control, for example, an enhanced type of teletext including video (cabletext)

In the first service, the user has no control over the starting times and can only "tune in" as one would a TV station. In the second, the information is relayed in a sequence of frames in cyclical repetition, so the user has some degree of control over the start of the presentation. This is analogous to the narrowband teletext service. Cabletext would be a broadband enhancement of this system, using cyclical transmission of text, images, video, and sound passages.

In the United States, significant large-volume installations of B-ISDN are just beginning to gather momentum, although full B-ISDN might not be available to large numbers of users until after the year 2000. B-ISDN might be achieved a few years earlier in Japan, which would make it the first country to attain full deployment.

10.8.2   B-ISDN implementation

There are three main facets of B-ISDN that must be in place before a resulting network can benefit from the synergistic combination of all three:

1. Optical FTTH
2. Synchronous digital hierarchy (SDH), or SONET
3. Asynchronous transfer mode (ATM)

Optical FTTH. Central to any serious implementation of B-ISDN is the information transporting medium. It is universally accepted that optical fiber is the solution. As mentioned in Chap. 9, there are several architectures for a fiber-based B-ISDN network. While some cost saving can be made depending on which architecture is used, the fact remains that eventually there must be fiber to every home, which is extremely expensive. Fiber to the curb is the interim solution, having already reached cost parity with copper wiring due to decreased fiber manufacturing costs. This means that in most developed countries the transmission medium choice for new installations will soon automatically be fiber. In the United States, fiber is already the economical choice for all growth projects. It is the replacement of copper cable with fiber that incurs the very large expense, so much so that there is a resurgence of the use of twisted copper pairs for data in the form of HDSL, ADSL, etc. Copper technology has been given a new lease on life that could last until the year 2010 or beyond.

Two possible architectures that allow steady growth and minimize costs are shown in Fig. 10.32. These two topologies have different properties and the choice depends on various factors such as service requirements, geographical situation, and reliability needs. By partitioning the subscriber loop into feeder and distribution loops, the cost is minimized. Figure 10.32a is a double star topology. High-bit-rate SONET transmission systems interlink the remote terminal (RT) and the CO (exchange) in the feeder loop. The RTs can do simple multiplexing or line concentration functions to use the available feeder band-width efficiently. Existing carrier system subscriber loops can be most easily upgraded using this topology.

Figure 10.32*b* is a ring and star topology. The feeder loop is in the form of a fiber ring and the RTs are attached to points on the ring. The ring can carry SONET variable transmission bit rates to accommodate a variety of traffic requirements. A drop and insert capability at the RTs allows variable band-width distribution as required around the ring. Again, traffic concentration can be done at the RTs, and some applications might require ring access protocols. The inherent route diversity of the ring, coupled with the RT bypass and loopback facilities, provides a highly reliable network.

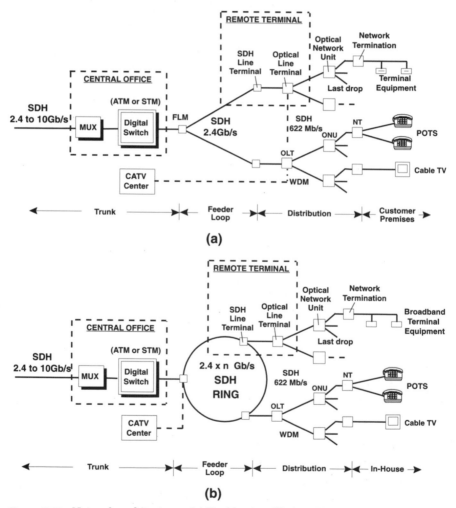

Figure 10.32 Network architectures. (*a*) Double star; (*b*) ring star.

The distribution loops both use a star-connected single-mode fiber to each customer. Full-duplex transmission can be achieved by using either two fibers, one for each direction, or just one fiber operating at a different wavelength for each direction of transmission.

It is essential to have the flexibility to be able to deliver a wide range of bandwidths to the customer. Some might want only conventional, digital POTS. Commercial customers might require high capacity, warranting the use of SONET. The distribution loop, often referred to as "the last mile (or kilometer)," will probably be the final link in the fiber chain.

There is another difficulty associated with FTTH that must be overcome. That is the simple, but not trivial, question of power. For the traditional pair of copper wires the power is supplied to the subscriber equipment by the battery bank in the CO, but fiber has no metal wires to carry this power. There are several possible solutions: (1) use copper wires in the fiber cable up to the curb or customer residence, (2) terminate the fiber at the curb and back-feed the optical equipment from the customer residence, or (3) use power in the customer residence to power the equipment. Solutions (2) and (3) mean relying on commercial power, which is not always as reliable as necessary, especially in an emergency situation. Possible alternatives are solar cells or chemical fuel cells. More elegant is the use of laser power to feed very low-power-consumption electronic sets. This issue is still not fully resolved, although all of the above methods might be used depending on the circumstances. Centralized power to the curb from the CO appears to be cost-competitive with copper-pair POTS.

SDH/SONET.   SDH/SONET, which was explained in Chap. 2, is a key feature of B-ISDN networks. Not only is it primarily a synchronous multiplexing scheme, but it also has the versatility to be able to handle asynchronous data within the synchronous system. This sounds like a contradiction of terms, but it merely states that both PDH and SHD multiplexing bit rates can coexist within the same SDH system.

SDH has the advantageous characteristics of modularity, flexible network management, self-healing via rings and crossconnects, and drop and insert capability. SDH allows ease of upgrading to higher-order bit rates. Also, cost benefits can be realized by skip multiplexing. This reduces the amount of equipment manufactured by internally multiplexing from a low bit rate to very high bit rates instead of having outputs at several specific hierarchical levels. SDH penetration into the network is a gradual process with a typical deployment process as in Fig. 10.33. Local exchanges have to make the transition from wire to fiber cable. In the interim, they must be able to offer either twisted-pair wire at the basic ISDN bit rates, or analog POTS and fiber at the B-ISDN rates. Initial SDH penetration for the trunks and feeders uses the 155.52-Mb/s (STM-1) and 622.08-Mb/s (STM-4) systems. To illustrate the versatility of SDH, the STM-1 could include, for example, 28 PDH-multiplexed DS1 (1.544 Mb/s) bit streams and two

PDH-multiplexed DS3 (44.736 Mb/s) bit streams, or three SDH-multiplexed SONET STS-1 (51.84 Mb/s) bit streams, etc. As SDH penetration advances, the STM-16 SDH level, which supports the 2.48832-Gb/s rate, is a reasonable data rate for the feeder loop. Drop and insert from this loop can be done as required. When the final phase of Fig. 10.33c is accomplished, customer premises' terminals will have their own SDH interfaces. The trunk section of these networks will have bit rates at 10 Gb/s or higher. No doubt, the business community will be the first customers to take advantage of high data rates. As the cost comes down with penetration, the home will be serviced by higher and higher bit rates. Already ITU-T Recommendation I.121 specifies that the data rate from subscriber to network need not be the same as network to subscriber. This takes into account the fact that customers in the home will probably want to receive more information than they transmit. Bandwidth on request at a reasonable price is the ultimate objective.

Asynchronous transfer mode (ATM).   ATM has been universally declared as the standard for B-ISDN. It was developed independently in the United States by Bellcore and standardized through ANSI, and in Europe where it was standardized through ETSI. The small differences between the two are gradually converging. The major objective of ATM is to integrate real-time information such as voice and video with non-real-time computer data, within the same transmission and switching medium. The conflicting requirements for these two categories can basically be stated as follows:

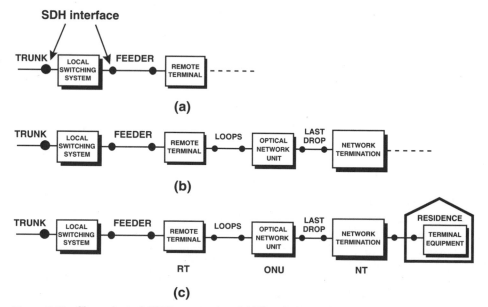

Figure 10.33   Chronological SDH penetration. (*a*) Fiber from trunks to feeder; (*b*) fiber into distribution loops; (*c*) fiber to the home.

- Data requires very low BER but can tolerate large propagation delays (seconds).

- Voice and video require small propagation delays (milliseconds) but can tolerate some errors or small losses of information.

These differences can be observed as the different sampling or scanning times. For voice, regular 8000 samples per second are necessary, whereas video might need a short burst of several megabytes followed by nothing for a few seconds. In contrast, computer data might need a high data rate for some applications, such as image transfer, but a low data rate for simple text transfer. TDM processes such as PCM are considered inefficient because time slots are allocated whether or not information is placed in them. PCM and therefore N-ISDN are examples of the synchronous transfer mode (STM).

In ITU-T Recommendation I.121, the asynchronous transfer mode (ATM) is used for the ISDN user network interface. The Recommendation states that although B-ISDN will support circuit mode applications, it will have a packet-based transport mechanism. That means N-ISDN will evolve from a circuit-switched telephone network to a packet-switched B-ISDN. ATM has some similarities to and some major differences from X.25. One difference is that ATM uses common channel signaling, whereas X.25 has control signaling on the same channel as the data transfer.

ATM improves on the inefficiencies of STM by packetizing the data into cells. Every cell is given a connection address and sent on a fixed route through the network. Packets are made small enough so that if they get lost or collide with other packets sharing the same route, they can be resent if necessary.

Algorithms are used to compensate for the variable delay inherent in packet transmission so that ATM can be used for continuous flow applications such as telephony and video. ATM has considerable flexibility in its use for all services. New services can be added without restructuring the network, or switching exchanges and services already in existence can be modified or expanded. For example:

- Broadband unrestricted bearer services

- High-quality broadband video telephony

- High-quality broadband videoconferencing

- Existing quality TV distribution

- HDTV distribution

10.8.3 ATM networks

ATM is a connection-oriented technique and ATM networks are essentially fast packet-switching networks. Each packet has a fixed length and is called a cell to distinguish it from its X.25 equivalent, which is a variable length packet. The ATM destination address contains a virtual circuit identifier (VCI) or virtual path identifier (VPI) carried in the fast packet header instead of the time slot used in the STM systems. The VCI or VPI identifies a virtual channel such that in a virtual connection all cells of the same connection are transferred through the same route. This ensures cells are received in the same order as transmitted. The VCI is established during the call setup and released on completion of the call. Signaling and information are transported on separate virtual channels.

The unused time slot inefficiency of PCM is overcome by the fast packet-switching process called statistical multiplexing. Several connections are made over the same link depending on their traffic details. Voice coexists with data by sending bursts of data during the inactive voice times. If simultaneous bursts are transmitted, buffering enables the voice to take priority while the data is held back for transmission possibly a few milliseconds later. The statistical multiplexing process evens out the different types of traffic over time to fill the transmission link. This process is not possible with the more rigid formatting of PCM in the synchronous transfer mode. Interestingly, an ATM system that is very heavily loaded with traffic degenerates into an STM-like system. Congestion is currently a hot topic in ATM research.

An important question to answer is, How will the ATM network exist with the present LAN networks such as Ethernet, token rings, or FDDI? As with all new technology, there must be a transition phase, or overlay period, during which the old and the new inevitably coexist. Users gain access to the ATM network through a user network interface (UNI). In the case of the ATM connection to a standard Ethernet or token ring LAN, the UNI converts the LAN frames into ATM cells on entry and ATM cells to LAN frames on exit. Figure 10.34 shows the ATM interface to different types of LANs through their respective gateways to the ATM UNI. The heavy investment in LAN technology dictates a gradual transition to ATM networks. The full power of the ATM network will be experienced only when telephones, videos, and computer terminals (DCE) connect directly to the ATM network through an ATM interface as in Fig. 10.35. Here, the ATM interface statistically multiplexes the different media of voice, video, and data and becomes the UNI for transmission over the ATM backbone network.

ATM cells. ANSI has defined each ATM cell to have 53 bytes as shown in Fig. 10.36. The header has 5 bytes and contains the VCI label (24 bits), control bits (8 bits), and checksum (8 bits). The rest of the cell (48 bytes) is for data, which can be partitioned, if desired, into a 4-byte adaptation layer and 44 bytes of data. The adaptation layer allows the flexibility to subdivide and reassemble cells prior to transmission and after reception. One of the 8 bits of header control is used to set data at 44 or 48 bytes. At certain intervals, the control bits will be used to identify a cell used as a flow control cell that does not contain customer data. The control also contains 1 bit to indicate that the cell can be deleted in situations of extreme congestion.

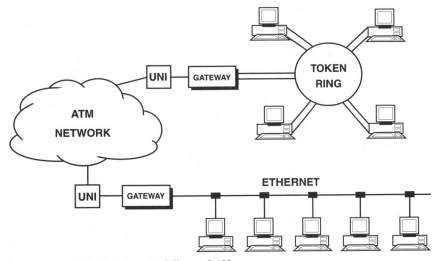

Figure 10.34   ATM interface to different LANs.

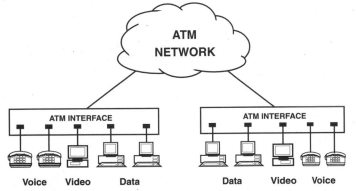

Figure 10.35   Direct connection to the ATM network.

The ETSI ATM cell is also defined as 53 bytes long, containing 48 bytes of data and 5 bytes of header. The differences lie in the composition of the VCI and the checksum fields.

ATM network connections.   When a user wants to access the ATM network, the UNI is contacted first. The UNI is given information about the type of traffic, destination, bit rate, cell loss acceptability, and delay acceptability, and forwards these details to the network. The network uses these parameters to establish the best available route for transmission. A connect setup request is sent to all nodes in the proposed route. A virtual connection is made right through to the destination. Remember, a virtual connection is one that is made only for the duration of the information transfer. Data is then transferred at

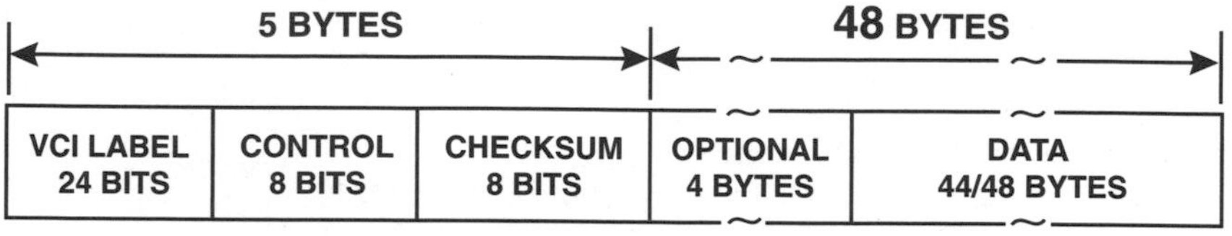

Figure 10.36 ATM cells.

the prenegotiated quality of service. On termination of the virtual connection the VCI is reallocated to another connection. This virtual circuit connection differs from that of the X.25 packet-switching process in that ATM has no acknowledgment sent back from the destination to source. This is a speed advantage at the expense of arrival verification. Also, ATM cells are not sequence numbered like X.25 packets. When a virtual circuit has been set up, there is only one possible path for cells to take, so they cannot be received in the wrong order. Cells might be buffered at some nodes to assist with flow control, but they still arrive at the destination in the same order in which they were transmitted. If cells are sent into the network using the adaptation layer without prearranging a virtual circuit, they can arrive at the destination in the wrong sequence. This adaptation layer requires a sequence number to be allocated to ensure correct reassembly at the destination.

ATM Standard. ATM can be related to the OSI scheme as in Fig. 10.37, which also shows its position with respect to the TCP/IP. The physical layer has an SDH lowest bit rate of 155 Mb/s (STS-3c or OC-3) for ATM traffic. Other SDH rates can be used, such as 622 (OC-12) or 2.488 (OC-48). The ATM cell can be considered part of the data link layer. It also contains some network layer activities (adaptation layer) such as route setup, connection formation, and data flow control.

The transmission control protocol/Internet protocol (TCP/IP) can interact with ATM even though the IP is a connectionless protocol. This is done by the fragmentation of IP packets into ATM cells using software at the UNI source and destination. In this manner, two LAN gateways can have a virtual connection to the ATM network.

In order to maximize throughput, ATM does not perform end-to-end flow control, but monitors the queues forming inside ATM switches at all nodes. In this manner, there is very early detection of congestion buildup. The streams causing the queue buildup are identified and flow control messages are sent to the UNIs causing the problem to reduce the rate, but not completely stop the input of data into the network. The outcome of maximizing throughput and not allowing requests for repeats or receipt acknowledgments is that there is no data arrival verification and, consequently, no quality of service check. However, the TCP/IP or higher protocol layers do have these higher functions.

The effectiveness and accompanying simplicity of the ATM protocol will probably ensure its survival and, together with the Internet protocol, might even render insignificant protocol stacks such as the OSI model. There is much debate about the future of many of the protocols that emerged in the late 1980s and early 1990s. Protocols are meant to facilitate, not hinder, the rate of technological progress. Some equipment manufacturers complain of being hamstrung by the need to comply with multiple protocol stacks.

10.8.4 Frame relay versus cell relay

These two terms have become more widespread as WANs move to higher operating rates. The main difference between the two is that frame relay uses variable length frames to transport data, whereas cell relay uses fixed-length frames.

Frame relay. Frame relay is a protocol, not a service. It is a fast, simplified method of transporting packetized data. Frame relay is effectively a streamlined version of the X.25 protocol. X.25 was originally designed for analog circuits, whereas frame relay was designed for digital technology, which covers only Level 1 and part of Level 2 of the OSI scheme. Frame relay can detect but not correct errors. The corrections are left to the higher levels within the application itself. Frames should always follow each other along the same route because this connection type of protocol provides no sequencing information. There is only destination routing and congestion control

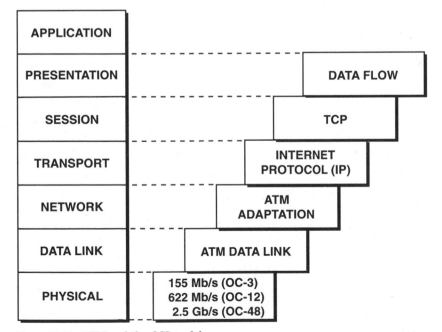

Figure 10.37 ATM and the OSI model.

information in the address. Because each frame is addressed individually, data from several applications can be multiplexed onto the same link. Bandwidth is not dedicated to any one application. This means that the full bandwidth of the circuit could be taken by one application for short periods of time.

The protocols relating to frame relay are the ITU-T Recommendations I.122 and Q.922. They add relay and routing functions to the data link layer (Layer 2). Frame relay is a form of multiplexing that transports frames through the network as quickly as possible.

Functions such as error detection and correction, flow control, etc., are now processed on an end-to-end user device basis instead of being done by the network. Although frame relay was initially designed for the primary rate interface up to 1.92 Mb/s, it is projected that bit rates up to 34 to 45 Mb/s might be introduced. Figure 10.38 shows the construction of frames, which starts with a 1-byte flag followed by a 2-byte frame relay header. The header incorporates a data link connection identification (DLCI) for routing each frame on a hop-by-hop basis through a virtual path established at call setup or at the time of subscription. Next comes the V.120 terminal adapter header followed by the control state information. The Layer 2 messages are then transmitted, after which there is a 2-byte cyclic redundancy check, and then the end of the frame flag.

Comparison with cell relay.   As a comparative summary, Fig. 10.39 illustrates the differences between TDM, frame relay, and cell relay. Figure 10.39*a* shows the conventional synchronous TDM frame structure. Figure 10.39*b* shows the frame relay format. This is asynchronous multiplexing with variable frame length as opposed to the fixed frame length of the ATM cell relay asynchronous multiplexing of Fig. 10.39*c*. For a view of the process of a frame relay data transfer, see Fig. 10.40*a*, which shows the data and signaling paths through the protocol architecture. For a file to be transferred from A to B, the user (application A) initially sends a request to establish a session via the

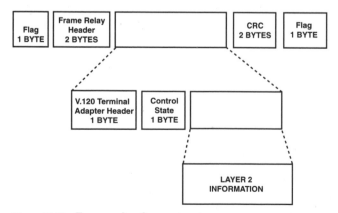

Figure 10.38 Frame relay frame structure.

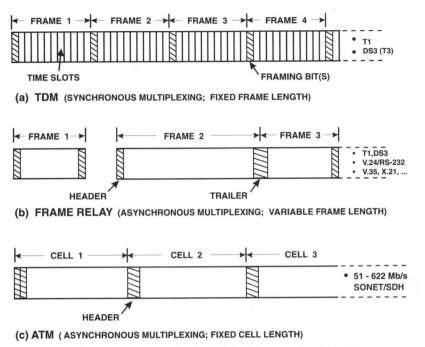

Figure 10.39 Multiplexing methods. (*a*) TDM; (*b*) frame relay; (*c*) ATM.

presentation and session layers to the transport layer. Call control information is forwarded through the ISDN D channel by the transport layer using ITU-T Recommendation Q.931. As also indicated in Fig. 10.40*a*, the signaling is used to define a virtual path and call control necessary to set up the data transfer. When the call has been set up, data is transferred from A to B using the DCLI in the frame header. The amount of frame processing by the network is reduced to a minimum for frame relay, which facilitates very fast data transfer. One problem with this technique is that in some circumstances reliability can be compromised for speed. In other words, some data can be lost, because frame relay has no error correction. This is in comparison with packet switching, which guarantees virtually 100 percent data transfer but takes a relatively long transfer time. Frame relay is suitable for high-volume data transfer such as imaging and visualization (e.g., 3-D displays of engineering designs). It is less suitable for services such as voice and video that are sensitive to time delays.

The ATM and IEEE 802.6 MAN are examples of cell relay technologies. ATM, as described previously, is designed to operate over optical fiber. The transmission bandwidth is organized into cells that are periodic sequences of undedicated, fixed-sized blocks of data. Figure 10.40*b* shows the signaling and data paths, as a comparison to frame relay. The ATM takes care of cell routing, cell multiplexing and demultiplexing, and header error control. The ATM

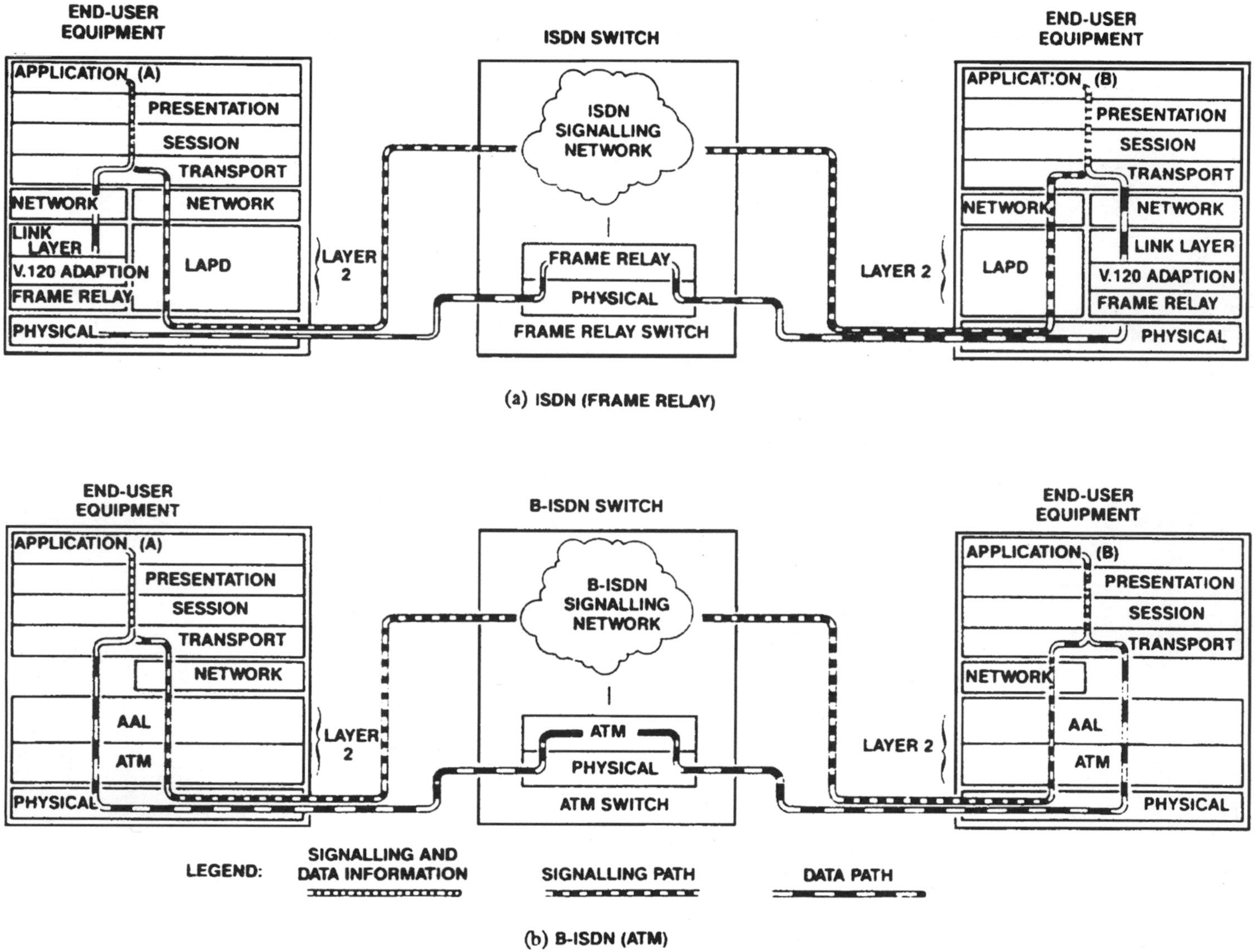

Figure 10.40 Protocol architectures. *(Reproduced with permission from Telecommunications. Roy, D., Frame Relay Technology: Complement or Substitute?, October 1990, pp. 39 –48.)*

adaptation layer (AAL) transforms the upper-layer message format and the fixed-size cell format of the ATM layer. The AAL also performs error detection, correction, and flow control.

Data transfer will be initiated by, for example, application A requesting session establishment to the transport layer, etc., as before. Call control information is carried through the network on a separate virtual circuit. A signaling message is sent by the transport layer using the signaling virtual channel. The AAL and ATM layers convert the signaling message into cells, after which they are sent to the B-ISDN signaling network. Path and call parameters will probably be defined by the signaling message in a similar manner to existing ISDN for basic and primary rate services. Figure 10.40*b* shows the data transfer path once the call has been established. During cell transfer, cells are routed through the ATM switches by a cell VPI and routing information organized during the call setup procedure. Very fast information transfer is established by minimizing the amount of processing performed by the network.

To summarize, cell relay has some advantages over frame relay. First, the switching functions can be performed more efficiently, which translates to a lower transmission cost per bit. Also, delay-sensitive applications such as voice and video have a more acceptable quality. This is because cell relay has no long delays caused by long data bursts from one user monopolizing the circuit, as can occur with frame relay.

10.9   The Internet

The term *internet* is simply an abbreviation of internetworking, which means the interconnection of networks. The name Internet is the internetworking of thousands, and soon millions, of subnetworks on a global scale that has been growing at a prodigious rate since the early 1990s. The subnetworks can be LANs, MANs, or WANs. The Internet can also be viewed as an enormous global WAN. The objective here is to explain the relationship between this huge computer network and the telecommunications network. It is probably fair to say that without the telecommunications network infrastructure the Internet would not exist. The biggest problem facing telecommunications network designers today is the fact that telephone networks were designed for telephone voice connections but are now required to send large amounts of data down those same lines. Adjusting to the incompatibilities of the two requirements is not a simple matter. Voice telephone traffic (POTS) still accounts for more than 50 percent of all telecommunications revenues worldwide. Telecommunications networks are nevertheless being used for networking, but a serious gap is already opening up between today's throughput requirements and the throughput that telecommunications companies can actually deliver. Furthermore, all this seemed to happen overnight without any single body being able to plan its evolution.

Fortunately, some of the "rules of the game" for Internet information flow are remarkably simple. The two protocols—transmission control protocol (TCP) and Internet protocol (IP), usually written as TCP/IP—are the key to allowing one subnetwork to communicate with another. They correspond to the OSI model, where the TCP is the transport layer and the IP part is the network layer. Because their functions are transparent to the physical and data link layers they can be applied to Ethernet, FDDI, or token ring networks.

The TCP/IP was created by the U.S. Defense and Advanced Research Projects Agency (DARPA), with the original intention of connecting research centers and universities to promote rapid information flow and dialogue. The costs of the interconnections through the telephone network were originally paid for by research grants or direct government funds. The commercial benefits are changing the Internet landscape and no doubt eventually the costs will all be passed on to Internet users.

The initial purpose of file transfer and remote computer log-on have now become a minor activity compared to information location by "browsing." Telecommunications service providers are finding it difficult to dimension their telephone networks adequately. A large part of Internet use is by individuals who pay monthly subscriptions to an Internet service provider for dial-up connection. Whereas the average phone call used to be 3 to 5 min, it is rising significantly as many browsers stay on the line for hours. Serious data flow bottlenecks are developing that will probably only be alleviated by the eventual deployment of *all-optical networks.*

The TCP/IP is used by the Internet to transfer data. Every Internet node has its own unique network address known as its IP address. A node is also known as a TCP/IP host and is the start or end point for communications over the Internet. Hosts are computers themselves which run software applications. TCP/IP gateway nodes connect different types of network together within the Internet, such as token ring and Ethernet networks. Both hardware and software are needed to perform this function. Gateways route TCP/IP packets between networks. Figure 10.41 shows several gateways. A router is a computer that connects similar types of networks together, such as two or more token rings.

The Internet uses a connectionless packet-based protocol, which means there is no established route for the series of packets to take. Every packet has its own addressing information to find its way to the destination independently of other packets. This method has pros and cons. A major advantage is that if one or more routers in the system fail, it does not affect the overall flow of packets because they simply find another path through the maze of routers. This benefit is offset by the fact that each router has to do a route determination for every packet it receives. The IP has no flow control mechanism, so when a router becomes congested it deletes excess packets until the congestion is reduced.

The Internet datagram.   Routing through the network is done by an addressing scheme written into each packet. All packets are constructed in the Internet datagram format which is composed of a header followed by information data

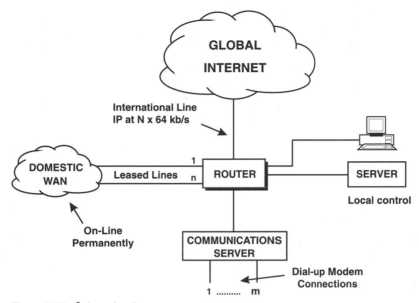

Figure 10.41   Internet gateways.

as in Fig. 10.42. A maximum of 65,536 bytes can be sent in one datagram and larger blocks of data are split into two or more datagrams. The header contains all the routing information. First, the TCP/IP *version number* allows nodes and gateways to interpret the datagram correctly as newer versions become available. The *type of service* 8 bits take the form PPPDTRXX and characterize the datagram as follows. PPP is a number from 0 to 7 to define the priority, D defines a low delay, T defines high throughput, R defines high reliability, and XX is for future use. The *header length* describes the length of the header to follow in multiples of 4 bytes normally up to 20 bytes (without *options*). The D-bit is to inform gateways that the data must not be fragmented into smaller data blocks. Gateways can split data into fragments and the M-bit denotes that data is fragmented. The *fragment offset* indicates the number of the fragment (not the fragment quantity). Depending on traffic circumstances, a datagram can be delayed anywhere from small fractions of a second to never arriving at all. The *time-to-live* 8-bit words place a maximum datagram transfer time value in seconds. At each gateway, the value is reduced automatically. On reaching zero, the datagram is deleted. This time-to-live word is very useful because it also gives the destination node a maximum time it should wait for the next datagram fragment. Other Internet protocol variants can be used in the datagram, and the 8-bit *protocol* word indicates which one is used. The *header checksum* is a special 16-bit word that is used for error detection. The *source IP address* and the *destination IP address* are both unique 32-bit words to allow the datagram to reach the desired location. Finally, the *options* field provides routing information and other items such as error control.

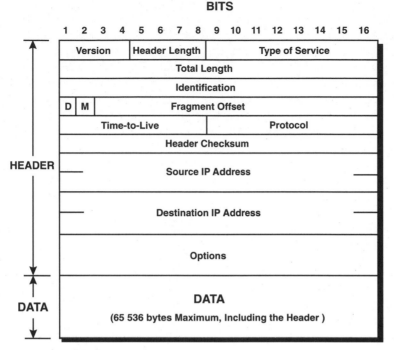

BITS

Figure 10.42   The Internet datagram.

The IP address.   The IP address is a 32-bit word. $2^{32} = 4,294,967,296$, which means over 4 billion hosts or physical networks can be addressed with this length of digital word. No doubt 32 bits will be inadequate a few years from now and will have to be increased in length. The 32-bit word is split into two parts, to identify a particular network number and a specific host within that network. Three address formats have been defined, depending on the quantity of networks and the quantity of hosts to be internetworked, as in Fig. 10.43.

The IP address can be specified as four decimal numbers, designated by the letters W, X, Y, and Z, where each decimal number represents 1 byte (8 bits) of IP address. Because $2^8 = 256$, the maximum decimal number for each of the four letters W, X, Y, or Z is 255. The network information center (NIC) assigns IP addresses. This address format allows quite a lot of flexibility and, for example, subnetworks can be defined within a large network.

The 32-bit IP address is difficult to remember so user names are usually given to colleagues and friends, and application programs are used to convert between the two. For example WWW.IEEE.COM denotes a World Wide Web server for the Institute of Electrical and Electronics Engineers, based in the United States, and the corresponding Internet IP address is 140.98.1.1.

Each institution within the Internet has its host which incorporates a domain name server (DNS) that has a directory of all hosts within the network. Figure 10.44 indicates a typical tree for domain names. In this manner, the Internet addressing scheme is the equivalent of a telephone numbering plan. Interestingly, the telephone network is used as part of the interconnection of networks, gateways, and hosts. The main difference in connectivity of addressing is created by the *leased line*. The computer community uses the term *being on-line*. That simply means there is a permanent connection between a user's computer and the others in the network. The network can be just one other computer in the next office with an Ethernet LAN connection or it can be a computer across the other side of the world within, say, a large university campus LAN. Telephone companies have set up numerous leased lines to allow, for example, a permanent connection between a gateway in San Francisco and a gateway in New York. New Internet users get the impression that such long distance connections are free. They are not. With speech over the Internet improving in quality and becoming ever more popular, telecom service providers are taking a serious look at the definition of both a long-distance and a local call.

10.9.1   ATM and the Internet

An ATM network and the Internet are fundamentally different in their modes of transmission. ATM is connection-oriented, whereas Internet data transfers use the connectionless Internet protocol (IP). The question has been asked,

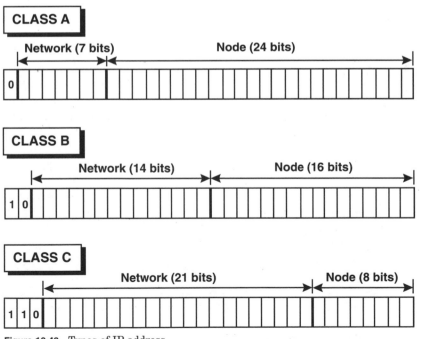

Figure 10.43   Types of IP address.

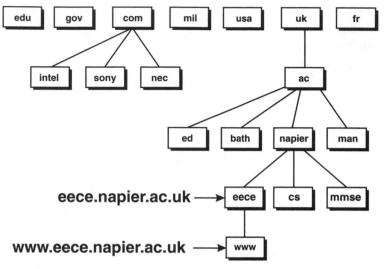

Figure 10.44 Domain names.

Why does the Internet need ATM, because it is functioning well as it is? The fact is the Internet will grind to a halt if new technology does not increase the rate of data flows as more and more users connect into it. The Internet is a telecom-network-based phenomenon, by virtue of its global connectivity using telecom-owned long-distance transmission lines. Because ATM is the chosen broadband ISDN telecom technology, the Internet of the future will be unavoidably bound to ATM technology. The problems of sending ATM data over the Internet are gradually being resolved. Eventually an integrated solution will no doubt evolve, and the differences in packet encapsulation, addressing, signaling, and routing are receiving considerable attention. There is plenty of time in the sense that there is only a relatively small, but rapidly growing, number of ATM switches in operation. Meanwhile, the Internet bandwidth bottlenecks are worsening daily.

10.10   All-Optical Networks

"It's the size of the pipe, stupid." This statement might be made by a senior water engineer addressing his junior colleague. It is equally applicable to the networking environment. Everyone is aware of the problem; the question is how to solve it. To appreciate the enormous bandwidth limitation problem that has been looming for some time, it is instructive to analyze the evolution of bit rate capabilities of transmission technologies within the local network environment and the long-haul environment (see Fig. 10.45). Starting with the modem rates, in 1997 about 34 kb/s seemed a reasonable modem speed simply because it was about 20 percent better than the year before. The next step up to 56 kb/s seemed a healthy advancement. Unfortunately, the annual growth

in bit rate by *each user* is a factor of about 8. Classical time-division electronic multiplexing, which has served so well in the past, is now a serious bottleneck. TDM demands that each port in a network take care of its own bits in addition to bits for most of the other ports. TDM schemes have worked well from the first days of PCM where only 24 individuals each use 64 kb/s, which requires a throughput of 1.544 Mb/s. The problem is that the speed of the electronics needed to cope with more users scales as the product of the number of ports and the bit rate per user, regardless of network topology. The rate of increase in users is by far outpacing the already impressive rate of increase in the speed of digital electronics.

Even worse, the local and long-distance backbone access bit rate gap is a factor of more than 10^4. Web access at 10 Mb/s is a totally different experience from using a 64-kb/s dial-up telephone line via a service provider. The response times are milliseconds compared to tens of seconds. ADSL promises to alleviate this situation by providing each user with several megabits per second, in what will be a quantum leap for user bit rate. But, it also shifts the problem to the backbone network, which will have difficulty in supplying all users with data at several megabits per second. Upgrading the long-distance equipment

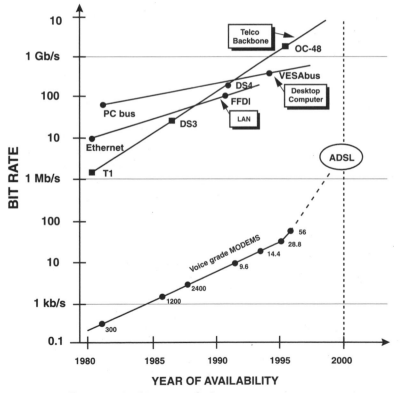

Figure 10.45 Transmission bit rate evolution.

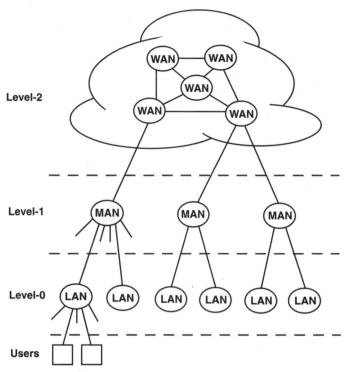

Figure 10.46 All-optical network hierarchy.

to 40 Gb/s and above will help, but other all-optical network components will need to be installed.

10.10.1 All-optical WDM

Fortunately, there is a solution to this bandwidth bottleneck problem, although the technology is not yet very mature in some crucial areas. Figure 10.46 is a view of the way in which LANs, MANs, and WANs are forming levels in an all-optical network hierarchy. WDM demands that each port process only its own bits. Optical fiber systems using WDM offer a solution that becomes totally effective only when the electronic-to-optical conversions are all eliminated. The resulting *all-optical networks* will no doubt be the way of the future because they would provide a universal solution to the present bandwidth inadequacies. A total single fiber bandwidth of 25,000 GHz would be unleashed in all-optical networks. The total available *radio* bandwidth of about 100 GHz looks very small in comparison. There is a consensus emerging in the research laboratories that WDM combined with wavelength conversion will alleviate the present networking bottleneck. WDM has a relatively small number of wavelengths available for use within a network; presently less than 100. While this is clearly too small for a broadcast and select network, it might be sufficient for a mesh network. This is analogous to the small pool of frequencies available for a multitude of users in a cellular network. Just as frequencies are reused in different cell clusters to enhance the cellular radio system capacity, so can wavelength conversion (frequency reuse) at the interconnection points of mesh networks scale up the capacity of all-optical networks. Figure 10.47 shows how the individual networks are analogous to radio cells, and how wavelength (frequency) reuse at the cell (network) borders expands the capacity without bounds. However, it must be acknowledged that this is an oversimplification of a very complex subject. The extent to which wavelength converters will increase an all-optical network capacity is still under intense debate. It is recognized that they will be used; the question is at which hierarchical level(s) and in what quantities.

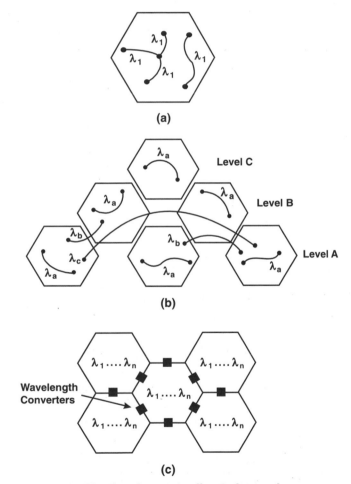

Figure 10.47   Wavelength reuse in all-optical networks.
(*a*) A single wavelength has limited reuse (one cell).
(*b*) Multiple wavelengths have more extensive reuse (cell-to-cell).
(*c*) Wavelength conversion of cell borders greatly expands the wavelength reuse.

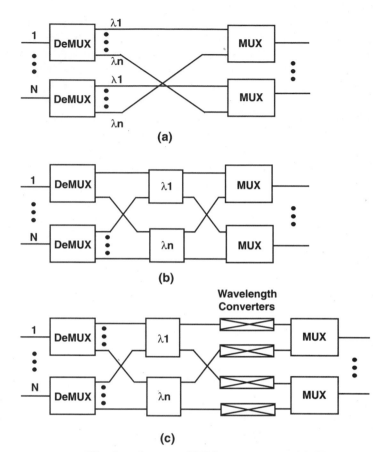

Figure 10.48 Wavelength router/WDM crossconnect. (*a*) Using two back-to-back (de)multiplexers; (*b*) incorporating space switches; (*c*) wavelength interchange crossconnect.

10.10.2 WDM wavelength routers

Figure 10.48*a* shows the wavelength router (also called WDM crossconnect) with N input and output port pairs having M wavelengths. This arrangement performs wavelength routing by using two back-to-back (de)multiplexers, so that wavelengths are interchanged between input and output fibers in a predetermined manner. This passive element, which could be used at the nodes of an all-optical network, has no automated reconfigurability.

Space switches need to be included as in Fig. 10.48*b* to achieve the reconfigurable freedom. This WDM crossconnect can interconnect any wavelength on any input fiber to any output fiber, provided the output fiber is not already using that wavelength. The total crossconnect bandwidth is proportional to the product of input fibers N, the number of wavelengths M, and the bit rate per wavelength B.

The crossconnects in Fig. 10.48a and b are wavelength selective because wavelengths are selected and reconfigured in the space domain. Figure 10.48c expands on the other two by having wavelength converters prior to the output multiplexer. These devices have the effect of allowing crossconnecting in the wavelength domain as well as the space domain. The resulting network elements are called wavelength interchanging crossconnects, and any configuration of input wavelengths on fibers 1 to N can be transferred to predetermined wavelengths on the 1 to N output fibers.

These network elements will probably form the basis of future transparent all-optical networks that can be reconfigured at will to optimize the network as traffic grows, to minimize congestion, and to neutralize the effects of component failures.

Final comment.   Data communications has come a long way in the past ten years but, to keep the progress in perspective, it is appropriate to make the following final observation on the subject: if networking and communications today were the motor industry, this would be about the year 1930.

Acronyms

AAL	ATM adaptation layer
ACG	Asymptotic coding gain
ADI	Alternate digit inversion (code)
ADM	Add/drop multiplexer
ADPCM	Adaptive differential pulse code modulation
ADSL	Asymmetric digital subscriber line
AGC	Automatic gain control (circuit)
AIS	Alarm indication signal
AM	Amplitude modulation
AMI	Alternate mark inversion (code)
AMPS	Advanced mobile phone service
AMR	Analog microwave radio
ANSI	American National Standards Institute
APD	Avalanche photodiode
APS	Automatic protection switching
ARQ	Automatic request for repeat
ASK	Amplitude shift keying (modulation)
ATM	Asynchronous transfer mode
AU	Administrative unit
AWGN	Additive white Gaussian noise
BB	Baseband
BBER	Background block error ratio
BER	Bit error rate
BIP	Bit-interleaved parity
B-ISDN	Broadband integrated services digital network
BNZS	Binary N zero substitution
BPSK	Binary phase shift keyed (modulation)
CAP	Carrierless amplitude/phase (modulation)
CATV	Community antenna TV
CBE	Chemical beam epitaxy
CBTR	Carrier and bit timing recovery
CCIR	International Radio Consultative Committee
CCITT	International Telegraph and Telephone Consultative Committee

CDMA	Code-division multiple access
CELP	Code-excited linear predictor
CEPT	Conference of European Posts and Telecommunications
CMI	Coded mark inversion
C/N	Carrier-to-noise ratio
CO	Central office
CP	Customer premises
CPE	Customer premises equipment
CSMA/CD	Carrier sense multiple access with collision detection
DAMA	Demand assigned multiple access
DBR	Distributed Bragg reflector
DCC	Digital crossconnect
DCE	Data circuit terminating equipment
DCME	Digital channel multiplication equipment
DCS	Digital cellular system
DCT	Discrete cosine transform
DECT	Digital European cordless telecommunications
DFB	Distributed feedback
DLCI	Data link connection identification
DMR	Digital microwave radio
DMT	Discrete multitone (modulation)
DNI	Digitally noninterpolated (data channel)
DPCM	Differential pulse code modulation
DQDB	Distributed queue dual bus
DQPSK	Differential quaternary phase shift keying
DRO	Dielectric resonator oscillator
DSB	Direct satellite broadcasting
DSB-SC-AM	Double-sideband suppressed carrier amplitude modulation
DS-CDMA	Direct-sequence code division multiple access
DSE	Data switching exchange
DSI	Digital speech interpolation
DSSS	Direct-sequence spread spectrum
DTE	Data terminal equipment
DVCC	Digital verification color code
ECL	Emitter-coupled logic
EDFA	Erbium-doped fiber amplifier
EIRP	Equivalent isotropic radiated power
ENG	Electronic news gathering
ESR	Errored second ratio

ET	Exchange termination
ETSI	European Telecommunications Standards Institute
FACCH	Fast associated control channel
FCC	Federal Communications Commission
FDD	Frequency-division duplexed
FDDI	Fiber distributed data interface
FDM	Frequency-division multiplexing
FDMA	Frequency-division multiple access
FEC	Forward error correction
FEXT	Far-end crosstalk
FFSK	Fast frequency shift keying
FFT	Fast Fourier transform
FH-CDMA	Frequency-hopping code division multiple access
FHSS	Frequency-hopping spread spectrum
FM	Frequency modulation
FSK	Frequency shift keying
FSR	Free spectral range
FTTH	Fiber to the home
GaAs FET	Gallium arsenide field-effect transistor
GMSK	Gaussian minimum shift keying
GSM	Global system for mobile communications
GSO	Geostationary satellite orbit
HBT	Heterostructure bipolar transistor
HDB3	High-density bipolar 3 (zero substitution)
HDSL	High-bit-rate digital subscriber line
HDTV	High-definition TV
HEMT	High electron mobility transistor
HLR	Home location register
HRDP	Hypothetical reference digital path
HRDS	Hypothetical reference digital section
HRP	Hypothetical reference path
HRX	Hypothetical reference connection
IBS	Intelsat Business Services
IDN	Integrated digital network
IDR	Intermediate data rate
IEEE	Institute of Electrical and Electronics Engineers
IF	Intermediate frequency
IP	Internet protocol
ISDN	Integrated services digital network

ISI	Intersymbol interference
ISO	International Standards Organization
ISPBX	Integrated services private branch exchange
ITU	International Telecommunication Union
LAN	Local area network
LAP	Link access procedure
LD	Laser diode
LD-CELP	Lower-delay code-excited linear predictor
LED	Light emitting diode
LEO	Low earth orbit (satellite)
LHCP	Left-hand circularly polarized (wave)
LNA	Low-noise amplifier
LNB	Low-noise block (downconverter)
LO	Local oscillator
LRE	Low-rate encoding
LT	Line termination
LTE	Line termination equipment
MAC	Medium access control
MAN	Metropolitan area network
MAP	Mobile application part
MCPC	Multiple channels per carrier
MEO	Medium earth orbit (satellite)
MIC	Microwave integrated circuit
MLA	Microwave link analyzer
MMIC	Monolithic microwave integrated circuit
MOS	Mean opinion score
MOVPE	Metal organic vapor phase epitaxy
MPEG	Motion Picture Experts Group
MQW	Multiple quantum well
MSK	Minimum shift keying
MTSO	Mobile telephone switching office
NEXT	Near-end crosstalk
N-ISDN	Narrowband integrated services digital network
NNI	Network node interface
NPR	Noise-to-power ratio
NRZ	Nonreturn to zero
NRZI	Nonreturn-to-zero invert
OC	Optical carrier
OEIC	Optoelectronic integrated circuit

OMJ	Orthomode junction
OMT	Orthomode transducer
OPLL	Optical phase-locked loop
OSI	Open systems interconnection
PACS	Personal Access Communications Systems
PAD	Packet assembly and disassembly
PAM	Pulse amplitude modulation
PAMA	Preassigned multiple access
PBX	Private branch exchange
PCM	Pulse code modulation
PCN	Personal communications network
PCS	Personal Communications Services
PDH	Plesiochronous digital hierarchy
PH	Packet handler
PHY	Physical protocol
PIC	Photonic integrated circuit
PLL	Phase-locked loop
PM	Phase modulation
PMD	Physical media dependent
POH	Path overhead
POTS	Plain old telephone sets, or services
PRBS	Pseudorandom bit sequence
PRMA	Packet reservation multiple access
PSK	Phase shift keying
PSTN	Public switched telephone network
PT	Path termination
PVC	Permanent virtual circuit
QAM	Quadrature amplitude modulation
QoS	Quality of service
QPSK	Quadrature or quaternary phase shift keying
RF	Radio frequency
RHCP	Right-hand circularly polarized wave
RMT	Ring management
RN	Remote node
RS	Reed-Solomon (block code)
RT	Remote terminal
RZ	Return to zero
SACCH	Slow associated control channel
SAP	Service access point

SAPI	Service access point identifier
SAW	Surface acoustic wave (filter)
SBB	Subbaseband
SCPC	Single channel per carrier
SDH	Synchronous digital hierarchy
SES	Severely errored second
SESR	Severely errored second ratio
S/I	Signal-to-interference ratio
SMT	Station management
S/N	Signal-to-noise ratio
SOA	Semiconductor optical amplifier
SOH	Section overhead
SONET	Synchronous optical network
SPC	Stored-program controlled
SPE	Synchronous payload envelope
SSB-AM	Single-sideband amplitude modulation
SS-TDMA	Satellite-switched time-division multiple access
STM	Synchronous transport module
STS	Synchronous transport signal
TA	Terminal adapter
TCM	Trellis-coded modulation
TCP	Transmission control protocol
TDM	Time-division multiplexing
TDMA	Time-division multiple access
TDMA-DA	Time-division multiple access demand assigned
TE	Transverse electric
TM	Transverse magnetic
TMI	Two-mode interference
TMN	Telecommunications Management Network
TRT	Token rotation time
TTRT	Target token rotation time
TU	Tributary unit
TUG	Tributary unit group
TWT	Traveling wave tube
UHF	Ultrahigh frequency
UMTS	Universal Mobile Telecommunication System
UNI	User network interface
USAT	Ultrasmall-aperture terminal
VAR	Voltage axial ratio

VC	Virtual call
VC	Virtual container
VCI	Virtual channel (or circuit) identifier
VCO	Voltage-controllable oscillator
VDSL	Very-high-bit-rate digital subscriber line
VHF	Very-high frequency
VLR	Visitor location register
VLSI	Very-large-scale integration
VPI	Virtual path identifier
VSAT	Very-small-aperture terminal
VSELP	Vector sum excited linear prediction
VSWR	Voltage standing-wave ratio
VT	Virtual tributary
WAN	Wide area network
WDM	Wave-division multiplexing
WLL	Wireless local loop
XPD	Cross-polarization discrimination
XPIC	Cross-polarization interference canceller

Index

ABOUT THE AUTHOR

Robert G. Winch obtained a doctoral degree in microwave electronics at the University of Oxford in 1974. He subsequently worked as staff scientist for Teledyne Microwave, Inc., California on the design of microwave communications subsystems and components for defense and space applications. He held faculty positions at the University of the West Indies and the University of the Virgin Islands during which time he worked as a consultant to local and California-based companies on telecommunications and aerospace projects. He then moved to ITT and continued full time work on intra- and interisland optical fiber and microwave telecommunication links in the Caribbean.

In 1984, he became a full time employee of the International Telecommunication Union, a member organization of the United Nations, for whom he worked as a digital telecommunications expert for 11 years in an advisory capacity to governments of the international community. Areas of specialization include the design and implementation of transmission system projects involving optical fiber, digital microwave radio, wireless, satellite, and data communications technology. Dr. Winch is now an independent international consultant based in Oxford, England.